12th South African Conference on Computational and Applied Mechanics (SACAM2020)

MATEC Web of Conferences Volume 347

Cape Town, South Africa
29 November - 1 December 2021

Editor:

Sebastian Skatulla

ISBN: 978-1-7138-4077-0

Printed from e-media with permission by:

Curran Associates, Inc.
57 Morehouse Lane
Red Hook, NY 12571

Some format issues inherent in the e-media version may also appear in this print version.

For additional information, please contact EDP Sciences – Web of Conferences at the address below.

EDP Sciences – Web of Conferences
17, Avenue du Hoggar
Parc d'Activité de Courtabœuf
BP 112
F-91944 Les Ulis Cedex A
France

Phone: +33 (0) 1 69 18 75 75
Fax: +33 (0) 1 69 28 84 91

contact-edps@webofconferences.org

Additional copies of this publication are available from:

Curran Associates, Inc.
57 Morehouse Lane
Red Hook, NY 12571 USA
Phone: 845-758-0400
Fax: 845-758-2633
Email: curran@proceedings.com
Web: www.proceedings.com

TABLE OF CONTENTS

Preface

The Proceedings of the 12th South African Conference on Computational and Applied Mechanics (SACAM2020) present findings from a large range of fundamental and applied research topics as well as industrial applications in the broad field of mechanics. SACAM is a biennially held conference series under the auspices of the South African Association for Theoretical and Applied Mechanics (SAAM) which aims at promoting activities concerning theoretical, experimental and computational mechanics in South Africa in order to enhance the development of expertise in mechanics in this country. As such, it provides a forum for postgraduate students, emerging and established researchers as well as professional engineers to discuss and answer questions on the subject of mechanics relating to solids and fluids.

SACAM2020 has received excellent support by researchers and practitioners all across South Africa, with authors being drawn from numerous research and industrial organisations. These Proceedings contain papers presented at the conference, classified into a total of 18 sub-themes which can be grouped under five main themes: (i.) Structural Mechanics and Dynamics, (ii.) Fluid Dynamics, (iii.) Thermodynamics, (iv.) Manufacturing and Process Engineering, and (v.) Control Theory and Optimisation. Only original contributions were considered for inclusion in the conference proceedings and all papers submitted were subjected to a full process of double-blind peer review. The review of manuscripts was undertaken by members of the Scientific Advisory Board and other identified leading experts, acting independently on one or more assigned manuscripts. This invaluable assistance, which has greatly enhanced the quality of the Proceedings, is gratefully acknowledged. The full conference papers are published by MATEC Web of Conferences, a provider of open-access proceedings in Materials Science, Engineering and Chemistry. This was done in order to give the papers better accessibility in the international research and industry communities.

Special acknowledgments are due to our supporters and partners in the industry:

- SIMTEQ Engineering
- QFINSOFT
- FEAS

Finally, the editor wishes to thank the authors for their efforts towards producing and delivering papers of high standard. This ensures that the Proceedings will be a valued reference for many working in this important field, and that they will form a suitable base for discussion and provide suggestions for future development and research.

S SKATULLA
Editor

Organising Committee

Prof Sebastian Skatulla (Chairman)
Prof Hans Beushausen (Co-Chairman)
Prof Malebogo Ngoepe (Secretary)

Scientific Committee and Advisory Board

Prof Thorsten Becker (University of Stellenbosch)
Prof Annie Bekker (University of Stellenbosch)
Prof Tunde Bello-Ochende (University of Cape Town)
Dr Francois Ebobisse Bille (University of Cape Town)
Prof Deborah Blaine (University of Stellenbosch)
Dr Alfred Bogaers (CSIR)
Dr Natasha Botha (CSIR)
Dr Tiri Chinyoka (University of Cape Town)
Dr Danie De Kock (Qfinsoft)
Dr Andie De Villiers (University of Stellenbosch)
Prof Jat Du Toit (North-West University)
Dr Reuben Govender (University of Cape Town)
Prof Helen Inglis (University of Pretoria)
Prof Modify Kaunda (Cape Peninsula University of Technology)
Dr Jan-Hendrik Kruger (North-West University)
Prof Ramesh Kuppuswamy (University of Cape Town)
Prof Schalk Kok (University of Pretoria)
Prof Genevieve Langdon (University of Cape Town)
Prof Craig Law (University of the Witwatersrand)
Prof Arnault Malan (University of Cape Town)
Prof Prof Maina Maringa (Central University of Technology)
Prof Chris Meyer (University of Stellenbosch)
Prof Pilate Moyo (University of Cape Town)
Prof Claudia Polese (University of the Witwatersrand)
Prof Daya Reddy (University of Cape Town)
Prof Robert Reid (University of the Witwatersrand)
Prof Pieter Rousseau (University of Cape Town)
Prof Gerhard Venter (University of Stellenbosch)
Dr Nico Wilke (University of Pretoria)
Prof Steeve Chung Kim Yuen (University of Cape Town)
Prof Alphose Zingoni (University of Cape Town)

Statement of Peer review

In submitting conference proceedings to *Web of Conferences*, the editors of the proceedings certify to the Publisher that

1. They adhere to its **Policy on Publishing Integrity** in order to safeguard good scientific practice in publishing.
2. All articles have been subjected to peer review administered by the proceedings editors.
3. Reviews have been conducted by expert referees, who have been requested to provide unbiased and constructive comments aimed, whenever possible, at improving the work.
4. Proceedings editors have taken all reasonable steps to ensure the quality of the materials they publish and their decision to accept or reject a paper for publication has been based only on the merits of the work and the relevance to the journal.

Title, date and place of the conference

12th South African Conference on Computational and Applied Mechanics, 29. November - 1. December 2021, Cape Town

Proceedings editor(s): Prof Sebastian Skatulla

Date and editor's signature

Representation of simulation errors in single step methods using state dependent noise

Edward Boje[1,*]

[1]Electrical Engineering, University of Cape Town, South Africa

Abstract. The local error of single step methods is modelled as a function of the state derivative multiplied by bias and zero-mean white noise terms. The deterministic Taylor series expansion of the local error depends on the state derivative meaning that the local error magnitude is zero in steady state and grows with the rate of change of the state vector. The stochastic model of the local error may include a constant, "catch-all" noise term.
A continuous time extension of the local error model is developed and this allows the original continuous time state differential equation to be represented by a combination of the simulation method and a stochastic term. This continuous time stochastic differential equation model can be used to study the propagation of the simulation error in Monte Carlo experiments, for step size control, or for propagating the mean and variance. This simulation error model can be embedded into continuous-discrete state estimation algorithms.
Two illustrative examples are included to highlight the application of the approach.

1 Introduction

In the solution of ordinary differential equations using numerical methods the problem of local and global errors arises because of finite order methods and finite step size. The problem has been considered exhaustively in the literature where, in simulation studies, the aim is usually to get a solution that has sufficiently small global error and over-estimating the error is wasteful but does no harm. In time- or resource-critical applications of numerical solution of ODEs, correct local error estimates are required and this motivates a more careful analysis of the local error. Of particular interest is the application of numerical simulation used to project state equations between measurement updates in continuous-discrete state estimation problems where a reliable estimate of the local error behaviour can be embedded into the state estimator to improve estimator optimality, for example, if the effect of the local error is over-estimated, the estimator will de-emphasise the state estimates in favour of the measurements and this will result in sub-optimal performance. (Two conundrums remain: (i) If one can estimate the local error, there is no reason not to add this to the computed solution to get an improved solution but then no local error estimate is available. (ii) Local errors accumulate to the global error making it hard to quantify the latter even with smoothness and stability assumptions on the underlying system model and good local error estimates.)

*e-mail: edward.boje@uct.ac.za

The development of the field of probabilistic solutions to differential equations is described in [1] and [2]. The ideas in [1] are developed in [3] to establish stronger convergence results for probabilistic integrators, and as their results apply to general additive noise models, they apply to the the state dependent noise with bias model used to model the local error in this paper. A general survey of probabilistic numerics is provided by [4].

In this paper, in Section 2 the local error for any single step method is modelled as a function of the state derivative multiplied by a bias term and a zero-mean white noise term, both appropriately scaled by the step size. Scaling the error model by the state derivative provides a richer model for the local error than used by [1], captures the fact that the local error depends on the rate of change of the state, and correctly gives zero local error effect in steady state. A stochastic difference equation is formed by combining the single step method with the local error model. To account for non-autonomous behaviour and as a "catch-all," an unscaled noise term may also be added to the model.

Following [1], in 3 a continuous time extension of the discrete time local error model is developed so that the original continuous time state differential equation can be represented by a combination of the simulation method and a stochastic term. This continuous time stochastic differential equation can be used to propagate the simulation error via Monte Carlo analysis or by statistical moments; can be used for step size control; and has specific application to incorporating local simulation errors in continuous-discrete state estimation.

Section 4 presents two examples to illustrate the utility of the approach presented in the paper.

2 Stochastic, state dependent local error model for single step methods

Consider the autonomous n^{th} order dynamic system,

$$\dot{\boldsymbol{x}} = \boldsymbol{f}(\boldsymbol{x}), \quad \boldsymbol{x}(t_0) = \boldsymbol{x}_0. \tag{1}$$

Non-autonomous systems can be dealt with by appending an additional state variable, $x_{n+1}(t) = t$, with $f_{n+1} = 1$, which makes the extended system autonomous ([5], Chapter 4). Apply any single-step method with p^{th} order global error, with step size h_i per time step, i,

$$\boldsymbol{x}^{\psi}_{i+1} = \boldsymbol{x}_i + h_i \boldsymbol{\psi}(\boldsymbol{x}_i, h_i). \tag{2}$$

If (1) is Lipchitz along the state trajectory, to achieve a global error $\boldsymbol{O}(h^p)$, the local error must be $\boldsymbol{O}(h^{p+1})$. The local error (with index i to denote local error in making the step from $\boldsymbol{x}_i$ to $\boldsymbol{x}_{i+1}$) is,

$$\begin{aligned}\boldsymbol{e}_i &= \boldsymbol{x}(t_{i+1}) - \boldsymbol{x}^{\psi}_{i+1}\big|_{\boldsymbol{x}_i=\boldsymbol{x}(t_i)} \\ &= h^{p+1}\left(\boldsymbol{D}^{(p)}(\boldsymbol{x}_i) + \boldsymbol{O}_e(h)\right)\boldsymbol{f}(\boldsymbol{x}_i) \\ &= \boldsymbol{O}(h^{p+1})\end{aligned} \tag{3}$$

where $\boldsymbol{D}^{(p)}(\boldsymbol{x}_i)\boldsymbol{f}(\boldsymbol{x}_i)$ represents the accumulation of the differences between $h^{(p+1)\text{th}}$ order elementary differentials ([5], [6]) of the Taylor series expansions of the true solution and the method with a common term, $\boldsymbol{f}(\boldsymbol{x}_i)$, factored out. Because there may be product terms, e.g. $f_a(\boldsymbol{x}_i)f_b(\boldsymbol{x}_i)$, in the expansion, there may be ambiguity in elements of matrix $\boldsymbol{D}^{(p)}(\boldsymbol{x}_i)$ but not in the vector, $\boldsymbol{D}^{(p)}(\boldsymbol{x}_i)\boldsymbol{f}(\boldsymbol{x}_i)$. For a pth order method, terms in the local error up to and including h^p cancel.

When applied to the linear time invariant (LTI) system, $\dot{x}(t) = \boldsymbol{A}x(t)$, (3) gives the local error as $\boldsymbol{e}_i = h^{p+1}(d_p\boldsymbol{A}^p + \boldsymbol{O}_e(h))\boldsymbol{A}\boldsymbol{x}_i$, where d_p is a coefficient. ($d_p = 1/(p+1)!$ for explicit methods with $p \leq 4$ so that the stage number equals the order, [6].)

Although the local error is deterministic, based on the above observations and developing on previous work in this area ([1], [2],[3]), we propose to model the local error for each row, $r \in \{1 \dots n\}$, of the state differential equation as a combination of a deterministic term and a stochastic term,

$$e_{r,i} \approx h^{p+1}\left(f^{\mathrm{T}}(x_i)\, m_r + \left[f^{\mathrm{T}}(x_i),\ 1\right] \xi_{r,i}\right) \tag{4}$$

where m_r is a (generally) non-zero mean vector that captures the average or deterministic component of the local error and $\xi_{r,i} \sim \mathcal{N}(\mathbf{0}, Q_r)$, a $(n+1)\times 1$, zero-mean white Gaussian random vector with constant covariance Q_r to capture the residual effect of the local error. (Equation (4) can be represented using a single, equivalent term, $\xi_{r,i} \sim \mathcal{N}(m_r, Q_r)$.) For LTI systems with p^{th} order method and a fixed step size, the local error multiplier, $m_r(h)$ is constant (depending on h), as the local error is, $e_i = h^{p+1} M(h) A x_i$ with $M(h) = \sum_{k=p+1}^{\infty} \frac{1}{k!} A^{k-1} h^{k-(p+1)}$. For general non-linear systems or varying step size, $D^{(p)}(x_i) + O(h)$ is not constant but may still have a non-zero mean over the operating region of the state space and range of step sizes. In (4), augmenting the state derivative vector with $f_{n+1} = 1$ takes care of time dependence for non-autonomous systems and provides a "catch-all" for errors that are not modelled via the state dependent terms. (This is the only term used in [1].) We do not recommend adding a constant offset to the deterministic part as this will change the steady state behaviour by adding bias to the equilibrium point.

The following approach is proposed for calibrating of the local error model.

1. Select a representative set of points over the typical operating state of the system, $x_{[k]} \in \{X\}, k = 1 \dots N$. Notice that the sample points selected do not need to form a solution to the differential equation, they only need to cover the points to be traversed in the state space adequately.

2. At each point, apply the intended numerical method to make a single step, $x^{\psi}_{i+1,[k]} = x_{i,[k]} + h_{i,[k]} \psi(x_{i,[k]}, h_{i,[k]})$. Also record $f_{i,[k]} = f(x_{i,[k]})$. Calculating the local error over a range of plausible step sizes, or making use of an automatic step size control algorithm to capture the step sizes appropriate to various points in the state space ensures that the local errors are properly represented in the data.

3. Use a higher order method or sub-steps (possibly with Richardson extrapolation) to get an accurate estimate of the state at the next time, $x_{i+1,[k]} \approx x_{[k]}(t_i + h)$, and hence estimate the local error, $e_{i,[k]} = x_{i+1,[k]} - x^{\psi}_{i+1,[k]}$. This step is only performed with high precision during calibration but typically there would be some form of (cheap) step size control to limit the worst-case local error during normal operation.

4. Use linear least squares (or any other method) to find elements of a fixed M (with rows, m_r^{T}) to minimise the mean square residual error, $\gamma_{r,i,[k]} = f^{\mathrm{T}}_{i,[k]} m_r - e_{r,i,[k]}/h^{(p+1)}_{i,[k]}$ over k. (M can be solved in a single least-squares step.) As $M f(x_i) h^{p+1}$ is scaled by h^{p+1} it does not affect the asymptotic error and elements of M may be forced to zero, if statistical analysis of variance supports the hypothesis that individual elements are zero within acceptable confidence levels, or if modelling of the problem indicates that this choice is appropriate. The justification for weighting the local error with $1/h^{p+1}$ is to follow the asymptotic error behaviour and will penalise large errors incurred with smaller step sizes. Different weighting could be used to emphasise areas of the state space that require tighter modelling of the error.

5. Using the white noise model of the residual errors, $\xi_{r,i} \sim \mathcal{N}(\mathbf{0}, Q_r)$ to give $\vartheta_{r,i} = \left[f^{\mathrm{T}}(x_i),\ 1\right] \xi_{r,i}$, find Q_r to minimise $\|\gamma^2_{r,i} - \mathcal{E}\{\vartheta^2_{r,i}\}\|_s$ over k for some norm, s. This problem can be solved in two steps: (i) For diagonal $Q_r = \operatorname{diag}\{q_{jj}\}$, $\mathcal{E}\{\vartheta^2_{r,i}\} = \sum_{j=1}^{n+1} q_{jj} f_j^2$

with $f_{n+1} = 1$) is linear in unknown q_{jj}, and can be found using a non-negative least squares solver (e.g. Matlab's lsqnonneg). (ii) If required, and having regard to the risk of over-fitting, find (e.g. upper) triangular Cholesky factor $\boldsymbol{R}$, ($\boldsymbol{R}^{\mathrm{T}}\boldsymbol{R} = \boldsymbol{Q}_r$) to solve the optimisation problem. The diagonal $\boldsymbol{R} = \mathrm{diag}\{\sqrt{q_{jj}}\}$ found above provide an appropriate starting value. The choice of metric, s (e.g. one- or two-norm) for the fit between $\gamma^2_{r,i}$ and $\vartheta^2_{r,i}$ will affect the result of the optimisation.

2.1 Variations on the theme

Step 5 can be interpreted/implemented in other ways:

1. It may be preferred to assume diagonal, $\boldsymbol{Q}_r$, both for simplicity and to avoid the risk of over-fitting. This assumes that the residual error covariance depends only on f_a^2 terms and not on product terms, $f_a f_b, a \neq b$ which may miss couplings that are physically relevant.

2. The residual error can be viewed/modelled per row as $\vartheta_{r,i} = \left[\boldsymbol{f}^{\mathrm{T}}(\boldsymbol{x}_i),\ 1\right]\boldsymbol{w}_r\ \xi_{r,i}$, with $\xi_{r,i} \sim \mathcal{N}(0,1)$ a scalar white noise term with unit variance, and $\boldsymbol{w}_r$ a weighting vector, found to minimise $\|\gamma^2_{r,i} - \mathcal{E}\{\vartheta^2_{r,i}\}\|_p$. The non-negative least squares solution discussed above for diagonal $\boldsymbol{Q}_r$ does not apply as $\mathcal{E}\{\vartheta^2_{r,i}\} = \left(\sum_{j=1}^{n+1} m_{r,j} f_j\right)^2 \neq \sum_{j=1}^{n+1} r^2_{jj} f^2_j$.

3. As noted in Section 3 below, if the residual error vector over the entire simulation is modelled as, $\boldsymbol{\vartheta}_i = \boldsymbol{W}\begin{bmatrix}\boldsymbol{f}(\boldsymbol{x}_i)\\ 1\end{bmatrix}\xi_i$ with $\xi_i \sim \mathcal{N}(0,1)$ a scalar stochastic element, the equivalent, continuous time stochastic differential equation can be simulated using the strong first order Milstein method for the special case of scalar noise. This requires finding the (n^2+n) elements of the matrix $\boldsymbol{W}$ to minimise the matrix norm $\|\boldsymbol{\gamma}_i\boldsymbol{\gamma}_i^{\mathrm{T}} - (\boldsymbol{W}\boldsymbol{f}_i)(\boldsymbol{W}\boldsymbol{f}_i)^{\mathrm{T}}\|_s$ over k. The row solutions in variation (2) above (corresponding to the case that the rows of $\boldsymbol{\gamma}_i$ are uncorrelated) might be used as a starting value for numerical optimisation.

4. We have tried an alternative approach of constructing (term by term), a least-length pseudo noise term from the residual errors via, ${\zeta_{r,i,[k]}}^{\mathrm{T}} = \gamma_{r,i,[k]}\begin{bmatrix}\boldsymbol{f}(\boldsymbol{x}_i)\\ 1\end{bmatrix}^{\dagger}$ and then finding the covariance, $\mathrm{cov}(\boldsymbol{Z}_r)$, with $\boldsymbol{Z}_r = [\zeta_{r,i,[1]}, \ldots, \zeta_{r,i,[N]}]^{\mathrm{T}}$. (Recall that the pseudo inverse of a vector, z, is very simple, $z^{\dagger} = z^{\mathrm{T}}/\|z\|_2^2$). We have found that this approach is optimistic: While (by construction) the pseudo-noise generates the residual error, because of the way it allocates the residual error to individual pseudo-noise terms the scheme typically gives too low a variance and since $\mathcal{E}\left\{\xi_{r,i,[k]}\ {\xi_{r,i,[k]}}^{\mathrm{T}}\right\} = \left[f_{i,[k]}\ {f_{i,[k]}}^{\dagger}\right]^{\mathrm{T}} Q_r \left[f_{i,[k]} {f_{i,[k]}}^{\dagger}\right]$, the resulting covariance cannot be expected to be constant. Finally, the dynamic range of $\boldsymbol{f}(\boldsymbol{x})$ depends on choice of units so weighting may be required.

The algorithm could be run on-line if computing resources allowed the local errors to be calculated accurately on the fly. Although this use-case seems to defeat the purpose of the exercise, the model parameters, $\boldsymbol{m}_r$, and $\boldsymbol{Q}_r$, could be adapted as new information becomes available while traversing of the operating region of the state space.

3 Continuous time extension of stochastic, local error model

There are applications where a continuous time equivalent of the local error model will be useful. These include continuous-discrete state estimation incorporating simulation error effects, cheap local and global error estimation, and step-size control. This section develops the continuous time extension of the results above by representing the continuous time system via a stochastic differential equation. [1] modelled the local error as a stochastic term added to the differential equation and calibrated this as a zero mean, iid Gaussian sequence, with a covariance scaled by h^{2p+1}. In this section, this idea is developed and applied to the more complicated local error model in (4).

If $\chi(t)$ is a zero mean, white Gaussian noise with covariance, $\boldsymbol{Q}_\chi(t)\delta(t)$, ($\delta(t)$ is the Dirac delta function) and $\boldsymbol{Q}_\chi(t)$ constant over the interval, $t \in (0, \tau]$. Then, the Wiener process, $\xi(\tau) = \int_0^\tau \chi(\lambda)d\lambda$, is zero mean with covariance, $\boldsymbol{Q}_\xi = \tau\boldsymbol{Q}_\chi$.

To develop a continuous time extension of the discrete time result in Section 2, the Taylor series expansions of the true solution and solution of the simulation method over the interval, $t = t_i + \tau,\ \tau \in (0, h]$, are written as follows. True solution:

$$\begin{aligned}\boldsymbol{x}(t_i+\tau) &= \boldsymbol{x}_i + \tau\boldsymbol{f}(\boldsymbol{x}_i) + \frac{\tau^2}{2}\frac{\partial \boldsymbol{f}}{\partial \boldsymbol{x}}\bigg|_{\boldsymbol{x}_i}\boldsymbol{f}(\boldsymbol{x}_i) + \frac{\tau^3}{3!}\boldsymbol{D}^{(2)}_{\boldsymbol{x}_i}\boldsymbol{f}(\boldsymbol{x}_i) + \dots \\ &= x_i + \int_{t_i}^{t_i+\tau} f_i(\lambda)d\lambda\end{aligned} \tag{5}$$

with all quantities evaluated at $(\boldsymbol{x}_i, t_i)$, e.g. $\boldsymbol{f}_i = \boldsymbol{f}(\boldsymbol{x}_i, t_i)$ and,

$$\boldsymbol{f}_i(\tau) = \boldsymbol{f}(\boldsymbol{x}_i) + \tau\frac{\partial \boldsymbol{f}}{\partial \boldsymbol{x}}\bigg|_{\boldsymbol{x}_i}\boldsymbol{f}(\boldsymbol{x}_i) + \frac{\tau^2}{2}\boldsymbol{D}^{(2)}_{\boldsymbol{x}_i}\boldsymbol{f}(\boldsymbol{x}_i) + \dots \tag{6}$$

Simulation method: The single step method, (2), is written out in the interval, giving a continuous time function, $\boldsymbol{y}(t)$,

$$\begin{aligned}\boldsymbol{y}(t_i+\tau) &= \boldsymbol{y}_i + \tau\boldsymbol{\psi}(\boldsymbol{y}_i,\tau) \\ &= \boldsymbol{y}_i + \tau\boldsymbol{\psi}(\boldsymbol{y}_i,0) + \tau^2\frac{d\boldsymbol{\psi}(\boldsymbol{y}_i,\tau))}{d\tau}\bigg|_{\tau=0} + \frac{\tau^3}{2!}\frac{d^2\boldsymbol{\psi}(\boldsymbol{y}_i,\tau)}{d\tau^2}\bigg|_{\tau=0} + \dots \\ &= \boldsymbol{y}_i + \int_{t_i}^{t_i+\tau} \boldsymbol{g}_i(\lambda)d\lambda\end{aligned} \tag{7}$$

$\boldsymbol{g}_i(\tau) = \boldsymbol{\psi}(\boldsymbol{y}_i,\tau) + \tau\frac{d\boldsymbol{\psi}(\boldsymbol{y}_i,\tau))}{d\tau}$ can be written as the Taylor series expansion of the time derivative of $\tau\boldsymbol{\psi}(\boldsymbol{y}_i,\tau)$ in the interval. If a deterministic correction is used in the simulation, the corresponding continuous time extension would be $\boldsymbol{y}(t_i+\tau) = \boldsymbol{y}_i + \tau\boldsymbol{\psi}(\boldsymbol{y}_i,\tau) + \tau^{p+1}\boldsymbol{M}\boldsymbol{f}(\boldsymbol{x}_i)$.

The continuous time extension of the local error is $\boldsymbol{e}(t) = \boldsymbol{x}(t) - \boldsymbol{y}(t)$,

$$\begin{aligned}\boldsymbol{e}(t_i+\tau) &= \boldsymbol{x}(t_i+\tau) - \boldsymbol{y}(t_i+\tau)|_{\boldsymbol{y}_i=\boldsymbol{x}_i} \\ &= \frac{\tau^{p+1}}{(p+1)!}(\boldsymbol{D}^{(p)}_{\boldsymbol{x}_i} - \boldsymbol{D}^{(p)}_{\boldsymbol{y}_i} + \boldsymbol{O}_e(\tau))\boldsymbol{f}(\boldsymbol{x}_i) \\ &= \int_{t_i}^{t_i+\tau} \boldsymbol{\ell}_i(\lambda)d\lambda.\end{aligned} \tag{8}$$

where $\boldsymbol{\ell}_i(s)$ is the time derivative of the local error's continuous time expansion,

$$\boldsymbol{\ell}(\tau) = \frac{\tau^p}{p!}\left(\boldsymbol{D}^{(p)}_{\boldsymbol{e}_i} + \boldsymbol{O}_\ell(\tau)\right)\boldsymbol{f}(\boldsymbol{x}_i). \tag{9}$$

Table 1. Continuous time extension for Heun rule

Heun	$x_{i+1} = x_i + h/2\,(f(x_i) + f(x_i + hf(x_i)))$
Continuous time extension	$g(t_i+\tau) = f + \tau\frac{\partial f}{\partial x}f + \frac{3\tau^2}{4}\left(\frac{\partial^2 f}{\partial x^2}f^2\right) + \ldots$
Local error	$e(t_i+\tau) = \tau^3\left(-\frac{1}{12}\frac{\partial^2 f}{\partial x^2}f + \frac{1}{6}\left(\frac{\partial f}{\partial x}\right)^2 + O_e(\tau)\right)f$
Local error derivative	$\ell(t_i+\tau) = \tau^2\left(-\frac{1}{4}\frac{\partial^2 f}{\partial x^2}f + \frac{1}{2}\left(\frac{\partial f}{\partial x}\right)^2 + O_\ell(\tau)\right)f$
LTI local error	$\boldsymbol{e}(t_i+\tau) = \tau^3\left(\frac{1}{6}A^2 + \frac{\tau}{4!}\boldsymbol{A}^3 + \boldsymbol{O}(\tau^2)\right)\boldsymbol{A}\boldsymbol{x}_i$
LTI local error derivative	$\boldsymbol{\ell}(t_i+\tau) = \tau^2\left(\frac{1}{2}\boldsymbol{A}^2 + \frac{\tau}{3!}\boldsymbol{A}^3 + \boldsymbol{O}(\tau^2)\right)\boldsymbol{A}\boldsymbol{x}_i$

Table 2. Continuous time extension for linearly implicit Taylor-Heun

Heun	$x_{i+1} = x_i + h\left(1 - \frac{h}{2}\left.\frac{\partial f}{\partial x}\right\|_i\right)^{-1} f(x_i)$
Continuous time extension	$g(t_i+\tau) = f + \tau\frac{\partial f}{\partial x}f + \frac{3\tau^2}{4}\left(\frac{\partial f}{\partial x}\right)^2 f + \ldots$
Local error	$e(t_i+\tau) = \tau^3\left(\frac{1}{6}\frac{\partial^2 f}{\partial x^2}f - \frac{1}{12}\left(\frac{\partial f}{\partial x}\right)^2 + O_e(\tau)\right)f$
Local error derivative	$\ell(t_i+\tau) = \tau^2\left(\frac{1}{2}\frac{\partial^2 f}{\partial x^2}f - \frac{1}{4}\left(\frac{\partial f}{\partial x}\right)^2 + O_\ell(\tau)\right)f$
LTI local error	$\boldsymbol{e}(t_i+\tau) = \tau^3\left(-\frac{1}{12}\boldsymbol{A}^2 - \frac{\tau}{12}\boldsymbol{A}^3 + \boldsymbol{O}\left(\tau^2\right)\right)\boldsymbol{A}\boldsymbol{x}_i$
LTI local error derivative	$\boldsymbol{\ell}(t_i+\tau) = \tau^2\left(-\frac{1}{4}\boldsymbol{A}^2 - \frac{\tau}{3}\boldsymbol{A}^3 + \boldsymbol{O}\left(\tau^2\right)\right)\boldsymbol{A}\boldsymbol{x}_i$

The above developments mean that, $\dot{\boldsymbol{x}}(t) = \boldsymbol{f}(t) = \boldsymbol{g}_i(t) + \boldsymbol{\ell}_i(t)$ in the interval $t = t_i + \tau$, $\tau \in (0, h]$. Both $\boldsymbol{e}(t_i + \tau)$ and $\boldsymbol{\ell}(t_i + \tau)$ depend on a principal term, $\boldsymbol{D}^{(p)}_{e_i} = \boldsymbol{D}^{(p)}_{x_i} - \boldsymbol{D}^{(p)}_{y_i}$, scaled by $\frac{\tau^{p+1}}{(p+1)!}\boldsymbol{f}(\boldsymbol{x}_i)$ and $\frac{\tau^p}{p!}\boldsymbol{f}(\boldsymbol{x}_i)$ respectively, and on higher order terms (also scaled by $\boldsymbol{f}(\boldsymbol{x}_i)$ but dependent on the step size). The higher order expansions in the local error, $\boldsymbol{O}_e(\tau)$, and local error derivative, $\boldsymbol{O}_\ell(\tau)$, do not have the same coefficients, for example, an LTI system with an explicit method (truncated Taylor series) has, $\boldsymbol{\ell}(t_i + \tau) = \left(\frac{\tau^p}{p!}\boldsymbol{A}^p + \boldsymbol{O}_\ell(\tau)\right)\boldsymbol{A}\boldsymbol{x}_i$.

As examples of the above discussion, the Heun (modified Euler) and linearly implicit Taylor-Heun schemes applied to the scalar, autonomous system, $\dot{x} = f(x)$, $x(t_0) = x_0$, with correct $x_i = x(t_i)$; and to the linear time invariant (LTI) system $\dot{\boldsymbol{x}} = \boldsymbol{A}\boldsymbol{x}$ are summarised in Tables 1 and 2 respectively. For the LTI examples, the next higher order term in the error expansion is shown to illustrate that $\boldsymbol{O}_e(\tau)$ and $\boldsymbol{O}_\ell(\tau)$ terms are not the same.

The above calculations are possible because $\boldsymbol{\ell}_i(t)$ is deterministic. If however, (after removing the deterministic component), $e_{r,i} = e_r(t_i + h)$ is modelled via the stochastic term, $h^{p+1}\vartheta_{r,i} = h^{p+1}\left[f_i^{\mathrm{T}},\ 1\right]\xi_{r,i}$, the stochastic version of $\ell_r(t_i + \tau) = \left[f_i^{\mathrm{T}}\ 1\right]\chi_r$ with $\chi_r \sim N(\boldsymbol{0}, h_i^{2p+1}\boldsymbol{Q}_r)$ so that $\operatorname{cov}\left(\int_0^{h_i}\chi(\tau)d\tau\right) = h_i^{2(p+1)}\boldsymbol{Q}_r = \operatorname{cov}\left(h_i^{p+1}\xi_{r,i}\right)$ as required. Although this arrangement ensures that the local error has the required covariance at the end of each step, it does not model the error correctly during the step because of the difference between deterministic and stochastic integrals. For the deterministic version, the principal component of the local error is constant during the evolution of the continuous time extension, while for the stochastic version, the principal error effect (standard deviation) grows with $\sqrt{\tau}$. As most studies will end up with numerical simulation of each step, the under-estimation of the local error covariance during the initial part of the continuous time extension will not matter.

Defining the stochastic increment per row as, $d\boldsymbol{\beta}_r = \boldsymbol{\chi}_r dt$, the simulation of $\dot{x}_r(t) = f_r(\boldsymbol{x})$, can be approximated and studied via the stochastic differential equation,

$$dx_r(t) = g_{r,i}(t)dt + \left[\boldsymbol{f}^{\mathrm{T}}(\boldsymbol{x}_i),\ 1\right] d\boldsymbol{\beta}_r \tag{10}$$

If (10) is simulated using the strong order 0.5 Euler-Maruyama approach ([7]), the original, generally higher order, deterministic method selected for the problem is retained as $\int_{t_i}^{t_i+h_i} g_{r,i}(\tau)d\tau = h_i\psi_{r,i}(\boldsymbol{x}_i)$ from (7) and the stochastic terms are propagated with low order. This scheme is justified by noting that the noise term $\boldsymbol{\chi}_r$ is pre-scaled by $h_i^{p+0.5}$ before integration. [3] emphasises the difference between solvers for stochastic differential equations and the present problem of probabilistic numerics where the variance of the local error model decays faster than h. No general claims about the resulting method order are made here and the cautions raised in [8] should be noted. The algorithm would be,

1. Draw independent random vectors, $\Delta\boldsymbol{\beta}_{r,i}$ from the distributions, $\Delta\boldsymbol{\beta}_{r,i} \sim \mathcal{N}(\boldsymbol{0}, h_i^{2(p+1)}\boldsymbol{Q}_r)$.
2. Propagate the state vector via,

$$x_{r,i+1} = x_{r,i} + h_i\psi_r(\boldsymbol{x}_i) + h^{p+1}\boldsymbol{f}_i^{\mathrm{T}}\boldsymbol{m}_r + \left[\boldsymbol{f}^{\mathrm{T}}(\boldsymbol{x}_i)\ 1\right]\Delta\boldsymbol{\beta}_{r,i} \tag{11}$$

If the local error is modelled with a scalar noise term that scales all rows of the SDE, $d\boldsymbol{x}(t) = \boldsymbol{g}_i(t)dt + \boldsymbol{W}\begin{bmatrix}\boldsymbol{f}(\boldsymbol{x}_i),\\ 1\end{bmatrix} d\beta$, the state can be propagated using a modification of the first order Milstein method ([7]), giving,

$$\boldsymbol{x}_{i+1} = \boldsymbol{x}_i + h_i\boldsymbol{\psi}(\boldsymbol{x}_i) + \boldsymbol{W}\left(\begin{bmatrix}\boldsymbol{f}(\boldsymbol{x}_i),\\ 1\end{bmatrix}\Delta\beta_i + \frac{1}{2}\begin{bmatrix}\frac{\partial \boldsymbol{f}}{\partial \boldsymbol{x}}\\ \boldsymbol{0}^{\mathrm{T}}\end{bmatrix}\boldsymbol{W}\begin{bmatrix}\boldsymbol{f}(\boldsymbol{x}_i),\\ 1\end{bmatrix}(\Delta\beta_i^2 - qh_i)\right) \tag{12}$$

Notice that the Jacobian required in (12) for the stochastic component is already available for linearly implicit methods.

3.1 Application of results to numerical simulation

The results developed above may be applied to numerical simulation in several ways:

3.1.1 Simulation of an ensemble of solutions:

For each solution the underlying numerical method is modified and simulated via (11).

3.1.2 Simulation of mean and covariance of solutions:

The solution of the mean, $\bar{\boldsymbol{x}}(t) = \mathcal{E}\{\boldsymbol{x}(t)\}$, and covariance, $\boldsymbol{P}(t) = \mathcal{E}\left\{(\boldsymbol{x}(t) - \bar{\boldsymbol{x}}(t))(\boldsymbol{x}(t) - \bar{\boldsymbol{x}}(t))^{\mathrm{T}}\right\}$ of the stochastic difference equation can be obtained using standard techniques of either linearizing the covariance equation, using unscented transforms or Monte Carlo particle approaches. The covariance solution provides an estimate of the uncertainty range of the global error. Ignoring high order terms the mean is,

$$\frac{d}{dt}\bar{\boldsymbol{x}} = \mathcal{E}\{\boldsymbol{f}(\boldsymbol{x})\} \approx \boldsymbol{f}(\bar{\boldsymbol{x}}) \tag{13}$$

giving,

$$\bar{x}_{i+1} \approx \bar{x}_i + h\psi(\bar{x}_i, h) + h^{p+1} M f(\bar{x}_i) \tag{14}$$

The covariance is,

$$\begin{aligned} \frac{d}{dt} P &= \mathcal{E}\{(x - \hat{x}) f^{\mathrm{T}}(x) + f(x)(x - \hat{x})^{\mathrm{T}} + L(x) Q L^{\mathrm{T}}(x)\} \\ &\approx \left.\frac{\partial f}{\partial x}\right|_{\hat{x}} P + P\left(\left.\frac{\partial f}{\partial x}\right|_{\hat{x}}\right)^{\mathrm{T}} + L(\bar{x}) Q L^{\mathrm{T}}(\bar{x}) \end{aligned} \tag{15}$$

with $L(\bar{x}) = \mathrm{diag}\left\{\left[f^{\mathrm{T}}(x_i),\ 1\right]\right\}$ and $Q = \mathrm{diag}\{Q_r\}$. The approximations in (13) and (15) assume sufficiently smooth behaviour of $f(x)$ around the mean value - see for instance [9], Section 4.9 for second order expansions of the mean and covariance, or [7]. To discretize (15) with a simple (recall that the noise covariance is already $O(h^{2p+1})$), A-stable method that preserves positive semi-definite behaviour of P, the following Padé type approximation is recommended ([10]),

$$P_{i+1} \approx M_i P_i M_i^{\mathrm{T}} + K_i L_i Q_i L_i^{\mathrm{T}} K_i^{\mathrm{T}} \tag{16}$$

with $L_i = L(\bar{x}_i)$, $K_i = \left(I - \frac{h}{2} A_i\right)^{-1}$, $M_i = K_i\left(I + \frac{h}{2} A_i\right)$, and $A_i = \frac{\partial f}{\partial x}$. ([10] uses a mid-time-step calculation to achieve second order accuracy. Also see [11] for further discussion.)

3.1.3 Step size control based on stochastic model of local error:

If the local error, (4), is well-characterised by the noise term over the operating state of the system, this can be used for cheap local error estimation and control as $f(x_i)$ is calculated at the start of every step. For any step, the stochastic model of the local error for row r will have standard deviation, $\sigma_{r,i} = h^{p+1} \left\| \left[f^{\mathrm{T}}(x_i),\ 1\right] R_r^{\mathrm{T}} \right\|_2$.

If the error standard deviation is required to be less than $\hat{\sigma}_r = g_r h$, (the additional h to take care of the step size effect on the global error accumulation), the step size should be

$$h \leq \sqrt[p]{g_r / \left\| [f^{\mathrm{T}}(x_i),\ 1] R_r^{\mathrm{T}} \right\|_2} \tag{17}$$

to ensure that the step size is small whenever there is a high rate of change in a direction that causes high local error sensitivity. This idea is illustrated in Example 2.

4 Examples

Both examples in this section use inappropriately large step size to give visual illustrations of the results in the paper. Example 1 uses the first order Euler method and Example 2 uses the second order Heun (modified Euler) method.

4.1 Example 1: Explicit Euler Method

The explicit Euler method, $x_{i+1} = x_i + hf(x_i)$ is applied to the FitzHugh-Nagumo oscillator used as an example in [1]: $\frac{dx_1}{dt} = c\left(x_1 - x_1^3/3 + x_2\right)$, $\frac{dx_2}{dt} = -\frac{1}{c}(x_1 - a + bx_2)$ with $(a,\ b,\ c) = (0.2,\ 0.2,\ 3.0)$ and $x_0 = (-1,\ \ 1)^{\mathrm{T}}$.

The principal term in the local error can be calculated readily as, $e(t_i + \tau) = \tau^2\left(\frac{1}{2}\frac{\partial f}{\partial x} + O(\tau)\right) f$ with $\frac{1}{2}\frac{\partial f}{\partial x} = \begin{pmatrix} 1.5(1 - x_1^2) & 1.5 \\ -0.167 & -0.033 \end{pmatrix}$. The approach outlined above is applied for $h = 0.01$ and $h = 0.1$ over the time interval $t \in (0,\ 20)$. Two cases of region of

Table 3. FitzHugh-Nagumo Oscillator

$h = 0.01$: On trajectory
$\boldsymbol{M} = \begin{pmatrix} 0.0321 \pm 0.0319 & 1.5705 \pm 0.0974 \\ -0.1667 \pm 0.0000 & -0.0351 \pm 0.0001 \end{pmatrix}$
$\begin{pmatrix} \sigma_{y1} & \sigma_{y2} \end{pmatrix} = \begin{pmatrix} 2.0198 & 0.0022 \end{pmatrix}$
$\boldsymbol{R}_1 = \text{diag}\left\{\begin{bmatrix} 1.1096 & 2.9702 & 0.0000 \end{bmatrix}^{\text{T}}\right\}$
$\boldsymbol{R}_2 = \text{diag}\left\{\begin{bmatrix} 0.0012 & 0.0033 & 0.0000 \end{bmatrix}^{\text{T}}\right\}$
$h = 0.01$: In disc
$\boldsymbol{M} = \begin{pmatrix} -0.1659 \pm 0.0846 & -0.2650 \pm 0.5369 \\ -0.1664 \pm 0.0001 & -0.0330 \pm 0.0006 \end{pmatrix}$
$\begin{pmatrix} \sigma_{y1} & \sigma_{y2} \end{pmatrix} = \begin{pmatrix} 4.0711 & 0.0045 \end{pmatrix}$
$\boldsymbol{R}_1 = \text{diag}\left\{\begin{bmatrix} 0.9423 & 7.7670 & 0.0000 \end{bmatrix}^{\text{T}}\right\}$
$\boldsymbol{R}_2 = \text{diag}\left\{\begin{bmatrix} 0.0010 & 0.0087 & 0.0000 \end{bmatrix}^{\text{T}}\right\}$
$h = 0.1$: On trajectory
$\boldsymbol{M} = \begin{pmatrix} -0.1021 \pm 0.0997 & 1.5282 \pm 0.3049 \\ -0.1656 \pm 0.0011 & -0.0504 \pm 0.0034 \end{pmatrix}$
$\begin{pmatrix} \sigma_{y1} & \sigma_{y2} \end{pmatrix} = \begin{pmatrix} 1.9966 & 0.0223 \end{pmatrix}$
$\boldsymbol{R}_1 = \text{diag}\left\{\begin{bmatrix} 1.1463 & 2.4561 & 0.0000 \end{bmatrix}^{\text{T}}\right\}$
$\boldsymbol{R}_2 = \text{diag}\left\{\begin{bmatrix} 0.0127 & 0.0289 & 0.0000 \end{bmatrix}^{\text{T}}\right\}$
$h = 0.1$: In disc
$\boldsymbol{M} = \begin{pmatrix} -0.2400 \pm 0.0763 & 0.9059 \pm 0.4839 \\ -0.1639 \pm 0.0009 & -0.0403 \pm 0.0055 \end{pmatrix}$
$\begin{pmatrix} \sigma_{y1} & \sigma_{y2} \end{pmatrix} = \begin{pmatrix} 3.6691 & 0.0418 \end{pmatrix}$
$\boldsymbol{R}_1 = \text{diag}\left\{\begin{bmatrix} 0.9272 & 5.3044 & 1.4480 \end{bmatrix}^{\text{T}}\right\}$
$\boldsymbol{R}_2 = \text{diag}\left\{\begin{bmatrix} 0.0105 & 0.0681 & 0.0098 \end{bmatrix}^{\text{T}}\right\}$

interest are considered: (i) points along the trajectory from a simulation and (ii) an ellipsoidal disc shaped region covering the anticipated trajectory, $0.65^2 \le \left(\frac{x_1}{2.1}\right)^2 + \left(\frac{x_2}{1.3}\right)^2 \le 1.0^2$. The results are tabulated in Table 3: For each linear least squares fit of the deterministic part of the local error term, $\boldsymbol{M}$, the square root of the diagonal elements of the covariance matrices of the solutions are shown to give an indication of the confidence regions. Also tabulated are the residual error standard deviations and the noise terms standard deviations required to solve the optimisation problem. Only diagonal noise terms are used because a full covariance matrix (Step (5.ii) in Section 2) only reduces the per-element residual error norms from (0.827 and 1.113×10^{-4}) to (0.704 and 9.328×10^{-5}) respectively in the (worst) case of disk region and $h = 0.1$ used below. Notice how the elements of $\boldsymbol{M}$ match $\frac{1}{2}\frac{\partial f}{\partial x}$ for the linear terms in the state space model, especially when the calculations are performed on trajectory points.

Fig. 1 illustrates the solution trajectory obtained with by comparing Matlab's ODE45 code with tight tolerances to the explicit Euler solution with $h = 0.1$, and using the diagonal-

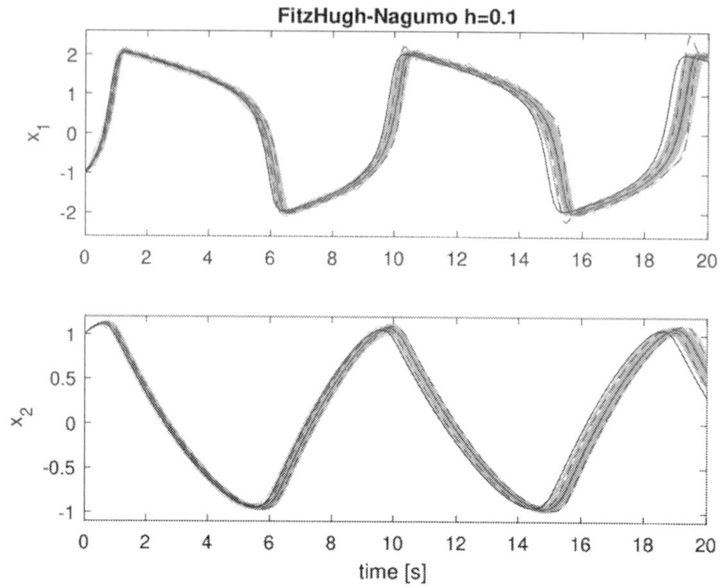

Figure 1. FitzHugh-Nagumo oscillator, $h = 0.1$: Cyan - ensemble of solutions; Blue - accurate solution; Green - low order solution; Red dashed - 1σ bounds;

only results for the disc-shaped region. To illustrate further, an ensemble of solutions to the SDE and the explicit Euler solution with one standard deviation, $y_i \pm \sqrt{\text{diag}(P_i)}$, are shown. Fig. 2 illustrates the development of the global error. Notice that in this simulation, after multiple cycles, the predicted global error is optimistic compared to the actual value.

4.2 Example 2: Heun Method

The Heun method, $x_{i+1} = x_i + h\psi(x_i, h)$ with $\psi(x_i, h) = \frac{1}{2}(f(x_i) + f(x_i + hf(x_i)))$ is applied to the classic two body problem discussed in [5] and elsewhere. The model for an orbit in the x-y plane with the origin as one of the foci, is $\ddot{x} = -x/r^3$, $\ddot{y} = -y/r^3$, with $r^2 = x^2 + y^2$, $(\dot{x}_0, \dot{y}_0, x_0, y_0)^{\mathrm{T}} = \left(0, \sqrt{\frac{1-e}{1+e}}, 0, 1-e\right)^{\mathrm{T}}$ and large eccentricity, $e = 0.9$. Heun is a second order explicit method and the purpose of the example is to illustrate the results presented in this paper rather than to advocate Heun as the best numerical method for this problem. (Higher order methods can exploit the smoothness, and implicit methods and step size control be used to deal with the stiffness.)

The value of $\boldsymbol{M}$ obtained is quite consistent over a range of step sizes but it is averaging the local error effect over the whole of an orbit, which has significant velocity differences at different points due to the eccentricity. Even though the simulation method, $x_{i+1} = x_i + h\psi(x_i, h) + h^3\boldsymbol{M}f(x_i)$ is still globally second order this understanding suggests excluding the offset term (i.e. $\boldsymbol{M} = \boldsymbol{0}$). As a result, the local error is modelled in its entirety via noise terms. The local error data is obtained by selecting points on the orbit found using Matlab's ODE23 with high tolerance.

To illustrate a simple step size control scheme, at each step, the noise covariance per row is calculated from (17) as $h = \min\left\{\left(\frac{p_{tol}}{\sigma_p}\right)^{1/2}, \left(\frac{p_{tol}}{\sigma_v}\right)^{1/3}\right\}$, with p_{tol} the allowed position tolerance per step and σ_p and σ_v the position and velocity standard deviation increments

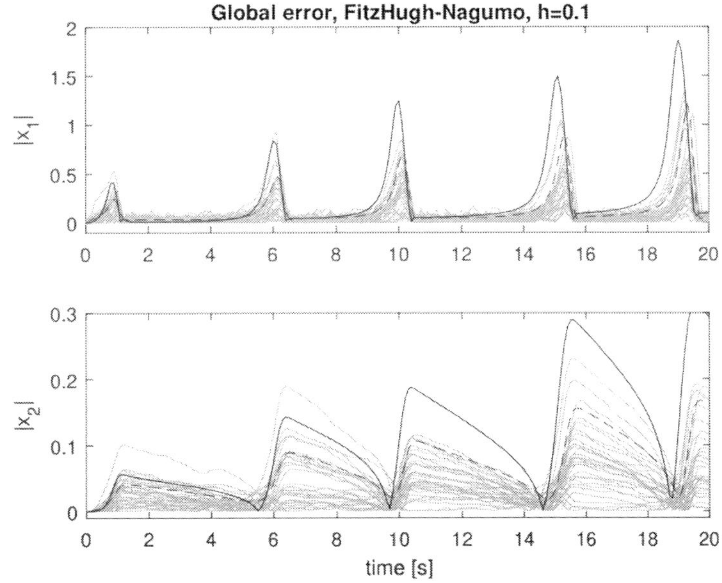

Figure 2. Global error magnitudes for FitzHugh-Nagumo oscillator, $h = 0.1$: Cyan - error for ensemble of solutions; Blue - error between accurate and low order solutions; Red dashed - 1σ bounds;

predicted using the noise model and (16). (Practically, the step size selection is also jacketed between minimum and maximum values but these are not active in the simulation results presented below.)

Fig. 3 shows the position along a single orbit with $p_{tol} = 5 \times 10^{-4}$ used for step-size control. The accurate solution is shown along with Heun method and stochastic differential equation solutions are shown every 30 steps. The step size control algorithm varies the step size between 0.0010 and 0.0158 along the trajectory. Because of the way the step size is calculated this is less reactive than a regular step size control algorithm would be - for example, Matlab's ODE23 (albeit a third order method) used step sizes ranging from 0.0003 to 0.071 for the same problem with tighter tolerance. The displayed points are fairly uniformly distributed in space because larger step sizes are used when the body is far from the origin and moving slowly. The solution mostly remains within the predicted one standard deviation ellipses of the accurate solution but becomes hopelessly uncertain within one orbit at this resolution. The important point is that the error ellipses give insight into the loss of reliability of the solution.

5 Conclusions

This paper has introduced the idea of representing local errors in numerical simulation as stochastic variables formed by the product of the state derivative and white noise, with the possibility of including a constant stochastic term to take care of unmodelled effects and time dependence. This richer representation of local error allows a less conservative model of the local error that, for instance, correctly models zero local simulation error when the system is in steady state. A continuous time extension of the results allows numerical simulation to be modelled using the framework of stochastic differential equations with the time derivative of the single step method as the deterministic part and a state dependent noise to represent the

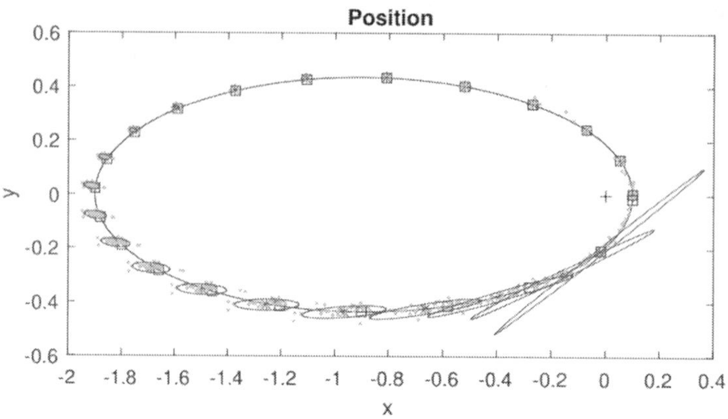

Figure 3. Position for two body problem: Blue - orbit, Red - Heun and uncertainty ellipses, Cyan - SDE solutions, every 30th step

continuous time effect of the local error. The stochastic term's covariance decays to zero with $h^{2p+1)}$. Two low order examples have illustrated various aspects of the proposed approach.

References

[1] Conrad, P.R., Girolami, M., Särkkä, S. Stuart, A. and Zygalakis, K., "Statistical analysis of differential equations: introducing probability measures on numerical solutions," Statistics and Computing 27, 1065–1082, (2017).

[2] Schober, M., Särkkä, S. and Hennig, P., "A probabilistic model for the numerical solution of initial value problems," Statistics and Computing 29, 99–122, (2019).

[3] Lie, H.C., Stuart, A.M. and Sullivan, T.J., "Strong convergence rates of probabilistic integrators for ordinary differential equations," Statistics and Computing, 29, 1265–1283 (2019).

[4] Oats, C.J., Sullivan, T.J., "A modern retrospective on probabilistic numerics," Statistics and Computing 29, 1335–1351 (2019).

[5] Shampine, L.F., Numerical Solution of Ordinary Differential Equations Chapman and Hall, (1994). ISBN 0-412-05151-6

[6] Butcher, J.C., "Numerical methods for ordinary differential equations in the 20th century," Journal of Computational and Applied Mathematics 125, 1-29, (2000).

[7] Särkkä, S. and Solin, A., Applied Stochastic Differential Equations, Cambridge University Press, (2019).

[8] Burrage, K; Burrage, P.; Higham, D. J.; Kloeden, P. E. and Platen, E., "Comment on 'Numerical methods for stochastic differential equations'," Physical Review Part E, 74, 68701, (2006).

[9] Jazwinski, A. Stochastic Processes and Filtering Theory Academic Press, (1970).

[10] Mazzoni, T., "Computational aspects of continuous–discrete extended Kalman-filtering," Computational Statistics, 23, 519–539, (2008).

[11] Paul Frogerais, P., Bellanger, J. and Senhadji, L., "Various Ways to Compute the Continuous-Discrete Extended Kalman Filter," IEEE Transactions on Automatic Control, 57 (4), 1000-1004, (2012).

Optimisation of the rotary friction welding process of titanium alloy rods

Mthobisi Zulu[1*], and *Madindwa* Mashinini[2]

[1]Mangosuthu University of Technology, Department of Mechanical Engineering, Durban, South Africa
[2]University of Johannesburg, Department of Mechanical and Industrial Engineering Technology, Doornfontein, South Africa

Abstract. This paper presents a study on the performance behaviour of friction welded titanium alloy rods. The Ti6Al4V alloy rods were friction welded with continuous-drive friction welding using different combinations of process parameters. The welding speed ranging from 1600 to 2700 RPM and the applied pressure ranging from 25 to 140 MPa were utilized for the welding process. The weld joints and the parent material were examined for microstructure characterisation, Vickers hardness and tensile properties. The microstructure of the weld joint revealed fine equiaxed grains with complete recrystallization. The presence of martensitic grains was obtained at process parameter combinations of low welding speed and high pressure. The tensile tests conducted revealed that the weld joints of low speed and high pressure had improved tensile properties when compared to that of the parent material. The ultimate tensile strength of 1040 MPa with the elongation of 26.5 % was obtained at a speed of 1900 RPM and 80 MPa pressure. The micro-hardness tests revealed an increase in hardness across all the weld zones when compared to the parent material, with the maximum hardness obtained at the weld nugget. Further study was conducted to evaluate the effect of parameters on the weld joint integrity.

1 Introduction

The use and the application of titanium alloys have increased worldwide and currently account for over 50 % of the total titanium usage worldwide [1]. This is due to the alloys having outstanding properties as compared to other materials and alloys in the same group. Titanium alloys have similar mechanical properties as steel alloys yet are approximately 45 % lower in specific gravity. These properties have brought about a wide range of applications that require a high level of reliability, corrosion resistance, and excellent performance in various conditions of applications such as in surgical and mechanical equipment, automotive industries, chemical plants, marine and aerospace industries, and biomedical body implants. In a wide range of engineering applications, titanium alloys have replaced a range of materials due to their ability to work continuously up to a temperature of 600 °C without the loss of mechanical properties, providing reliable components and prolonging the service life

* Corresponding author: zulu.mthobisi@mut.ac.za

of many equipment [2, 3]. The availability of titanium alloys enables engineering designers and fabricators to use them for different forms of critical application, such as in orthopaedics and dental implants [4].

Ti6Al4V alloy is the most used titanium alloy. Its application in various industries has brought about positive changes in the serving life of many pieces of equipment and employees' injury rate due to failure of critical components. This is due to its excellent properties that enable the material to withstand various forms of demanding applications. Although Ti6Al4V alloy has outstanding mechanical properties and obvious economic advantages, the alloy is difficult to weld through conventional welding techniques [5, 6]. This is caused by the high embrittlement and high reactivity of Ti6Al4V when exposed to elevated temperatures. Such a challenge results in its weld joints having poor mechanical properties, and thus become the failure point during testing or application.

The friction welding technique has been the most economical and remarkable process in the welding of various materials including Ti6Al4V alloys. The technique produces the weld joints with improved properties and minimum volume fraction of defects [2, 7]. The friction welding process is a solid-state joining technique that produces the weld joint at a temperature below the solidus temperature of the parent material. The application of rotary friction welding (RFW) process (as one type of friction welding technique) in the joining of various materials has provided a better alternative to the conventional welding techniques with an ability to weld materials of different properties and those regarded as difficult to weld such as Ti6Al4V [2, 8]. Since the process is solid-state, many defects caused by the melting of the interface are not found, which acts as an added advantage over other fusion welding processes [9, 10]. RFW process was found by Bevington in 1891 after recognising that tubes can be welded by frictional heat generated when spinning them in different directions while applying axial pressure [11]. American Welding Society [12] further developed the process and formally defined the recommended practices for the welding industries. Over fifty years, TWI and the expertise have been at the forefront in the development of the process. Various experiments and tests have been conducted to improve the efficiency of the process [13].

Over the past decades, RFW has been gaining familiarity in various industrial applications for critical components. This is due to the process being the most economical, efficient, and successful welding technique in joining of hard materials, and its joints having excellent mechanical properties obtained from the recrystallized, fine grains in the weld zone [14, 15]. Although the process has been widely used in various industries, there have been limited studies that have focused and reported on the performance behaviour of rotary friction welded titanium alloys. Due to an increase in the use of titanium alloys worldwide and an increasing application of RFW process in the joining of various industrial components, the information on the performance behaviour of rotary friction welded titanium alloys has been in great demand. Therefore, this study aims to focus and report on the behaviour and the performance of rotary friction welded Ti6Al4V. The experimental method will be utilized to obtain different properties of the friction welded titanium alloy.

2 Friction welding of Ti6Al4V

The welding of Ti6Al4V has been the biggest challenge for conventional welding processes. Although it can be welded, several challenges require special attention to make the welding process successful and further produce a high-quality weld joint. These challenges may be time-consuming and costly for the production of a single weld joint. For example, when Ti6Al4V is heated (welded) it tends to react with the environment, hence absorbing nitrogen and oxygen in the atmosphere. This causes weld oxidation, embrittlement, and discoloration that eventually deteriorates weld joint properties. Therefore, proper weld shielding is required to isolate the welding process from the environment [2, 16]. Also, the fusion welds of Ti6Al4V alloy are susceptible to many forms of cracking, embrittlement through contaminations, and segregation [17]. Due to these challenges, various fabricators believe that the alloy is hard to weld.

RFW process has been the most efficient and cost-effective welding technique for the production of components of similar and dissimilar materials. Since the process involves rotation and application of pressure, much attention is paid to the selection of process parameters taking into consideration the geometry and the chemical properties of the materials to be joined. When Ti6Al4V is friction welded, the joining occurs below the solidus temperature. Defects and stresses caused by bulk melting do not occur [2, 18]. Embrittlement through contamination and segregation does not occur as well because the process is self-cleaning and filler-free [19, 20].

Numerous studies of friction welding of Ti-6Al4V have been conducted and various researchers have reported that the weld joint of the alloy had remarkable properties. These results were obtained with the use of suitable combinations of process parameters. Munchen [21] reported that friction welding of Ti6Al4v requires one to determine the friction characteristics of the joining material to be able to select the welding speed as the friction coefficient changes with rotational speed. Yates [19] conducted a study on friction welding of Ti6Al4V and the author reports that process parameters play a significant role in the properties of the weld joint. The author reported an increase in tensile strength and hardness when the pressure is increased or the speed is reduced [19]. Bohme et al [7] reported that the friction welding process produces weld joints with improved properties. The authors reported a maximum strength of 1026 MPa and a hardness of 385 HV [7]. Wisbey et al [22] reported an elongation of 15% and the strength of 1030 MPa on the friction welded titanium tubes. Threadgill [20] reported three distinct weld zone in friction welded titanium alloy. These zones are weld nugget (WN), heat affected zone (HAZ), and thermo-mechanically affected zone (TMAZ) [20]. Mashinini and Hattingh [23] found only the WN and HAZ in the friction welding of titanium plates. Although researchers reported different results in friction welding of titanium alloys, it is of great importance to further study the behaviour of friction welded Ti6Al4V in order to improve the efficiency of the process in titanium alloys and to further understand the evolution of different microstructures and properties of the friction welds.

3 Material and experimental procedure

3.1 Material

The material used in the experimental work is the commercially available titanium grade 5 (Ti6Al4V rods) of 25.4 mm diameter. The rods of Ti6Al4V were machined to a diameter of 12.5 mm and a length of 82.5 mm. The chemical composition of the workpiece was V (4 % wt), Al (6 % wt), Fe (0.25 % wt), H (0.015 % wt), C (0,01 % wt), N (0,02 % wt) and O (0.155 % wt). Before the welding process, the workpieces were cleaned to remove organic contaminants, such as grease and oil.

3.2 Experimental procedure

A continuous-drive friction welding machine, Process Development System (PDS), available at Nelson Mandela University was used. The machine has a maximum spindle speed of 10000 RPM, spindle torque of 200 Nm, and a vertical load output of 100 kN. The welding platform is illustrated in Fig. 1.

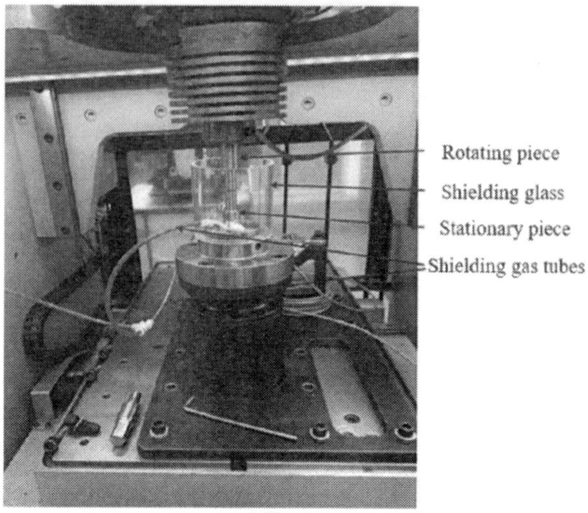

Fig. 1. The welding platform.

The combination of process parameters used to weld Ti6Al4V workpieces are shown in Table 1. The heating and breaking times of the process were obtained using the process cycle produced by each welding process. The range of pressures and rotational speeds were selected based on the preliminary welds analysis and the recommendations made by various authors.

Table 1. Combination of process parameters.

Weld Number	Rotational speed [N] (RPM)	Friction Pressure (MPa)	Forging Pressure (MPa)	Forging Time (sec)
RFW1	1600	25	95	25
RFW2	1900	25	95	25
RFW3	2300	25	95	25
RFW4	2700	25	95	25
RFW5	1600	40	95	25
RFW6	1900	40	95	25
RFW7	2300	40	95	25
RFW8	2700	40	95	25
RFW9	1600	60	95	25
RFW10	1900	60	95	25
RFW11	2300	60	95	25
RFW12	2700	60	95	25
RFW13	1600	80	140	25
RFW14	1900	80	140	25
RFW15	2300	80	140	25
RFW16	2700	80	140	25

The characterisation of the weld joint microstructure was performed using an optical microscope. The samples were cut from the welded specimens using the electrical discharge machining (EDM), hot mounted in a PolyFast resin using CitoPress-1 mounting machine, wet grounded with progressively finer grades of silicon carbide emery paper using water to lubricate and cool and polished with DiaPro Allegro diamond paste and Collodial silica of 9 and 0.04 μm grain size respectively. Finally, the samples were etched using Kroll's reagent for 30 seconds. The surfaces of the etched samples were then viewed using an Olympus SZX16 microscope.

The tensile testing process was performed at a laboratory temperature according to the ASTM E8/E8M-13a testing standard [24], using Zwick/Roell Z250 test machine. Both, the Parent material and welded samples, were tested using a pre-load of 2 MPa and a test speed of 0.002 m/min. The percentage elongation of the tensile test specimens was determined using the initial and final lengths. The microhardness testing was performed on the microstructure samples using the ASTM E92 - 82 hardness test method [25]. The tests were carried out using Time Vickers microhardness testers with a load of 0.3 kg force for a dwell time of 10 seconds. An indentation spacing of 0.5 mm over a length of 10 mm was performed on the samples.

4 Results and discussion

4.1 Process cycle response graph

Fig. 2 illustrates the process response graph that was obtained at a rotational speed of 1600 RPM and 40 MPa friction pressure. The process parameters are presented against the welding time. This process cycle has similarities to that reported by Yates [19] and Palanivel et al [26] in the friction welding of titanium alloys. The response graph is made out of four stages marked by letters A to D, where:

- A – Dry friction stage
- B – Friction stage
- C – Braking Stage, and
- D – Forging stage

The dry friction stage has the workpiece rotating in the air for about 5 seconds, while the speed increases to a predetermined value of 1600 RPM. The friction stage consists of initial contact and the heating phases. The initial contact of the workpieces caused the torque to reach its first peak, as shown in Fig. 2. This was caused by instant contact of stationary and rotating workpieces, and a high abrasion rate as a result of low temperature at the interface. As the friction stage progressed, the torque decreased with an increase in the interface temperature. Due to an increase in the interface temperature to maximum weld temperature and material self-cooling process, the torque dropped to some value considered to be an equilibrium value. The torque oscillated within the equilibrium value for the remainder of the friction stage. There was an axial shortening that was observed as the friction stage progressed. This was caused by the soft material being ejected in a form of flash as a part of the self-cleaning process (all oxides layers and impurities are ejected). The disengagement of the drive motor and the application of brakes at the braking stage, caused the torque to spike up once again and reach the maximum torque of the process. This was caused by the inertia of rotating components and an increase in rotational resistance.

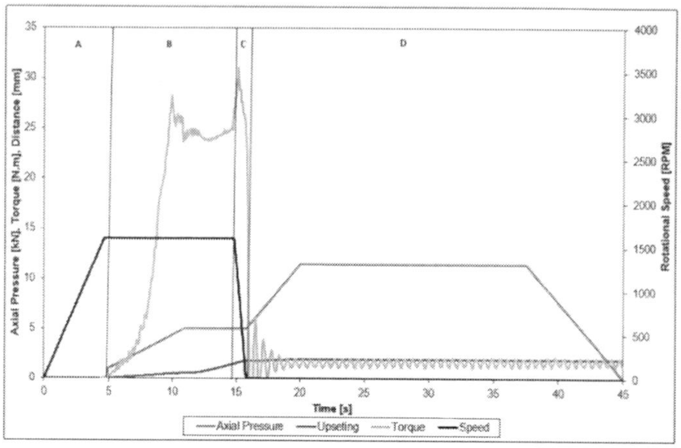

Fig. 2. Process cycle response graph obtained at 1600 RPM and 40 MPa

The application of forging pressure of 95 MPa followed after the breaking stage to consolidate the weld joint before the interface material cools down. This resulted in a further increase in axial shortening to 2 mm. The forging pressure was maintained for 25 seconds to allow the weld joint to cool under pressure. In the end, the pressure was removed and the welding process was complete. The welding time recorded was 45 seconds, as shown in Fig. 2.

4.2 Microstructure of the weld joints

The material's microstructure is a significant element that determines the mechanical behaviour of the component in any form of application. The microstructure Ti6Al4V used for this research study is shown in Fig. 3. The alpha grains (light) are interspersed within the transformed beta (dark). In some regions of the microstructure, coarse acicular alpha grains were observed. This type of microstructure can be linked to different processes of Ti6Al4V processing that give the material its unique properties and allow Aluminium and Vanadium stabilizers to diffuse through the structure.

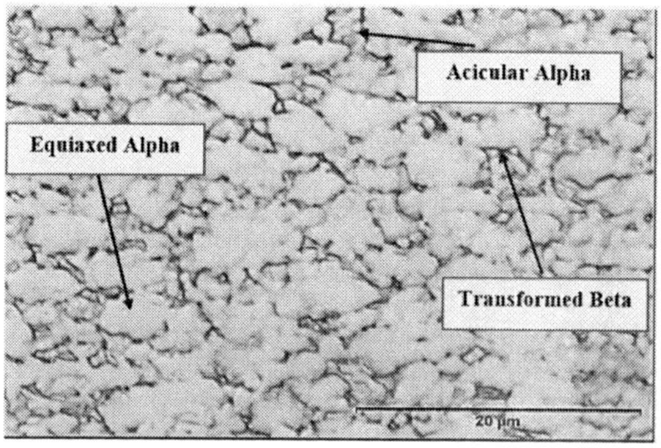

Fig. 3. Microstructure of Ti6Al4V

The microstructure of friction welded Ti6Al4V revealed 3 distinct weld zones namely, weld nugget (WN), thermo-mechanically affected zone (TMAZ), and the heat-affected zone (HAZ). These weld zones were identified based on the variation of the microstructure of the centre of the weld to the Parent material. Fig. 4 illustrates the weld zones of the Ti6Al4V weld joint.

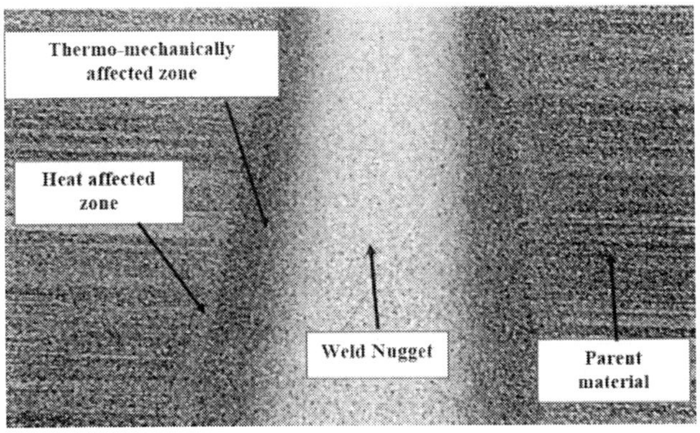

Fig. 4. Distinct weld zones of Ti6Al4V weld joint

The microstructure of the WN revealed fine equiaxed alpha grains within a matrix of Widmanstattten beta grains, as shown in Fig. 5. This was linked to complete recrystallization that occurred as a result of the temperature elevating beyond beta-transus temperature. A basket-weave microstructure with fine acicula alpha particles was also observed at the WN and may be linked to cooling from an extreme temperature, as shown in Fig. 5. Mashinini et al [3] and Munchen [21] reported similar results in their studies on friction welding of Ti6Al4V. Also, very fine 'needle-like' martensitic grains were observed at weld joints of low energy input, as shown in Fig. 5. This was attributed to the heating and cooling processes in a short period as a result of a high-temperature gradient. The use of low welding speed and/or high friction pressure results in the production of the WN microstructure with very fine grains and a high volume fraction of martensitic grains.

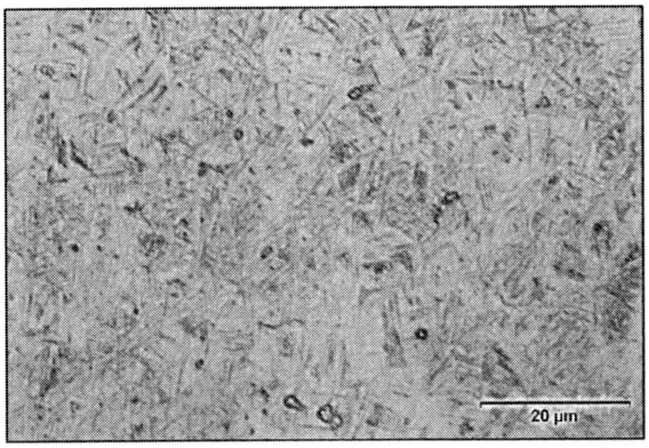

Fig. 5. WN Microstructure

The microstructure of the TMAZ is almost similar to that of the WN because the region gets exposed to extreme temperatures that may result in recrystallization, as shown in Fig. 6. In most cases, it is very difficult to identify and distinguish this region from the WN as its existence is negligibly small in friction welded titanium alloys, resulting in other researchers treating it as part of the WN. However, it is very important to understand and clarify the evolution of this weld zone as it may be a point of poor mechanical properties. As observed, the region contains very fine equiaxed alpha grains with a very small volume fraction of martensitic grains, as shown in Fig. 6. This may be linked to the region being exposed to extreme temperature even though it is not physically involved in the joining process.

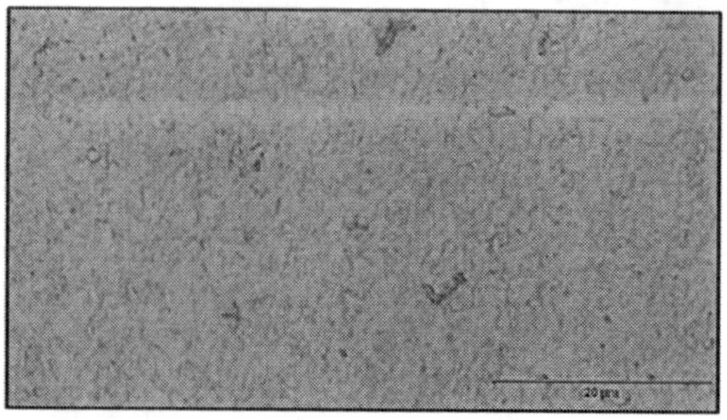

table

Fig. 6. TMAZ microstructure

The microstructure of a HAZ was observed to have elongated equiaxed grains aligned in a flow pattern, as shown in Fig. 7. This was linked to the rotational nature of FRW and the high heat flow rate as this was more visible on weld joints of high heat input, i.e., high welding speed and/or low friction pressure. The volume fraction of beta grains was higher than that of the parent material due to the exposure to elevated temperatures. The same observation was reported by My Nu et al [2] when friction welding titanium alloy.

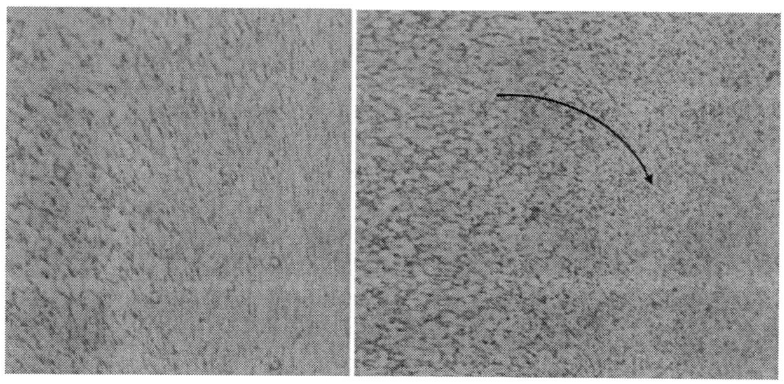

Fig. 7. HAZ Microstructure

4.3 Mechanical properties of welded Ti6Al4V

Table 2. illustrates the tensile test results of weld joints obtained at different process parameters. The tensile test of the parent material revealed the Ultimate Tensile Strength (UTS) of 1026 MPa, the Yield Strength (YS) of 950 MPa, and the elongation of 18 %. The tensile tests conducted on the friction welded Ti6Al4V specimens revealed that the specimens produced at speeds of 1600 and 2700 RPM, and friction pressure of 25 MPa (RFW1 and RFW4) had poor tensile properties. This was concluded after obtaining the failure points within the weld zone, as shown in Fig. 8A. Such results were linked to the use of low friction pressure that may have demoted the rate of heat generation, the inter-atomic bond at the interface, and the efficiency of the self-cleaning process, resulting in the presence of weld defects. The YS, UTS, and the elongation of these specimens were almost equivalent to that of the parent material.

Table 2. Tensile test results

Weld Number	Rotational speed [N] (RPM)	Friction Pressure (MPa)	Ultimate Tensile Strength	Yield Strength	Strain (%)

			[UTS] (MPa)	[YS] (MPa)	
PM			1026	950	18.26
RFW1	1600	25	1029	948	19.87
RFW2	1900	25	1030	951	22.48
RFW3	2300	25	1029	949	22.06
RFW4	2700	25	1026	951	17.73
RFW5	1600	40	1033	947	22.64
RFW6	1900	40	1032	945	24.90
RFW7	2300	40	1036	952	25.34
RFW8	2700	40	1032	946	25.88
RFW9	1600	60	1036	950	25.12
RFW10	1900	60	1037	952	24.08
RFW11	2300	60	1036	947	25.52
RFW12	2700	60	1037	944	26.00
RFW13	1600	80	1038	953	24.73
RFW14	1900	80	1040	948	26.68
RFW15	2300	80	1038	950	26.36
RFW16	2700	80	1035	946	27.32

Fig. 8. Tensile test specimens

The weld joints made at 40, 60, and 80 MPa friction pressures respectively had the failure point away from the weld zone, as shown in Fig. 8B. This revealed that their weld joints had improved tensile properties compared to the parent material. This is attributed to the recrystallized fine equiaxed grains obtained in the weld joints. This was also linked to narrower weld joints achieved as a result of increased friction pressure resulting in low energy input for the process. The maximum UTS of 1040 MPa and the elongation of 26,68 % were achieved at the specimens welded with 1900 RPM rotational speed and 80 MPa (RFW14) friction pressure. Yates [19] and Da Silva [8] reported similar results in their studies.

Fig. 9 illustrates the micro-hardness results obtained at weld joints of 1900 RPM and 80 MPa. The hardness was tested in the outer, middle, and inner diameters of the weld joints. The results revealed an increase in the hardness properties from the HAZ to the WN. The maximum hardness of 376 HV0.5 Ti6Al4V was obtained at the WN, as shown in Fig. 9. This was attributed to the very fine grains achieved at this zone. Other weld joints obtained at different process parameters revealed an increase in the hardness properties on the HAZ, TMAZ, and WN. Such behaviour was linked to the heating and cooling processes that occurred and the fine grains observed. Similar results were reported by Dalgaard [27] in friction welding of titanium alloys.

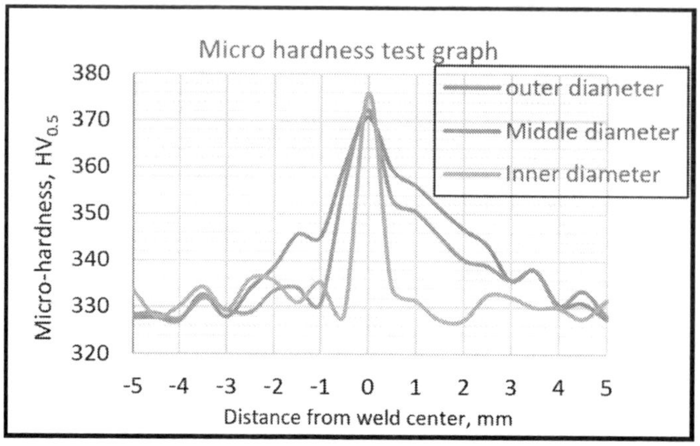

Fig. 9. Micro-hardness test graph

4.4 The effect of speed and pressure on the mechanical properties of friction welded Ti6Al4V

Fig. 10 presents the effect of varying speed and pressure on the tensile and hardness properties of the welded specimens. Rotational speed and friction pressure are the primary and sensitive process parameters of the RFW process. The main function of the speed is to generate the relative motion to enhance heat generation, while the function of friction pressure is to maintain intimate contact of the surfaces, prevent oxidation and enhance consolidation [16]. Fig. 10A and B present the effect of varying speed on the UTS and hardness of welded Ti6Al4V. The results revealed that an increase in rotational speed decreases the UTS and hardness properties. This may be linked to a reduced friction coefficient at high-speed resulting in a low rate of heat generation, and thus wider weld width. Therefore, rotational speed has an inverse proportionality to the UTS and hardness.

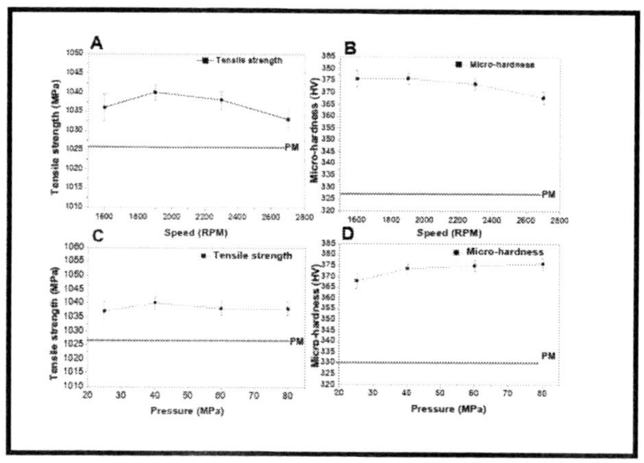

Fig. 10. The effect of speed and pressure on the weld joint properties

Fig. 10C and D illustrate the effect of varying friction pressure on the UTS and hardness of welded Ti6Al4V. The results revealed that the increase in friction pressure increases the UTS and hardness properties of the weld joints. This can be linked to a high heat generation rate at high pressure, promoting recrystallization at the weld interface. The weld joints of high friction pressure were observed to have very fine grains which may be the reason for high UTS and hardness. Friction pressure is directly proportional to the UTS and hardness properties of the weld joint. Friction pressure and welding speed do not have a significant effect on the weld joint properties if varied within the optimum pressure and speed respectively.

5 Conclusion

Titanium alloy Ti6Al4V was successfully welded with the RFW process. The weld joint microstructures achieved had three distinct weld zones. The WN had fine equiaxed alpha grains within the transformed beta grains. The martensitic alpha grains appeared well on the weld joints produced at high friction pressure and low rotational speed. The use of very high or low speed with low friction pressure increases the probability of obtaining welds with poor tensile properties. The maximum UTS, strain, and hardness were achieved at welds of 1900 RPM speed and 80 MPa pressure. Speed and pressure had no significant impact on the weld joint properties when were varied with the optimum range.

Acknowledgement

The success of this research study was made possible by Mr. Brown and Prof. Hattingh of Nelson Mandela University, the staff of MIET at the University of Johannesburg, and Mr. Zondi of the Mangosuthu University of Technology. The authors wish to thank them for their contributions and for providing the platform to perform the experimental work. Lastly, this work is based on the research supported by the National Research Foundation (NRF) of South Africa. The authors wish to thank the support for the success of this study.

References

1. H. Bhadeshia, *Metallurgy of Titanium and its Alloys*, University of Cambridge, launch site forums metals: videos (2008)
2. H. T. My Nu, L. P. Minh, N. H. Loc, *A study on rotary friction welding of titanium alloy (Ti6Al4V)*, Adv. Mater. Sci. Eng., **vol, 2019** (2019)
3. P. Mashinini, I. Dinaharan, J. D. R. Selvam, D. Hattingh, *Microstructure evolution and mechanical characterization of friction stir welded titanium alloy Ti–6Al–4V using lanthanated tungsten tool*, Mater. Charact., **vol. 139, pp. 328-336** (2018)
4. A. McAndrew, P. Colegrove, C. Buhr, B. Flipo, *A literature review of Ti-6Al-4V linear friction welding*, Prog. Mater. Sci., **vol. 1, no. 92, pp. 225-257** (2018)
5. L. M. Gammon, R. D. Briggs, J. M. Packard, K. W. Batson, R. Boyer, C. W. Domby, *Metallography and microstructures of titanium and its alloys*, ASM handbook, **vol. 9, pp. 899-917** (2004)
6. S. Tolvanen, *Microstructure and mechanical properties of Ti-6Al-4V welds produced with different processes*, Chalmersa University of Technology, Gothenburg, Sweden (2016)
7. D. Bohme, L. Appel, H. Cramer, *Joint of titanium/titanium/steel by friction welding with continuous drive*, TIMA, Timisoara, Romania, **pp. 11-12** (2009)
8. A. A. M. Da Silva, *An Investigation on the Structure, Property Relationships of Solid State Welding Processes in a Titanium Matrix Composite Alloy (Ti6Al4V+ 10 Wt.% of TiC)*, GKSS-Forschungszentrum (2006)

9. M. Grujicic, G. Arakere, H. Yalavarthy, T. He, C.-F. Yen, B. Cheeseman, *Modeling of AA5083 material-microstructure evolution during butt friction-stir welding*, J. Mater. Eng. Perform., **vol. 19, no. 5, pp. 672-684** (2010)
10. R. Palanivel, R. Laubscher, I. Dinaharan, *An investigation into the effect of friction welding parameters on tensile strength of titanium tubes by utilizing an empirical relationship,* Meas. Tech., **vol. 98, pp. 77-91** (2017)
11. A. W. Society, *Friction welding process: how does rotary or rotational friction welding works*, AFW (2016)
12. A. W. Society, *Recommended practices for Friction welding*, New York, (ANSI) (1989)
13. TWI, *What is friction welding*, TWI (2018)
14. Y. Morisada, T. Imaizumi, H. Fujii, *Clarification of material flow and defect formation during friction stir welding*, Sci. Technol. Weld. Join., **vol. 20, no. 2, pp. 130-137** (2015)
15. P. Podržaj, B. Jerman, D. Klobčar, *Welding defects at friction stir welding*, Metalurgija, **vol. 54, no. 2, pp. 387-389** (2015)
16. V. I. Vill, *Friction welding of metals*, American Welding Society, trade distributor: Reinhold Publishing Company (1962)
17. E. P. Alves, C. Y. An, F. P. Neto, E. F. dos Santos, *Experimental determination of temperature during rotary friction welding of dissimilar materials*, FAE, **vol. 1, no. 1, pp. 20-26** (2012)
18. G. Pinheiro, A. Meyer, J. dos Santos, *A Literature Review on Friction Welding*, GKSS, J. Mater. Res. Technol. (2000)
19. A. Yates, *The effect of microstructure on mechanical properties in inertia welded mtitanium 6-4*, University of Birmingham (2015)
20. P. L. Threadgill, *The potential for solid state welding of titanium pipes in offshore industries*, Symposium on the Right Use of Titanium, November 4-5 (1997)
21. V. Munchen, *Joint of Titanium/Titanium/Steel by Friction Welding with Continuous Drive*, TIMA, **vol. 1, no. 1** (2000)
22. A. Wisbey, I. Wallis, H. Ubhi, P. Sketchley, C. Ward-Close, P. Thereadgill, *Mechanical properties of friction welds in high strength titanium alloys*, **pp. 1718-1725** (2009)
23. P. Mashinini, D. Hattingh, *Influence of laser heat input on weld zone width and fatigue performance of Ti-6Al-4V sheet*, Mater. Sci. Eng. B., **vol. 262, pp. 114699** (2020)
24. A. Standard, E8/E8M-13a, *Standard Test Methods for Tension Testing of Metallic Materials,* ASTM Int., West Conshohocken, PA, (2013)
25. A. Standard, E92 - 82, *Standard Test Method for Vickers Hardness of Metallic Materials,* ASTM Int., West Conshohocken, **PA 19428** (1997)
26. R. Palanivel, R. Laubscher, I. Dinaharan, D. Hattingh, *Microstructure and mechanical characterization of continuous drive friction welded grade 2 seamless titanium tubes at different rotational speeds*, Int. J. Press. Vessel., **vol. 154, pp. 17-28** (2017)
27. E. C. Dalgaard, *Evolution of microstructure, microtexture and mechanical properties in linear friction welded titanium alloys*, McGill University Library (2011)

The off-design modelling of a combined-cycle power plant

Rushavya Naidu[1,*], and *Wim* Fuls[2]

[1]ATProM Research Unit, University of Cape Town, Cape Town, South Africa
[2]Department of Mechanical Engineering, University of Cape Town, Cape Town, South Africa

Abstract. The objective of this project was to develop a model of a combined-cycle power plant in Flownex which can be solved in off-design conditions in order to compare it to plant data. The verification of this model will show that Flownex can be used to effectively and efficiently model a combined-cycle power plant. The process of development of the final Flownex model was achieved using various additional software. Initially, an analytical model was developed in Mathcad (software used for engineering calculations). Thereafter, a model was built in Virtual Plant, a thermodynamic modelling software for assessing plant performance. Finally, the Flownex model was designed. For the single, double, and triple pressure combined-cycle power plant systems, the analytical, Virtual Plant and Flownex models were compared. The results of all the models agreed closely with one another. The triple-pressure design and off-design Virtual Plant and Flownex models were also compared to plant data and it was concluded that Flownex was successful in modelling the design and off-design conditions of a combined-cycle power plant.

1 List of Nomenclature

Table 1. General Symbols.

h	Enthalpy (kJ/kg)
$\dot{m}$	Mass flow rate (kg/s)
Q	Heat transfer (W)
T	Temperature (°C)
ΔT_{lm}	Log mean temperature difference (K)
UA	Overall conductance (W/K)
Subscripts	
ec	Economiser
ev	Evaporator
g	Gas phase
$g1, g2, g3...$	Gas properties at points of the HRSG

* Rushavya Naidu : rushavyan@gmail.com

s	Water/steam phase
sh	Superheater
so	Steam exiting superheater
v	Saturated steam
w1	Water entering economiser
w2	Water exiting economiser

Table 2. Acronyms and Abbreviations.

ECON	Economiser
EVAP	Evaporator
HP	High pressure
HRSG	Heat Recovery Steam Generator
HT	High temperature
IP	Intermediate pressure
IT	Intermediate temperature
LP	Low pressure
LT	Low temperature
NTU	Number of transfer units
RH	Reheater
SH	Superheater

2 Introduction

Renewable energy presents itself as a solution to the downfalls of fossil fuels, however, these resources are only able to account for less than 10% of the world's total energy demand [1]. Unfortunately, this statistic is not expected to change significantly in the near future. As a result of this, and in order to ensure the conservation of fossil fuels, the increase of efficiency in present power generation systems is critical. Combined-cycle power plants have the highest efficiency in the world in comparison to other power plants that operate on power grids, being between 45% and 57% [2].

These plants consist of a gas turbine as the topping cycle, forming the core of the plant, and a Rankine cycle with a steam turbine as the bottoming cycle [2]. A component called the Heat Recovery Steam Generator (HRSG) forms a connection point between the two cycles. It uses the heat released from the gas turbine to produce high pressure and temperature steam to be sent to the steam turbine. The schematic of a combined-cycle power plant is shown in Figure 1.

The objective of this project was to develop a model of a combined-cycle power plant in Flownex which can be solved in off-design conditions in order to compare it to results from other software as well as plant data. The verification of this model will show that Flownex can be used to effectively and efficiently model a combined-cycle power plant.

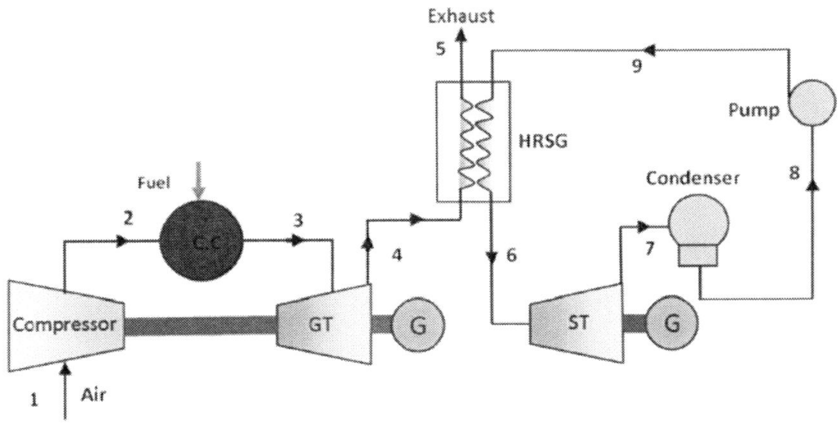

Fig. 1. Basic Combined-Cycle Power Plant [3].

3 Heat Recovery Steam Generator

The HRSG acts as a heat exchanger which uses exhaust gases from the gas turbine to generate steam. When designing an HRSG, the evaluation of steam generation and steam temperature profiles are important starting points [4]. The steam production and gas and steam temperature profiles are dependent on the assumed pinch point and approach point values. Therefore, HRSG's can be thermally designed without having to physically size them [5]. A basic single pressure HRSG is composed of an economiser, evaporator and superheater. The components work together to convert thermal energy from the exhaust gas into steam. Feedwater is heated in the economiser after which it enters the drum at slightly subcooled conditions. The water then circulates in the evaporator, re-entering the drum as a water/steam mixture [6]. The water and steam separate in the drum and the saturated steam travels to the superheater where it experiences the maximum heat exchange temperature (inlet temperature of the exhaust gas from the gas turbine) and is superheated to the desired temperature.

4 Previous Work

There are numerous models of combined-cycle power plants that have been developed previously. This paper is based mainly on the works of V. Ganapathy [5] and Alobaid et al. [7]. Ganapathy developed a program for HRSG simulation in order to produce the thermal design and off-design performance of unfired, fired, simple or complex, multimodule, multi-pressure gas turbine HRSGs without having to physically size them [5].

Alobaid et al. presented a comparative study of different dynamic process simulation codes for combined-cycle power plants at design and off-design conditions [7]. A combined-cycle power plant utilising a sub-critical triple-pressure HRSG was developed using Aspen Plus Dynamics and Apros. These models were validated against plant data at 100%, 80% and 60% load.

Tică et al. developed a combined-cycle power plant for optimisation [8]. The model was developed using Dymola/Modelica, with the use of ThermoPower which is a library for modelling thermal power plants, and the components were validated by experimental data.

Rauch et al. developed a model in Matlab which determined the maximum thermal efficiency of a combined-cycle power plant limited by gas temperature entering and exiting the gas turbine, condenser temperature and the dryness fraction at the outlet of the steam turbine [9].

Srinivas et al. used computer simulation software to assess the performance of a combined-cycle power plant at part loads [10]. The study presented the optimum process parameters of steam exiting the HRSG at part loads.

5 Methodology

5.1 Single Pressure – Analytical

First, a single pressure system was analysed and modelled in the design mode. Figure 2 provides a schematic of the system. The design values used for this model were obtained from a textbook written by V. Ganapathy [11]. The architecture and inputs used for the model were therefore based on that of the textbook.

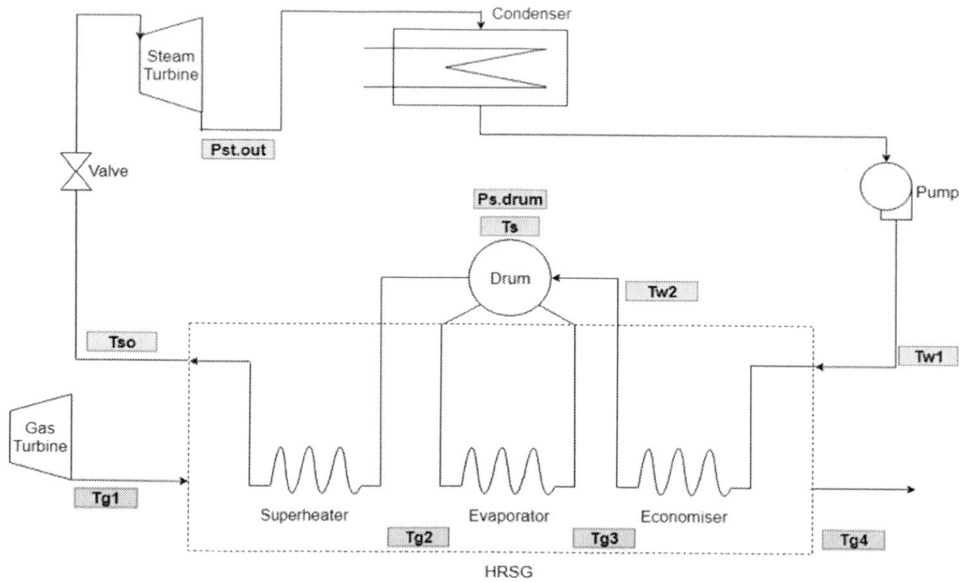

Fig. 2. Schematic of a single-pressure combined-cycle power plant.

A temperature profile of both the gas and water/steam in the HRSG was developed so that the duties of each component could be calculated and the foundation of the HRSG could be established.

The known values for the system were the exhaust gas mass flow and temperature, desired pressure of the water/steam in the HRSG, temperature of the feedwater into the HRSG and the desired temperature of steam to be generated.

Assumptions for the pinch and approach points were made (8.3°C each) in order to form the basis of the HRSG design. Energy balance equations were used to evaluate the system which is shown in equation (1).

$$Q = \dot{m} \times \Delta h \quad (1)$$

The overall conductance (UA) of each component was calculated using equation (2) in order for the system to be evaluated in the off-design model.

$$UA = \frac{Q}{\Delta T_{lm}} \quad (2)$$

Finally, the pressure drop over the steam turbine was calculated. The feedwater temperature was used to find the condenser pressure. This was for simplicity reasons and led to a high condenser temperature and pressure. This did not decrease the accuracy of the model as it was just for example sake.

5.5 Triple Pressure – Analytical

This model was based on that of Alobaid et al [7]. The approach to modelling the triple pressure system was different to the single and double pressure systems. Firstly, the gas flow path no longer flowed through the pressure levels from high- to low-pressure consecutively. Secondly, this model contained a reheating (RH) section following the intermediate-pressure section. Also, it included multiple economisers and superheaters in the high-pressure section, as well as locations where the gas would split through different heat exchangers that were at the same gas path location.

The schematic in Figure 4 shows the layout of the model with the initial known temperatures of water/steam represented by green circles.

Certain assumptions made in this model were critical. These included the pinch point of each pressure level (3), the ratio of the 2 gas splits in the parallel components and 2 water temperatures that had to be assumed. These 7 values had to be fine-tuned in the highly optimised system until there was no steaming or temperature cross situations in the system.

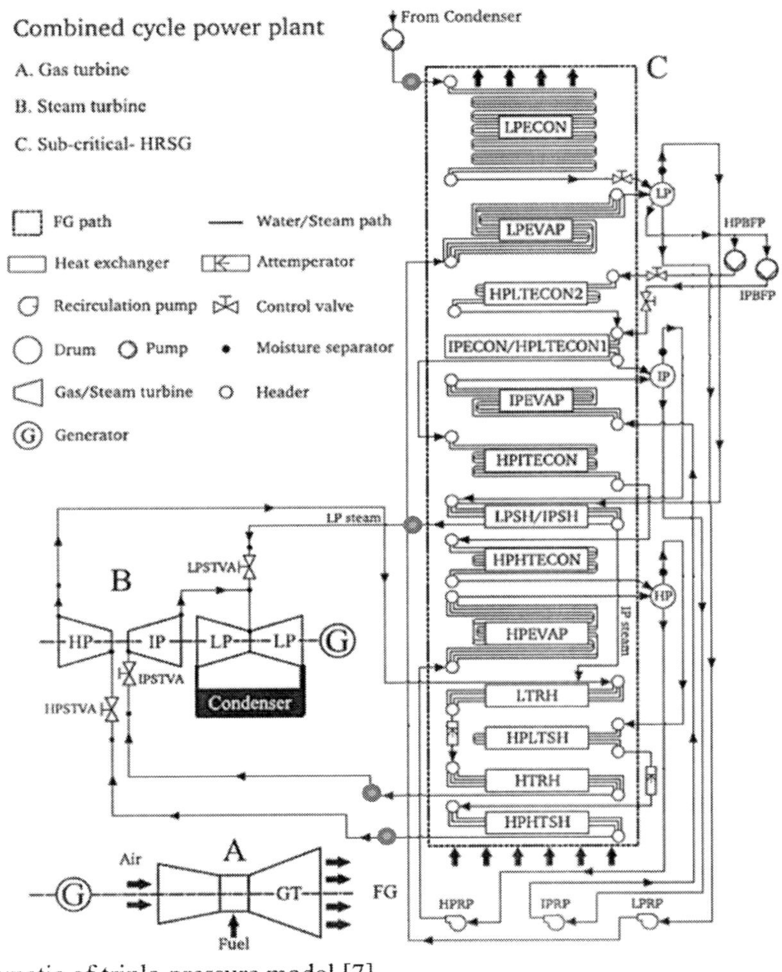

Fig. 4. Schematic of triple-pressure model [7].

5.6 Triple Pressure – Flownex

The triple pressure Flownex model is shown in Figure 5. Due to the complexity of the model, it was separated onto 3 drawing pages and interconnected using projected nodes called View Nodes.

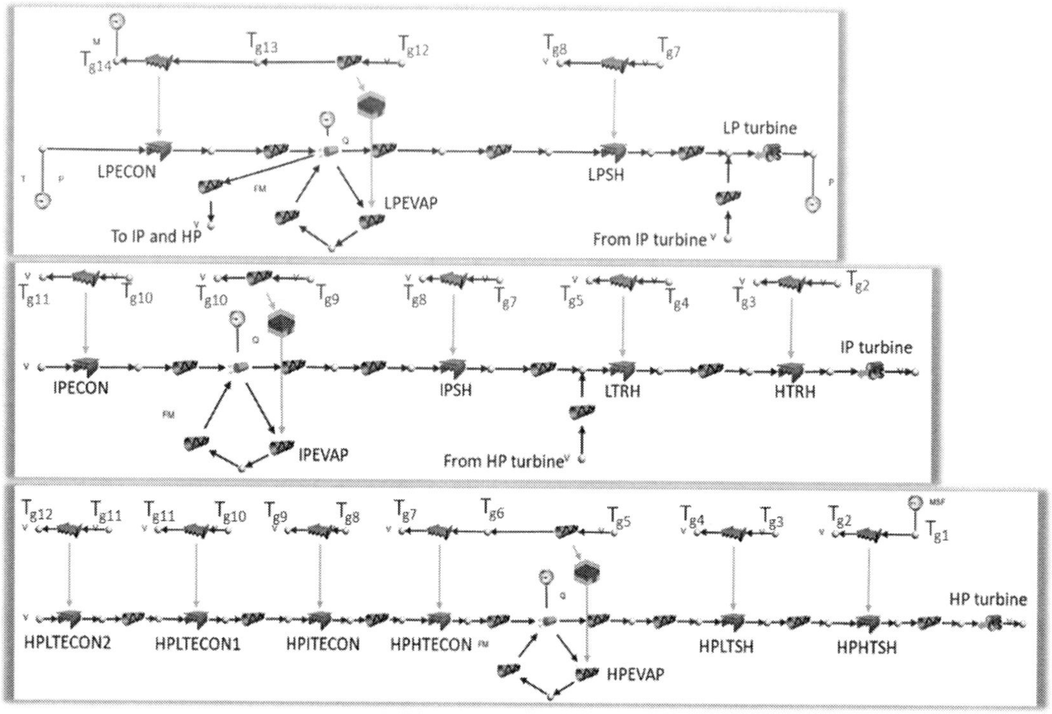

Fig. 5. Triple pressure Flownex model.

The modelling of the economisers, evaporators and superheaters were similar to that of the single and double pressure models.

For the off-design model, the input low-pressure value and both pump heads had to be manually adjusted until all three drums had an energy source close to zero, without causing any steaming in the economisers. Only two low load conditions were analysed (60% and 80%) as these were the only low loads at which information was available from Alobaid et al [7].

It was not feasible to solve for an off-design analytical model of the triple pressure system due to its complexity. Therefore, the triple pressure off-design Flownex model was verified using the triple pressure Virtual Plant model.

6 Results

The triple pressure model was analysed in design and off-design mode. The design case was analysed using a comparison between each model. The design and off-design cases were then analysed by comparing results of the Virtual Plant and Flownex models to published plant data obtained from Alobaid et al [7].

6.1 Design Model

The comparisons between plant data and the analytical results for steam are shown in Table 3.

Table 3. Comparison between design base plant data and analytical results for steam.

Com-ponent	Steam mass flow (kg/s)			Outlet temp. (°C)			Pressure (bar)		
	Plant	Ana-lytical	Error %	Plant	Ana-lytical	Error %	Plant	Ana-lytical	Error %
HPSH outlet	78.2	78.2	0	567	567	0	97.7	97.7	0
RH outlet	83.2	83.2	0	567	567	0	21.4	21.4	0
LPSH outlet	9.8	9.8	0	293	280	4.44	4.1	4.1	0

The comparisons between plant data and the analytical results for gas variables are shown in Table 4.

Table 4. Comparisons between design base plant data and analytical results for gas.

Gas Variable	Plant	Analytical	Error %
Mass flow (kg/s)	587	587	0
Tg1 – Inlet (°C)	628	624.8	0.51
Tg14 – Outlet (°C)	81	100.7	-24.32

The low-, intermediate- and high-pressure levels were plotted separately for the triple pressure models. The low-pressure water/steam and gas temperature profiles for each model is shown in Figure 6. The upper line represents the gas flow and the lower one represents the flow of water/steam. The break in the gas flow line signifies the omittance of components from other pressure levels.

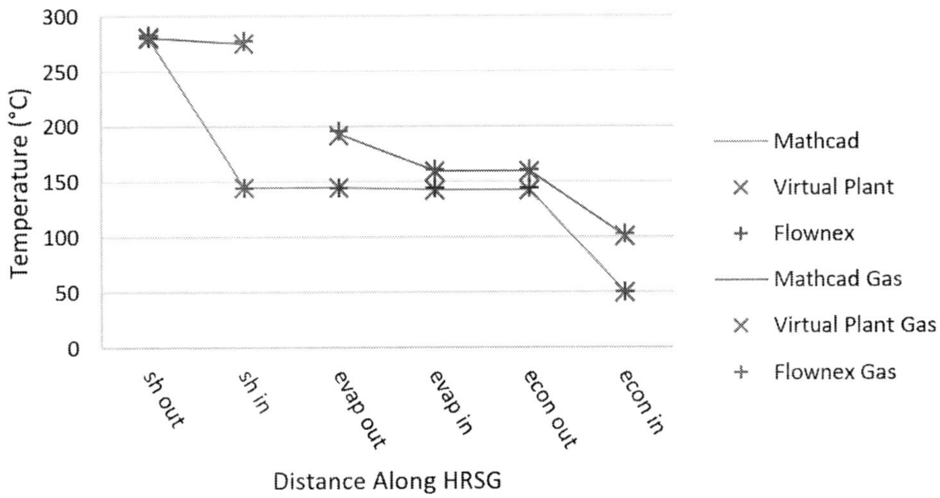

Fig. 6. Graph showing comparison of triple pressure models (low-pressure).

The intermediate-pressure water/steam and gas temperature profiles for each model is shown in Figure 7.

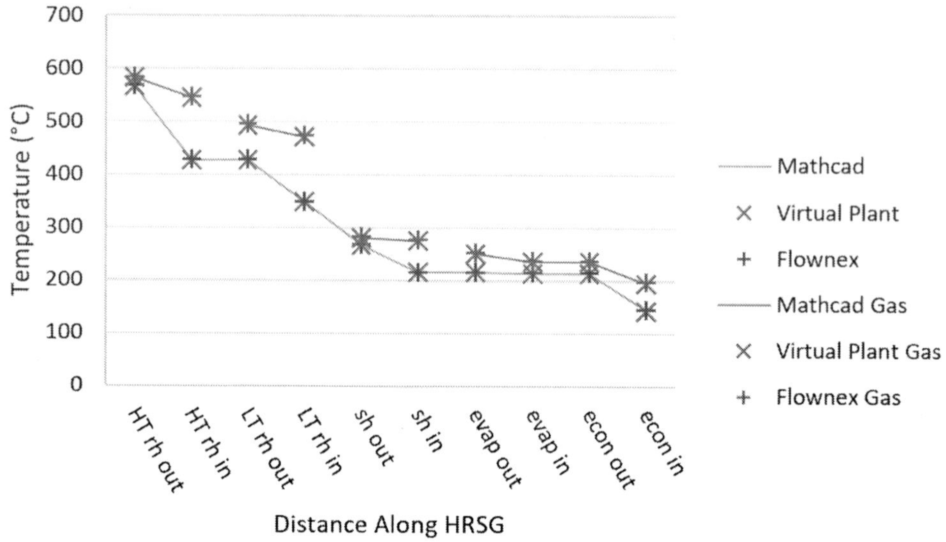

Fig. 7. Graph showing comparison of triple pressure models (intermediate-pressure).

The high-pressure water/steam and gas temperature profiles for each model is shown in Figure 8.

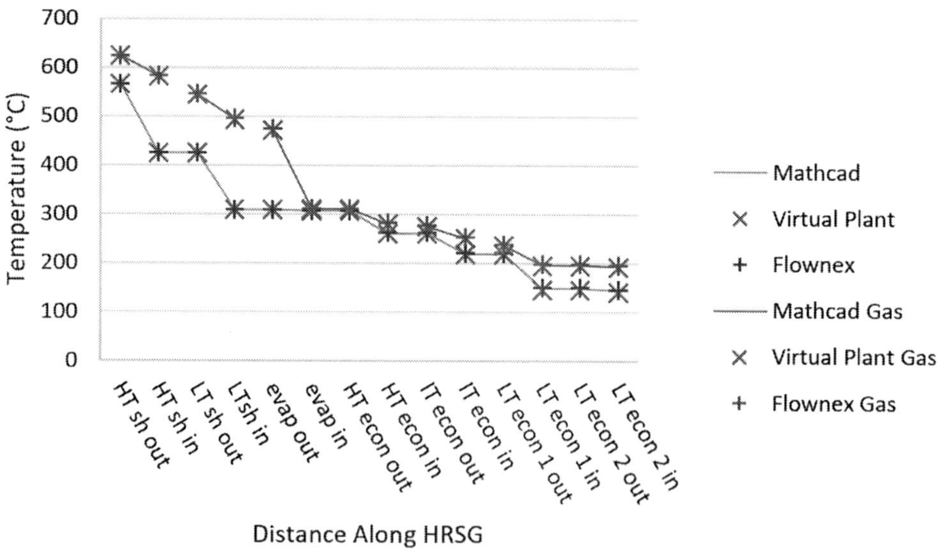

Fig. 8. Graph showing comparison of triple pressure models (high-pressure).

Despite the fact that numerous assumptions had to be made for the triple pressure system, the results agree quite closely with one another. Considering the complexity of the system and the fact that a small difference at one point of the system could cause a larger difference down the line, the models produced successful results.

6.2 Off-Design Model

The results available from Alobaid et al. were the mass flow rate, temperature and pressure of the exiting steam for each pressure level i.e. the outlet of the *HPHTSH*, *HTRH* and *LPSH* [7]. Results for the outlet of the *HPHTSH* are shown in Table 5 with 'Expected' representing published plant data.

Table 5. Triple pressure off-design comparisons for HPHTSH outlet.

Load %	HPHTSH steam mass flow (kg/s)			HPHTSH outlet temp. (°C)			HPHTSH pressure (bar)		
	Ex-pected	Vir-tual Plant	Flown ex	Ex-pected	Vir-tual Plant	Flown ex	Ex-pected	Vir-tual Plant	Flown ex
80	66.9	65.6	67.3	566.8	576.4	572.5	88	82.9	84.7
60	58.3	56.3	57.8	566.9	583.2	579.4	76.2	71.7	73.3

Results for the outlet of the *HTRH* are shown in Table 6.

Table 6. Triple pressure off-design comparisons for HTRH outlet.

Load %	HTRH steam mass flow (kg/s)			HTRH outlet temp. (°C)			HTRH pressure (bar)		
	Ex-pected	Vir-tual Plant	Flown ex	Ex-pected	Vir-tual Plant	Flown ex	Ex-pected	Vir-tual Plant	Flown ex
80	74.7	69.7	71.6	567	571.3	569.4	20.6	18	18.5
60	64.9	59.8	61.6	566.9	574.5	572.4	17.7	15.5	15.9

Results for the outlet of the *LPSH* are shown in Table 7. The design outlet steam temperature had to be decreased from 293°C to 280°C, therefore, a considerable difference in these results were expected.

Table 7. Triple pressure off-design comparisons for LPSH outlet.

Load %	LPSH steam mass flow (kg/s)			LPSH outlet temp. (°C)			LPSH pressure (bar)		
	Ex-pected	Vir-tual Plant	Flown ex	Ex-pected	Vir-tual Plant	Flown ex	Ex-pected	Vir-tual Plant	Flown ex
80	7.7	7.7	8.1	289.9	271.4	271	3.9	3.4	3.5
60	6.4	6.4	6.8	281	264	263.3	3.3	2.9	3

7 Discussion

For the low-pressure steam outlet, there was a larger error in outlet temperature due to the temperature change that had to be made at this point to avoid a temperature cross situation. Additionally, the error in temperature for all pressure levels at low load is expected due to the absence of attemperation in the Flownex and Virtual Plant models as this was not modelled for. These variations meant that the Flownex and Virtual Plant models would behave differently to the plant under off-design conditions. There was also the possibility of various other control mechanisms, such as supplementary firing; control valves; moisture separators; and bypass controls, used in the plant.

8 Conclusion

8.1 Design

The verification of the models was a success for the single, double and triple pressure models. Results obtained from each software were in close agreement with each other which showed that the modelling methodology for each software was consistent. The results of the single and double pressure model were also consistent with that of V. Ganapathy from which the inputs for each were obtained and were, therefore, validated [11].

In order to validate the triple-pressure model, it was compared to plant data obtained from Alobaid et al. at design and off-design conditions [7]. The design model showed close agreement to that of the plant data with the exception of changes that had to be made. The changes made could not be avoided, which could have been due to the amount of assumptions that had to be made or a variation in gas composition which affects the calculation of gas enthalpy. Because of the complexity of the system, errors could not be easily pin pointed.

8.2 Off-Design

For the triple pressure model, the off-design case was able to be compared to plant data. The maximum error of the Virtual Plant model compared to plant data was 13% for off-design conditions. That of the Flownex model was 10%. This is acceptable due to the fact that multiple assumptions had to be made while modelling the triple pressure system as well as the changes in design temperatures.

The results obtained show that Flownex is capable of modelling a combined-cycle power plant in design and off-design conditions.

References

[1] M. S. Y. Ebaid and Q. Z. Al-hamdan, *"Thermodynamic Analysis of Different Configurations of Combined Cycle Power Plants,"* Mech. Eng. Res., **vol. 5**, no. 2, p. 89, (2015).

[2] M. P. Boyce, *"Combined cycle power plants,"* Comb. Cycle Syst. Near-Zero Emiss. Power Gener., pp. 1–43, (2012).

[3] T. K. Ibrahim and M. M. Rahman, *"Effect of Compression Ratio on Performance of Combined Cycle Gas Turbine,"* Int. J. Energy Eng., **vol. 2**, no. 1, pp. 9–14, (2012).

[4] V. Ganapathy, *"Heat-recovery steam generators: Understand the basics,"* Chem. Eng. Prog., **vol. 92**, no. 8, pp. 32–45, (1996).

[5] V. Ganapathy, *Steam Generators and Waste Heat Boilers for Process and Plant Engineers.* (2015).

[6] P. K. Nag, *Power Plant Engineering.* (2008).

[7] F. Alobaid, R. Starkloff, S. Pfeiffer, K. Karner, B. Epple, and H. Kim, *"A comparative study of different dynamic process simulation codes for combined cycle power plants – Part A : Part loads and off-design operation,"* Fuel, **vol. 153**, pp. 692–706, (2015).

[8] A. Tica, H. Guéguen, D. Dumur, D. Faille, and F. Davelaar, *"Design of a combined cycle power plant model for optimization,"* **vol. 98**, pp. 256–265, (2012).

[9] M. Rauch, A. Galović, and Z. Virag, *"Optimization of combined brayton-rankine cycle with respect to the total thermal efficiency,"* Trans. Famena, **vol. 40**, no. 1, pp. 1–10, (2016).

[10] T. Srinivas, A. V. S. S. K. S. Gupta, and B. V. Reddy, *"Thermodynamic simulation of a combined cycle power plant at part-load operation,"* Cogener. Distrib. Gener. J., **vol. 23**, no. 4, pp. 50–63, (2008).

[11] V. Ganapathy, *Industrial Boilers and Heat Recovery Steam Generators.* CRC Press, (2002).

Validation of a thermal non-equilibrium Eulerian-Eulerian multiphase model of a 620 MWe pulverized fuel power boiler.

Brad Rawlins[1,*], *Ryno* Laubscher[1,**], and *Pieter* Rousseau[1,***]

[1]University of Cape Town, Department of Mechanical Engineering

Abstract. The use of a thermal non-equilibrium Eulerian-Eulerian model for the simulation of a 620 MWe power boiler is proposed for capturing the combustion and radiative heat transfer found in the pulverized fuel systems. The models eliminates the use of a Lagrangian reference frame in tracking solid fuel particles thereby reducing the computational expense and time. The model solves the scalar transport for the particle mass, energy and radiation interactions between the pseudo-particle and continuous phases. The goal is to apply the modelling approach to generate a simulation database for different load cases and firing conditions which in turn will be used to study flexible operation. The model is validated against both numerical and applicable site data measurements. It is shown that the model is able to adequately resolve the furnace and superheater wall heat fluxes. Additionally the resolution of the flow field, combustion dynamics and wall fluxes are demonstrated for both an 80% and 60% operational loads. Moreover, it is shown that the Eulerian-Eulerian model results in approximately a 30% computational resource reduction when compared to traditional modelling approaches.

1 Introduction

The mid-merit operation of Coal-Fired Power Plants (CFPP) arises from the intermittent supply of electricity produced by renewable sources. Research into the flexible/mid-merit operation of power plants is needed to better understand the dynamic operations arising from these off-design conditions if renewable energy sources are to be integrated onto electrical grids.

Three-dimensional computational fluid dynamics (CFD) can capture the thermal performance of CFPPs at different load conditions with the necessary accuracy. These 3D CFD models include the modelling of combustion, radiation, and fluid flow phenomena. The traditional method for modelling solid-fuel combustion systems, such as CFPPs, is acknowledged as the Eulerian-Lagrangian (EL) approach. Successful implementations, using the EL approach include the following studies [1],[2] and [3], where the latter focused on the effect that the burner swirl direction has on the heat uptake to the furnace and

*e-mail: rwlbra001@myuct.ac.za
**e-mail: ryno.laubscher@uct.ac.za
***e-mail: pieter.rousseau@uct.ac.za

superheater walls of a 620 MWe boiler.

Another approach is to use a Eulerian multi-fluid model as used by the following studies [6], [4] and [5]. Knaus et al [7] compared the Eulerian-Eulerian (EE) and EL approaches for a 550 MWe coal-fired utility boiler. The predicted combustion products yield only small differences, with carbon monoxide being better predicted when using the Eulerian-Lagrangian approach. The computational effort of the Eulerian approach proved more efficient and faster convergence times were noted. It should be stated, that these studies were conducted using in-house CFD codes which is difficult to apply by engineers and scientists outside the research group.

The current work proposes the use of an EE modelling strategy that incorporates thermal non-equilibrium between the gas and particle phases which allows for adequate particle concentration tracking and heat transfer interactions, with a reduced computational cost when compared to the EL simulation approach. The model is validated for a full scale utility power boiler for various load cases. The validated modelling technique is to be used for future studies in developing a CFD data-driven surrogate model of a 620 MWe CFPP's furnace and radiant superheaters.

2 Mathematical model

2.1 Numerical modelling methodology

An outline of the modelling methodology, utilized by the proposed EE model, is provided in this section highlighting the general conservation equations, pseudo-solid phase tracking, and the combustion and radiation modelling techniques employed. The basis of the model assumes that the gas and solid phases behave as inter penetrating continua with both phases being described in a Eulerian reference frame. A further assumption is that the fluid is in dynamical equilibrium with the solid phase, (similar to the work of [6]), meaning that velocities of both phases are locally the same.

2.1.1 Transport equations

The time-averaged steady-state continuity, momentum, k-ϵ turbulence and energy equations for the gas-phase can all expressed using the following generalised transport equation for a variable ϕ expressed as:

$$\underbrace{\frac{\partial}{\partial x_i}(\rho u_i \phi)}_{\text{Advection}} = \underbrace{\frac{\partial}{\partial x_i}(\Gamma_\phi \frac{\partial \phi}{\partial x_i})}_{\text{Diffusion}} + \underbrace{S_\phi}_{\text{Source}} \tag{1}$$

where ρ is the gas density and Γ_ϕ is the diffusion coefficient for variable ϕ.

In the present work the flue gas phase is modelled using the species-transport approach where a transport equation is solved for each constituent mass fraction in the gas mixture. The governing equations for the gas phase can be written as follows;

Mass conservation:

$$\frac{\partial}{\partial x_i}(\rho \bar{u}_i) = S_m. \tag{2}$$

Momentum conservation:

$$\frac{\partial}{\partial x_i}(\rho_{eff}u_iu_j) + \frac{\partial \overline{P}}{\partial x_j} = \frac{\partial}{\partial x_i}\left[\mu\left\{\frac{\partial u_j}{\partial x_i} + \frac{\partial u_i}{\partial x_j} - \frac{2}{3}\delta_{ij}\frac{\partial u_i}{\partial x_i}\right\}\right] + \frac{\partial}{\partial x_i}(-\rho\overline{u_i^{'}u_j^{'}}) + S_m \quad (3)$$

The momentum equation is closed by approximating the Reynolds stresses using the Boussinesq relationship as shown in equation (4).

$$-\rho\overline{u_i^{'}u_j^{'}} = \mu_t\left(\frac{\partial u_i}{\partial x_j} + \frac{\partial u_j}{\partial x_i}\right) - \frac{2}{3}\left(\rho k + \mu_t\delta_{ij}\frac{\partial u_k}{\partial x_k}\right) \quad (4)$$

This approximation is used in conjunction with the realizable k-ϵ turbulence model of Shih et al [8]

Energy conservation:

$$\frac{\partial}{\partial x_i}(u_i[\rho E + P]) = \frac{\partial}{\partial x_j}\left[\lambda\frac{\partial T_g}{\partial x_j}\right] + S_h \quad (5)$$

where E in the total energy term, P the pressure, λ is the effective thermal conductivity and T_g is the gas temperature.

Species transport:

$$\frac{\partial}{\partial x_i}(\rho u_j Y_k) = -\frac{\partial}{\partial x_j}(\vec{J_k}) + \sum_r R_{j,r} + S_k$$
$$k = 1,2,3...N \quad (6)$$

where k is the number of species present in the domain and $\vec{J_k}$ is the species diffusion flux. $\sum_r R_{j,k}$ is the net rate of production of species j due to r reactions.

To correctly account for the particle inertial effects on the gas phase convection an effective density is defined as follows and is used in the above equations.

$$\rho_{eff} = \frac{\rho\rho_p\left(\phi_{mp} + 1\right)}{\rho\phi_{mp} + \rho_p} \quad (7)$$

In equation (7), ρ_p is the particulate matter density and ϕ_{mp} $[kg_p/kg]$ is the local mass concentration of particulate matter present in the gas volume.

2.1.2 Solid phase modelling

The pseudo particles transported into the domain are modelled using a scalar field and transported through the computational domain according to equation (1). The pseudo particle scalar fields are used to define the fuel characteristics based on the proximate analysis composition (refer to Table 2), namely consisting of moisture, volatile matter, fixed carbon and ash. Each of the scalar field equations solved for are given in Table 1.

Table 1. Scalar field equation descriptions

Variable	Description	Transport equation
ϕ_{mp0}	Original/initial mass of particles	$\frac{\partial}{\partial x_i}(\rho u_i \phi_{mp0}) = 0$
ϕ_M	Moisture present in particles	$\frac{\partial}{\partial x_i}(\rho u_i \phi_M) = \frac{1}{V}\frac{dM_{evap}}{dt}$
ϕ_{VM}	Volatile matter present in particles	$\frac{\partial}{\partial x_i}(\rho u_i \phi_{VM}) = \frac{1}{V}\frac{dM_{devol}}{dt}$
ϕ_{FC}	Fixed carbon present in particles	$\frac{\partial}{\partial x_i}(\rho u_i \phi_{FC}) = \frac{1}{V}\frac{dM_c}{dt}$
ϕ_{ASH}	Ash present in particles	$\frac{\partial}{\partial x_i}(\rho u_i \phi_{ASH}) = 0$
ϕ_{hp}	Enthalpy of particle	Equation (8)

The energy transport of the pseudo particle, is transported by defining the particle enthalpy using the following equation:

$$\frac{\partial}{\partial x_i}(\rho u_i \phi_{hp}) = \left(f_{heat}\frac{dM_c}{dt}h_{rxn} + \dot{Q}_{rad} + \dot{Q}_{conv} - \frac{dM_{evap}}{dt}h_{fg}\right)\frac{1}{V} \tag{8}$$

The equation accounts for all the processes associated with energy transport to the particle, namely convection $\left(\dot{Q}_{conv}\right)$, radiation $\left(\dot{Q}_{rad}\right)$, latent heat $\left(\frac{dM_{evap}}{dt}h_{fg}\right)$ and near surface char oxidation $\left(f_{heat}\frac{dM_c}{dt}h_{rxn}\right)$. This gives the model the ability to track the particle temperature in the domain, moving the model away from the thermal equilibrium approach incorporated by previous studies using an EE approach ([6], [7]). The particle temperature is important in describing the sequential steps found in modelling combustion processes, especially at low boiler loads where mixing and ignition become problematic.

2.1.3 Combustion modelling

The combustion modelling of coal comprises four sequential steps, namely the heating/cooling of the particles, evaporation, devolatilization and char oxidation. The process starts with the pseudo-particles entering the domain and heating up until the moisture's boiling temperature of 373 K is reached [9]. The realization of the boiling temperature would release moisture based on a boiling rate which is defined as:

$$\frac{dM_{evap}}{dt}h_{fg} = \dot{Q}_{rad} + \dot{Q}_{conv}. \tag{9}$$

Following this is the devolatilization process whereby the volatiles present in the fuel are liberated. The initialization temperature of the devolatilization process was found to vary between 553 – 623 K for a variety of coals by Ranade and Gupta [10], thus particle heating would need to ensue after the moisture evaporation to bring the particles to the initialization temperature. A single-step Arrhenius kinetic rate model is used to approximate the devolatilization process since these models can predict the devolatilization of coal and are simpler to implement, according to Sankar et al [11]). The rate is expressed in equation (10) as:

$$\frac{dM_{vol}}{dt} = R_{vol}(m_{0,vol} - m_{vol}). \tag{10}$$

where $R_{vol} = A_{vol}\exp(-E_{a,vol}/RT_p)$, with $A_{vol} = 2E5\ [s^{-1}]$ and $E_{a,vol} = 6.7E7\ [J/kmol]$ from the works of Sheng et al [12]. Here m_{vol} is the mass of the volatile released during combustion and $m_{0,vol}$ is the initial volatile mass present before combustion.

The final step of the solid fuel combustion process is realized when all the volatiles have been driven off from the pseudo-particles. The product species of the char oxidation process is CO, with 100% of the resultant heat being absorbed by the pseudo-particle contributing to the energy rise of the particle matter in a cell. The char oxidation rate is derived from the works of Baum and Street [13] and Spalding [14]. The model is commonly referred to as the diffusion-kinetics limited model, written as:

$$\frac{dM_c}{dt} = -A_p \frac{R_c R_{diff}}{R_c + R_{diff}} P_{O_2} \tag{11}$$

The interested reader is directed to [15] for an in-depth derivation of the above equation. The kinetic parameter $R_c = A_c \exp(-E_{a,c}/RT_p)$, where $A_c = 0.0053$ $[kg/m^3]$ and $E_{a,c} = 8.37E7$ $[J/kmol]$ from the works of Sheng et al [12] and Hanjalić et al [16]. Here A_p is the particle surface area, $R_{diff} = (5E-12)(0.5(T_p + T_g))^{0.75})/\bar{d}_p$ is the diffusion rate coefficient, which is a function of the gas and pseudo-particle temperature and average diameter, and P_{O_2} is the partial pressure of oxygen.

According to Sankar et al [11], the gas phase reactions in boilers are fast, with the chemical time scales being orders of magnitude faster than turbulent mixing time scales. This leads to a mixed-is burnt assumption [17], which assumes the chemistry is infinitely fast and irreversible, being adequate for species and temperature distribution predictions. The gas-phase volatile and CO reactions are approximated using the two-step global reaction shown in equation (12).

$$\underbrace{CH_{3.51}O_{0.78}N_{0.1106}S_{0.0466}}_{Volatiles} + 1.03O_2 \rightarrow CO + 1.75H_2O + 0.0466SO_2 + 0.0553N_2$$
$$CO + 0.5O_2 \rightarrow CO_2 \tag{12}$$

This study makes use of the two-step finite-rate/eddy-dissipation model (FR/EDM) in ANSYS Fluent® 19.3 [15] to calculate the gaseous reaction rates. The model calculates three rates and uses the minimum of the three values for the source term calculations for the species transport equations and subsequently for the energy release source terms. Chemical reaction rate:

$$R_{j,r,chemical} = A_r exp\left(\frac{-E_{a,r}}{RT_g}\right)\left[\vartheta'_{k,r} - \vartheta_{k,r}\right]\prod_L [C_{l,r}]^{m,r} \tag{13}$$

Rate of turbulent production eddies dissipation:

$$R_{j,r,prodcution} = \vartheta_{k,r} M_{w,k} AB\rho \frac{\varepsilon}{k} min\left(\frac{\sum_p Y_p}{\sum_j \vartheta_{j,r} M_{w,j}}\right) \tag{14}$$

Rate of reaction eddies dissipation:

$$R_{j,r,reactant} = \vartheta_{k,r} M_{w,k} A\rho \frac{\varepsilon}{k} min\left(\frac{Y_R}{\varepsilon_{R,r} M_{w,R}}\right) \tag{15}$$

In the above equations Y_p is the product species mass fraction, Y_R is the mass fraction of any reaction species, A and B are model constants, $M_{w,k}$ is the molecular weight of jth species, $\vartheta_{k,r}$ is the stoichiometric coefficient of reactant k in reaction r, $[C_{l,r}]^{m,r}$ is the molar concentration of species l in reaction r with rate exponent m, $A_r = 2.56E11$ $[s^{-1}]$ and $E_{a,r} = 1.081E8$ $[J/kmol]$ taken from the study of He et al [18].

2.1.4 Radiation modelling

The heat transfer to the furnace, platen and final superheater walls of the validation case are mainly due to radiation. The radiation transport for the current work was resolved using the P1 radiation model which includes the effects of particle absorption (α_p) and scattering (σ_p) as well as gas mixture absorption (α_g). The P1 model transport variable is the incident radiation (G - W/m^2), and can be written for a particle laden domain as:

$$\frac{\partial}{\partial x_i}\left(\Gamma\frac{\partial G}{\partial x_i}\right) = \left(\alpha_g + \alpha_p\right)G - 4\left(\alpha_g\sigma_{SB}T_g^4 - \pi E_p\right)$$
$$\Gamma = \frac{1}{\alpha_g + \alpha_p + \sigma_p} \tag{16}$$

The gas mixture absorption coefficient was calculated using the weighted sum of gray gases model (WSGGM) which accounts for the tri-atomic gases CO_2 and H_2O as proposed by Smith et al [19]. The solid phase radiative properties (namely α_p, σ_p and E_p) were determined using an equivalent Eulerian description for each expressed as:

$$\alpha_p \equiv \frac{\epsilon_{p,n}A_{proj}N_p}{V} \tag{17}$$

$$\sigma_p \equiv (1-\epsilon_{p,n})(1-f_p)\frac{A_{proj}N_p}{V} \tag{18}$$

$$E_p \equiv \frac{\epsilon_{p,n}A_{proj}\sigma_{SB}T_p^4N_p}{\pi V} \tag{19}$$

where $N_p = \rho\phi_{mp0}V/\left(\rho_p\pi\bar{d}_p^{\,3}/6\right)$, is the number of particles present based on the average diameter pseudo particle size $\bar{d}_p$, $A_{proj} = N_p\pi d_p^2/4V$ is the projected area, $\sigma_{SB} = 5.67E-8; [W/m^2K^4]$, T_p is the particle temperature and V is the cell volume. The particles emissivity (ϵ_p) and scattering factor (f_p) were implemented using a variable property correlation as used in the work of Laubscher and Rousseau [20].

2.2 Numerical setup and model inputs

The modelling methodology, described previously, was successfully incorporated in the modelling of a 620 MWe subcritical power boiler. Figure 1 illustrates the computational domain used in this validation study highlighting the various heat receiving surfaces and inlets. The burner feeds the combustion chamber with primary and secondary air through the burners inner and outer annuli respectively. At 100% maximum continuous rating (MCR) the primary air (PA) supplies the fuel and air at a temperature of 373 K and secondary air (SA) enters at a temperature of 577 K.

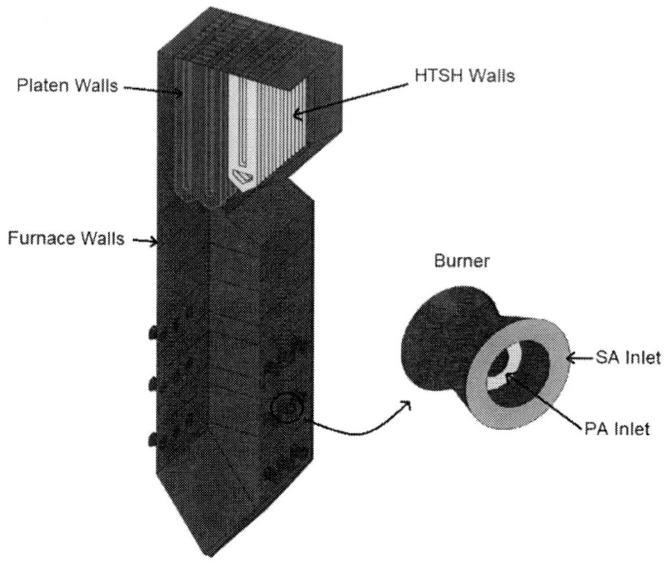

Figure 1. Computational domain layout

The fuel characteristics used for CFD modelling purposes are given in Table 2.

Table 2. Fuel characteristics used in CFD model

Fuel constituent	Fraction - (kg/kg_{fuel})
Ultimate analysis - [Dry-Ash-Free]	-
C	0.775
H	0.042
N	0.018
O	0.147
S	0.018
Proximate analysis - [As-received]	-
Fixed Carbon	0.34
Volatiles	0.196
Ash	0.409
H_2O	0.055
Energy Content	As-received - (kJ/kg_{fuel})
Higher Heating Value - (HHV)	15070

The plant in designed to operate at a 100% MCR with a excess air ratio of 15.5%. As the plant is turned down from 100% MCR the excess air ratio is increased to ensure sufficient air and fuel mixing occurs. At MCR ratings of 81% and 60% the excess ratios were calculated as 20.9% and 26.3% respectively, based on the design schedule data. The inlet mass flow rates and temperatures for the plant's MCR ratings of 100%, 81% and 60% are listed in Table 3.

Table 3. Inlet conditions for simulated load cases (per burner)

Load Rating	100 %	80 %	60 %	Unit
PA flow rate	4.75	4.01	2.81	kg/s
SA flow rate	14.27	12.04	8.43	kg/s
Fuel flow rate	3.12	2.52	1.67	kg/s
PA inlet temperature	373.15	373.15	373.15	K
SA inlet temperature	577.15	558.15	535.15	K

Other required wall boundary conditions such as wall temperatures and emissivities are listed in the work of Laubscher and Rousseau [2].

2.2.1 Numerical solution strategy

The current simulations were performed using the ANSYS Fluent® 19.2 pressure-based solver. Pressure-momentum coupling was set to the SIMPLE technique, with momentum, energy and species equations being discretized using the second-order upwind method. The pressure equation was discretized using PRESTO!. This setup holds for both the EE model and the detailed EL model. The difference arises in the use of scalar fields in the EE configuration and using the discrete phase modelling approach in the EL setup. Simulations were performed on a numerical mesh consisting of roughly 6.2 million cells. A mesh independence study was conducted for mesh sizes consisting of 4.2 and 10.2 millions with the 6.2 million cell model deemed acceptable for this study. To ensure an accurate numerical solution the mesh aspect ratio was kept below 15 and the minimum orthogonal quality above 0.2. The discrete phase equations, for the EL model, were solved once every 30 fluid phase iterations. The number of particles injected per burner was set to about 7800, totalling to 140 000 particles in the entire domain. To ensure a stable converged solution the spatial discretization for all fields were set to first order upwind (except pressure) and solved for 1500 iterations before increasing the discretization order. The simulations were then run for a further 10000 iterations. For all cases the maximum mass imbalance was 0.046 kg/s for a total gas flow rate of 376 kg/s. The maximum heat imbalance was 2450 kW for a total heat input of 855 MW.

3 Results and discussion

The discussion herewith compares the results of an EL numerical model with the results of the developed EE modelling approach, (discussed in section 2) for a 100%, 81% and 60% MCR loads cases. Overall a 30% (9 hours to approximately 6 hours) decrease in simulation time was observed across the simulated load cases.

Table 4 highlights the relative errors obtained between the EL and EE numerical models. A maximum error of 8.15% occurs at the platen superheater for the 100% load case. The relative errors are deemed acceptable for the decrease in computational time required by the EE model.

Table 4. Relative percentage errors of key parameters between the EL and EE models

Load Rating	100%	80%	60%
Furnace heat load	4.05%	8.15%	1.23%
Platen heat load	6.93%	7.36%	7.21%
HTSH heat load	2.81%	3.78%	4.23%
Furnace exit temperature	3.56%	4.76%	5.62%
Exit X_{CO_2} fraction	2.12%	1.56%	0.98%
Exit X_{O_2} fraction	1.86%	3.21%	2.84%

The graphs shown in figure 2 illustrate the calculated (based off operational plant measurements), EL and EE models results, for the heat loads to the furnace, platen and high temperature superheater (HTSH) walls.

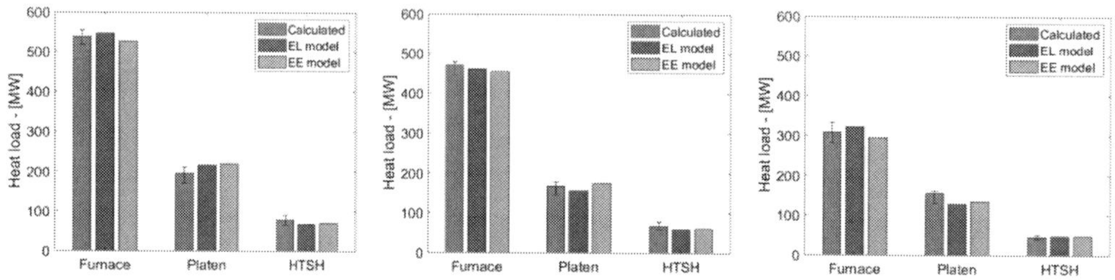

(a) 100% load case (b) 81% load case (c) 60% load case

Figure 2. Overall heat load performance for 100%, 81% and 60% MCR load cases

The general accuracy of the EE model is seen to be able to sufficiently capture the overall heat loads when compared to the calculated and EL model results. A notable difference is seen in the 60% load case of figure 2 (c), where both the EL and EE numerical models under-predict the platens heat load. Figure 3 compares the gas velocity fields for the EL and EE model. It can be seen that results are in good agreement. The EE model tends to under predict the lower burner velocity transport resulting in combustion occurring closer to the burner mouth and lower furnace walls. This effect is seen in both the temperature contour plots of figure 4 (d-f) and wall heat flux contours of figure 5 (d-f).

The temperature contour plots of figure 4 (d-f) illustrate the EE model's ability to sufficiently resolve the temperature field when compared to the EL model. The lower burners tend to initiate combustion closer to the burner leading to high temperatures. Due to the lack of velocity of the gas phase in this area, the high temperatures are closer to the walls of the furnace leading to a higher heat flux observed as seen in figure 5 (d-f). In general the temperature and velocity fields are deemed sufficiently accurate for the purposes of surrogate model development, since the parameter of interest is the resolution of the wall flux field. The heat flux profiles are given in figure 5 (a-f). The EE model over predicts the wall fluxes in the lower half of the boiler, namely due to the high temperatures experienced in these zones in and around the bottom burners. The platen and HTSH radiant superheaters show similar flux profiles with the top half of the furnace illustrating sufficient agreement between the EL and EE models. The spatial distribution of the particles (expressed in kg/m^3) shown in figure 6 (a-f) highlights the EE model's ability to resolve the particle concentration throughout the domain for the various load cases. The general trend of the EE model is with sufficient accuracy when compared to the EL model.

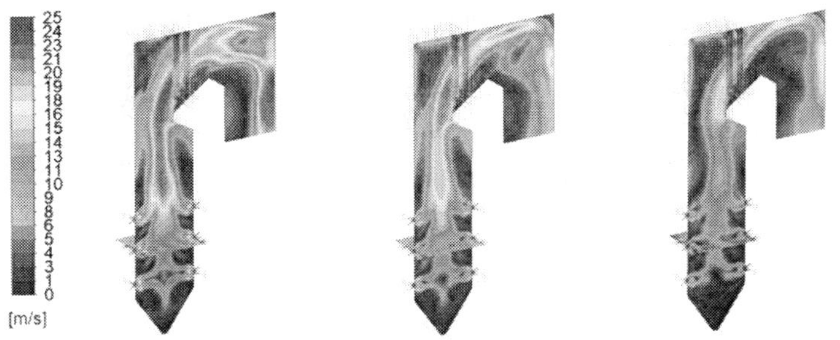

(a) 100% case (b) 81% case (c) 60% case

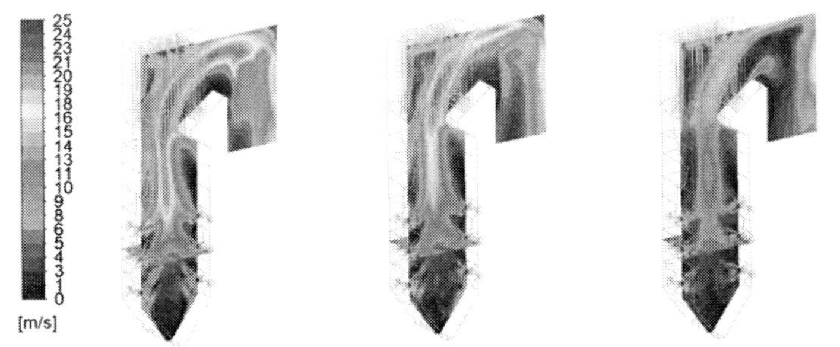

(d) 100% case (e) 81% case (f) 60% case

Figure 3. Gas velocity contour plots for the EL (a-c) and EE model (d-f) simulated load cases

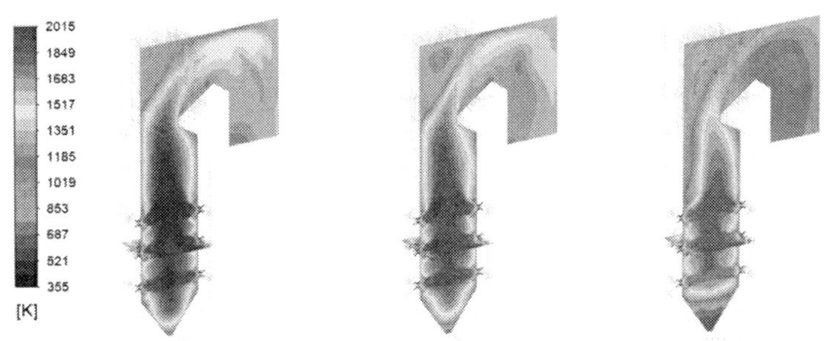

(a) 100% case (b) 81% case (c) 60% case

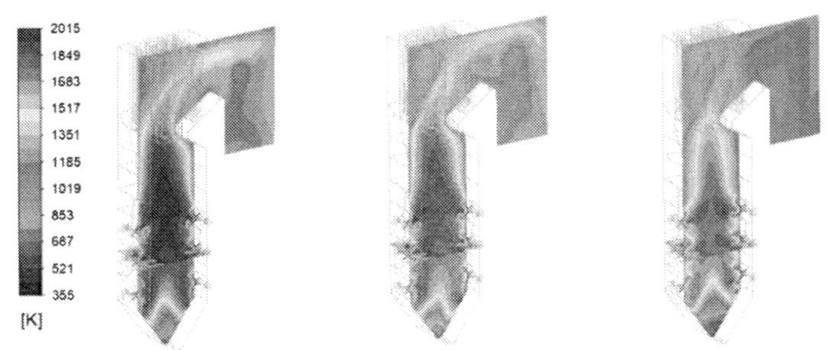

(d) 100% case (e) 81% case (f) 60% case

Figure 4. Temperature plots for the EL (a-c) and EE model (d-f) simulated load cases

https://doi.org/10.1051/matecconf/202134700004

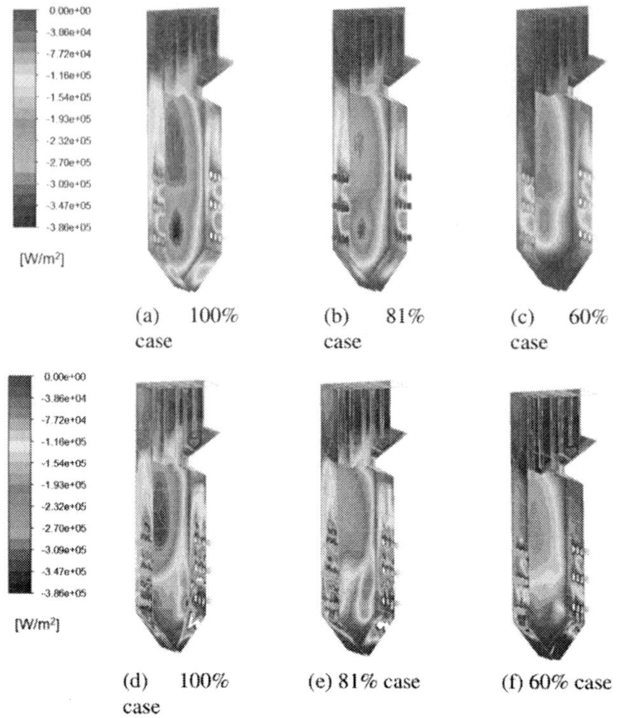

Figure 5. Wall heat flux profiles for the EL (a-c) and EE model (d-f) simulated load cases

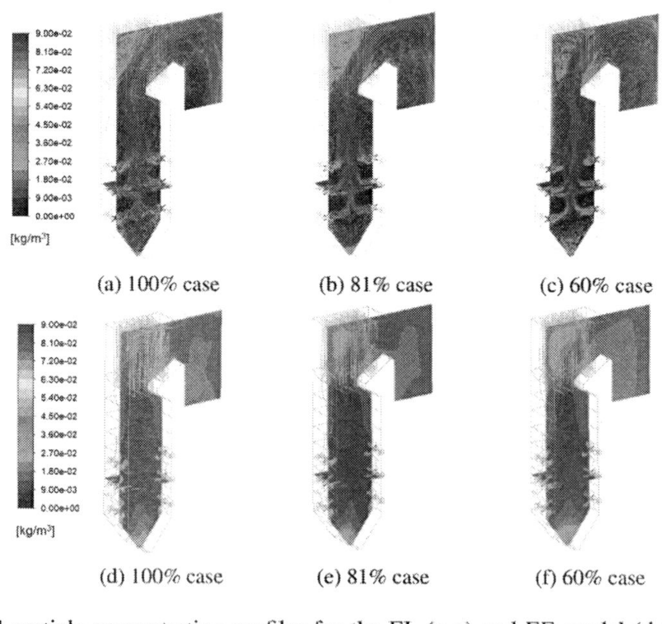

Figure 6. Spatial particle concentration profiles for the EL (a-c) and EE model (d-f) simulated load cases

4 Conclusion

The validation of a full scale 620 MWe boiler was conducted using a thermal non-equilibrium Eulerian-Eulerian model for three load cases, namely a 100%, 81% and 60% steady-state loads. The validation cases include a comparison with results obtained by a numerical model using a Eulerian-Lagrangian approach and where applicable measured site data. The developed EE model demonstrated adequate performance in predicting the flow field, wall heat fluxes and fluid property distributions throughout the domain.The relative accuracy of the EE model ranged from 2 to 8% for key parameters used in the study. The computational speedup, of 30%, by using the EE model is a beneficial attribute when considering the EE models intended future use in the development of a CFD data-driven surrogate model.

Acknowledgements

The authors would like to thank the Eskom EPPEI program for funding this project and the Centre for High Performance Computing (CHPC, South Africa) for the computational resources provided for completing this work.

References

[1] M. von Bohnstein, M. Richter, P. Schiemann, J. Ströhle, and B. Epple, Renewable and Sustainable Energy Reviews, **137**, (2021)
[2] R. Laubscher, and P. Rousseau, Applied Thermal Engineering, **160**, (2019)
[3] R. Laubscher, and P. Rousseau, International Journal of Heat and Mass Transfer, **137**, 506-522 (2019)
[4] J. Cai, M. Handa, and M.F. Modest, Combustion and Flame, **162**, 1550-1565 (2015)
[5] Y. Wu, D. Liu, J. Ma, and X. Chen, Powder Technology, **342**, 48-61 (2018)
[6] A.C. Benim, B. Epple, and B. Krohmer, Progress in Computational Fluid Dynamics, **5**, 345-361 (2005)
[7] H. Knaus, U. Schnell, and R.G. Klaus , Progress in Computational Fluid Dynamics , **1**, 194–207 (2001)
[8] T.H. Shih, W.W. Liou, A. Shabbir, Z. Wang, and J. Zhu, Computers and fluids, **24**, (1995)
[9] Y.A. Cengel, and M.A. Boles, *Thermodynamics: An engineering approach*, (McGraw Hill, Singapore, 2015)
[10] V..V. Ranade, and F.D. Gupta, *Computational Modeling of Pulverized Coal Fired Boilers*, (CRC Press - Taylor & Francis Group, Boca Raton, 2015) 101-105
[11] G. Sankar, D. Santhosh Kumar, and K.R. Balasubramanian, Fuel, **236**, 643-665 (2019)
[12] C. Sheng, B. Moghtaderi, R. Gupta, and T. F. Wall, Fuel, **83**, 1543-1552 (2004)
[13] M.M. Baum, and P.J. Street, Combustion Science and Technology, **3**, 231-243 (1971)
[14] D.B. Spalding, Symposium (International) on Combustion, **13**, (1970)
[15] *ANSYS Fluent Theory Guide*, (Ansys Inc., 2021)
[16] A. Đugum and K. Hanjalić, Energy, (2018)
[17] V..V. Ranade, and F.D. Gupta, *Combustion: Physical and chemical fundamentals, modelling and simulation, experiments, pollutant formation.*, (USA: Springer, 2006)
[18] B. He, Computational Fluids, **36**, 435-444 (2007)
[19] T. Smith , Z. Shen , and J. Firedman, Journal of Heat Transfer, **104**, 602-608 (1982)
[20] R. Laubscher, and P. Rousseau, Energy, **195**, (2020)

Characteristics of the asymmetric Stockbridge damper

Zakhele Zondi [1*], *Modify* Kaunda [2], and *Tiyamike* Ngonda[2]

[1]Mangosuthu University of Technology, Department of Mechanical Engineering, Durban, South Africa
[2]Cape Peninsula University of Technology, Department of Mechanical Engineering, Cape Town, South Africa

Abstract. The Stockbridge damper (tuned mass absorber) is used on overhead transmission power lines to suppress wind-induced vibrations. These power lines get exposed to various types of wind motions that cause them to vibrate and this causes fatigue failure at the suspension clamp where the maximum stress occurs. Aeolian vibration is the most common wind motion that causes fatigue and eventually, failure to transmission lines. This paper presents the study of an asymmetric Stockbridge damper. A set of experiments were conducted on the damper according to the Institute of Electrical and Electronics Engineers (IEEE) 664 standards to evaluate the characteristics of the asymmetric Stockbridge damper. The results obtained revealed that the asymmetric damper is four degrees of freedom. The results also revealed a relationship between the mass and the magnitude of the resonance frequency.

1. Introduction

Overhead power transmission lines get exposed to wind motions that cause them to vibrate. Wind motions cause vortex shedding and eventually leads to a fatigue failure, usually at the suspension clamp where the most significant stresses of the power lines occur [1]. This phenomenon leads to a power outage. There are three wind-induced mechanical vibrations of power lines, wake-induced oscillation, conductor galloping, and Aeolian vibration [2]. The wake-induced oscillation of a bundle conductor transmission line is caused by the wake interference between sub-conductors [2]. The spacing between sub-conductors of a bundle conductor is usually in the range of 15–20 times the sub-conductor diameter. During vibrations, the sub-conductors vibrate within sub-spans, which leads to clashing and fatigue failure of conductors. In cold climates, the build-up of ice or snow causes the conductor galloping, which is characterised by a low frequency, high amplitude, mainly vertical conductor motion [2, 3]. In most cases, ice builds up on the windward surface of the conductor. As a result, the conductor creates an aerodynamically unstable shape that produces large forces and moments on the conductor. The motion of the conductor due to wind can lead to oscillations in a vertical direction. Aeolian vibration is the most common and familiar

* Corresponding author: zakhele@mut.ac.za

wind motion that affects power lines and produces fatigue failure. It is characterised by a frequency ranging from 3 to 150 Hz produced by a wind speed of 1 m/s to 7 m/s [4].

Various researchers have investigated measures to minimise failure of power lines arising from wind induced conductor vibrations. They reported on different designs of wind motion vibration dampers that can be used to mitigate fatigue failure of power lines, one of which is a Stockbridge damper. The main function of a Stockbridge damper is to absorb Aeolian vibration by creating destructive interference of resonance frequencies. Over the past years, there have been several research studies that focused on improving Stockbridge dampers [5]. Barbieri and Barbieri [6] examined the linear and nonlinear dynamical behaviour of asymmetric Stockbridge damper with excitation frequencies in the range of 5 to 17 Hz using a machine cam with five different profiles. The authors reported that the motion amplitude of vibration is inversely proportional to the natural frequency. Vaja et al. [7] developed a mathematical model with minimum assumptions of the Stockbridge damper that suits the conductor cable. The bending stiffness of the cable and the effect of this stiffness on the natural frequencies were examined. The authors found equivalent stiffness compensated for the assumption of modeling the messenger. The results obtained revealed that the moment of inertia of the power line was directly proportional to the inertia of the messenger cable. Vaja et al. [8] developed an analytical model of a fresh vibration damper with an increased number of resonant frequencies. A 3D finite element model was developed to validate the analytical model. Markiewicz [9] presented a method and computational model for the evaluation of the optimum dynamic characteristics of Stockbridge dampers to be mounted near dead! end towers. The method was applied to a general model of a tension system with insulator strings and antiarc protection elements. It revealed how the efficiency of a standard damper used in such spans may be enhanced by its proper location on a cable.

Although several research studies have been conducted to improve the efficiency of the Stockbridge damper, there have been limited research studies that have focused specifically on the asymmetric Stockbridge damper. This paper aims at examining the characteristics of the asymmetric Stockbridge damper to improve the knowledge pool about this damper. This will assist in the future improvements of the asymmetric Stockbridge damper. Experiments will be conducted on different sizes of an asymmetric Stockbridge damper.

2. Design, functioning, and history of the Stockbridge damper

Fig. 1 and Fig. 2. Illustrate the types of Stockbridge dampers, namely asymmetric and symmetric dampers respectively. In general, the asymmetric Stockbridge damper (4-R damper) consists of two unequal masses attached to the end of two unequal lengths of stranded cables, which are known as messenger wires, as shown in Fig. 1 [10]. It also consists of a clamp that is used to clamp damper on the transmission line. The symmetric Stockbridge damper (2-R damper) differs from an asymmetric damper with two equal masses and lengths of stranded cables, as shown in Fig. 2.

The symmetric damper is known as the acronym 2-R, it is a two-resonance system. The two resonance frequencies produced by the damper are the ones that counteract the resonance frequency produced by the transmission line when is it exposed to wind motion. The asymmetric damper is known as acronym 4-R, the four resonance frequencies produced by the damper counteract the resonance frequencies produced by the transmission line during operation.

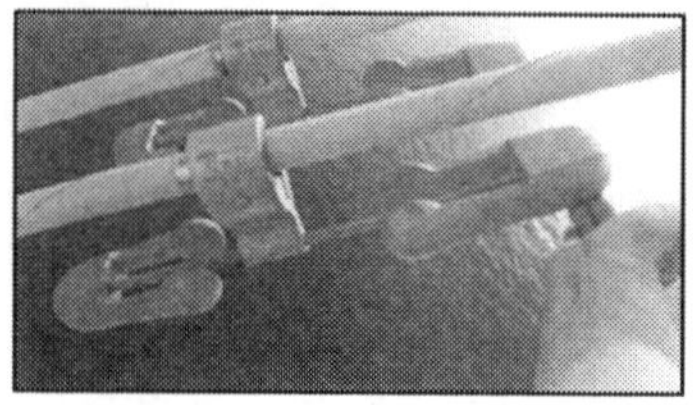

Fig. 1. Asymmetric Stockbridge damper [2].

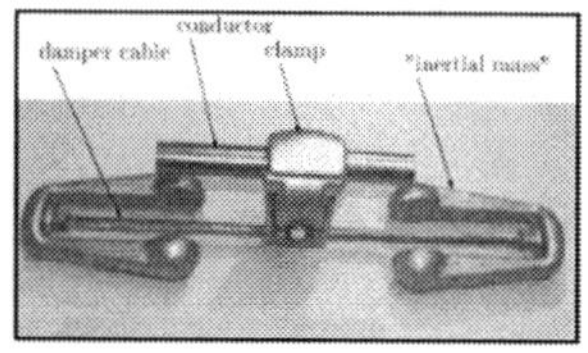

Fig. 2. Symmetric Stockbridge damper [2].

The first Stockbridge-type damper was invented by H. Stockbridge in 1924 after discovering the failure of overhead transmission lines as a result of wind vibrations, as shown in Fig. 3 [2]. Developments and modifications from the original design were conducted and invented to improve the efficiency of the damper. In 1932, Monroe and Templin improved the damper to two degrees of freedom, as shown in Fig. 4 [2]. Both the moment of inertia and the shape were improved to counteract the second vibration mode of the cantilever beam within the frequency range of operation of the damper [2, 10]. In 1968 Claren and Diana came with a new idea in which the two halves of the Stockbridge damper were made asymmetrical leading to two different masses with different moments of inertia and unequal lengths of the messenger cable[2]. A damper known as the Haro damper, as shown in Fig. 5a, was developed by Lauri Haro and Tapani Seppa based on the Stockbridge damper principle in 1970 [2]. An Australian Dogbone damper that develops a torsional resonance was designed in addition to the two flexural resonances, as shown in Fig. 5b. A torsional damper, shown in Fig. 5c, was designed by Tebo of Ontario Hydro and it was mostly used in Canada and other countries worldwide for many years. Elgra is a Swedish damper that consists of a vertical stem having three cast masses loosely fitted to a vertical shaft, as shown in Fig. 5d. The Bretelle damper, shown in Fig. 5e, is widely used in France. It is made from pieces of scrap conductors that are the same size as the line on which it is used [2]. Festoon dampers, shown in Fig. 6, had been used on various long span. Practically no design rules had been published for their use [2].

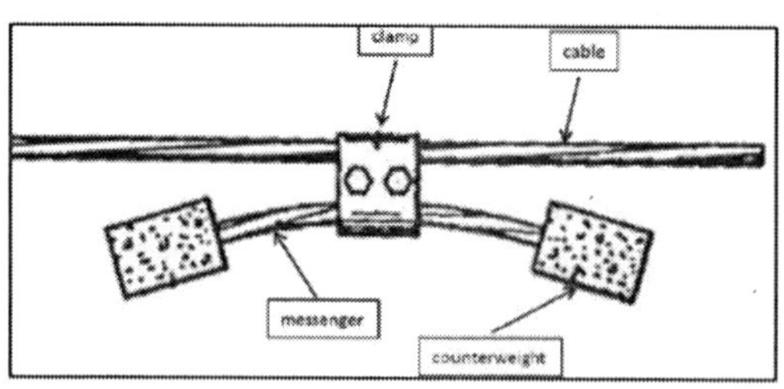

Fig. 3. The original type of Stockbridge damper [2].

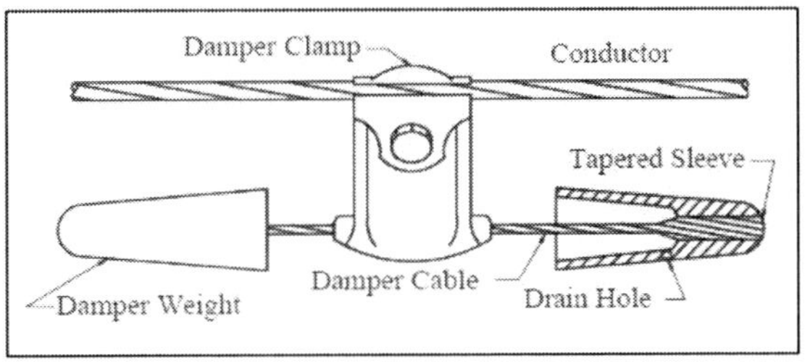

Fig. 4. Monroe and Templin damper [2].

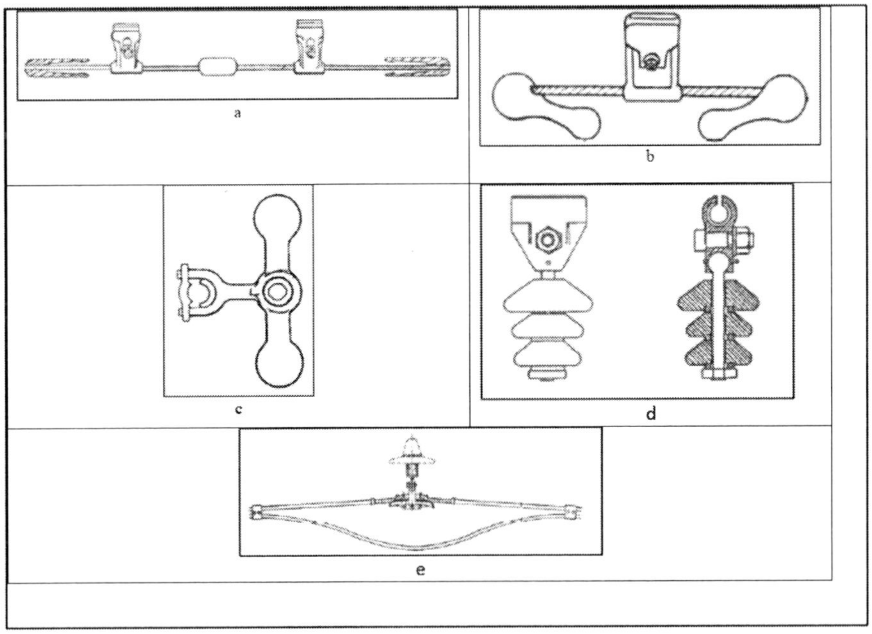

Fig. 5. Different types of dampers: (a) Haro damper. (b) Dogbone damper. (c) torsional damper. (d) Elgra damper. (e) Bretelle damper [2].

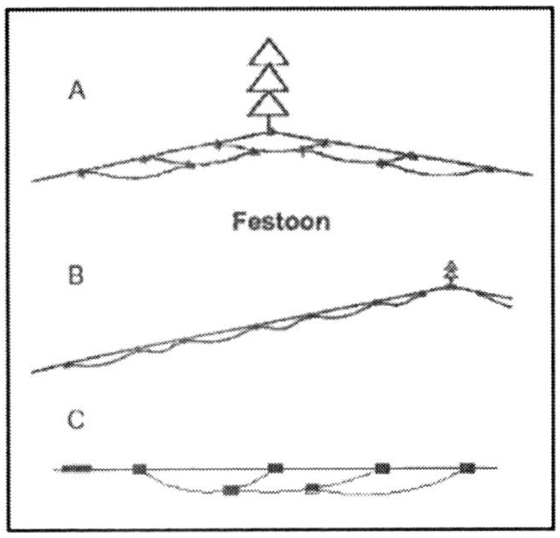

Fig. 6. Festoon dampers. (A) and (B) are festoon dampers for suspension; (C) is a festoon damper for tension points [2].

3. Method

The equipment used to conduct the experimental work is shown in Fig. 7. The asymmetric Stockbridge damper is not clamped on the power transmission line, it is mounted on the shaker shown in Fig. 8 to represent the transmission line exposed to Aeolian vibration. The speed of the shaker was kept with a constant velocity of 0.1 m/s while the frequency was increasing in a range of 5 – 300 Hz. Three (3) asymmetric Stockbridge dampers of different sizes were tested on a shaker and the test was repeated three times per sample. Two accelerometers were stuck on each damper's mass and one on the shaker base using glue as shown in Fig. 8. The important components of a vibration system consist of:

- An electro-dynamic shaker (TIRA Model, Type TV 56263/LS-340) for providing an input force to the damper.
- An amplifier to control the voltage or current to the shaker.
- A control system (known as a computer-controlled data acquisition system)
- A compressor for load protection
- Three (3) asymmetric Stockbridge dampers of various sizes.

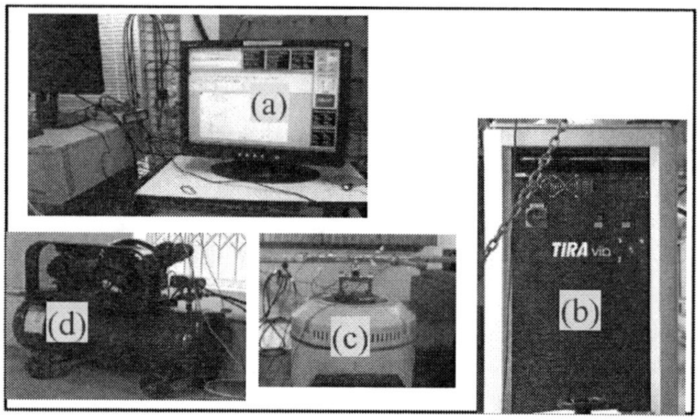

Fig. 7. Components of the vibration test system on the shaker base: (a) controller. (b) Amplifier. (c) Shaker, asymmetric damper, and accelerometers. (d) Compressor.

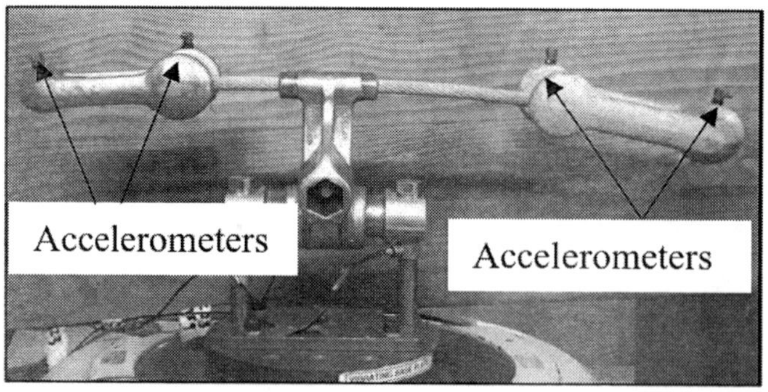

Fig. 8. Stockbridge damper with three accelerometers on each mass [2].

4. Results and discussions

Fig. 9 below shows the positions at which the resonance frequencies were obtained as per the arrangement of accelerometers. During the operation, the physical observation made on the behaviour of the Stockbridge damper is strictly characterised by a sequential vibration pattern. The outer part ($\mathbf{f_1}$) of a bigger mass vibrates then translates the vibration signal to the outer part of a smaller mass ($\mathbf{f_4}$) as the frequency was increasing. Thereafter the inner part ($\mathbf{f_2}$) of the bigger mass vibrates which is ultimately followed by the inner part ($\mathbf{f_3}$) of the smaller mass as the frequency was increasing. Overall, the sequential movement is the one leading the vibration of the damper. The sequence will start from $\mathbf{f_1}$, $\mathbf{f_4}$, $\mathbf{f_2}$, and $\mathbf{f_3}$. There are effectively two positions per mass on the Stockbridge damper responsible to produce resonance frequencies, as shown in Fig. 9.

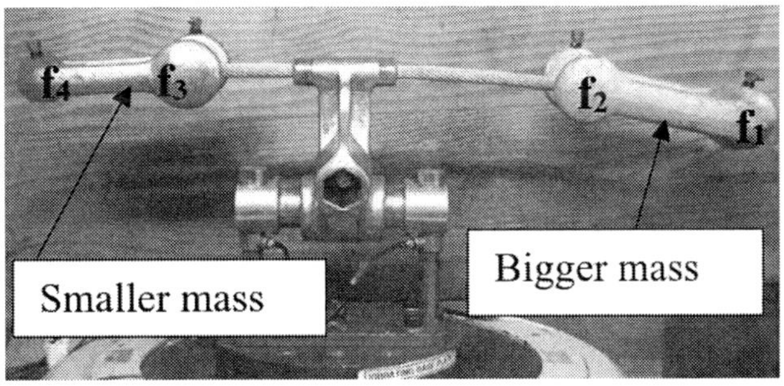

Fig. 9. The asymmetric Stockbridge damper

The graph of acceleration versus frequency shown in Fig. 10 presents the results of a damper (182 025-301) Ø31 – 39 mm. From the big mass (at point $\mathbf{f_1}$), as shown in Fig. 9, it was observed that the frequency of vibration is 8 Hz with an acceleration amplitude of 0.6 G and 1G as shown in Fig. 10. From the same big mass (at point $\mathbf{f_2}$) the frequency of vibration was observed to be 50 Hz with an acceleration amplitude of 1.9 G and 5 G, as shown in Fig. 10. Based on these results the bigger mass has an output of two resonance frequencies and four acceleration amplitudes (inner and outer part). In all three tests conducted, similar results were achieved concerning resonance frequencies and acceleration amplitudes.

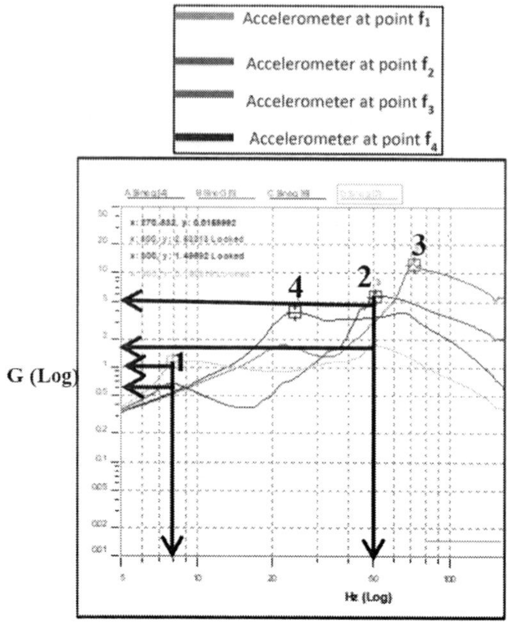

Fig. 10. The graph of acceleration amplitude G(log) vs Hz(log)

Fig. 11 presents the results of a damper (182 025-301) Ø31 – 39 mm. From the small mass (at point **f**$_4$), as shown in Fig. 9, it was observed that the frequency of vibration is 22 Hz with an acceleration amplitude of 1.9 G and 4G as shown in Fig. 11. From the same small mass (at point **f**$_3$), the frequency of vibration was observed to be 70 Hz with an acceleration amplitude of 4 G and 10 G, as shown in Fig. 11. Based on these results the smaller mass has an output of two resonance frequencies and four acceleration amplitudes (inner and outer part). In all three tests conducted, similar results were achieved concerning resonance frequencies and acceleration amplitudes.

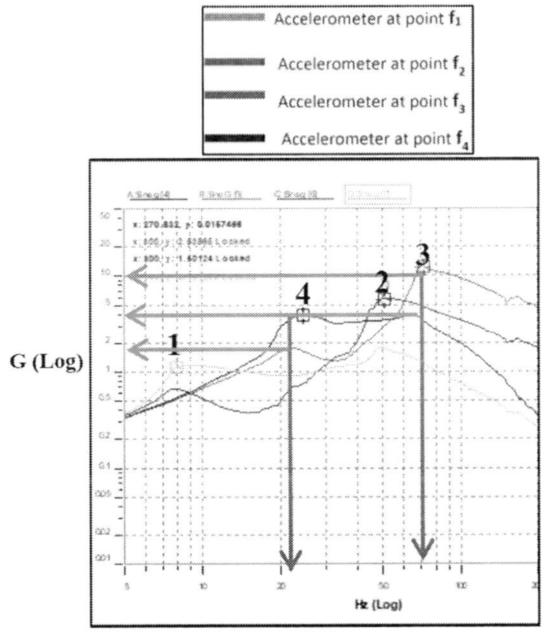

Fig. 11. The graph of acceleration amplitude G(log) vs Hz(log)

At the outer parts of both weights (at points $\mathbf{f_1}$ and $\mathbf{f_4}$) the resonance frequencies produced are smaller than the inner parts of the masses (at points $\mathbf{f_2}$ and $\mathbf{f_3}$). This can be linked to the vibration amplitudes being greater at the outer part of the masses than inner parts, as a result of mass size. Based on the results obtained, the asymmetric Stockbridge damper is four degrees of freedom. The change in the sizes of the asymmetric Stockbridge dampers produces the same number of frequencies and acceleration amplitudes but different magnitudes.

5. Conclusions

The experiments were successfully conducted on different sizes of asymmetric Stockbridge dampers. The results revealed that an asymmetric Stockbridge damper can absorb four resonance frequencies. The magnitude of the resonance frequency was inversely proportional to the acceleration amplitudes and the size of the mass. The outer part of the mass produced smaller frequencies than the inner part of the mass due to mass size. The number of resonance frequencies produced by the asymmetric Stockbridge dampers is the same no matter how big or small the asymmetric Stockbridge damper is. At any other resonance frequencies except those ($\mathbf{f_1}$, $\mathbf{f_2}$, $\mathbf{f_3}$, and $\mathbf{f_4}$) produced by the asymmetric Stockbridge damper, the damper is not a mass absorber. For the future, the work that will be done involves doing experiments when the damper is altered on the transmission line and measured again, in an attempt to see how cable lengths and masses affect the measured response.

Acknowledgment

This work was supported by the Cape Peninsula University of Technology, Mangosuthu University of Technology, and the VRTC laboratory at UKZN. The authors thank Mr. Pravesh Moodley and Mr. Mthobisi Zulu for their contribution.

References

1. O. Griffin, G. Koopmann, "The vortex-excited lift and reaction forces on resonantly vibrating cylinders," *Journal of Sound and Vibration,* **vol. 54, pp. 435-448, (**1977)
2. J. Chan, *EPRI transmission line reference book: Wind induced-induced conductor motion*. Palo Alto, California: Electric Power Research Institute (2006)
3. L. Cloutier, G. Diana, A. Goel, C. Hardy, J. Lilien, J. Wang, "EPRI Transmission Line Reference Book: Wind-Induced Conductor Motion" (2006)
4. A. L. Braun, A. M. Awruch, "Aerodynamic and aeroelastic analyses on the CAARC standard tall building model using numerical simulation," *Computers & Structures,* **vol. 87, pp. 564-581** (2009)
5. D. Havard, "Assessment of the Cowal JCT x Longwood TS for vibration control," *Toronto, Ontario (*2008)
6. N. Barbieri, R. Barbieri, "Dynamic analysis of Stockbridge damper," *Advances in Acoustics and Vibration,* **vol. 2012, (2012)**
7. N. K. Vaja, O. Barry, B. DeJong, "Finite Element Modeling of Stockbridge Damper and Vibration Analysis: Equivalent Cable Stiffness," in *ASME 2017 International Design Engineering Technical Conferences and Computers and Information in Engineering Conference* (2017)
8. N. Vaja, O. Barry, E. Tanbour, "On the modeling and analysis of a vibration absorber for overhead powerlines with multiple resonant frequencies," *Engineering Structures,* **vol. 175, pp. 711-720(**2018)
9. M. Markiewicz, "Optimum dynamic characteristics of Stockbridge dampers for dead-end spans," *Journal of sound and vibration,* vol. 188, **pp. 243-256, (**1995)
10. J. Chan, D. Havard, C. Rawlins, G. Diana, L. Cloutier, J.-L. Lilien, *et al.*, "EPRI Transmission Line Reference Book: wind-induced Conductor Motion" (2009)

Steady-state heat transfer analysis in a spherical domain revisited

Jat du Toit[1,*] , and *Christiaan* Pretorius[2]

[1]North-West University, Private Bag X6001, Potchefstroom 2520, South Africa
[2]Nelson Mandela University, PO Box 77000, Gqeberha 6031, South Africa

Abstract. The paper discusses the numerical solution of the one-dimensional radially axi-symmetric non-linear second-order differential equation to model the conduction and radiation transfer through a spherical domain as a result of an exothermic heat source. The equation is transformed to a non-dimensional form. The dimensionless numbers emanating from the transformation represent the effect of the reaction rate, reaction type, activation energy, radiation and the convection on the temperature. The non-dimensional differential equation for the temperature distribution was previously solved using the Runge-Kutta-Fehlberg method coupled with a Shooting technique. In this paper the solution of the non-dimensional differential equation using an iterative Galerkin finite element method approach employing the Picard method is described. The commercial finite element code Comsol is also employed to solve the non-dimensional differential equation. The current study was motivated by inconsistencies that were observed in the previous results that were presented. Although the assumed underlying physics is used to evaluate the results, the study focuses purely on the numerical solution of the non-dimensional differential equation. The results obtained by the Galerkin finite element code and Comsol were found to be in exact agreement and also exhibit no inconsistencies.

1 Introduction

Lebelo *et al.* [1] discussed a one-dimensional radially axi-symmetric non-linear second-order differential equation to model the conduction and radiation transfer through a spherical domain as a result of an exothermic heat source. A zero heat flux was applied at the centre of the domain and at the boundary a convection heat flux. The equation was transformed to a non-dimensional form. The dimensionless numbers emanating from the transformation represent the effect of the reaction rate, reaction type, activation energy, radiation and the convection on the temperature. The non-dimensional differential equation for the temperature distribution was solved using the Runge-Kutta-Fehlberg method coupled with a Shooting technique. The results for solutions obtained for selected combinations of the dimensionless numbers were presented by Lebelo *et al.* [1].

* Corresponding author: jat.dutoit@nwu.ac.za

In this paper the solution of the non-dimensional differential equation using an iterative Galerkin finite element method approach employing the Picard method is described [2-3]. The commercial finite element code Comsol [4] was also employed to solve the non-dimensional differential equation. Solutions for the same selected combinations of the dimensionless numbers as used by Lebelo *et al.* [1] were obtained.

The current study was motivated by inconsistencies that were observed in the results presented by Lebelo *et al.* [1]. Although the assumed underlying physics is used to evaluate the results, no attempt was made to determine the validity of the mathematical formulation of the physical problem. The study focused purely on the numerical solution of the non-dimensional differential equation.

The results obtained by the Galerkin finite element code and Comsol were found to be in exact agreement and also exhibit no inconsistencies. The paper will provide a full comparison and evaluation of the results presented by Lebelo *et al.* [1] and the results obtained by the finite element codes.

2 Governing equations

The conduction and radiation heat transfer through a spherical domain due to an exothermic heat source can be described by the one-dimensional radially axi-symmetric non-linear second-order differential equation [1]:

$$\frac{k}{r^2}\frac{d}{dr}\left(r^2\frac{dT}{dr}\right)+QA\left(\frac{\sigma T}{\upsilon l}\right)^m\exp\left(\frac{-E}{RT}\right)-\mu\sigma\left(T^4-T_w^4\right)=0 \qquad (1)$$

Where k is the thermal conductivity, r the radial distance, T the temperature, Q the heat of reaction, A the rate constant, σ is the Stefan-Boltzmann constant, υ the vibration frequency, l the Planck number, E the activation energy, R universal gas constant, μ a dimensionless activation energy parameter, and T_w the ambient temperature. The parameter m represents the chemical reaction kinetics type. The equation is subject to the boundary conditions [1]:

$$\left.\frac{dT}{dr}\right|_{r=0}=0 \quad \text{and} \quad \left.\frac{dT}{dr}\right|_{r=a}=\frac{h}{k}\left(T-T_w\right) \qquad (2)$$

Where a is the radius of the sphere and h the heat transfer coefficient. Eq. (1) can be written in dimensionless form to become [1]:

$$\frac{1}{\bar{r}^2}\frac{d}{d\bar{r}}\left(\bar{r}^2\frac{d\theta}{d\bar{r}}\right)+\lambda\left(1+\varepsilon\theta\right)^m\exp\left(\frac{\theta}{1+\varepsilon\theta}\right)-Ra\left[\left(1+\varepsilon\theta\right)^4-1\right]=0 \qquad (3)$$

With the dimensionless temperature $\theta=\frac{E\left(T-T_w\right)}{RT_w^2}$, dimensionless radius $\bar{r}=\frac{r}{a}$, activation energy parameter $\varepsilon=\frac{RT_w}{E}$, radiation parameter $Ra=\frac{\mu\sigma Ea^2T_w^2}{kR}$, and rate of reaction parameter $\lambda=\left(\frac{\sigma T_w}{\upsilon l}\right)^m\frac{QAEa^2}{kRT_w^2}\exp\left(\frac{-E}{RT_w}\right)$. The boundary conditions become:

$$\left.\frac{d\theta}{d\bar{r}}\right|_{\bar{r}=0}=0 \quad \text{and} \quad \left.\frac{d\theta}{d\bar{r}}\right|_{\bar{r}=1}=-Bi\theta \qquad (4)$$

Where Bi is the Biot number and $0 \leq \bar{r} \leq 1$. Lebelo *et al.* [1] solved eq. (3) numerically subject to the boundary conditions given in eq. (4) employing the Runge-Kutta-Fehlberg method coupled with the Shooting technique [5].

3 Finite element formulation

The weak formulation of eq. (3) is formed by multiplying the differential equation with $\bar{r}^2$ and an arbitrary but suitable weighting function v and integrating over the domain [2]. This then gives:

$$\int_0^1 v\frac{d}{d\bar{r}}\left(\bar{r}^2\frac{d\theta}{d\bar{r}}\right)d\bar{r}+\int_0^1 v\bar{r}^2\left\{\lambda(1+\varepsilon\theta)^m\exp\left(\frac{\theta}{1+\varepsilon\theta}\right)-Ra\left[(1+\varepsilon\theta)^4-1\right]\right\}d\bar{r}=0 \qquad (5)$$

Integrating eq. (5) by parts, rearranging the result, expanding the boundary terms and substituting the boundary conditions then leads to :

$$\int_0^1\frac{dv}{d\bar{r}}\bar{r}^2\frac{d\theta}{dr}d\bar{r}=\int_0^1 v\bar{r}^2\left\{\lambda(1+\varepsilon\theta)^m\exp\left(\frac{\theta}{1+\varepsilon\theta}\right)-Ra\left[(1+\varepsilon\theta)^4-1\right]\right\}d\bar{r}+v(1)\left[-Bi\theta(1)\right] \qquad (6)$$

The equation can be solved using an iterative Galerkin finite element (GFE) approach employing the Picard method [2]. For the (i+1)-th iteration with the solution for the i-th iteration available we get :

$$[K]\{\theta^*\}=\{F(\theta^i)\} \qquad (7)$$

Where

$$K_{IJ}=\int_0^1\left(\sum\bar{r}_J\Phi_J\right)^2\frac{d\Phi_I}{d\bar{r}}\frac{d\Phi_J}{d\bar{r}}d\bar{r}$$

$$F_I=\int_0^1\Phi_I\left(\sum r_J\Phi_J\right)^2\left\{\lambda\left(1+\varepsilon\left[\sum\theta_J^i\Phi_J\right]\right)^m\exp\left(\frac{\left[\sum\theta_J^i\Phi_J\right]}{1+\varepsilon\left[\sum\theta_J^i\Phi_J\right]}\right)-Ra\left[\left(1+\varepsilon\left[\sum\theta_J^i\Phi_J\right]\right)^4-1\right]\right\}d\bar{r}$$

$$K_{NN}=K_{NN}+Bi$$

The last equation accounts for the boundary condition on the right-hand side. The solution for the (i+1)-th iteration is then given by :

$$\{\theta^{i+1}\} = \omega\{\theta^*\} + (1-\omega)\{\theta^i\} \qquad (8)$$

Where ω is a relaxation parameter with $0 \le \omega \le 1$. The solution is initialized by taking $\{\theta^0\} = \{0.0\}$. The solution is considered to be converged when :

$$\frac{\sum(\theta_J^{i+1})^2 - \sum(\theta_J^i)^2}{\sum(\theta_J^{i+1})^2} < 10^{-6} \qquad (9)$$

For all the simulations 25 quadratic elements were employed with the relaxation parameter taken as $\omega = 0.5$.

The dimensionless non-linear second-order differential equation was also solved using the finite element method as implemented in the Comsol Multiphysics software (version 3.5a) [4]. The domain $0 \le \bar{r} \le 1$ was divided into 60 uniformly sized elements on which quadratic Lagrange interpolation was employed resulting in a total is 121 nodal points. Numerical integration was performed using Guassian quadrature with an integration order of four. The problem was solved iteratively by making use of a Damped Newton method with a relative error tolerance of 10^{-6}.

4 Results

Lebelo *et al.* [1] studied the effect on the dimensionless temperature distribution of selected values for λ, m, ε, Ra and Bi. The following arbitrary reference values were selected for the parameters, namely $\lambda = 0.1$, $m = 0.5$, $\varepsilon = 0.1$, $Ra = 1.0$ and $Bi = 1.0$. Lebelo *et al.* [1] stated that all the results are due to exothermic reactions that are taking place in the domain and where heat is thus generated inside the domain. In steady-state conditions conservation of energy requires that heat be released to the surrounding environment at the boundary of the domain. In order for the generated heat to be transferred to the boundary and released to the surrounding environment, the temperature gradient in the domain must be $d\theta/d\bar{r} < 0$.

Five sets of simulations were performed by Lebelo *et al.* [1] and repeated in the current study. In each set the value of one of the parameters was varied whilst the values of the other parameters were kept constant.

In the first set of simulations the values of m, ε, Ra and Bi are kept constant at the reference values, whilst the values 0.1, 0.2, 0.3 and 0.4 were selected for λ. According to Lebelo *et al.* [1] increasing the value of λ represents an acceleration in the rate of the reaction that supports the exothermic chemical reaction.

The results obtained by Lebelo *et al.* [1] are depicted in Fig. 1, whilst the results obtained in the current study are depicted in Fig. 2.

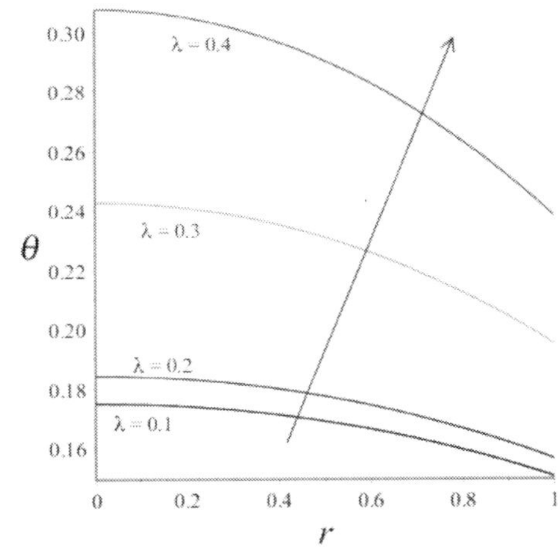

Fig. 1. Effect of λ on temperature as function of radial position, Lebelo *et al.* [1].

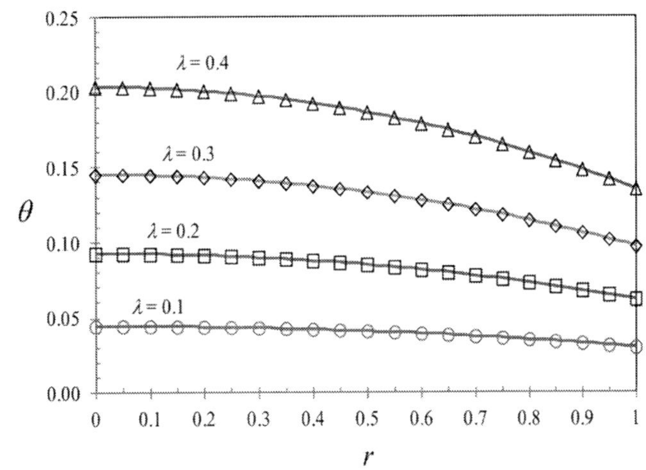

Fig. 2. Effect of λ on temperature as function of radial position, current study. Solid lines = GFE, symbols = Comsol.

It can be seen in both Figs. 1 and 2 that the gradients of the temperature profiles are negative as expected for an exothermic chemical reaction. Also in both Figs. 1 and 2 the level of the temperature increases as the rate of reaction increases and more heat is released. In Fig. 2 it can be seen that the GFE and Comsol results are in exact agreement. However, the temperature levels obtained by Lebelo *et al.* [1] are considerably higher than the corresponding temperature levels obtained by the GFE and Comsol simulations.

In the second set of simulations the values of λ, ε, Ra and Bi are kept constant at the reference values, whilst the values 0.5, 0.0 and -2.0 were selected for m. According to Lebelo *et al.* [1] the value of 0.5 represents biomolecular reaction kinetics, 0.0 Arrhenius reaction kinetics and -2.0 sensitized or light induced reaction kinetics.

Fig. 3 shows the results obtained by Lebelo *et al.* [1] for the selected values of m, whilst Fig. 4 shows the corresponding results that were obtained in the current study. Again it can be observed in Fig. 4 that the GFE and Comsol results are in exact agreement. However, the GFE and Comsol results differ markedly from the corresponding results shown in Fig. 3.

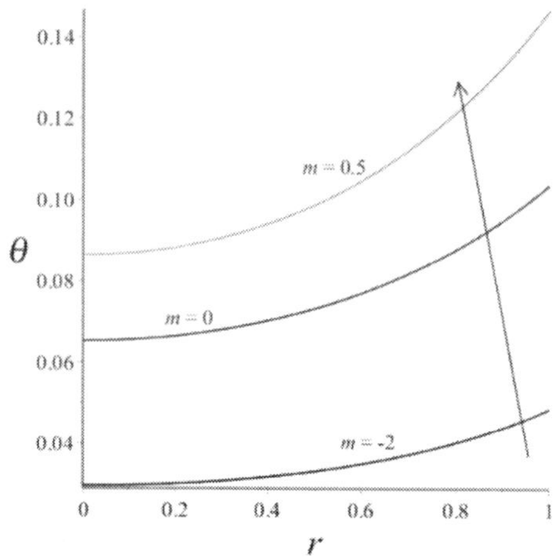

Fig. 3. Effect of m on temperature as function of radial position, Lebelo *et al.* [1].

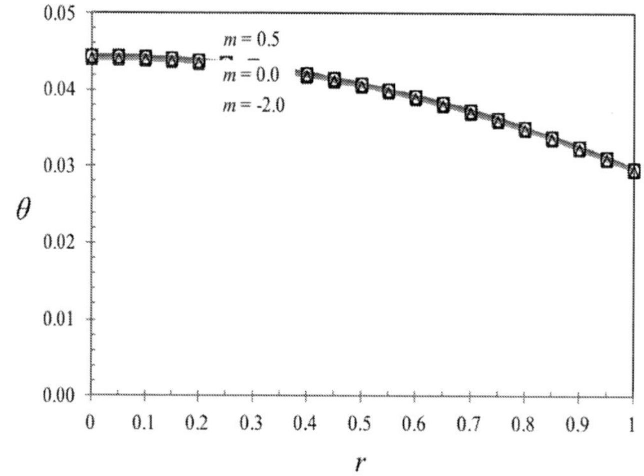

Fig. 4. Effect of m on temperature as function of radial position, current study. Solid lines = GFE, symbols = Comsol.

The results obtained by Lebelo *et al.* [1] depicted in Fig. 3 are indicative of an endothermic chemical reaction where heat is absorbed from the surrounding environment. In contrast the GFE and Comsol results shown in Fig. 4 correspond to the assumed exothermic chemical reaction. It can be observed in Fig. 4 that the reaction kinetics type has an almost negligible effect on the temperature level. However, according to the results shown in Fig. 3 the reaction kinetics type has a marked effect on the temperature level.

In the third set of simulations the values of λ, m, Ra and Bi are kept constant at the reference values, whilst the values 0.1, 0.2, 0.3 and 0.4 were selected for ε. According to Lebelo *et al.* [1] ε is the activation energy parameter and a reduction in the temperature level due to ε implies a reduction in the activation energy to stimulate the exothermic chemical reaction.

The results obtained by Lebelo *et al.* [1] are shown in Fig. 5, whilst the results obtained from the GFE and Comsol simulations are shown in Fig. 6. It can again be observed in Fig. 6 that the GFE and Comsol results are in exact agreement.

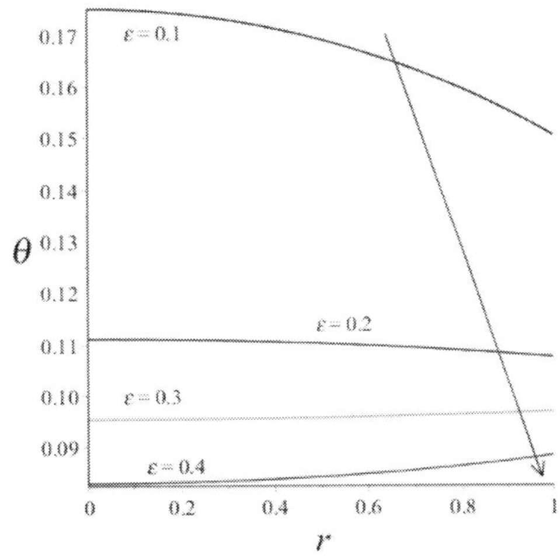

Fig. 5. Effect of ε on temperature as function of radial position, Lebelo *et al.* [1].

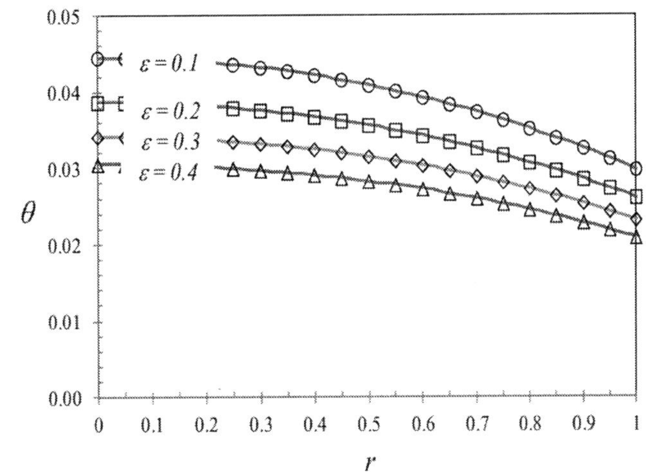

Fig. 6. Effect of ε on temperature as function of radial position, current study. Solid lines = GFE, symbols = Comsol.

In both Figs. 5 and 6 it can be seen that the temperature level decreases as the value of ε increases. However, in Fig. 5 the results for $\varepsilon = 0.1$ and 0.2 indicated that the chemical reaction is exothermic, whilst the results for $\varepsilon = 0.3$ and 0.4 indicate that the chemical reaction is endothermic. In Fig. 6 the results indicated that the chemical reaction is exothermic for all the values of ε. The values of the results shown in Fig. 6 are also markedly lower than the values of the corresponding results shown in Fig. 5.

In the fourth set of simulations the values of λ, m, ε and Bi are kept constant at the reference values, whilst the values 1.0, 2.0, 3.0 and 4.0 were selected for Ra. According to Lebelo *et al.* [1] Ra deals with the rate at which heat is released from the surface of the sphere to the surrounding environment due to radiation. When Ra increases the rate at which heat is released also increases and results in the reduction of the temperature level.

The results obtained by Lebelo *et al.* [1] are shown in Fig. 7, whilst the results obtained from the GFE and Comsol simulations are shown in Fig. 8. It can again be observed in Fig. 8 that the GFE and Comsol results are in exact agreement.

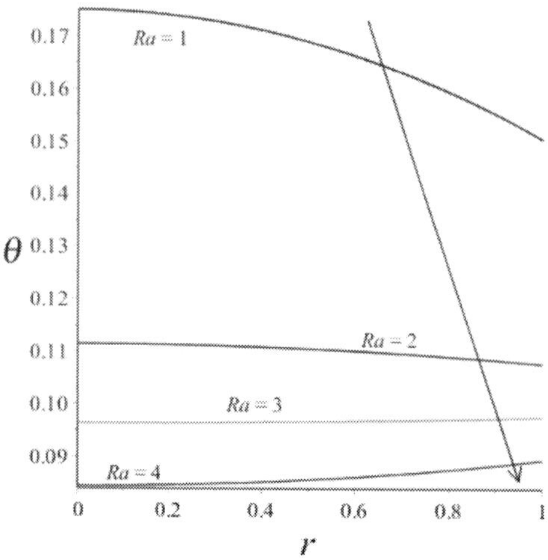

Fig. 7. Effect of Ra on temperature as function of radial position, Lebelo *et al.* [1].

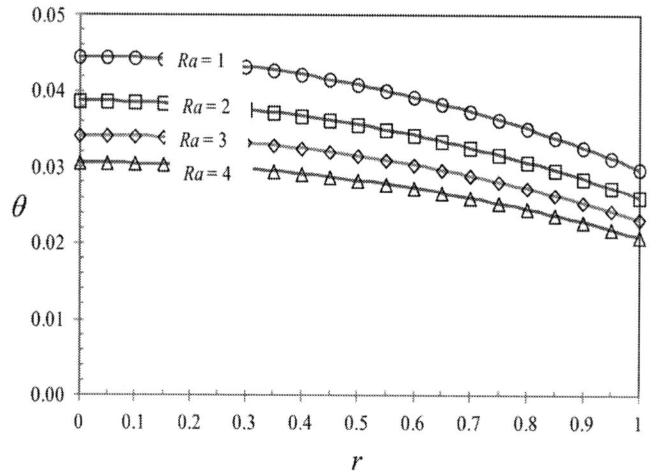

Fig. 8. Effect of Ra on temperature as function of radial position, current study. Solid lines = GFE, symbols = Comsol.

It can be seen in both Figs. 7 and 8 that the temperature level decreases as the value of Ra increases. However, in Fig. 7 the results for $Ra = 1.0$ and 2.0 imply that the chemical reaction is exothermic, whilst the results for $Ra = 3.0$ and 4.0 imply that the chemical reaction is endothermic. In Fig. 8 the results indicated that the chemical reaction is exothermic for all the values of Ra. The values of the results shown in Fig. 8 are also markedly lower than the values of the corresponding results shown in Fig. 7. Comparing the results in Figs. 5 and 7 and comparing the results in Figs. 6 and 8 it can be seen that ε and Ra appear to have exactly the same effect and produce the same temperature profiles.

In the fifth set of simulations the values of λ, m, ε and Ra are kept constant at the reference values, whilst the values 1.0, 2.0, 3.0 and 4.0 were selected from Bi. According to Lebelo *et al.* [1] Bi deals with the rate at which heat is released for the surface of the sphere to the surrounding environment due to convection. When Bi increases the rate at which heat is released also increases and results in the reduction of the temperature level.

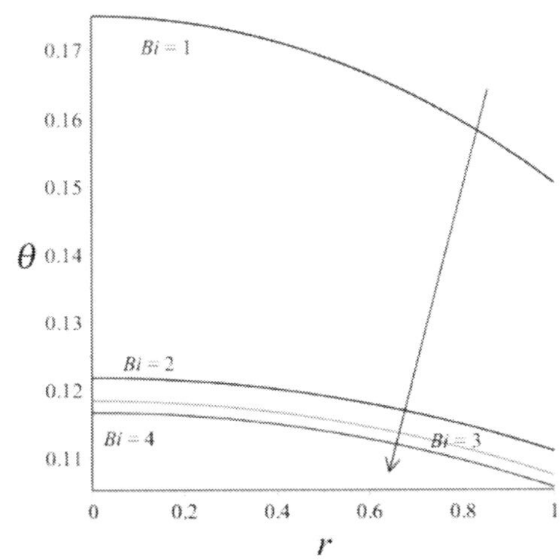

Fig. 9. Effect of Bi on temperature as function of radial position, Lebelo *et al.* [1].

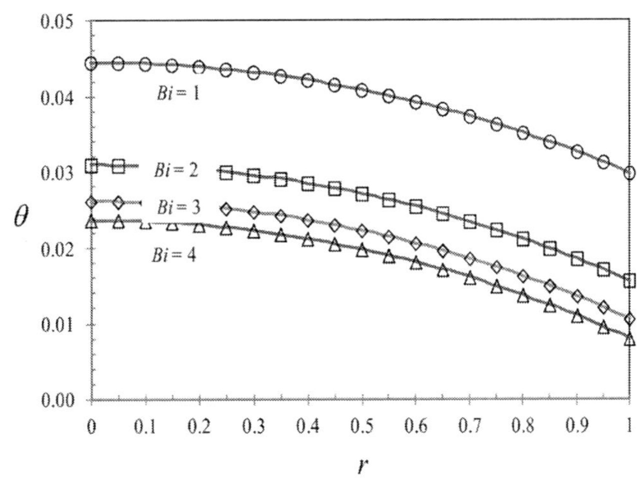

Fig. 10. Effect of Bi on temperature as function of radial position, current study. Solid lines = GFE, symbols = Comsol.

The results obtained by Lebelo *et al.* [1] are shown in Fig. 9, whilst the results obtained from the GFE and Comsol simulations are shown in Fig. 10.

It can be seen in both Figs. 9 and 10 that the gradients of the temperature profiles are negative as expected for an exothermic chemical reaction. Also in both Fig. 9 and 10 the level of the temperature decreases as the rate of convection heat transfer to the surrounding environment increases. In Fig. 10 it can be seen that the GFE and Comsol results are in exact agreement. However, the temperature levels obtained by Lebelo *et al.* [1] are considerably higher than the corresponding temperature levels obtained by the GFE and Comsol simulations.

In an attempt to resolve the difference between the temperature levels of the solutions obtained by Lebelo *et al.* [1] and the GFE and Comsol solutions, additional GFE simulations were performed where the simulations were initialized by taking $\{\theta^0\} = \{1.0\}$

and $\{\theta^0\} = \{10.0\}$. However, both cases still lead to the same solution as obtained with $\{\theta^0\} = \{0.0\}$.

In each of the sets of simulations one of the simulations corresponded to the reference values for λ, m, ε, Ra and Bi. When considering Figs. 1, 5, 7 and 9 it can be observed that the relevant curve is the same in all the figures. However, in Fig. 3 the relevant curve differs significantly from the corresponding curves in Figs. 1, 5, 7 and 9. Considering Figs. 2, 4, 6, 8 and 10 it can be seen that the relevant curves are the same in all the figures.

Finally the solutions for θ depicted in Figs. 1 to 10 all show that $\theta > 0$.Thus the boundary condition at $\bar{r} = 1$ given in eq. (4) dictates that the temperature gradient at the boundary must be negative for $Bi > 0$.Thus the solutions for θ shown in Fig. 3 and for $\varepsilon = 0.3$ and 0.4 in Fig. 5 and $Ra = 3.0$ and 4.0 in Fig. 7 are inconsistent with the boundary condition.

5 Conclusions

In this paper the numerical solution of the one-dimensional radially axi-symmetric non-linear second-order differential equation to model the conduction and radiation transfer through a spherical domain as a result of an exothermic heat source presented by Lebelo *et al.* [1] was evaluated. The evaluation was performed by generating corresponding solutions using a Galerkin Finite Element approach and the Comsol Multiphysics code [4].

In all the cases for which simulations were performed the GFE and Comsol solutions were in exact agreement and consistent with the described underlying physics of the problem. In particular the results indicated that the chemical reaction is exothermic with heat being released to the surrounding environment.

The solutions presented by Lebelo *et al.* [1] contain a number of inconsistencies. The set of simulations demonstrating the effect of the reaction kinetics on the temperature profile lead to results that are inconsistent with the assumed exothermic chemical reactions and are also inconsistent with the boundary condition prescribed at the surface of the sphere. The sets of results associated with the effects of the activation energy and the radiation parameter on the temperature profile also contain inconsistencies. Half of the results in each set are indicative of exothermic chemical reactions, whilst the other half of the results are indicative of endothermic chemical reactions. In the latter cases the results are also inconsistent with the prescribed boundary condition at the surface of the sphere.

It is recommended that the implementation of the Runge-Kutta-Fehlberg method coupled with the Shooting technique [5] by Lebelo *et al.* [1] be revisited and clarified.

This work is based on research supported by the South African Research Chairs Initiative of the Department of Science and Technology and National Research Foundation of South Africa (Grant No 61059). Any opinion, finding and conclusion or recommendation expressed in this material is that of the authors and the NRF of South Africa do not accept any liability in this regard.

References

1. R.S. Lebelo, K.C. Moloi, C. Chitumwa, M. Sadiki, P. Baloyi, *Steady-state heat transit analysis in a spherical domain*, in Proceedings of the 11th South African Conference on Computational and Applied Mechanics, SACAM 2018, 17-19 September 2018, Vanderbijlpark, South Africa (2018).

2. E.B. Becker, G.F. Carey, J.T. Oden, Finite elements: An introduction, Prentice-Hall (1981).
3. J.N. Reddy, An introduction to the finite element method (McGraw-Hill , 2006).
4. COMSOL Multiphysics Software. Online, URL: https://www.comsol.com/comsol-multiphysics. Accessed 16 June 2021.
5. E. Kreyszig, Advanced Engineering Mathematics (John Wiley & Sons, 1999).

An accurate lumped convective plus radiative heat exchanger model for performance-based modelling

Wim Fuls[1,*]

[1]ATProM Research Unit, University of Cape Town, Private Bag X3, Rondebosch, 7701, South Africa

Abstract. There are several thermo-fluid process modelling tools available on the market which can be used to analyze the off-design performance of thermal plants. These tools all offer the user with a simple convective heat exchanger component that requires the design-base process conditions as inputs. The tools would then calculate an effective overall heat transfer factor (UA) and make use of gas flow mass ratios to scale the UA value for off-design conditions. The models employed in these tools assume that the contribution of gas radiation is insignificant, hence only applies convection scaling laws. This paper presents an improved model which considers the contribution of the gas and particle radiation, as is often encountered in the first few heaters in coal fired boilers and heat recovery steam generators. A more fundamental scaling law is applied for the convection scaling and incorporates a cleanliness factor which allows for the consideration of fouling of the heater surfaces. The model's performance was validated against a discretized tube-level heater model that solves the fundamental convection and radiation terms. The model is accurate within 1% for the cases considered, as compared to more than 20% error if radiation contribution is not considered.

1 Introduction

The ability to accurately predict the heat transfer occurring in a thermal application is key to a successful design. It is therefore not unusual to find a vast volume of literature dealing with heat exchanger performance modelling. The methods range from lumped heuristic and empirical models, to detailed discretised analysis, and more recently high resolution CFD analyses. The choice of method employed by the designer or analyst is a function of the expected outcome as well as the available information about the specific heat exchanger.

There are various commercial software tools available which implements one or multiple methods. The tools generally offer three approaches: a) determine the physical design requirements of a heat exchanger, given certain performance needs; b) determine the performance of a heat exchanger given detailed design information; c) determine the performance of a heat exchanger given a known operating point. This paper deals with the last approach where the analyst is usually interested in the off-design performance of an existing heater without detailed knowledge of the physical design or physical process

* Corresponding author: wim.fuls@uct.ac.za

occurring during the heat transfer. More specifically, the focus is on gas-to-fluid heat exchangers such as superheaters found in typical steam generating plants.

2 Literature

The following is a brief discussion of the methods employed by some power plant modelling software tools and researchers. It is by no means comprehensive to cover all tools and methods available, but a common trend is observable. Also, the method described here is not necessarily the only option available to the user of the mentioned software.

In the case where the user only specifies the design point performance of a heater, the method used by the software is sometimes called a "black box" model, or "performance" model and is essentially a lumped model. The inputs for these are typically the inlet thermal properties and mass flow of both streams, as well as one stream's exit condition. The software will then use these, along with a user specified heat exchanger type (cross-flow, parallel flow or counter flow), and determine the design base overall heat transfer factor (UA_D) such that:

$$\dot{Q}_D = \varepsilon_D C_{min}\left(T_{H.in} - T_{C.in}\right)$$
$$\text{with } \varepsilon_D = f(Ntu,...) \quad (1)$$
$$\text{and } Ntu = \frac{UA_D}{C_{min}}$$

This is the well-known ε-NTU method which defines the heat exchanger's effectiveness (ε) in terms of the Number of Transfer Units (*NTU*) which in turn is a function of the UA value and the minimum stream's thermal capacitance, with $C_{\min} = \min(C_H\dot{m}_H, C_C\dot{m}_C)$. The subscript *D* denotes the Design conditions, and H and C points to the Hot and Cold stream respectively. In some cases, the equivalent log mean temperature difference (LMTD) method is used, which requires a flow correction factor *F* in the case of cross flow.

In the off-design condition, the design UA value is scaled based on the change of process conditions to arrive at an off-design UA value, which can then be used to calculate the heat transfer using the same ε–NTU method. The method of scaling varies slightly between various tools, but the most significant factor is the ratio of the off-design mass flow vs. the design flow, raised to some exponent.

$$UA \approx UA_D\left(\frac{\dot{m}}{\dot{m}_D}\right)^a \quad (2)$$

This ratio originates from the common shape of the Nusselt number correlation when considering forced convection of fluids, which is:

$$Nu = C \cdot Re^a Pr^b \quad (3)$$

If one assumes that the gas stream is the dominant thermal resistance, equation (2) an (3) can be combined and rewritten using the known definitions for Nusselt, Reynolds and Prandtl number. The coefficient *C* cancels out when considering only convection ratios. The exponent *a* varies based on the tube arrangement for tubular heat exchangers, but is typically in the range of 0.5 to 0.8 [1], while the exponent *b* is generally taken as $\frac{1}{3}$. The off-design UA value is thus:

$$UA = UA_D\left(\frac{\dot{m}_g}{\dot{m}_{g.D}}\right)^a\left(\frac{k_g}{k_{g.D}}\right)^{2/3}\left(\frac{\mu_g}{\mu_{g.D}}\right)^{1/3-a}\left(\frac{cp_g}{cp_{g.D}}\right)^{1/3} \quad (4)$$

The commercial software Ebsilon offers a Component 61 which splits the UA into an internal (liquid side) and external (gas side) value, and then applies the following scaling laws [2]:

$$UA_l = UA_{l.D}\left(\frac{\dot{m}_l}{\dot{m}_{l.D}}\right)^{a_l}$$
$$UA_g = UA_{g.D}\left(\frac{\dot{m}_g}{\dot{m}_{g.D}}\right)^{a_g}\left(1-\frac{\left(T_{g.av}-T_{g.av.D}\right)}{2000K}\right) \tag{5}$$

The user is required to enter values for $UA_{l.D}$ and $UA_{g.D}$. The help file does not suggest a suitable value for the exponents. The temperature difference term added to the gas side scaling is probably trying to compensate for fluid property changes.

The commercial tool PEPSE has a performance mode for all its heat exchanger components, however the manual does not state how the calculation is done [3]. Based on the required inputs, it is likely that it also uses a mass flow ratio scaling.

The software Thermoflex uses a multi-parameter thermal resistance scaling, and translated into UA values results in [4]:

$$UA_l = UA_{l.D}\left(\frac{\dot{m}_l}{\dot{m}_{l.D}}\right)^{a_l}\left(\frac{T_l}{T_{l.D}}\right)^{c_l}\left(\frac{p_l}{p_{l.D}}\right)^{d_l}$$
$$UA_g = UA_{g.D}\left(\frac{\dot{m}_g}{\dot{m}_{g.D}}\right)^{a_g}\left(\frac{T_g}{T_{g.D}}\right)^{c_g}\left(\frac{M_g}{M_{g.D}}\right)^{d_g} \tag{6}$$

The user enters the ratios for liquid, gas and wall resistance vs overall resistance, and this is used to determine the individual stream's design UA. No guidelines are given for the value of the exponents. It is interesting to note that for steam (liquid side)† the fluid property scaling is related to temperature and pressure, while the gas side is scaled using temperature and the fluid's average molecular weight M. The only reason why the molecular weight would change is if the gas composition changes, which probably has the greatest effect on the specific heat c_P.

The VirtualPlant software is part of the EtaPro performance and condition monitoring platform [5]. For its boiler and steam generator heat exchangers, it uses the following scaling law:

$$UA = UA_D\left(\frac{\dot{m}_g}{\dot{m}_{g.D}}\right)^{a}\left(\frac{k_g}{k_{g.D}}\right)^{2/3}\left(\frac{\mu_g}{\mu_{g.D}}\right)^{2/3}\left(\frac{cp_g}{cp_{g.D}}\right)^{1/3} \tag{7}$$

The default suggestion is $a = 0.8$. It closely resembles the fundamental equation (4), except for a small difference in the viscosity ratio exponent.

Nord used Aspen HYSYS to model the steam generator of a combined cycle, and applied the following scaling [6]:

$$UA = UA_D\left(\frac{\dot{m}_g}{\dot{m}_{g.D}}\right)^{0.57} \tag{8}$$

† Although the steam in power plant superheaters are in the gaseous phase, its pressure and flowrate are typically so high that the heat transfer coefficients resembles that of a liquid.

In his book, Ganapathy suggests the following scaling for modelling heat recovery steam generators [7]:

$$UA = UA_D \left(\frac{\dot{m}_g}{\dot{m}_{g.D}} \right)^{0.65} \left(\frac{F_g}{F_{g.D}} \right) \left(\frac{\dot{m}_l}{\dot{m}_{l.D}} \right)^{0.15}$$
$$with\ \ F = \frac{c_P^{\,0.33} k^{0.67}}{\mu^{0.32}} \tag{9}$$

It is unsure why the liquid mass flow ratio is included in this fashion, as the combined effect of inner and outer heat transfer is not the product of the two, but rather the reciprocal sum.

3 Model definition

The primary drawback with all the methods employed in the presented literature is that it ignores the contribution of thermal radiation to the overall heat transfer and how it changes with plant load. Although this may be suitable for low temperature heat exchangers, it is insufficient for those that are close to the fuel combustion, such as superheaters in a coal fired boiler. The ratio of thermal radiation compared to the total heat transfer can be more than 50% [8]. If the overall heat transfer is scaled only on mass flow, one may make a substantial error for off-design conditions, especially in the case where the temperature remains relatively high while the mass flow reduces.

Let us define a ratio of the heat transfer due to thermal radiation relative to the total heat transfer at design conditions as R_r. Using the general equation for thermal radiation, one can write the thermal radiation component as:

$$\dot{Q}_{rD} = R_r \cdot \dot{Q}_D = \sigma \varepsilon AF \left(T_{g.D}{}^4 - T_{l.D}{}^4 \right) \tag{10}$$

$T_{g.D}$ and $T_{l.D}$ is the average temperature of the gas and liquid stream between inlets and outlets. This is a simplification of the actual process occurring, most significantly because one assumes radiation occurs between the gas and fluid, instead of the gas and the outer wall surface. This assumption implies that the thermal resistance on the fluid-side, as well as the wall resistance is substantially lower than the resistance on the gas side, making the wall temperature on the gas side to be approximately the same as the fluid temperature. This is generally an adequate assumption for gas to liquid heat exchangers.

The term $\sigma \varepsilon AF$ can be called the radiation factor of the heat exchanger. It contains the collective effect of thermal emissivity, radiation surface area and radiation view factor. Apart from minor changes to the effective gas emissivity, this radiation factor can be assumed to remain constant irrespective of the process conditions.

One can also write the radiation heat transfer as an effective UA value such that:

$$\dot{Q}_{r.D} = UA_{r.D} \left(T_{g.D} - T_{l.D} \right) \tag{11}$$

The overall design UA_D is calculated with the design process conditions and the appropriate ε-NTU correlation. This enables one to determine the convective $UA_{c.D}$ value at design conditions, still assuming the wall resistance is insignificant:

$$UA_{c.D} = UA_D - UA_{r.D} \tag{12}$$

For off-design conditions, the convection portion UA_c can be calculated using equation (4). The radiation-convection is determined using the constant $\sigma \varepsilon AF$ factor and the off-design temperatures by combining equation (10) and (11):

$$UA_r = \sigma\varepsilon AF \cdot \frac{\left(T_g^{\,4} - T_l^{\,4}\right)}{\left(T_g - T_l\right)} \tag{13}$$

This results in the effective off-design UA value:

$$UA = UA_c + UA_r \tag{14}$$

The design conditions are typically for a clean heat exchanger. Many software tools provide the user with a correction factor option to adjust the overall heat transfer to account for fouling, hence:

$$UA_{fouled} = CF \cdot UA_{clean} \tag{15}$$

The *CF* factor can also be termed a Cleanliness Factor. The effective thermal resistance due to the added fouling R_{foul} can be obtained from:

$$UA_{fouled} = \left(UA_{clean}^{\;-1} + R_{foul}\right)^{-1} \tag{16}$$

Thus, for a given cleanliness factor, the resulting fouling resistance is:

$$R_{foul} = \frac{1/_{CF} - 1}{UA_D} \tag{17}$$

The overall UA value for a fouled heat exchanger at off-design conditions is thus given as:

$$UA = \left(\frac{1}{UA_r + UA_c} + R_{foul}\right)^{-1} \tag{18}$$

This method is different to what most software tools use, as they would typically scale the off-design UA factor with the cleanliness factor (equation (15)). This is not entirely correct, as it implies that the convection and radiation is proportionally affected by fouling. Fouling rather results in a constant thermal resistance in the heat transfer path. As can be seen from equation (18).

The off-design UA factor can now be used with the appropriate ε-NTU or LMTD method to calculate the accurate off-design performance.

4 Model validation

Gwebu and Rousseau published results from modelling a superheater in a discretized pipe-per-pipe network approach [9]. The significant contribution was a special pipe component developed in the Flownex Simulation Environment [10] which includes accurate correlations for internal and external convection, pipe and fouling layer conductivity, as well as thermal radiation from the surrounding gas. The method is similar to the work done by Trojan and Taler [11].

This special pipe component was used in this study to model a hypothetical multi-pass heat exchanger which resembles a typical final superheater of a subcritical coal fired boiler. **Figure 1** shows an image of the heater model in Flownex SE. The model considers one tube bank only, and consists of 8 longitudinal passes, each discretized into 6 vertical pipe elements. The geometrical parameters of each pipe element as well as the design base process conditions and performance are given in **Table 1**. It is worth-while to note that this specific heater transfers 51.9% of the heat via thermal radiation.

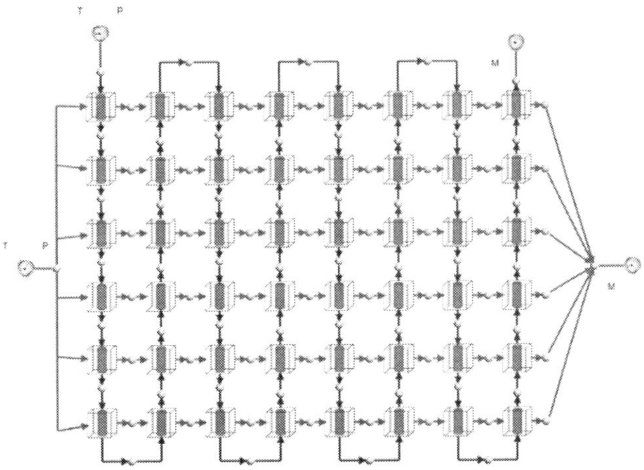

Figure 1. Flownex Model of the validation heat exchanger.

The IF97 formulation of water and steam as per IAPWS [12] was used for the liquid side, and ideal gas air for the gas side. It is assumed that there is no maldistribution of temperature or mass flow per pipe element. The discretized model was used to predict the "true" result of the heater at off-design conditions and was then compared to the newly proposed scaling method implemented in a lumped model.

Table 1. Inputs and Process Results for the heat exchanger.

Geometry	Properties
Pipe increment length : 1.0 m	Wall emissivity : 0.7
Pipe outer dia. : 45 mm	Gas emissivity : 0.1
Pipe inner dia. : 30 mm	
Gas volume length : 200 mm	
Gas volume width : 500 mm	
Process inputs	**Process results**
Steam inlet :17 MPa, 450 °C	Steam outlet : 530.08 °C
Steam mass flow : 0.3 kg/s	Gas outlet : 795.58 °C
Gas inlet : 100 kPa, 850 °C	Total heat : 75.631 kW
Gas mass flow : 1.2 kg/s	Radiation heat : 39.274 kW
	Design UA : 0.230 kW/K

The lumped model used the ε-NTU formulation of a pure parallel flow heat exchanger. This is not entirely correct, as there are better models to use for multi-pass cross flow heat exchangers, but this would require more detailed information from the user. Recalling that the main goal of the lumped model is to predict off-design performance with little to no physical inputs, these models typically only provide for pure parallel flow, counter flow, or cross flow with both streams un-mixed.

The first set of validation calculations focussed on the ability of the model to properly account for the thermal radiation contribution. Two scenarios were studied: a) change in gas mass flow; and b) change in gas temperature. For both cases the lumped model was solved with the actual design base radiation ratio, and a radiation ratio of zero. The zero condition would resemble the results produced by traditional methods. The total heat transfer and the steam outlet temperature were used as primary indicators. The curves in **Figure 2** show the percentage error for the heat transfer, as well as the absolute steam temperature error for the various cases.

The results clearly show how the new model performs exceptionally well in capturing the actual off-design performance, even in the very extreme conditions as was tested. A maximum heat error of 1.6% and 0.9 K is reported when radiation is included, as compared to the very large errors from traditional models which do not include the effect of thermal radiation.

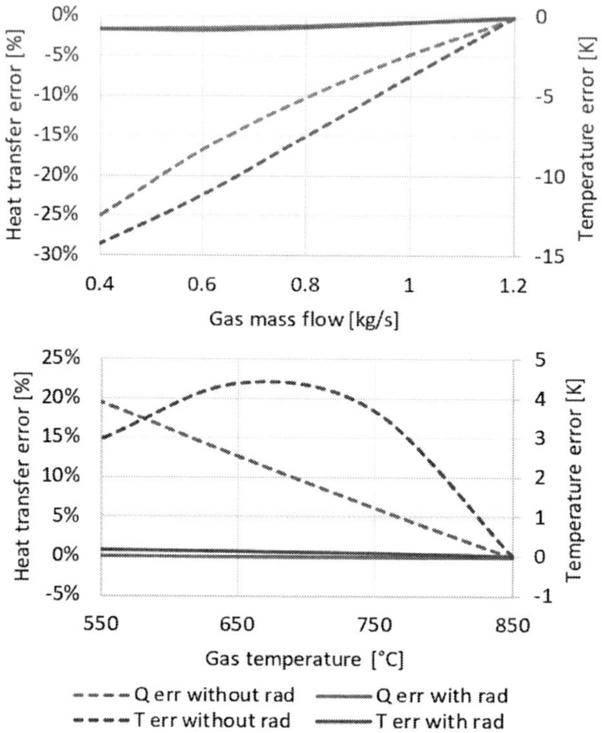

Figure 2. Difference between models with and without the Radiation Effect for off-design gas mass flow (top) and inlet temperature (bottom).

The next case considers the inclusion of a fouling layer. For this, a 1mm thick layer of fouling with arbitrary thermal conductivity of 0.07 W/mK was added to the outside of all the pipes. When solved for the design base input conditions, an exit steam temperature of 510.6 °C and 58.3 kW of heat was reported. This translates into a fouled overall $UA = 0.1686$ kW/K, and thus a cleanliness factor $CF = 0.733$.

Two sets of results are reported in **Figure 3** where the inlet temperature and mass flow of the gas was varied together. It is clear how the traditional method of scaling the overall UA for fouling results in substantial errors as compared to the new method which rather includes a fouling resistance. Using the fouling resistance, the maximum heat error is about 3% at the lowest flow rate considered, with a temperature error of only 0.4 K. Any change in wall thermal emissivity caused by the fouling is not considered in the model.

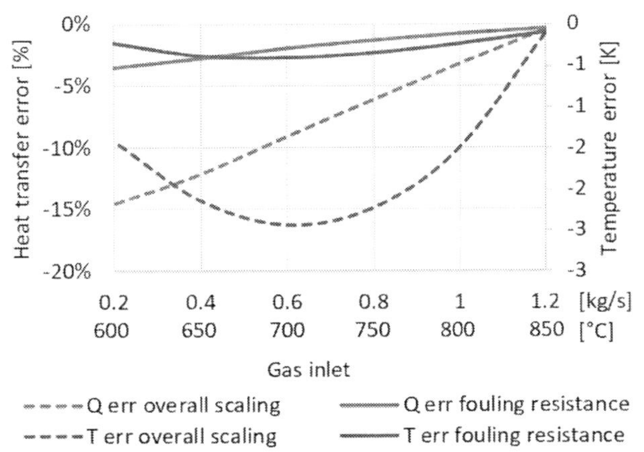

Figure 3. Difference between models using Overall Scaling or a Fouling Resistance for off-design gas inlet conditions.

The primary reason for the new model producing relatively large errors as compared to the previous case, is the fact that the 1mm fouling thickness added to the pipe resulted in a noticeable change in the gas volume around the pipe. This changed the mean beam length used for radiation as well as the stream length used for convection around the pipe. The actual thermal radiation ratio changed from 0.519 to 0.565, but this change was not included into the lumped model, as one would typically ignore the geometric effect that fouling has on a heat exchanger.

The assumption that the thermal resistance on the liquid side is insignificant was tested by varying the steam mass flow from 0.3 kg/s down to 0.05 kg/s. The maximum heat error was only 0.42%, and temperature error of 0.8 K. At the lowest mass flow, the convection coefficient on the steam side was calculated with the detail model as 550 W/m^{2}K, compared to only 16 W/m^{2}K on the gas side. This therefore confirms that one may ignore the effect of the convection on the liquid side.

5 Practical example

Taler and colleagues have modelled a complex superheater using a discretised method, as well as using CFD analysis [13]. The exact geometry and inputs that was published were used to model the heater using the special pipe component in Flownex as is shown in **Figure 4**. The inputs and process parameters are given in **Table 2**. The paper does not report the wall or gas emissivity used, hence typical values were used.

For the design case process conditions with a fouling layer of 1.8 mm at 0.18 W/mK conductivity, the discretized model reported results which are very close to those from Taler. This can be seen in **Table 3**. The heat error is 2.4%. The good accuracy of the results adds further confidence to the validation done previously, as it demonstrates that the special pipe component does capture the fundamental physical processes correctly.

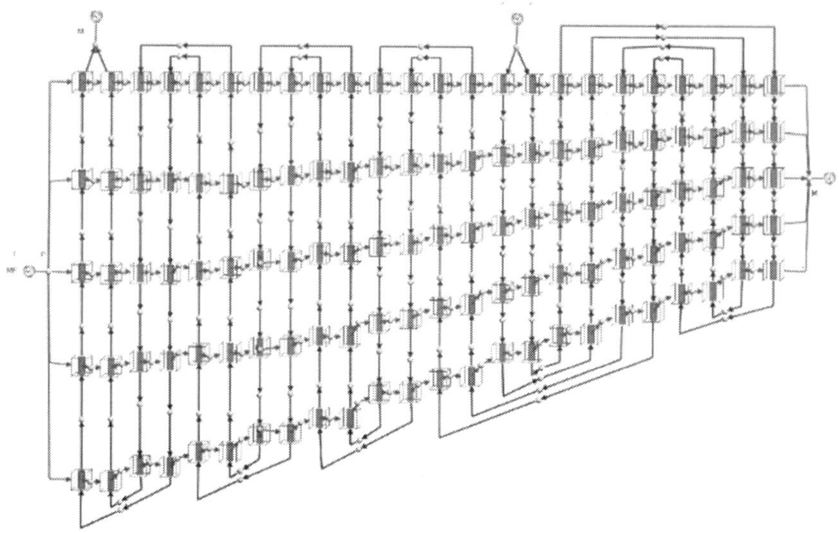

Figure 4. Discretized Flownex model of the heater exchanger analysed by Taler et al. [13]

Table 2. Inputs and Process results for the clean heat exchanger used by Taler.

Geometry	Properties
Pipe increment length : 1.068 m – 0.892 m Pipe outer dia. : 42 mm Pipe inner dia. : 32 mm Gas volume length : 180 mm Gas volume width : 104 mm	Wall emissivity : 0.7 Gas emissivity : 0.1 Ideal gas Air
Process inputs	**Process results**
Steam inlet : 9.61 MPa, 317.2 °C Steam mass flow : 0.668 kg/s Gas inlet : 100 kPa, 862 °C Gas mass flow : 0.815 kg/s	Steam outlet : 420.6 °C Gas outlet : 589.1 °C Total heat : 253.56 kW Radiation heat : 67.12 kW Design UA : 0.7209 kW/K

Table 3. Result comparison for the fouled heat exchanger at Design conditions.

Property	Taler's	This study
Steam outlet Temp. [°C]	399.1	402.2
Gas outlet Temp. [°C]	651.9	648.5
Total heat transfer [kW]	212.5	217.5

Because of the strange flow arrangement of the superheater, it is not obvious which ε-NTU correlation should be used in the lumped model. The model was solved using the three different options and compared with the discretized model results. To capture all changing effects, an artificial "start-up" scenario was implemented where the steam and gas mass flow increases with the gas inlet temperature, but the steam inlet temperature was kept constant. The detail model steam exit temperature and heat transfer results are shown in **Figure 5**. Also shown is the heat transfer error from the three types of flow options using the lumped model.

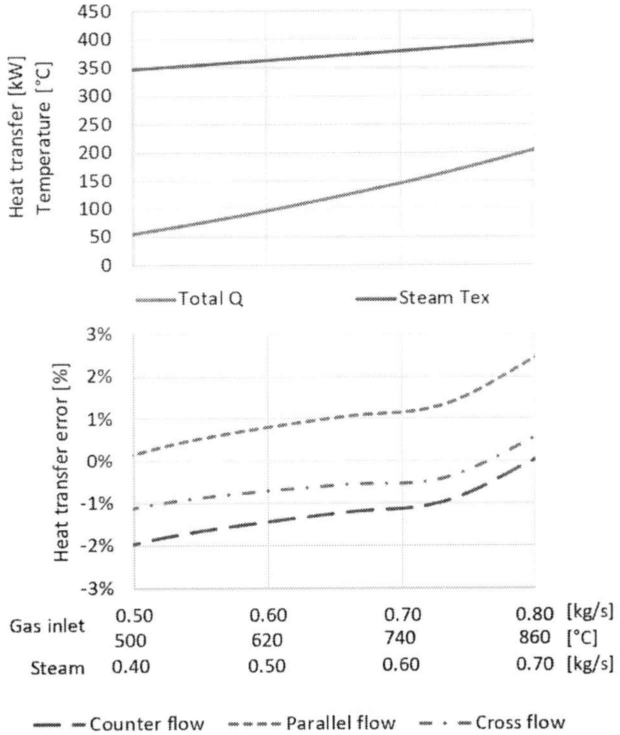

Figure 5. Results for off-design detail analysis of the chosen heat exchanger, along with the heat transfer error for the three flow options using the lumped model.

The errors are quite small for all the flow configuration options, but it does appear as-if the cross-flow with both streams un-mixed option would be the best type to use for this specific heater, producing errors within 1% of the accurate result.

6 Conclusion

It has been shown that many commercial software tools make use of only a mass flow ratio to scale the heat transfer coefficient for gas-to-liquid heat exchangers in the off-design performance calculations. A new approach was presented where the overall UA term is split into a convection and a radiation term and treated separately. Furthermore, a fouling factor is correctly introduced into the overall UA calculation based on a user-supplied cleanliness factor.

The new model has been tested against a discretized pipe-per-pipe heat exchanger model which demonstrated the improved accuracy for off-design performance prediction. The model remains very accurate, even for substantially large deviations from design conditions.

The only additional parameter required from the user is the radiation ratio. This would require some informed estimate of the radiation contribution, either from analogy of similar heaters, a detailed CFD result, or a hand calculation of the radiation and convection terms for the first tube in the heat exchanger using the method described by Trojan and Taler [11]. It is therefore recommended that this model be incorporated in commercial software to produce a more accurate result, without the need for detailed geometry.

The method is applicable for any gas-to-liquid heat exchanger where the gas side's heat transfer dominates the overall heat transfer such as in a water heater / economiser, a convective evaporator, or a superheater at typical power plant pressures. It is believed that this model would even produce reasonable results for a platen superheater of a fired boiler, even though this type of heater also experiences direct radiation from the flame ball. Because the inlet gas temperature to the platen will also change if the flame temperature changes, the model should be able to capture the effect of a change in flame temperature. This is the subject for a future study.

Acknowledgements

The author would like to thank the Applied Thermofluid Process Modelling (ATProM) research group for access to the software and other modelling tools.

Nomenclature

a	Mass flow ratio exponent []
b, c, d	Fluid property ratio exponents []
ε	Effectiveness, Emissivity []
CF	Cleanliness factor []
M	Molecular weight [mol/g]
Q	Rate of heat transfer [W]
R_r	Radiation ratio []
R_{foul}	Fouling factor [K/W]
T	Temperature [°C]
UA	Heat transfer factor [W/K]

Subscripts

av	Average
c	Convection
D	Design
r	Radiation
H	Hot
C	Cold

References

[1] A. S. Incropera, F.P., Dewit, D.P., Bergman, T.L., Lavine, *Fundamentals of heat and mass transfer*, 7th ed. John Wiley & Sons, 2007.

[2] STEAG Energy Services GmbH, "EBSILON Professional Online Help," 2018. [Online]. Available: http://help.ebsilon.com/EN/webframe.html. [Accessed: 30-May-2019].

[3] P. G. L. Minner, P. E. D. R. Fleming, and P. E. W. C. Kettenacker, *PEPSE Volume II Engineering Model Description*. 2004.

[4] Thermoflow, "THERMOFLEX Fully-Flexible Heat Balance Engineering Software Help," 2018. [Online]. Available: https://www.thermoflow.com/products_generalpurpose.html.

[5] GP Strategies, "EtaPRO VirtualPlant Editor Help." [Online]. Available: http://www.etapro.com/platform/virtualPlant.aspx.

[6] L. O. Nord, R. Anantharaman, and O. Bolland, "Design and off-design analyses of a pre-combustion CO2 capture process in a natural gas combined cycle power plant," *Int. J. Greenh. Gas Control*, vol. 3, no. 4, pp. 385–392, 2009.
[7] V. Ganapathy, *Engineering Steam Generators and Waste Heat Boilers For Process and Plant Engineers*. CRC Press, 2015.
[8] P. Basu, C. Kefa, and L. Jestin, *Boilers and Burners - Design and Theory*. New York: Springer, 2000.
[9] E. Z. Gwebu and P. G. Rousseau, "A Network Approach Applied in Modelling the Heat Transfer and Fluid Flow in a Superheater Heat Exchanger," in *Proceedings of the IASTED International Conference Modelling, Simulation and Identification*, 2017, pp. 14–25.
[10] M-Tech Industrial, "Flownex Simulation Environment." [Online]. Available: https://www.flownex.com/.
[11] M. Trojan and D. Taler, "Thermal simulation of superheaters taking into account the processes occurring on the side of the steam and flue gas," *Fuel*, vol. 150, pp. 75–87, 2015.
[12] W. Wagner and H. Kretzschmar, *Properties of Water and Steam Based on the Industrial Formulation IAPWS-IF97*, 2nd ed. Springer-Verlag, 2008.
[13] D. Taler, M. Trojan, P. Dzierwa, K. Kaczmarski, and J. Taler, "Numerical simulation of convective superheaters in steam boilers," *Int. J. Therm. Sci.*, vol. 129, no. March, pp. 320–333, 2018.

Aerodynamic Effect of Wheel Fairings on the Wake of a Formula One Car

Liam Davkin[1,*] and *Craig* Law[1,**]

[1]School of Mechanical, Industrial and Aeronautical Engineering
University of the Witwatersrand, Johannesburg, South Africa

Abstract. The open-wheeled configuration of Formula One cars has been known to create unfavourable aerodynamic conditions for overtaking during races. New aerodynamics regulations are set to come into effect by the 2022 season. Early concepts suggested that covering the wheels of a car are amongst the aerodynamic design changes to be implemented. This is in contrast to the traditional open-wheeled configuration. This study aims to evaluate the effect that these fairings have on the behaviour of the wake behind the car. This has been achieved through computational fluid dynamics (CFD) studies which surveyed the flow in the wake, that determined any changes that are added by these fairings. The fairings were found to have altered the structure of the wake, by changing the size of the regions of low-speed airflow at different positions behind the car.

1 Introduction

1.1 Background

The aerodynamic wake behind a car has been commonly attributed as a cause of the lack of overtaking in Formula One racing. This is because the aerodynamic wake is characterised by highly turbulent air flow. Any trailing car attempting to perform an overtaking manoeuvre is forced to operate in this area. The result is a loss of aerodynamic downforce experienced by the trailing car. This trailing car is prone to understeering into corners, making it difficult to successfully perform an overtake [1]. It has been found that the open-wheeled configuration of Formula One cars greatly influences the turbulent wake [2]. The high turbulence is further increased by the rotation of the wheels. Early car concepts for the 2022 F1 season included wake-control devices (fairings) on the front and rear wheels (as seen in Figure 1). These were supposedly introduced to clean up the air flow downstream of a car. This would aid a trailing car in successfully performing an overtaking manoeuvre.

*Liam.Davkin@students.wits.ac.za
**Craig.Law@wits.ac.za

Figure 1: Concept car featuring partially covered rear wheels [3]

The car design for the 2022 F1 season was confirmed in August of 2019, with only the front wheels making use of these wake-control fairings. There is little information in the public domain regarding the effect of wheel fairings in Formula One racing.

Previous studies have successfully characterised the turbulent wake behind a Formula One car. Watts [4] conducted an experimental study utilising a wind tunnel scale model. Newbon [5] examined the wake with the use of a numerical model. Both studies described a 'mushroom-like' wake, with strong centreline upwash and counter-rotating vortices originating from each edge of the rear wing. A prominent velocity/dynamic pressure deficit was attributed to the loss of downforce by a trailing car. However, these studies only focused on open-wheeled configurations and did not implement rotating wheels. Wäschle [6] successfully demonstrated through a numerical study, the difference in wake structure for both stationary and rotating wheels. The rotating wheels reduced the size of the wake behind a car by allowing for additional energy to be transferred. These previous studies thus showed that the wake behind an open-wheeled car is highly turbulent, with regions of low flow velocity.

1.2 Objective of Study

This study aimed to evaluate the effect rear wheel fairings have on the wake behind a Formula One car. A survey of the velocity components in the flow field behind a Formula One car model, with various rear wheel configurations, provided data that was used to understand the effect that rear wheel fairings have on the flow field.

2 Method

The objective was achieved with CFD models, that surveyed the velocity components in the wake of a Formula One car, with and without wheel fairings. The method, results and conclusions pertaining to the CFD study are outlined here.

2.1 Car Model

A generic Formula One car model was used in the CFD study (illustrated in Figure 2(a)). This model was based on an openly published computer-aided design (CAD) model that was simplified in order to allow for easier scale model manufacturing. All modifications were made utilising Autodesk Inventor 2020. The model was without any unique aerodynamic features that may pertain to a car designed by a particular team. The model contained all primary downforce producing components, namely the front wing, rear wing and diffuser.

The front and rear wings were simplified for manufacturing purposes and thus do not produce the same magnitude of downforce as would be seen in a competition vehicle. However the flow-field that results from the simplified wings should still be qualitatively similar.

Two fairing concepts were attached to the rear wheel of the car. These fairings began at the front of the wheel, at a clockwise angle of 30° from the horizontal mid-plane. The fairings then extended circumferentially to a clockwise angle of 180° to the back of the wheel. The dimensions for each fairing concept remained the same. A partial fairing concept had no cover over the sidewall of the wheel, whereas the full fairing concept did (shown in Figure 2(b)).

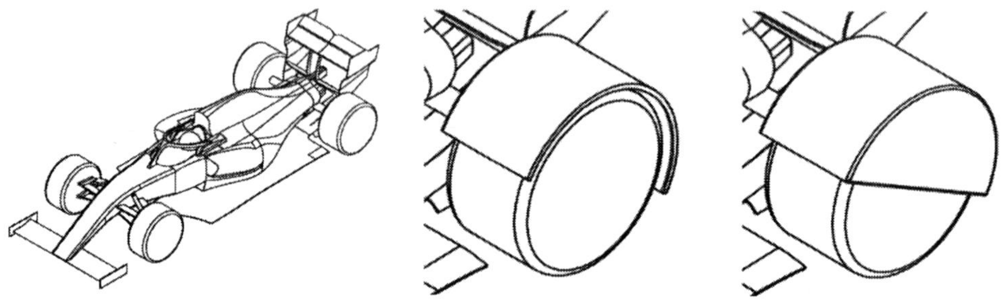

(a) (b)

Figure 2: (a) Car model used for CFD study (without wheel fairings), (b) Partial (left) and full (right) rear wheel fairing concepts used in presented study

2.2 Domain

The mid-plane of the car was modelled as a symmetry plane, combined with the far fields to the top and side of the car. A velocity inlet and pressure outlet were set ahead of, and behind the car respectively. The ground plane and the surfaces of the car were modelled as smooth walls. The length of the domain was set as multiples of the length of the car (L), chosen as the model characteristic length. The domain extended ahead of the car by a distance of $4L$, and behind the car by $6L$. The height of the domain was set at $6h_c$ (equivalent to $\sim 1.28L$), with h_c being the height of the car. The width of the domain was set at $5w_c$ (equivalent to $\sim 0.94L$), with w_c being the width of the vehicle. A domain with the implemented dimensions is recommended by Siemens [7] to produce a blockage ratio of less than 1%, thus preventing the flow-field around the car from being affected by the domain boundary conditions.

2.3 Mesh

The mesh (shown in Figure 3) was composed solely of hexahedral cells, with 5 layers of prism cells at the car surfaces. A wall y^+ value of 50 was chosen (i.e. the log-law region of the turbulent boundary layer) for the height of the first cell. The resolution of the boundary layer was not considered to be of primary importance in the study, as the main focus was on the wake behind the vehicle. This wall treatment was thus chosen in an attempt to minimise the cell count and hence reduce computational times. The chosen value lies within the recommended range for the selected turbulence model. Prism layer cells were disabled for the domain boundaries. Refinement zones were added immediately downstream of the car in order to accurately capture the flow properties of the wake. These zones are identified in Figure 3. The resultant meshes were relatively coarse, consisting of ~ 4 million cells.

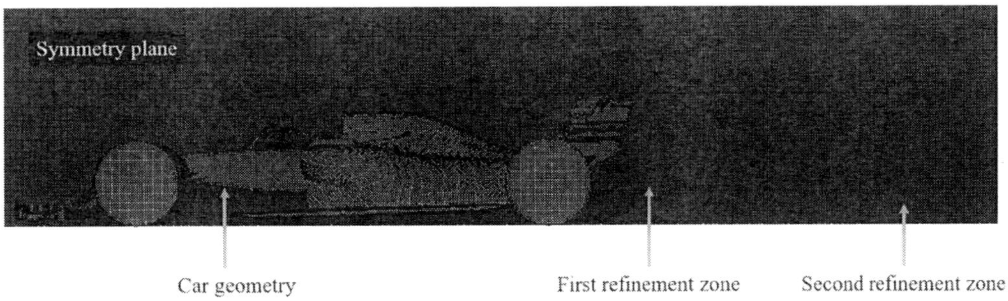

Figure 3: Side view of hexahedral mesh, showing symmetry plane and refinement zones

2.4 CFD Model Setup

The CFD software package used was Simcenter STAR-CCM+ 2020.3.1 (15.06.008). Only steady state simulations are presented in this study. A coupled solver was chosen in order to reach convergence faster, at the expense of more computational resources. Air density was assumed to be constant at all times. The SST $k-\omega$ turbulence model was employed as this is the recommended turbulence model for studies pertaining to external vehicle aerodynamics [8]. The value of a_1 was changed from the default of 0.31 to 1. The realisability coefficient was set at 1.2. This setup originates from a recent study that showed improved stability and convergence using these values [9]. The CFL number was initially linearly ramped to avoid the development of an unstable solution. Grid sequencing initialisation was activated for all simulations in order to speed up convergence. The inlet air velocity was set to 60 m/s. This was based on a rounded off average speed of 59.8 m/s amongst all F1 circuits in the 2019 season [10]. The outlet was set at a gauge pressure of 0 Pa. The ground was assigned a vector speed of 60 m/s in the same direction as the airflow. This was to prevent the formation of a boundary layer on the ground plane. Both wheels were assigned a tangential velocity that corresponds to the linear speed of 60 m/s. Normalised residuals were monitored throughout the simulations, along with monitors of drag coefficient, lift coefficient, mass flow rate, and maximum velocity. The solution was considered to be converged once all normalised residuals had dropped to at least 10^{-3} [11], and all monitored quantities had approached steady values. These criteria were satisfied for all conducted simulations, with all normalised residuals dropping below 10^{-4}.

3 Results and Discussion

Normalised values are presented for measurements of characteristic length, height and width of the car. Three planes perpendicular to the mid-plane of the car were considered for investigation, namely $x = 1.0L$, $x = 1.5L$ and $x = 2.0L$. The normalised length $x = 1.0L$ corresponds to a plane immediately aft of the car. The normalised length $x = 1.5L$ plane corresponds to a plane half a car length aft of the car. The normalised length $x = 2.0L$ plane corresponds to a plane a full car length aft of the car.

3.1 Wake at $x = 1.0L$

Figures 4 - 6 show that for all three fairing concepts, there existed an area of low normalised axial velocity at the rear wing of the car ($z/h_c \approx 1.0$, labelled as Figures *A* in 4 - 6). This led to a region of low dynamic pressure aft of the rear wing. This can be attributed to the low-speed air from the upper surface of the wing.

To the sides of the end plates, there existed a slightly increased region of normalised axial velocity. This was indicative of the formation of vortices emanating from the rear wing tips (labelled as *B* in Figures 4 - 6). These vortices were caused by the circulation generated by the rear wing, which creates downforce. The formation of these wake vortices was further shown by the direction of the secondary flow vectors at the tips of the rear wing. These could be seen to be circulating at the rear wing tips. This circulation was in opposite directions on either side of the rear wing (i.e. the vortices were counter-rotating). The properties of these vortices were consistent with a lifting surface generating downforce. The length of the secondary flow vectors were directly proportional to the magnitude of the planar velocity. Longer vectors thus correlated with vortices with higher disturbance. These secondary flow vectors were longest for the car without any wheel fairings and shortest for the full fairing concept. The addition of wheel fairings therefore reduced the diameter of the vortices originating from the rear wing tips, by reducing the effect of the wheel disturbance. These less disturbed vortices contribute less to the upwash in the wake of a trailing car, creating favourable downforce generating conditions.

Circulation was again seen to the outside of the rear wheels ($y/w_c \approx \pm 0.7$, labelled as *C* in Figures 4 - 6) for all three concepts. These vortices originated from the rear wheels during rotation. Air close to the ground was forced outwards as the wheel rotates. The centres of these vortices contained a dynamic pressure deficit as a result of the low axial velocity of the airflow. The size of the low speed core (normalised axial velocities between 0.4 and 0.6) was seen to increase with the addition of both faring concepts. However, the planar velocity magnitude of these vortices were highest for the car without any wheel fairings attached. This was due to stronger inwash that was present for this configuration. These areas of airflow on the outside of the wheel are of primary concern for any trailing car. A trailing car would have to move through these wheel vortices as it attempts to pass the leading car on the side. The upwash induced by the wheel vortices reduces the effective angle of attack of the trailing car's front wing, thus reducing downforce. The front wing of the trailing car would be at such a height that it moves directly through these vortices. If a dynamic pressure deficit existed in this region, the front wing of the trailing car would lose downforce, and thus understeer into corners.

A low velocity region (normalised axial velocity ≈ 0.0) existed at the exit of the diffuser of the car for all three cases (labelled as *D* in Figures 4 - 6). This was indicative of airflow from underneath the car expanding through the diffuser. This low velocity region was smallest for the partial wheel fairing concept, and largest for the car without any wheel fairings. The addition of wheel fairings would increase the ground effect experienced by a car by reducing the size of this low velocity region. A low velocity region at this height would result in a higher pressure on the underside of a trailing car (the space between the car and the ground), thus reducing the amount of downforce generated, while also reducing the amount of drag force experienced.

The most obvious difference between the three cases was the area immediately aft of the rear wheel (y/w_c between ± 0.3 and± 0.5, labelled as *E* in Figures 4 - 6). For the car with no wheel fairings, an area of low normalised velocity (≈ 0.0) existed between $z/h_c \approx 0.0$ and 0.5 (i.e. half the car height). With the addition of wheel fairings, this low normalised velocity region was reduced by approximately 50%. This region also moved slightly inwards. These phenomena were again observed for the full fairing configuration. Before wheel fairings were applied to the car, there existed an airflow separation point towards the top of the rear wheel. This separated flow caused a region of low dynamic pressure aft of the wheel. By adding fairings, any separated airflow was forced to flow along the path created by the fairing.

Airflow thus exited this enclosed compartment at the bottom of the fairing. Here the air experienced possible rapid expansion, which led to a loss of dynamic pressure and thus low velocity. The addition of the wheel fairings also created a stronger downwash aft of the wheel. The addition of wheel fairings also resulted in inward secondary flow vectors close to the ground, aft of the wheel. A strong centreline upwash was present for all three cases. For all three cases, the height of the wake remained at a normalised height of approximately $z/h_c \approx 1.0$.

Thus, the addition of fairings has changed the size and positions of the low-speed regions of airflow immediately aft of the car ($x = 1.0L$). By reducing the size of this region towards the centreline of the car, favourable downforce generating conditions are created for a trailing car immediately behind. However, as a trailing car moves to the side of a leading car during the overtaking manoeuvre, it will encounter larger regions of low-speed airflow when fairings are applied. This creates unfavourable conditions for downforce generation as a trailing car moves closer, and thus diminishes the chance of successful overtaking in corners.

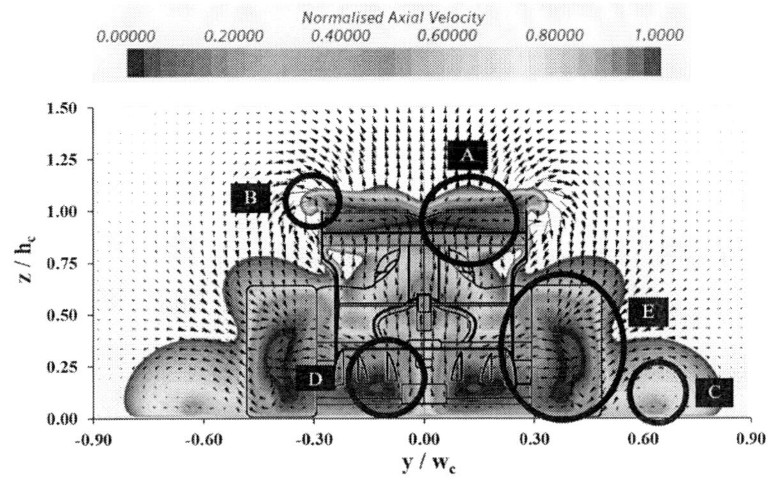

Figure 4: Normalised axial velocity contours and secondary flow vectors for no wheel fairings at $x = 1.0L$

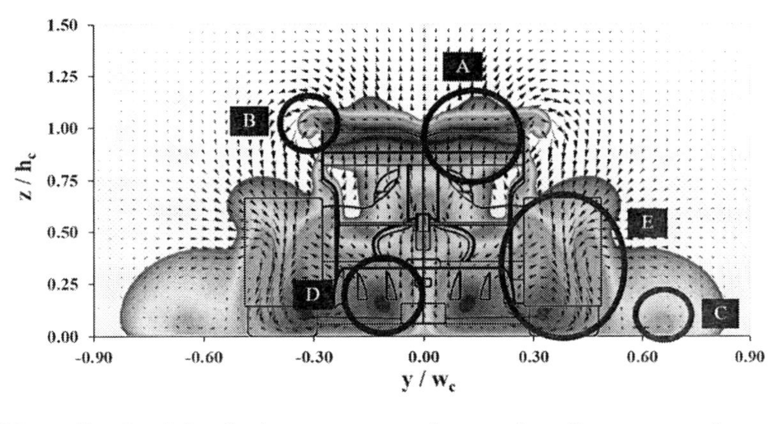

Figure 5: Normalised axial velocity contours and secondary flow vectors for partial wheel fairings at $x = 1.0L$

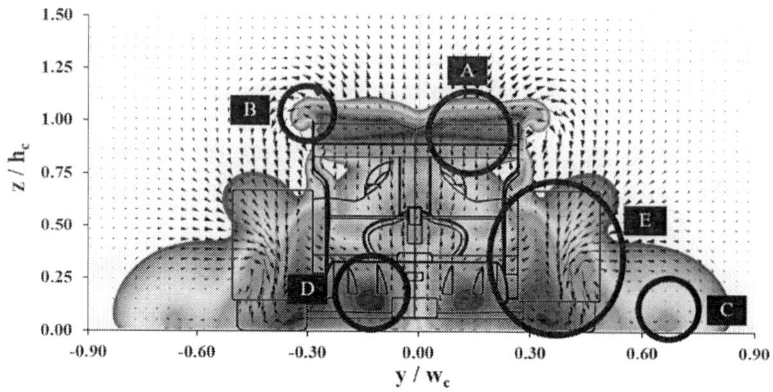

Figure 6: Normalised axial velocity contours and secondary flow vectors for full wheel fairings at $x = 1.0L$

3.2 Wake at *x* = 1.5*L*

Figures 7 - 9 show that the height of the wake increased for all three cases (labelled as *A*), and that the wing tip vortices had increased in size. These counter-rotating vortices (labelled as *B*) had formed the previously observed 'mushroom-like' wake.

Flow vectors representing the centreline upwash were influenced by the wing tip vortices and were directed diagonally upwards away from the mid-plane of the car (labelled as *C* in Figures 7 - 9). The secondary flow vectors on the outer edge of the wake tended to be directed diagonally downwards to the mid-plane of the car (labelled as *D* in Figures 7 - 9).

Regions of low dynamic pressure were still present to the sides of the rear wheels ($y/w_c \approx \pm 0.6$, labelled as *E* in Figures 7 - 9). These regions moved slightly inwards to the mid-plane of the car. The car without any wheel fairings produced this smallest region. The addition of wheel fairings extended the thickness of this low pressure region to the region aft of the rear wheels. The addition of partial wheel fairings increased the thickness of this region by approximately 67.5%. This increase remained approximately constant between the two fairing concepts. The presence of wheel fairings also decreased the normalised axial velocity in this region.

However, along the centreline, there existed regions of low velocity airflow for all three cases (labelled as *F* in Figures 7 - 9). The position of this region was different between the three cases. For a car with no wheel fairings, the centre of this low velocity region existed at a normalised height of $z/h_c \approx 0.45$. With the addition of partial wheel fairings, the centre height was increased to a normalised height of $z/h_c \approx 0.575$ (an increase in height of 27.8%). With the addition of full wheel fairings, the centre height was increased to a normalised height of $z/h_c \approx 0.75$ (an increase in height of 66.7%). An increase in height may have adverse conditions for downforce generation by a trailing car. Oncoming airflow to the rear wing of the trailing car would be at a lower velocity when a leading car has full wheel fairings.

The region around the centreline of the car formed a low velocity column for the entire height of the car (labelled as *G* in Figures 7 - 9). This column was thickest for a car without wheel fairings, with a normalised thickness of approximately 0.9. This column thickness was particularly prominent at a height corresponding to the upper half of the rear wheel. With partial and full wheel fairings, this normalised thickness was reduced to approximately 0.6. This represented an approximate thickness decrease of approximately 30%. The fairings prevented interaction between the airflow around the rear wheel and the airflow around the rest of the car. This interaction was thus delayed and occurred further aft of the car, resulting in low velocity regions.

Therefore, at the distance of x = 1.5L, the wake began to develop the previously seen 'mushroom-like' wake. Columns of low-speed air existed along the centreline of the car, with the region of lowest velocity having increased with the addition of wheel fairings. The region of low velocity airflow have moved inwards, which would reduce the dynamic pressure deficit of a trailing car attempting an overtaking manoeuvre, thus creating favourable downforce conditions.

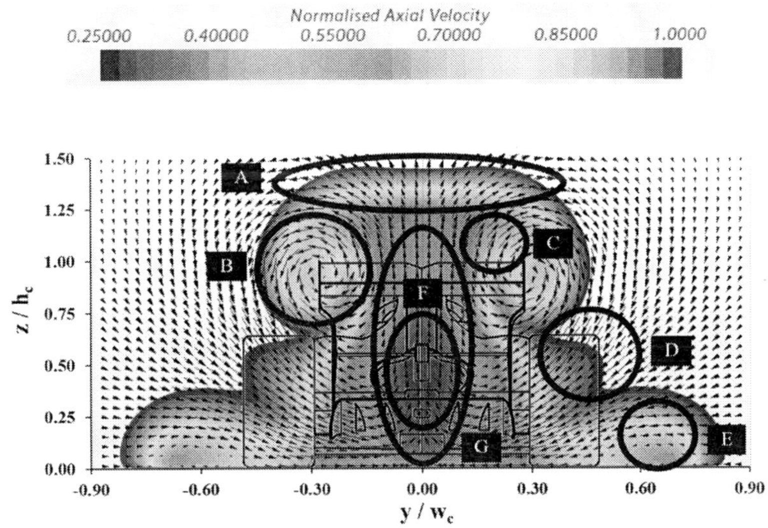

Figure 7: Normalised axial velocity contours and secondary flow vectors for no wheel fairings at $x = 1.5L$

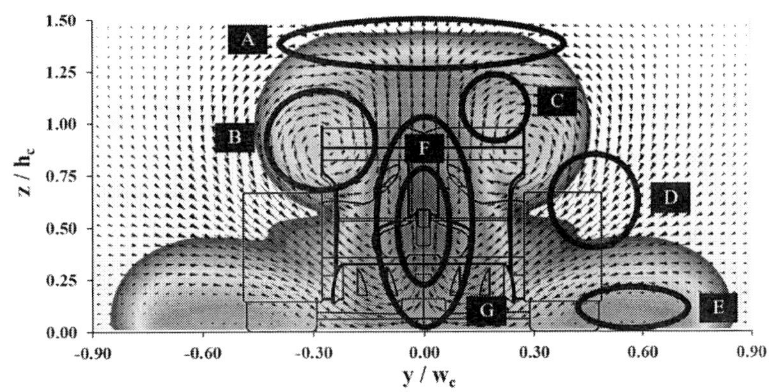

Figure 8: Normalised axial velocity contours and secondary flow vectors for partial wheel fairings at $x = 1.5L$

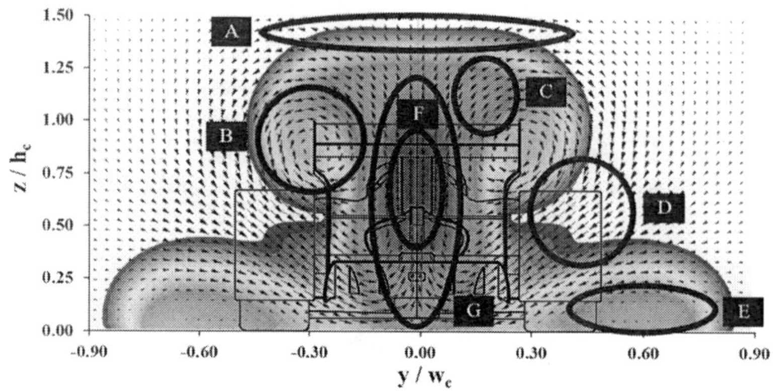

Figure 9: Normalised axial velocity contours and secondary flow vectors for full wheel fairings at $x = 1.5L$

3.3 Wake at $x = 2.0L$

Figures 10 - 12 show that the normalised height of the 'mushroom-like' wake increased beyond $z/h_c \approx 1.50$ (labelled as A).

Regions of low normalised axial velocity still existed to the sides of the wheels for all three cases (labelled as B in Figures 10 - 12). The velocities in this region were lowest for the car with wheel fairings attached.

The main difference between the three cases was the difference in structure of the low-velocity column at the centreline of the car (labelled as C in Figures 10 - 12). For all three cases, the height of the lowest velocity region (≈ 0.45) extended to a normalised height of $z/h_c \approx 1.25$. However, this region extended to the ground plane when no wheel fairings were attached. This resulted in a low velocity region for the entire height of the car. The addition of wheel fairings caused this column of low velocity air to increase in magnitude. For the partial wheel fairing configuration, this region correlated to a normalised axial velocity of approximately 0.615 (an increase of 37%). For the full wheel fairing configuration, this region correlated to a normalised axial velocity of approximately 0.725 (an increase of 61%). If a trailing car was positioned at this distance aft of the leading car, this low velocity region would have considerable effects on downforce production. The lower the air velocity in this region, the less the downforce generated by the trailing car would be. The addition of wheel fairings moved this region to above the height of the car. The airflow above the height of the car would have less of an impact on the downforce production as it would not come into contact with any downforce generating components of the car.

The 'mushroom-like' wake has therefore continued to grow at a distance of $x = 1.5L$. The addition of wheel fairings has reduced the size of the column of low-speed airflow, which would create favourable downforce generating conditions immediately aft of the car. However, the regions to the side of the wheels still has a decrease in normalised axial velocity that would continue to hinder overtaking opportunities.

The results at these three planes indicate that a trailing car would be able to generate more downforce when moving immediately behind the car (with the addition of wheel fairings resulting in higher downforce production). However, as the trailing car moves to the side of the leading car, downforce production is reduced by the presence of low-speed axial velocity. As the trailing car gets closer to the car ahead, the axial velocity decreases, and downforce generation is diminished, resulting in understeer and an unsuccessful overtaking manoeuvre.

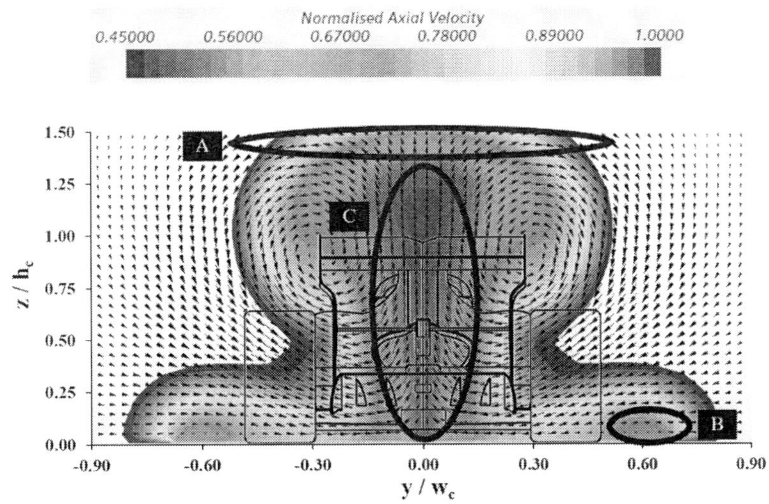

Figure 10: Normalised axial velocity contours and secondary flow vectors for no wheel fairings at $x = 2.0L$

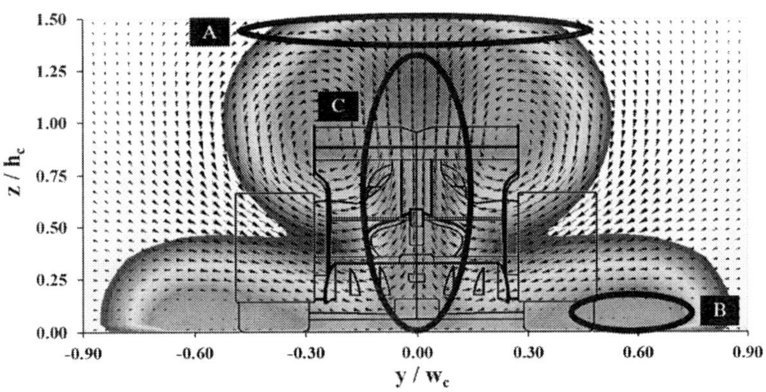

Figure 11: Normalised axial velocity contours and secondary flow vectors for partial wheel fairings at $x = 2.0L$

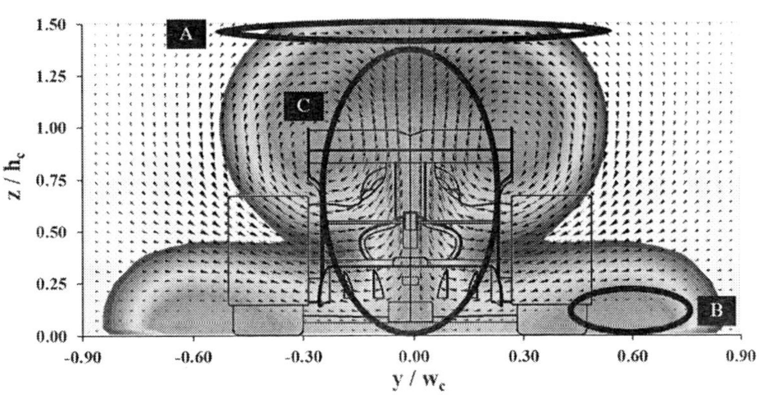

Figure 12: Normalised axial velocity contours and secondary flow vectors for full wheel fairings at $x = 2.0L$

3.4 Streamlines at rear wheel

The following streamlines are coloured by velocity magnitude and are seeded from the rear wheel.

Figure 13(a) shows that as air flowed around the wheel, it did not remain attached , and separated at some point downstream of the vertical mid-plane (labelled as *A*). This was the expected behaviour and is analogous with the airflow around a cylinder. Aft of the separation point, there existed a large wake of low speed airflow and backflow (labelled as *B*). The airflow in this region was highly turbulent and chaotic as indicated by the random pattern of the streamline paths. This would create unfavourable conditions for a trailing car to generate downforce. Figure 13(b) shows that by covering the rear wheels with partial fairings, the wake region of highly chaotic airflow has been reduced (labelled as *C*) as the airflow was forced to follow the enclosed path created by the fairing. The airflow escaped from the sidewall of the fairing (labelled as *D*). This air was at higher velocity than the wake created when no wheel fairings were attached. Figure 13(c) shows that the airflow exited from the bottom of the wheel fairing (labelled as *E*). This resulted in rapid expansion of the air. The result was that the airflow slowed down and created a low velocity wake aft of the wheel.

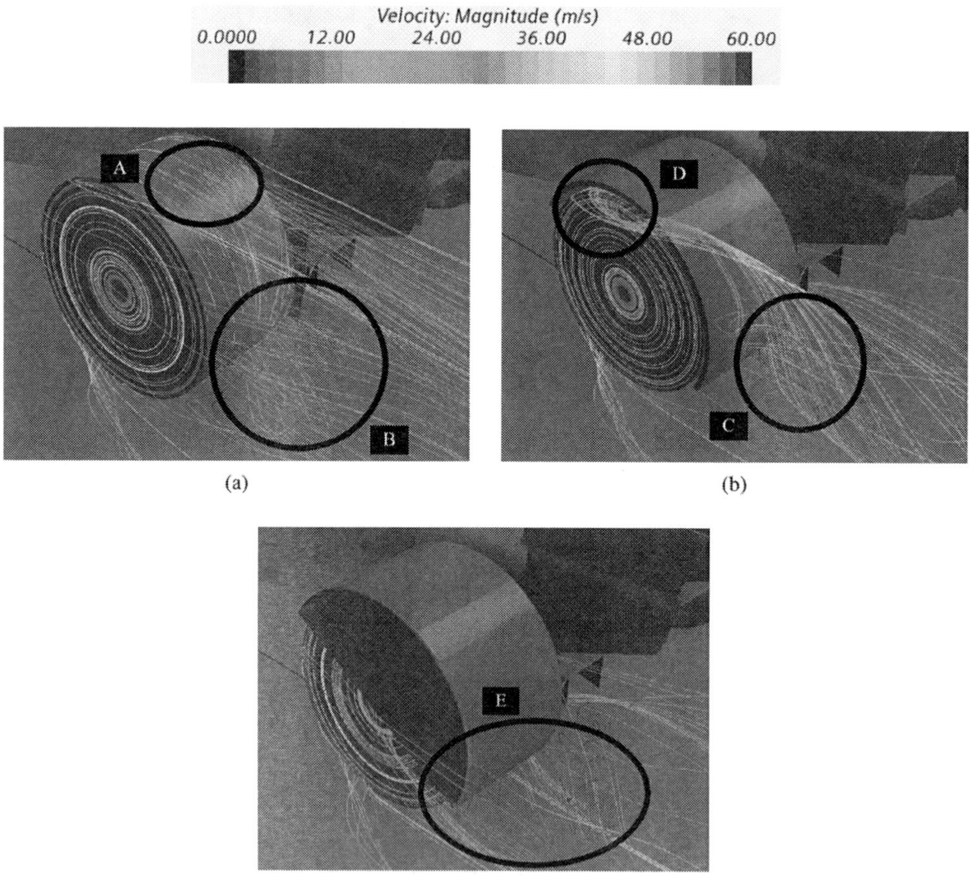

(a) (b)

(c)

Figure 13: Normalised axial velocity contours and secondary flow vectors for no wheel fairings (a), partial fairings (b), and full fairings (c)

4 Conclusion

This study has shown that the addition of wheel fairings has noticeable effects on the wake behind a Formula One car. The addition of wheel fairings reduced the size of the low speed airflow immediately aft of the rear wheels. The wheel fairings also concentrated the counter-rotating vortices originating from the tips of the rear wing. The addition of wheel fairings increased the size of the low speed airflow to the sides of the vehicle and created unfavourable conditions for downforce production here. However, the fairings have reduced the low axial velocity region at the centreline of the leading car, while also reducing the upwash, which created favourable downforce generating conditions for a trailing car that is immediately behind. The fairings have an adverse affect on overtaking opportunities due to the need for a trailing car to always pass through the low-speed airflow (worsened by the addition of wheel fairings) to the side of the wheels of the leading car whilst overtaking. This suggests a possible explanation as to why the confirmed 2022 design only makes use of partial fairings on the front wheels of the car. Higher resolution studies are being conducted, along with wind tunnel measurements in order to provide validation data.

References

[1] C. Scarborough. F1: Why No Overtaking and Are Bigger Wings the Answer? 2017.URL: https://drivetribe.com/p/f1-why-no-overtaking-and-are-bigger-auZby7jDQhuGAz6N9TEP-g?iid=SP0pLFGvTa-R62Klb6r7Ig (Last accessed on 4 April 2019).

[2] J. Noble. The key questions raised by Formula 1's 2021 car concepts. 2018. URL: https://www.autosport.com/f1/news/138720/f1-2021-concepts-the-key-questi ons-answered (Last accessed on 4 April 2019).

[3] A. Cooper and V. Khorounzhiy. 2021 Formula 1 concept claimed to produce five-times less dirty air. 2019. URL: https://www.autosport.com/f1/news/14 2014/2021-concepts-produce-fivetimes-less-dirty-air (Last accessed on 4 April 2020).

[4] Watts, M. and Watkins, S., "Aerodynamic Structure and Development of Formula 1 Racing Car Wakes," SAE Int. J. Passeng. Cars - Mech. Syst. 7(3):2014, doi:10.4271/2014-01-0600.

[5] Newbon, J., Sims-Williams, D., and Dominy, R., "Aerodynamic Analysis of Grand Prix Cars Operating in Wake Flows," SAE Int. J. Passeng. Cars - Mech. Syst. 10(1):2017, doi:10.4271/2017-01-1546.

[6] A. Wäschle. "The Influence of Rotating Wheels on Vehicle Aerodynamics - Numerical and Experimental Investigations". In: SAE Technical Paper Series. 2010. doi:10.4271/2007-01-0107.

[7] Siemens. Best Practices for Vehicle Aerodynamics. 2020. URL: https://thesteveportal.plm.automation.siemens.com/articles/en_US/FAQ/Best-Practices-for-Vehicle-Aerodynamics (Last accessed on 16 May 2020).

[8] Simcenter STAR-CCM+ User Guide. Siemens. 2020.

[9] S. Evans, S. Lardeau and C. Pettinelli. 'Validation of a turbulence methodology using the SST $k - \omega$ model for adjoint calculation'. In: 54th AIAA Aerospace Sciences Meeting. 2016. isbn: 9781624103933. doi: 10.2514/6.2016-0585.

[10] H. Cobb. The fastest F1 tracks on the calendar. 2020. URL: https://www.crash.net/f1/feature/937592/1/fastest-f1-tracks-calendar (Last accessed 18 May 2020)

[11] Celik, I. B., Ghia, U. Roache, P. J., Freitas, C. J., Coleman, H., and Raad, P. E. 2008. "Procedure for estimation and reporting of uncertainty due to discretization in CFD applications". Journal of Fluids Engineering, Transactions of the ASME, 130(7), 0780011–0780014. https://doi.org/10.1115/1.2960953

Train Wheels Eigenmodes on Effects of Vibration and Track Deterioration in the Belfast to Steelpoort Railway Line

Bingo Balekwa*[1], *Daramy Vandi Von* Kallon[1]

[1]Department of Mechanical and Industrial Engineering Technology, University of Johannesburg, Level 7, John Orr Building, 47 Nind Street, Doornfontein, Johannesburg, 2092, South Africa,

Abstract. In South Africa, one of the main lines situated in its Limpopo Province presents symptoms of excessive vehicle-track vibrations. A modal method is adopted to study the difference in dynamic response of corrugated rails resting on the Hytrel/6385 and High Density Polyethylene (HDPE) rail pads. The results for dynamic response of rails on these two types of rail pads are analysed in form of Frequency Response Functions (FRFs). Two positions are considered in obtaining these point FRFs; those are mid-span and on top of sleeper. Resonance modes of significance are excited in a frequency range of 0 – 500Hz on rails resting on both types of rail pads. In view of the entire vibration response window, the FRFs show a more damped vibration response by rail resting on the HDPE than that resting on the Hytrel/6358 rail pad. A Finite Element (FE) modal analysis method is used to investigate vibration mode shapes and corresponding frequencies of the wheelset used by the locomotive operating in the line. In the high frequency range of 350 – 500Hz, the gear-side wheel of a locomotive wheelset proves to be vibrating more erratically than the non-gear side wheel. Clearly pronounced mode shape excited at a frequency of 350Hz and higher are presented.

1 Introduction

Rail corrugation is a damage that forms on the running surface of the rail due to dynamic of the wheel-rail system. Short-pitch corrugation on the inner rail of track curves in the railway line from Belfast to Steelpoort drew interest in the current research [1]. Rail corrugation is considered a global enigma in the railway space and has been pursued for over a century [2]. This problem also forms in metro and high speed systems [3–4]. High speed systems are generally associated with higher frequencies of vibration, mostly beyond 1kHz [5]. In some environments corrugation can form on railway wheel threads and is referred to as wheel polygonisation [6]. Causes of rail corrugation formation may differ per environment, but it is understood from the literature to form due to the damage mechanism and fixed on rails by wavelength fixing mechanisms [7]. Currently there is no primary cure for rail corrugation and rail grinding remains the common treatment [8–9]. Researchers in the recent years are developing methods to detect, identify and study rail corrugation through computer vision [10–12]. People who live closer to affected railway sites are exposed to mechanical vibrations that may negatively impact on their well-being. Figure 1. shows rail corrugation captured at one of the track curves in the railway line of concern.

* Corresponding author: bingo.balekwa@yahoo.com

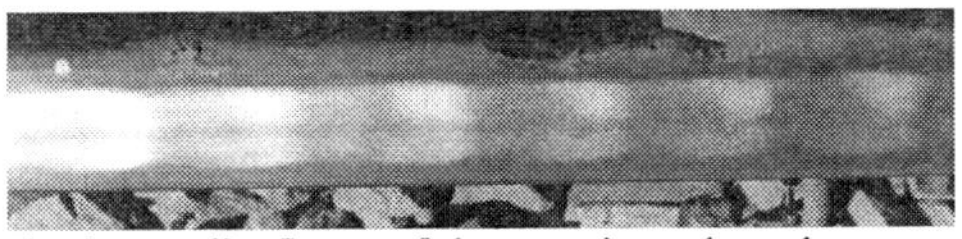

Fig. 1. Rail corrugation on the low rail of one of the monitored track curves.

A periodic wavelength average of 80 – 90mm was measured [1]. Together with train speeds, this average wavelength per corrugated track curve was used to calculate the "corrugation frequency" for track curves. After which the resonance frequencies of a wheelset used by a D39200 locomotive were measured using the modal method. The results showed reasonable agreement of correlation of the corrugation frequencies, with resonance frequencies of the wheelset on track curves. It was established that a stronger correlation occurs at 108Hz. Rail resonance modes were not considered in the previous investigation [13–14]. The current study focuses on a high frequency wheel-rail resonance modes. It also focuses mainly on the dynamic differences between the responses of rails resting on two types of rail pads i.e. HDPE and Hytrel/6358. Rail response to vibration excitation was measured in the vertical and horizontal directions. Figures 2a and 2b. show the HDPE and the Hytrel/6358 rail pads. The HDPE rail pad (Figure. 2a) is used on the FY-type concrete sleeper. The Hytrel rail pad is used on the PY-type concrete sleepers.

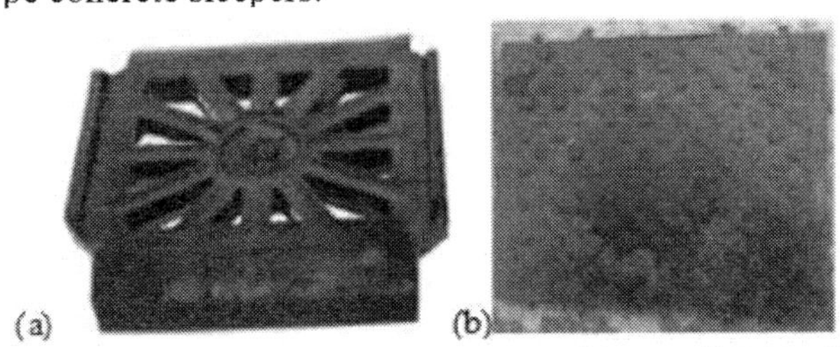

Fig. 2. (a) HDPE rail pad used on the FY-type concrete sleeper (b) Hytrel/6358 rail pad used on the PY-type concrete sleepers.

The design of the FY-type concrete sleeper restricts the horizontal displacement of the pad. The Hytrel/6358 is designed with studs on both sides of the pad. Figure 3. illustrates the arrangement of three studs along the cross section of a rail pad.

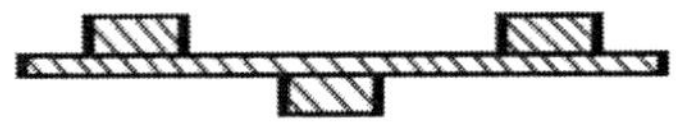

Fig. 3. Cross section of three studs arranged on the two opposite faces of the Hytrel/6358 rail pad.

Rail pads protect sleepers against impact and wear, ensure resiliency during vibration, ensures a higher creep resistance and provides electrical resistance [15]. In his illustration, Thompson [16] considered the dynamic difference of rail pads with a difference in stiffness. Figure 4. shows the illustration.

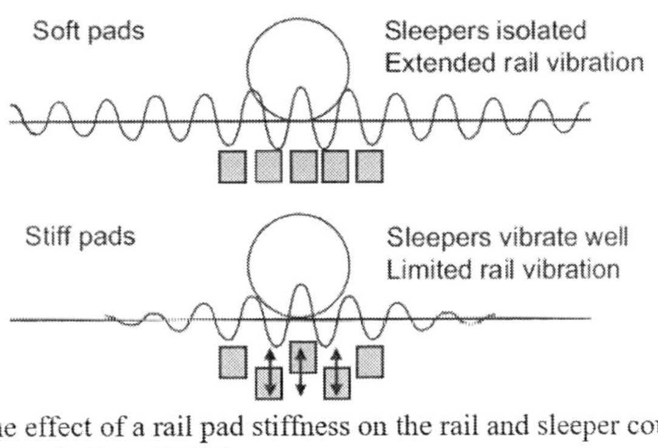

Fig. 4. Illustration of the effect of a rail pad stiffness on the rail and sleeper coupling; and on the damping of vibration waves in the rail [16].

Evident in Figure 4. is that for softer rail pads, vibration of the rail is isolated from that of sleepers but propagation of vibration can happen freely along the rail. Contrary to this, for stiffer pads, vibration by the rail is constrained by the fastening system and damping of the pads, ultimately causing a shorter rail length to vibrate. The rail pad stiffness significantly affects rail vibration damping and the amount of coupling between rail sleepers [16]. In general the damping of a wheel is higher than that of a track structure, however, this does not significantly affect the point FRF of the track structure at 500-1000Hz [16]. Nonetheless, the importance of damping introduced through the track structure is given that it has an influence on vibration decay rate along the rail track [16]. In the recent work a wavelength fixing mechanism for short pitch rail corrugation is established [17–18].

1.1 Effects of Rail Corrugation-Borne Vibration

Damages attributed to excessive vibrations were noted during in the Belfast to Steelpoort railway line. Figure 5a. shows the damage on rail pads due to excessive vibration and Figure 5b. shows the effect of the damage on rail pads. The rail ultimately makes direct contact with the concrete sleeper, ultimately exposes a sleeper to high dynamic loads and creepage. The clamping force of the fastening system is understood to be adequate when every component is at its right place, hence once the rail pad breaks or find its way out of the system there is no more sufficient clamping, this ultimately causes loosening of the fastening system at its entirety and dislocation of its components. is no more sufficient clamping, this ultimately causes loosening of the fastening system at its entirety and dislocation of its components.

(a) (b)

Fig. 5. (a) Rail pad damage (b) effect of rail pads damage due to excessive vibration.

Figure 6. shows deteriorated ballast quality due to excessive rail track vibration.

(a) (b)

Fig. 6. (a) Desirable ballast condition (b) undesirable ballast condition.

Other functions of the ballast in a track are to bear the load from the sleepers and provide horizontal stability to the track [19–20]. For the ballast to perform this function well, friction should be adequate between surfaces of the ballast rocks, to prevent ease of ballast settlement. Figure 6a. shows a new, desirable condition of a ballast for an ideal rail track and Figure 6b. an old undesirable condition. Under the ballast condition in Figure 6b. the ballast rocks may easily displace, leaving the sleepers exposed as seen in Figure 7a.

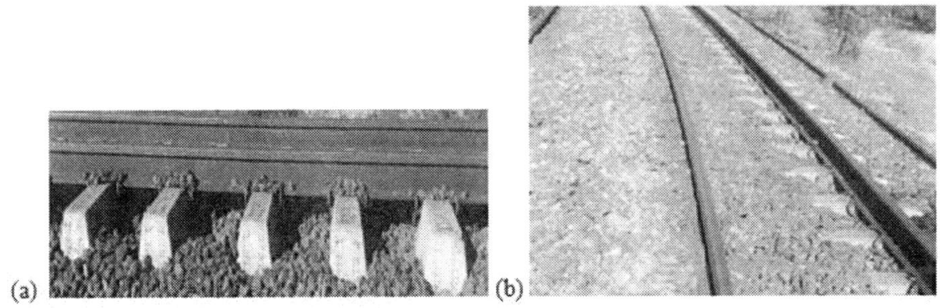

Fig. 7. (a) Excessively exposed concrete sleepers (b) Concrete sleepers adequately covered by ballast.

Figure 7b. shows the adequately ballasted part of the track where concrete sleepers are not excessively exposed. Exposed sleepers (as seen in Figure 7a.) can vibrate the most and for a longer period with lack of ballast friction to damp the energy.

2 Method

Two methods are used in the study i.e. the experimental and FEA modal analysis methods.

2.1 Setup for Experimental Modal Analysis on Rails

Rail vibration is excited using an impact hammer. The response is measured in the vertical and horizontal directions using an accelerometer placed on the field side of the rail crown as seen in Figure 8. The equipment utilised to supply power for the experiment was the battery bank, Uninterrupted Power Supply (UPS), power generator, Signal Express software (to compute measured FRFs), National Instruments PXi (hardware in which the computing software is installed). Rails were first cleaned using a pencil grinder and some chemicals for preparation of modal testing. Two test sites were used for modal analysis on rails; the track curves installed with the FY and PY-types of concrete sleepers. These are respectively situated at kilometer points 181/6 and 187/13 in the rail track. Figure 8 shows a prepared rail and an accelerometer placed on the crown for modal testing. The test site in Figure 8. installed with the FY-type concrete sleepers.

Fig. 8. Impact hammer, accelerometer and surface preparation for experiment on rail resting on FY-type concrete sleepers.

The rail is clamped on the FY-type concrete sleepers using the FISTclip fastening system. The impact hammer used for the experiment is also shown in Figure 8. The rail resting on the PY-type concrete sleepers was cleaned following the same procedure. The e-Clip fastening system is used for clamping the rail onto pads and the PY-type concrete sleepers. Vertical vibration modes were excited by applying impacts downwards on the rail working surface. Horizontal modes were excited by applying impacts on the rail crown, in the field side. The accelerometer remained in a single position, whilst impacts were applied as close as possible (approximately 5mm from the measuring point). Five excitation impacts were applied continuously to better define a single FRF in each direction. Three FRF measurements were taken in each direction to ensure accuracy and reporting of the most pronounced FRFs. Figure 9. shows an image of an e-Clip fastening system clamping the rail resting on the PY-type concrete sleeper.

Fig. 9. Rail on PY-type concrete sleeper with an e-Clip fastening system.

The e-Clip fastening system uses components embedded on concrete sleepers to restrict displacement of the rail, pad and clips.

2.2 Finite Element Modal Analysis Setup

A 3D Finite Element Model (FEM) of the D39 200 locomotive traction wheelset was developed as shown in Figure 10.

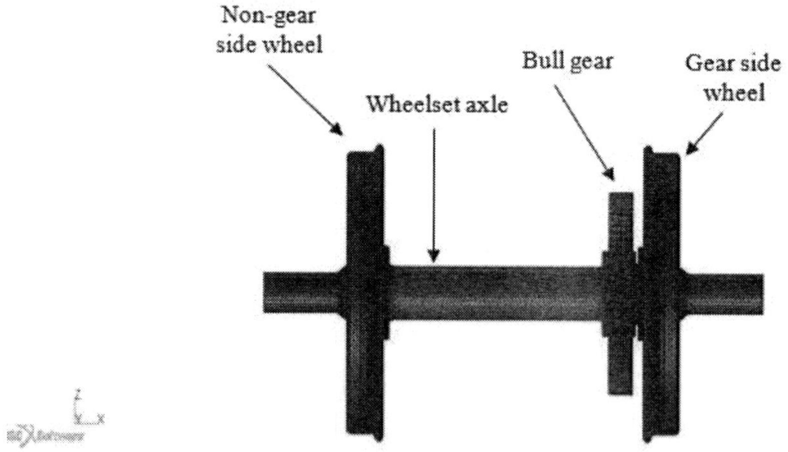

Fig. 10. FEA model for a D39200 locomotive traction wheelset.

The traction wheelset model in Figure 10. includes a bull gear mounted on a wheelset axle. The bull gear was modelled to accurate dimensions with teeth profile. The coupling between the bull gear and traction motor pinion is not modelled. The gear case is also not modelled. The mesh details for the FEM are tabulated in Table 1.

Table 1. Mesh details for the traction wheelset FEM.

Mesh Item	Mesh Details
Mesh type	Solid
Element type	Tetrahedron
Mesher	TetMesh
Topology	Tet 10

Tet10 solid elements were used on all solids of the wheelset model to ensure better contact definition. The element nodes were equivalenced to 0.1mm to ensure better alignment of the nodes on created contacts, and also to get rid of unnecessary nodes in the model. The element nodes were equivalenced to 0.1 mm to ensure better alignment of the nodes on created contacts, and also to get rid of unnecessary nodes in the model. Table 2 shows mesh convergence for the wheelset model. Mesh convergence was measured on the first bending mode only, for all mesh sizes.

Table 2. Mesh convergence for wheelset model.

Element Global Edge Length (mm)	Number of Nodes	Maximum Displacement (mm)
250	48,188	42.3
50	182,346	43.8
10	763,491	54.7
5	823,157	57.3
4	834,568	57.7
3.5	869,498	58.1
3	905,567	58.1

Materials are understood to be isotropic and the material properties of steel used in the FEA model are tabulated in Table 3.

Table 3. Material properties of the traction wheelset FEA model.

Material	Young's Modulus (GPa)	Density (kg/m3)	Poisson's Ratio
Steel	210	7850	0.3

Constraints and loads were not applied in the model in all Degrees of Freedom (DoF). Vibration frequency is set to a range of 0 – 500Hz to capture the vibration modes that correspond to a short pitch corrugation on rails.

Vibration modes requested are sixteen, which include the first six rigid body modes. Hence the vibration mode number seven is in principle the first mode. The bull gear material properties used are the same as those of the entire wheelset.

3 Results

In the current study only the low vibration frequency range (0 – 500Hz) is considered. Short-pitch rail corrugation found in the railway line between Belfast and Steelpoort corresponds to a low frequency range of vibration [1]. Furthermore, it is in a low frequency range that the horizontal vibration is predominantly transmitted downwards [21]. At higher frequencies the vibrations are understood to be transmitted along the rail. The stiffer the rail pad, the more vibration is transmitted downward to the sleepers [21–22]. This is in agreement with [16] as illustrated in Figure 4. Vibrations transmitted downwards to the sleepers are at their greatest around the antiresonance of the rail [21]. However in order to test this, FRFs from rail and sleepers would have to be tested simultaneously. The difference in the design of these two types of rail pads may play a role in allowing and or constraining horizontal vibration.

3.1 Experimental Modal Analysis Results

Modal analysis was conducted to investigate the difference in dynamics of rails resting on HDPE and Hytrel/6358 rubber rail pads. This was done to investigate how the rails respond to vibration excitation when resting on these two types of sleepers. Part of the ultimate aim of the study is to establish how these dynamics can be associated with existence of corrugation on rails. The data is presented in form of FRFs. Figure 11. shows vibration response of a rail in the vertical direction when the rail is excited on top of sleeper point and mid-span. Figure 12. shows vibration response of a rail in the horizontal direction when the rail is excited on top of sleeper point and mid-span.

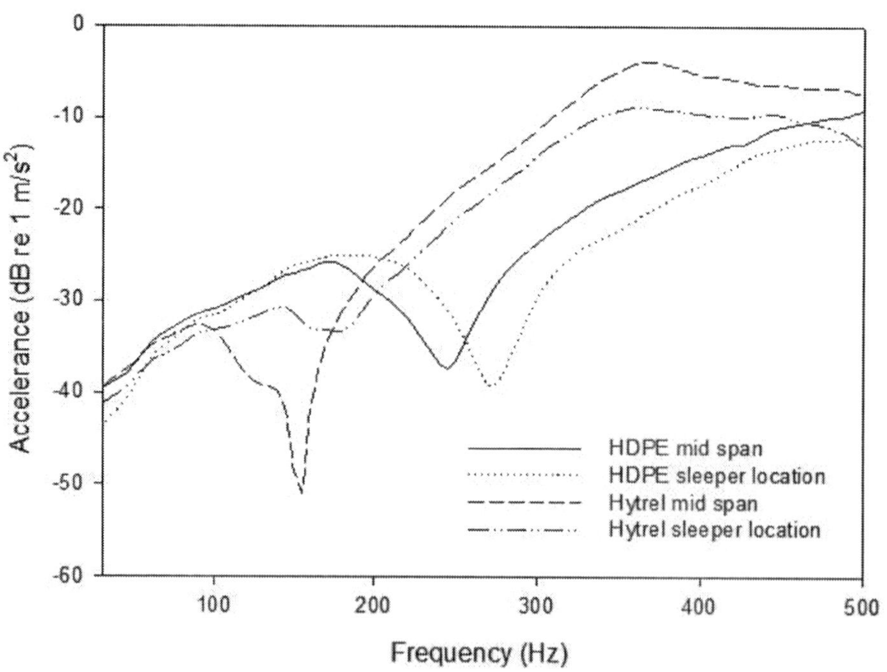

Fig. 11. Vibration response of rail resting on HDPE and Hytrel/6358 in the vertical direction. Rail is excited on two points - on top of sleeper point and mid-span.

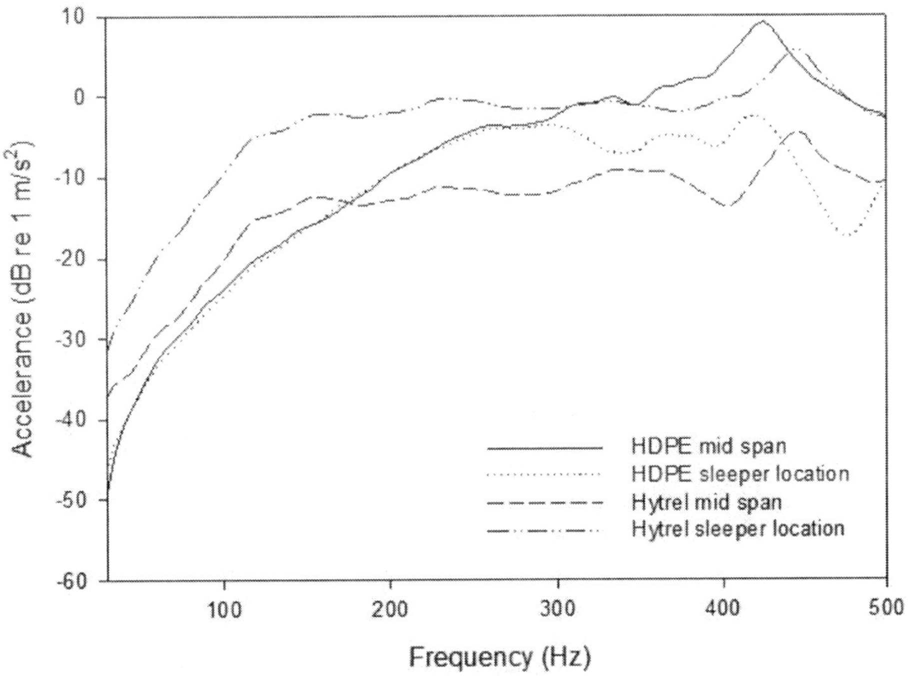

Fig. 12. Vibration response of rail resting on HDPE and Hytrel/6358 in the horizontal direction. Rail is excited on two points. On top of sleeper point and mid-span.

Seen in Figure 11. is that rails resting on both types of pads show a constant increase in vibration in the frequency range of 0 – 100Hz. In the previous study, rail corrugation fixed in the track curves of the railway line with which the current study is concerned showed evidence of development by D39200 locomotive class wheels at 108Hz. That frequency falls not far off from 100Hz (where response of rail is in the rise in the horizontal direction). At 150Hz the rail resting on a Hytrel/6358 pad vibrates with antiresonance when excited mid-span. This is seen by a narrow ditch in the FRF representing an antiresonance by the rail.

No antiresonance vibration is experienced by the rail when the rail excitation impact is applied directly on the rail crown point that is direct with the rail pad. The rail resting on HDPE rail pads experienced antiresonance twice at low frequency vibration i.e. when the rail is excited either mid-span and or on the rail crown point that is direct with the rail pad. This antiresonance vibration occurs at a frequency range of about 250 – 300Hz. In the horizontal direction, on top of a sleeper position, the rail on HDPE pads responds with more accelerance (acceleration per unit force) for most of the vibration response in a low frequency rang. This may be because the rail is much stiffer on top of a sleeper than mid-span due to the clamping force applied by the fastening system. Both rails on HDPE and on Hytrel/6358 respond at resonance at a frequency range of 400 – 500Hz. Only heavily damped vibration modes are identified in the frequency range of 100 – 150Hz, which corresponds to a short-pitch rail corrugation frequency.

3.2 FEA Modal Analysis Results

Resonance modes in the high frequency range of 350 – 500Hz are studied. Four resonance modes are captured with their shapes in the frequency range of 350 – 377Hz. Figures 13 to 16. show the mode shapes that occur at each resonance (in the high frequency range of 350 – 500Hz) during vibration.

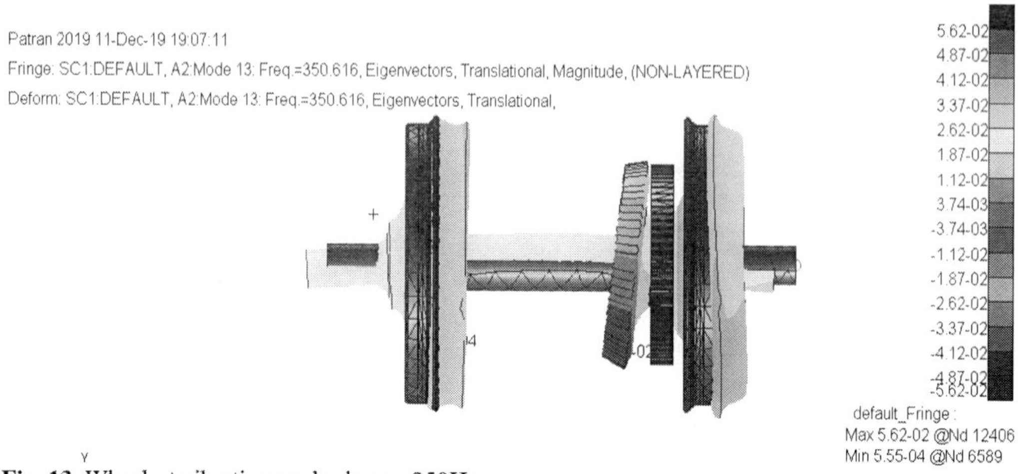

Fig. 13. Wheelset vibration mode shape - 350Hz

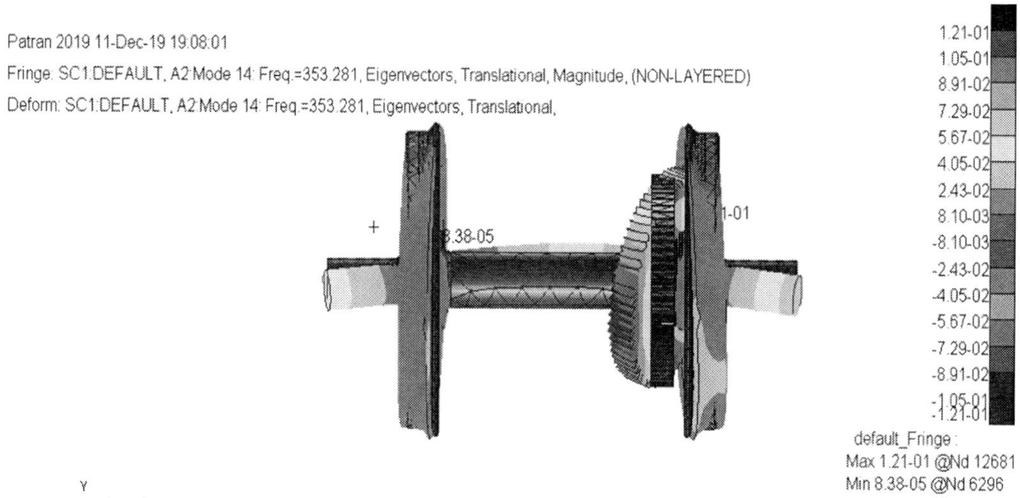

Fig. 14. Wheelset vibration mode shape - 353Hz

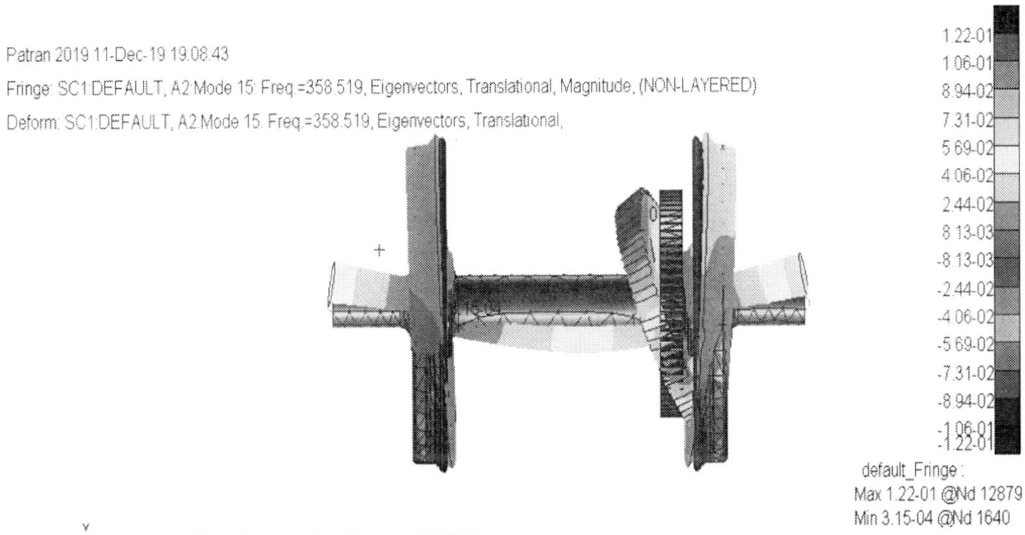

Fig. 15. Wheelset vibration mode shape - 358Hz

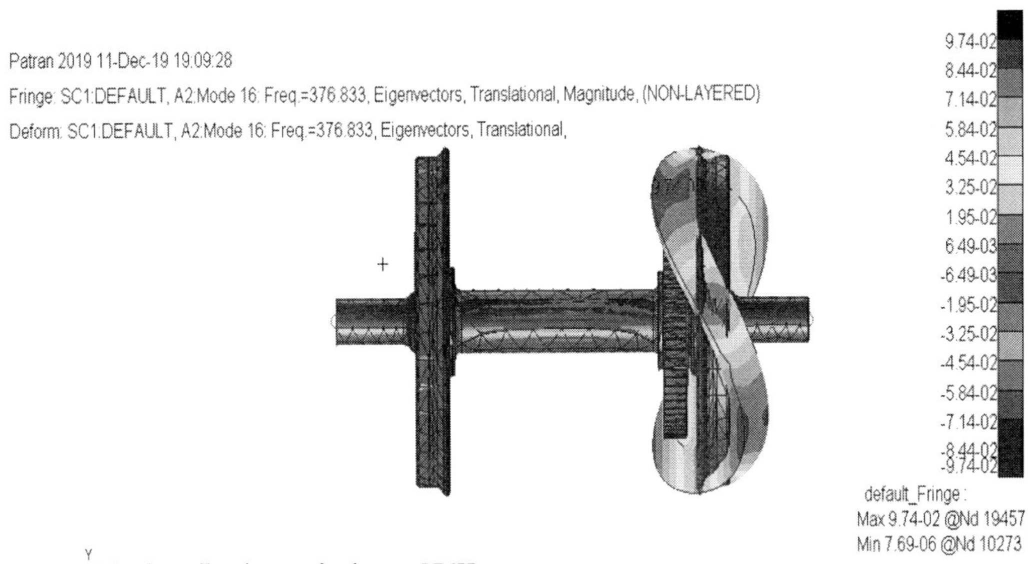

Fig. 16. Wheelset vibration mode shape - 376Hz

In Figures 13 to 16. the traction wheelset used by a D39200 locomotive class has a traction gear on one side of the axle. The two wheels of the wheelset are referred to as the gear and non-gear side wheels. Evident in Figures 13 to 16. is that for a selected frequency range of 350 – 500Hz the wheels vibrate at resonance at frequencies 350Hz, 353Hz, 358Hz and 376Hz. During vibration, in these resonance frequencies the gear side wheel proves to experience more erratical vibration than the non-gear side wheel. Only at a frequency of 350Hz does the non-gear side wheel show erratical oscillation in the lateral direction as the axle mode is on bending. For a selected frequency range, the axle vibrates with a bending mode shape for all for vibration modes except for the resonance mode that occurs at 376Hz. Erratical vibration is also experienced by the traction gear, which is pressed onto the axle much closer to the gear side wheel. The vibration mode shapes of the wheels become more complicated as the vibration frequency increases.

4 Implications, Conclusions and Recommendations

Rail pads design engineers may utilise findings in this work to look at designing rubber rail pads that will sufficiently damp excessive rail vibrations. Railway engineers may utilise findings in this work to look at different mechanisms that may be affected on railway train wheels to ultimately suppress excessive vibration. The HDPE and Hytrel/6358 rubber rail pads present different characteristics in the same environment. In the vertical direction, for mid-span rail excitation, the rail on the Hytrel/6358 pad experiences antiresonance vibration around 150Hz. However, the opposite occurs when the force is applied on top of sleeper point. For the rail on HDPE rail pad, this antiresonance is experienced when either position is excited, more in particular in the frequency range of 250 – 300Hz. In the horizontal direction, rails on both types of rubber rail pads vibrate at resonance, particularly in the frequency range of 400 – 500Hz. At lower frequencies below 300Hz, rails vibrate with antiresonance in the vertical direction. The first order axle bending; and the first and second order wheel bending modes are found to be of significance. There is a significant need of tuned wheelset vibration dampers for the wheelset used by the D39200 locomotives in the lines. One damper shall be tuned to damp resonance modes along the wheelset axle, whilst the other shall be applied around the wheel web of the gear-side wheel.

References:

1. B. M. Balekwa, Masters Dissertation: *Corrugation Wavelength Fixing Mechanism and its Relationship to Train and Track Geometry Parameters*, Johannesburg: University of Johannesburg, (2017).
2. S. L. Grassie and J. Kalousek, "*Rail Corrugation: Characteristics, Causes and Treatments*," Part F: J. Rail. Rapid. Trans, pp. 57 - 68, **207** (1993).
3. J. Han, X. Xiao, Y. Wu, Z. Wen and G. Zhao, "*Effect of Rail Corrugation on Metro Interior Noise and its Control*," App. Acous, https://doi.org/10.1016/j.apacoust.2017.09.007, pp. 63 - 70, **130** (2018).
4. X. Cui, Z. Cheng, Z. Yang, B. Huang and Z. Du, "*Study on the Phenomenon of Rail Corrugation on High-Speed Rail Based on the Friction-Induced Vibration and Feedback Vibration," International Journal of Vehicle Mechanics and Mobility*, DOI: 10.1080/00423114.2020.1817507, (2020).
5. X. Zhao, P. Zhang and Z. Wen, "On the Coupling of the Vertical, Lateral and Longitudinal Wheel-Rail Interactions at High Frequencies and the Resulting Irregular Wear," Wear, Vols. 430 - 431, https://doi.org/10.1016/j.wear.2019.05.017, pp. 317 - 326, (2019).
6. W. Cai, M. Chi, P. Shackleton , D. Crosbee and Y. Zhao, "*Experimental and Numerical Investigation into Formation of Metro Wheel Polygonization*," Shock and Vibration, Article ID 1538273, https://doi.org/10.1155/2019/1538273, (2019).
7. S. Grassie and J. Kalousek, "*Rail Corrugation: Causes and Cures*," Wheel-Rail Technology, pp. 24 - 26, July, (2000).
8. D. T. Eadie, J. Kalousek and K. C. Chiddick, "The Role of High Positive Friction Modifier in the Control of Short Pitch Corrugations and Related Phenomena," Wear, 253, pp. 185 - 192, **1** (2000).
9. T. Xin , S. Wang, L. Gao, H. Huo, Y. Ding, P. wang, P. Chen and P. Liu, "*Field Measurement of Rail Corrugation Influence on Environmental Noise and Vibration*: A Case Study in China," Measurement, ISSN 0263-2241, p. https://doi.org/10.1016/j.measurement.2020.108084, **164** (2020).
10. K. Alten and A. Fuchs, "Detecting and Classifying Rail Corrugation Based on Axle Bearing Vibration," International Conference on Acoustics, Speech and Signal Processing (ICASSP), DOI: 10.1109/ICASSP.2019.8683317, 12 - 17 May, (2019).
11. D. Wei , X. Wei, Y. Liu , L. Jia and W. Zhang, "*The Identification and Assessment of Rail Corrugation Based on Computer Vision*," App. Sci, no 18, p. https://doi.org/10.3390/app9183913, **9** (2019).
12. C. Dong, Q. Mao, X. Ren, D. Kou, J. Qin and W. Hu, "*Algorithms and Instrument for Rapid Detection of Rail Surface Defects and Vertical Short-Wave Irregularities Based on FOG and Odometer*," IEEE Access, doi: 10.1109/ACCESS.2019.2903488, pp. 31558 - 31572, **7** (2019).
13. B. M. Balekwa, A. Mashamba and D. Kallon, "*Application of Modal Analysis to Establish the Wavelength Fixing Mechanism for Rail Corrugation*," Proceedings of the 11th SACAM, pp. 212 - 223, (2018).
14. B. Balekwa, D. V. V. Kallon and D. J. Fourie, "*Vibration Response for class D44 and D39 200 Locomotive Wheelsets due to Dynamic Loads Excitation*," Proceedings of OIC 2019, pp. 418 - 421, 2 - 4 October, (2019).

15. Indianmart, "Studded Rail Rubber Pads, Railway, Railroad, Train Components," Indianmart, [Online]. Available: https://www.indiamart.com/proddetail/studded-rail-rubber-pads-10501823362.html. [Accessed 05 October 2018].

16. D. J. Thompson, Railway Noise and Vibration Handbook: Mechanisms, Modelling and means of Control, Oxford: Elsevier, (2009).

17. B. Balekwa, D. V. V. Kallon and D. J. Fourie, "*Vertical and Horizontal Vibration Response for Corrugated Track Curves Supported on Steel and PY-Type Concrete Sleepers*," Proceedings of OIC 2019, pp. 474 - 482, 2-4 October, (2019).

18. B. Balekwa and D. V. V. Kallon, "Correlation of Short Pitch Rail Corrugation with Railway Wheel-Track Resonance at Low Frequencies of Excitation," Vibration, no 4, pp. 491 - 520, **3** (2020).

19. E. Articles, "Engineering Articles," [Online]. Available: http://www.engineeringarticles.org/what-is-ballast-railway-ballast-functions-of-ballast/. [Accessed 05 October 2018], 20 May, (2015).

20. J. Ali Zakeri and S. Ali Mosayebi, "*Study of Ballast Layer Stiffness in Railway Tracks*," Gradevinar. DOI: 10.14256/JCE.1232.2015, no 68, pp. 311 - 318, **4** (2016).

21. X. Sheng, C. Zhao, P. Wang and D. Liu, "Study on Transmission Characteristics of Vertical Rail Vibrations in Ballast Track," Mathematical Problems in Engineering, pp. 13 pages, Article ID 5872419, https://doi.org/10.1155/2017/5872419, 19 December, (2017).

22. X. Song, Y. Qian, K. Wang and P. Liu, "Effect of Rail Pad Stiffness on Vehicle–Track Dynamic Interaction Excited by Rail Corrugation in Metro," Journal of the Transportation Research Board, no 6, pp. 225 - 243, **2674** (2020).

High frequency guided wave ultrasound measurements using low-cost equipment

Philip Loveday[1*]

[1]School of Mechanical, Industrial and Aeronautical Engineering, University of the Witwatersrand, Johannesburg, Wits 2050, South Africa

Abstract. Guided wave ultrasound can be applied to inspect large volumes of structures from a single transducer location and is particularly effective in one-dimensional structures such as rods, pipes and rail. An application, which is of interest in South Africa, is the inspection rock bolts in the mining industry. The difficulty in inspecting embedded rods is that there is rapid attenuation of the signals with distance. While guided wave systems generally operate at low frequencies (20 kHz to 200 kHz), high frequency operation (1 MHz to 10 MHz) can offer some advantages when inspecting rock bolts. Guided wave ultrasound testing is complicated by multi-modal propagation and dispersion, which require the use of sophisticated excitation signals, which are not available in equipment designed for conventional ultrasonic testing. It was therefore necessary to develop a measurement setup and the approach taken was to use standard lab instruments wherever possible and to only purchase the items that are unique to this measurement, using the very limited budget available. The measurement setup was used to perform measurements on a rod and confirmed theoretical predictions. These measurements identified the frequencies where modes propagate with low attenuation and allow the attenuation to be quantified.

1 Introduction

Guided wave ultrasound offers the possibility of inspecting large volumes of structures from a single transducer location. It is particularly effective in one-dimensional structures such as rods, pipes and rail. This has resulted in inspection and monitoring systems being developed for pipelines in the oil and gas industry internationally [1] and for a heavy haul train line in South Africa [2]. Another application, which has local relevance and interest, is the inspection of rock bolts in the mining industry [3]. The difficulty in inspecting embedded rods is that the ultrasound leaks into the surrounding media causing rapid attenuation of the signals with distance. An instrument for inspecting rock bolts was described in [4], which used high energy, wideband excitation followed by pulse compression digital processing of the received signals to increase the signal to noise ratio. While guided wave systems generally operate at low frequencies (20 kHz to 200 kHz) it was shown by Beard [5] that high frequency operation (1 MHz to 10 MHz) can offer some advantages when inspecting

* Corresponding author: Philip.Loveday@wits.ac.za

embedded structures such as rock bolts and could be combined with the lower frequency measurements. At the higher frequencies, there are some modes of propagation and frequency ranges where the ultrasonic energy is concentrated mainly inside the cross - section of the waveguide with little motion at the surface and this results in reduced leakage of energy.

Guided wave ultrasound testing is more complicated than conventional ultrasonic testing due to multi-modal propagation and dispersion of guided waves. The development of guided wave ultrasound based inspection or monitoring systems requires both numerical modelling capabilities and experimental measurement capabilities. While equipment for conventional ultrasonic testing can operate at high frequencies this equipment is not suitable for guided wave testing, which requires the use of more sophisticated excitation signals. Laboratory equipment for guided wave ultrasound research is available commercially (see for example www.ritecinc.com). A power amplifier for driving a transmit transducer may cost between R100 000 and R 200 000, while a pulser-receiver for pulse-echo measurements could cost between R300 000 and R500 000. Such a system would still require additional instruments such as a two channel arbitrary waveform generator and an oscilloscope and ultrasonic probes and cables. Unfortunately, such a system is not affordable for most South African researchers. This paper aims to show what can be achieved on a very limited budget. The approach taken was to use standard lab instruments wherever possible and to only purchase the items that are unique to this measurement.

Some of the features of guided wave ultrasound are introduced in section 2 as these determine how measurements must be performed and therefore influence the design of the measurement setup. Measurement results obtained from the measurement setup are presented in section 4. The capabilities and limitations of the setup are described in section 5 along with some recommendations for future improvements.

2 Principles of guided wave ultrasound

Conventional ultrasonic testing makes use of relatively simple elastic waves that propagate in an infinite medium. Generally, the compressional wave is used although shear waves may also be used. When these waves encounter a free boundary of a structure they reflect and conversion between compressional and shear waves occurs. There are numerous combinations of waves that can satisfy the wave equation inside the structure and the stress free boundary conditions on the perimeter of the waveguide. These solutions are modes that can occur in the waveguide. Some of these modes are travelling waves that propagate and transfer energy along the waveguide. At higher frequencies, where the wavelength is smaller than the cross – section of the waveguide, there are many modes of propagation. These modes of propagation can be dispersive in nature, meaning that energy at different frequencies propagates at different speeds (group velocities). This means that a pulse of energy will become stretched out as it propagates and therefore the amplitude decreases making it more difficult to detect.

Prediction of the guided wave modes and the properties of these modes requires mathematical modelling. Numerical techniques, such as the semi-analytical finite element method, are required for waveguides having complex cross – sections such as a rail. Analytical solutions are available for the cylindrical waveguide (rock bolt) considered in this paper. The dispersion curves for a 20 mm diameter steel rock bolt are shown in figure 1. This graph was generated using the free software PCDISP which is described in [6]. This software implements the Pochhammer-Chree solution of the 3D elastic wave equation in a cylinder, which is based on Bessel functions. The final stage of these analytical solutions involves finding the roots of characteristic equations and this only became possible with the arrival of digital computers. A Matlab code that implements the analytical solution and finds the roots of the characteristic equation was developed and made available for free by the authors of [6].

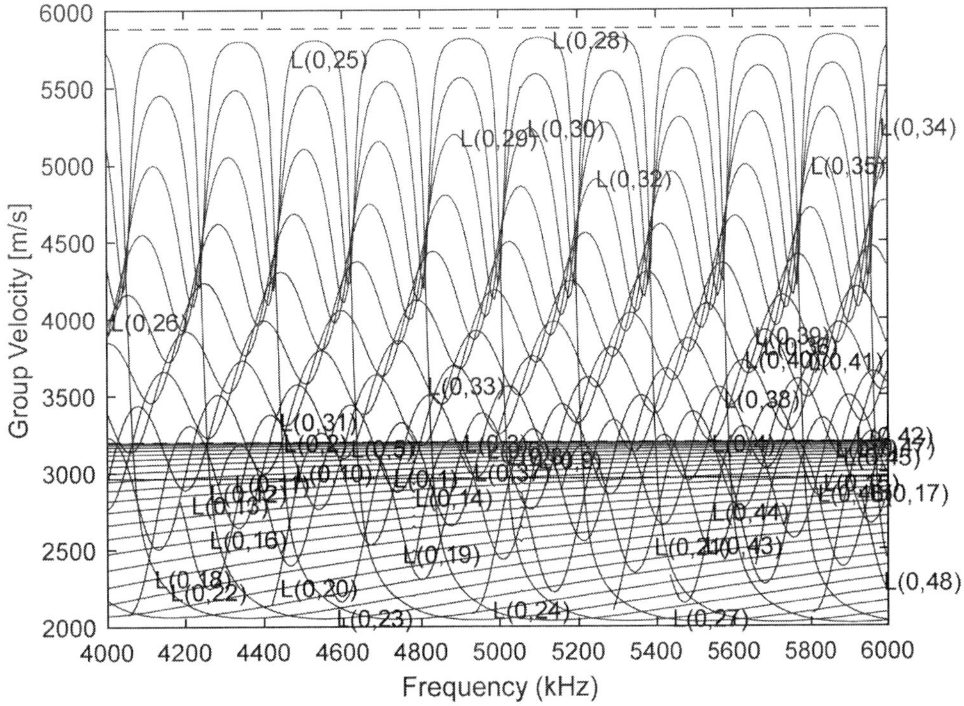

Fig. 1. Group velocity dispersion curves for a 20 mm diameter steel rod.

The figure shows that at the frequencies being considered there are dozens of modes of propagation. The modes and frequencies where the maximum group velocities are attained are the combinations that are of interest for inspecting embedded rock bolt because it was shown by Beard [5] that these points represent modes where most of the energy is concentrated in the centre of the rock bolt which implies that less energy leaks from the rock bolt into the surrounding medium. Modes and frequencies with higher leakage will attenuate quickly and be filtered out. The points with highest group velocity are also advantageous as energy reflected in these modes will arrive first. Tone burst excitation signals are often used in guided wave ultrasound as these signals have relatively narrow frequency bandwidth for a given length of signal. By increasing the length of the excitation signal (number of cycles) it is possible to reduce the bandwidth of the signal to excite only one mode with a peak group velocity. The tone burst signals are obtained by applying a Hanning window to a sine wave. Example signals with 10 and 50 cycles respectively are shown in figure 2. These signals were generated by applying the Hanning window to a sine wave with frequency of 5 MHz. Figure 2 shows how increasing the number of cycles in the signal causes the bandwidth of the signal to be reduced and the energy at the centre frequency to be increased. Long signals may be used when the distance resolution required is not a concern. There are also other types of signal that could be used including coded signals. It is therefore required that the measurement setup includes the capability to generate arbitrary signals that can be programmed on a computer.

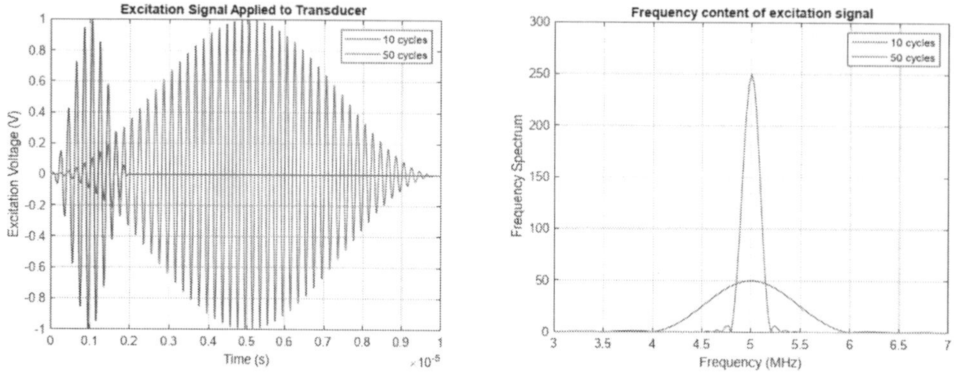

Fig. 2. Tone burst excitation with different signal lengths.

3 Measurement system

Measurements may be performed to measure the transmission or reflection of ultrasound. The transmission mode requires the use of two transducers, one to transmit the ultrasound and the second to receive the ultrasound. This mode of measurement is illustrated in figure 3.

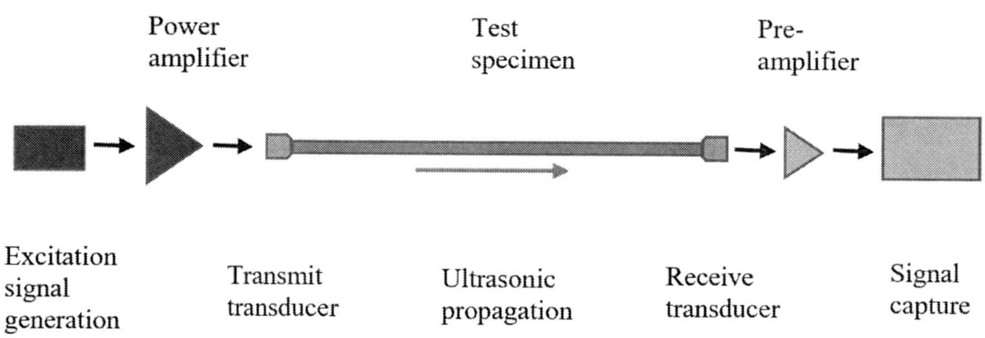

Fig. 3. Transmission mode measurement system configuration.

In the reflection mode a single transducer is used to transmit the ultrasound and then to receive reflected ultrasound. The reflection mode would be required for the inspection of installed rock bolts where only one end of the rock bolt can be accessed. The reflection mode requires additional electronics to protect the receive pre-amplifier from the high voltage excitation signal and then to switch out the power amplifier after the excitation so that this does not cause noise or a low impedance. It was therefore decided to implement the transmission mode measurement first.

The measurement setup is shown in figure 4 and the different components are described in the following subsections. Measurements performed with this system are presented in section 4.

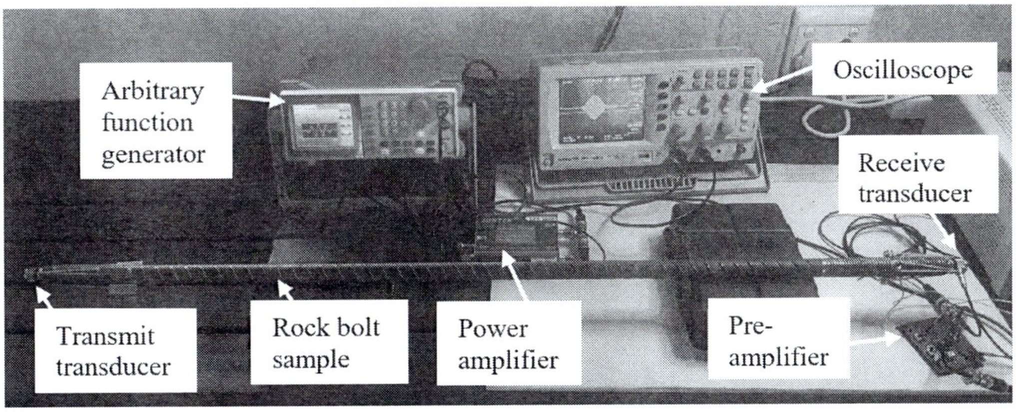

Fig. 4. Measurement system photograph.

3.1 Piezoelectric ultrasonic transducers

Two SIUI P5-6L transducers were purchased for this measurement. These are 5 MHz transducers having an active element diameter of 6 mm. The two transducers with cables (BNC to Lemo connectors) and ultrasonic coupling gel cost approximately R8200 (US$ 600). It is believed that these transducers are relatively inexpensive but have good performance. The transducers were held in place during testing by rubber bands and magnets.

3.2 Arbitrary function generator

An arbitrary function generator (GWInstek AFG-2225) was borrowed and used to generate the excitation signals. The signals were programed in Matlab and then saved as a .csv file which was loaded into the memory of the arbitrary function generator. This instrument can output a signal up to 4000 points long, with 10 bit resolution and with a sampling rate up to 120 MSa/s (mega samples per second). The burst mode was used to output the arbitrary signal and then pause before outputting the signal again.

3.3 Digital storage oscilloscope

A digital storage oscilloscope (GWInstek GDS-1102-U) was borrowed and used to capture the signals. This instrument has a 250 MSa/s real-time sampling rate and a 100 MHz bandwidth. Like almost all digital oscilloscopes this instrument had a vertical resolution of 8 bits and a minimum vertical scale of 2mV/division. The maximum record length that could be captured was 4000 points, which is low for this type of measurement. This meant that only parts of the received signal could be captured at a time.

3.4 Power amplifier

The power amplifier is required to amplify the voltage of the signal produced by the function generator and to supply the current required to drive the piezoelectric transducer. Commercially available power amplifiers for lab measurements in the frequency range of interest tend to be large and expensive. It was therefore necessary to compromise on performance. An alternative was to use a small amplifier that is used to amplify the signals from a function generator. A Juntek DPA-2698 10 MHz 2 channel amplifier was purchased for approximately R1300 from www.banggood.com. This amplifier could only apply a maximum excitation amplitude of 25 Vpp to the transducer while it is expected that the transducer could be driven at 100 Vpp or 200 Vpp depending on the duty cycle.

3.5 Pre-amplifier

The pre-amplifier is required to amplify the small signal generated by the receive transducer. Again, a good quality ultrasonic pre-amplifier was not affordable. The alternative was to use an evaluation board and a AD8331 Evaluation Board was purchased for approximately R1300. This board is intended for evaluation of the AD8331 IC which is a low noise preamplifier that includes ultrasound as one of the applications. The pre-amplifier board requires a 5 V DC supply and a power supply (borrowed) was used for this.

4 Measurement results

In order to investigate the capabilities of the measurement system a series of measurements were performed on a 1 m (more accurately, 0.996 m) long portion of a rock bolt, which had a diameter of 20 mm.

The first measurement was performed using a 10 cycle tone burst signal with a centre frequency of 5 MHz. This is a fairly large bandwidth signal (see figure 2) having energy spread across the peak velocities of numerous modes shown in figure 1. The pre-amplifier was not used in this measurement. The received signal is shown in figure 5. The first figure shows the entire captured signal while the second figure shows a zoomed in view of part of the signal. The first wave packet arrived at approximately 0.17 ms and there appear to be a succession of wave packets following the first one. It is tempting to think that the different wave packets observed correspond to different modes of propagation each having different group velocities. This is actually not the case as the peak group velocities are all very similar (less than 1% difference). Instead, this phenomenon is due to the presence of numerous modes propagating with dispersion causing the wave packets to become stretched in time. These different modes interfere constructively in certain time intervals and destructively in other time intervals and this causes the signal to appear like a succession of wave packets. This phenomenon was described by Beard [5] who showed that if the signal is filtered to contain only the frequency range where one mode has a peak group velocity (for example 4.8 MHz to 5 MHz) then the response is that of the single mode with dispersion. The frequency spectrum of the signal is shown in figure 6. It is seen that there is more energy in the frequency bands corresponding to the peak group velocities shown in figure 1.

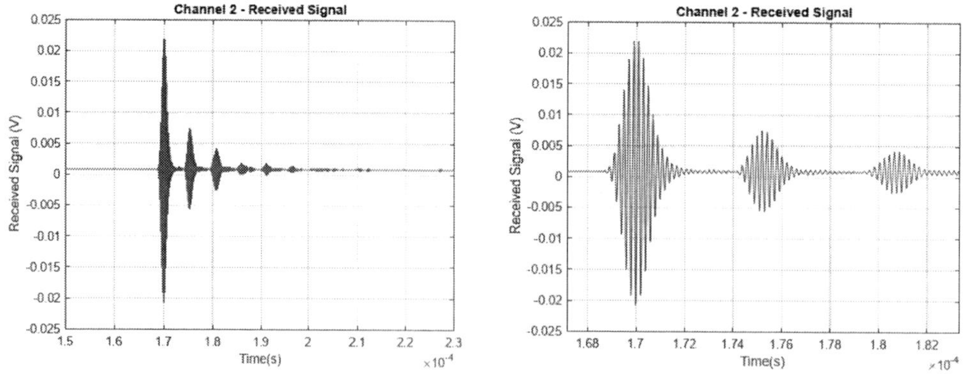

Fig. 5. Received signal for a 10 cycle tone burst with centre frequency of 5 MHz.

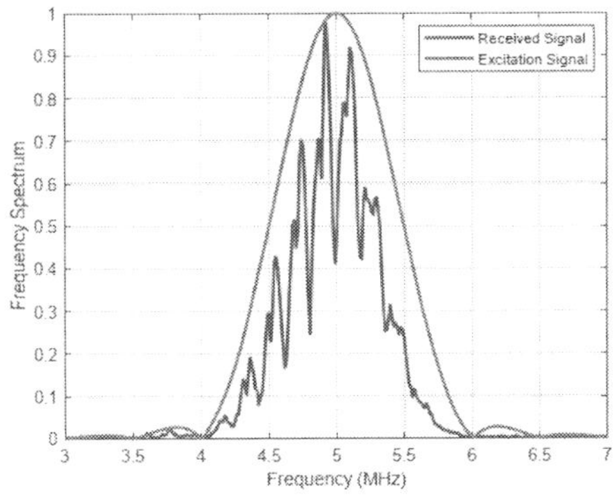

Fig. 6. Frequency spectrum of received signal and excitation signal.

This first result illustrates that in guided wave ultrasound the multi-modal nature of the propagation can make interpretation of the results complicated. To excite or sense individual modes, use is sometimes made of transducer arrays in which the spacing of the elements and phasing of the excitation signals is selected to preferentially excite a particular mode. A second option, which is more appropriate here, is to reduce the frequency bandwidth of the excitation signal. This can be achieved by increasing the number of cycles in the excitation tone burst signal. A tone burst signal containing 200 cycles was used for the remainder of the tests.

The centre frequency of the excitation signal was varied and the frequencies were maximum receive responses were obtained were recorded. Table 1 lists the peak group velocities and the frequencies at which these were predicted from the model. The measured frequencies, where peak amplitude responses were observed are compared to these frequencies. It is seen that there is generally very good agreement between the predicted and measured frequencies. This provides a simple way to measure these frequencies. These are the candidate frequencies where long range testing could be performed.

Table 1. Predicted peak velocity frequencies and measured peak amplitude frequencies

Mode	L0,24	L0,25	L0,26	L0,27	L0,28	L0,29	L0,30	L0,31	L0,32
Predicted Peak Group Velocity (m/s)	5800	5806	5812	5818	5822	5827	5830	5834	5838
Predicted Peak GV Frequency (MHz)	4.340	4.535	4.723	4.910	5.105	5.293	5.480	5.676	5.863
Measured Peak Amp Frequency (MHz)	4.360	4.550	4.735	4.925	5.110	5.290	5.470	5.640	5.835

Next we investigate the signals obtained from multiple arrivals after multiple reflections. For these measurements a 200 cycle tone burst excitation signal with centre frequency of 4.925 MHz was used. The first arrival corresponds to 1 m of travel and took approximately 0.17 ms to arrive. The second and third arrivals correspond to 3 m and 5 m of travel and arrive at proportionally later times. The first three arrivals are shown in figure 7. If the oscilloscope had more memory it would be possible to show all the arrivals in one captured signal. The time between arrivals can be used to estimate the group velocity while the decrease in amplitude can be used to estimate the attenuation. In this measurement the successive arrivals appear to be approximately 10 times smaller (-20 dB) and therefore the attenuation per metre of wave propagation is approximately -10 dB/m.

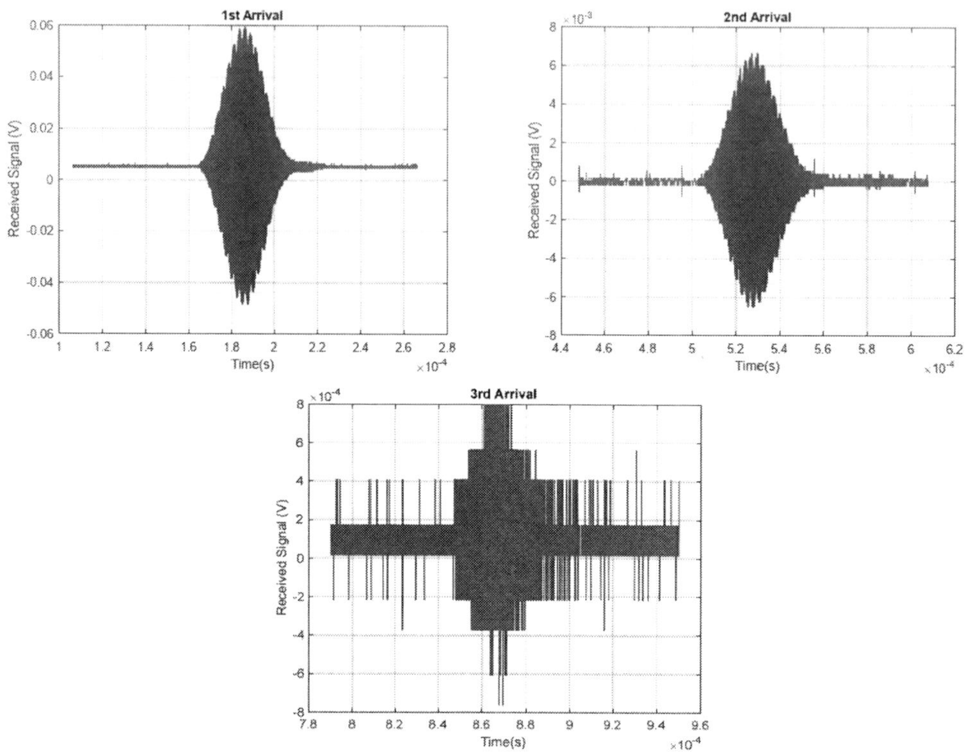

Fig. 7. First three arrivals.

It can be seen that the third arrival signal is very small and discrete levels are seen in the signal. This occurs because the oscilloscope is set on the smallest vertical scale of 2 mV/div and has only 8 bit resolution. In order to measure these small signals more accurately it was necessary to use the pre-amplifier. Figure 8 shows the same arrival but with the pre-amplifier used to amplify the received signal. This signal was measured using a vertical scale of 10 mV/div and is shown in the first graph in figure 8. The voltage signal is approximately 40 times larger than the signal without the pre-amplifier (third graph in figure 7). There is also considerable noise on this signal, which can be reduced by averaging as illustrated in the second and third graphs in figure 8 where 16 and 256 averages were used respectively. This averaging was performed on the oscilloscope. When random noise is present in a signal averaging may be applied to decrease the random noise without reducing the actual signal. The reduction in noise is related to the square root of the number of averages. Therefore 16 and 256 averages should reduce the noise to 1/4th and 1/16th of what is present in one measurement respectively.

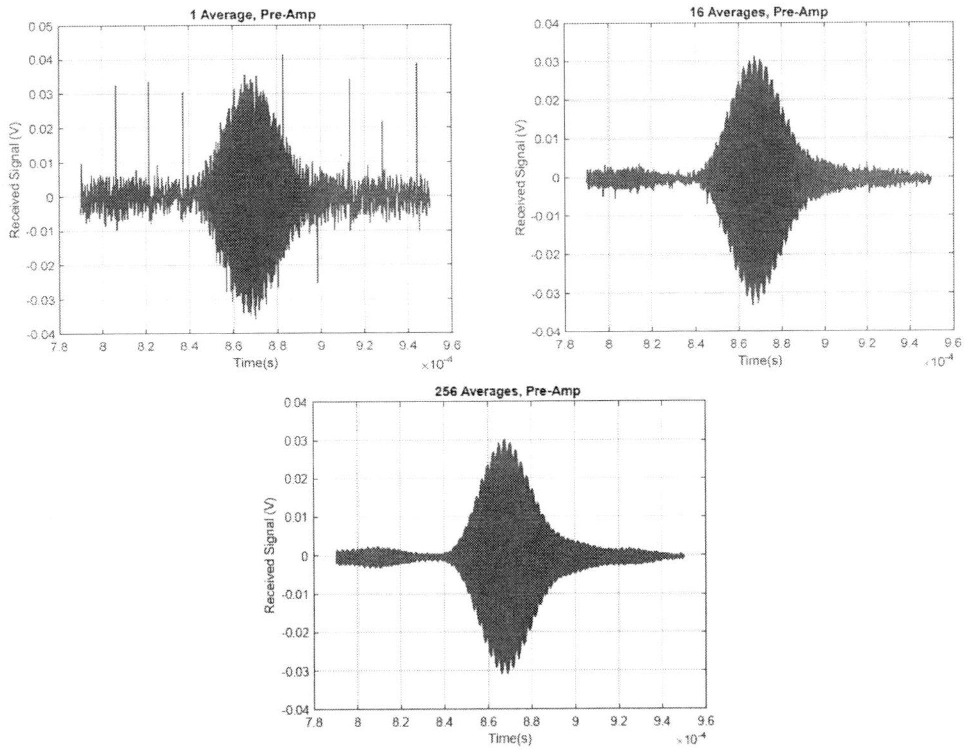

Fig. 8. Third arrival measured with pre-amplifier and averaging.

With this measurement setup it was possible to measure two more arrivals. The fourth and fifth arrivals shown in figure 9 involve propagation of the wave over distances of 7 m and 9 m respectively. Again it is observed that the signals decrease by a factor of ten for each additional arrival. The fifth arrival signal is very small and was the smallest signal that could be observed.

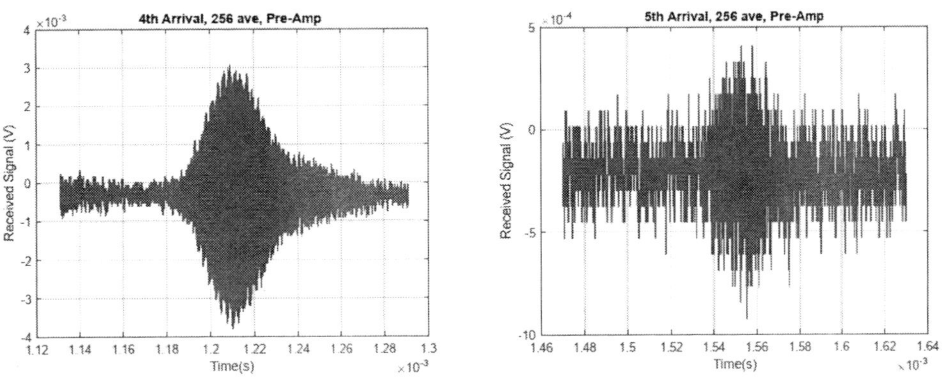

Fig. 9. 4th and 5th arrivals measured with pre-amplifier and averaging.

Once the signals have been captured and transferred to a computer it is possible to perform post processing. The signal has energy only in a small frequency range and if only frequencies between 4.8 and 5 MHz are retained then the signal becomes clearer as shown in figure 10.

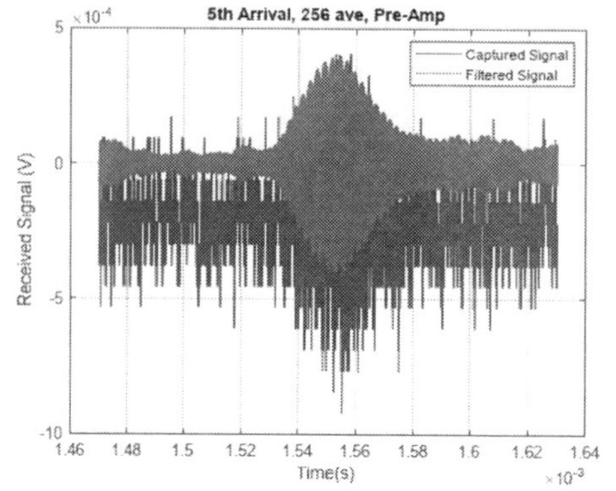

Fig. 10. Filtering to reduce noise.

5 Conclusions and future work

The system is capable of performing high frequency guided wave ultrasound measurements. Measurements were performed on a section of a rock bolt between 4 and 6 MHz and various guided wave modes were detected in good agreement with theoretical predictions. A wave that had propagated a total distance of 9 m in the rock bolt could be detected. The system can now be used, for example, to quantify the additional attenuation introduced when the rock bolt is embedded in concrete.

The measurement system was created using borrowed lab instruments and a few relatively inexpensive purchases. The total cost of the power amplifier, pre-amplifier, two transducers and cables was approximately R11000. The value of the borrowed instruments would be around R20 000. The function generator and the oscilloscope had limited memory and this restriction will be removed by purchasing new instruments which will cost a total of R22000 bringing the total cost of the system to R33 000 (US$ 2 200). High frequency guided wave ultrasound measurements can therefore be performed on a limited budget.

PC based oscilloscopes were considered as an alternative to the bench top instruments. A product that met all the performance requirements and had a significant cost saving was not identified and the ease of use of bench top instruments was preferred. PC based oscilloscopes may be reconsidered at a later stage if field measurements are to be performed.

The drive amplifier does not supply the maximum voltage levels that can be used and this could be improved. Current efforts in this area involve the development of a MOSFET based amplifier. Reflection mode testing should be implemented. This requires additional electronics and these are being developed.

References

1. P. Cawley, Struct. Heal. Monit. **17**, 1225–1244 (2018)
2. P.W. Loveday, J. Nondestruct. Eval. **31**, 303–309 (2012)
3. B.J. Buys, P.S. Heyns, P.W. Loveday, J. South. Afr. Inst. Min. Metall. **109**, 2 (2009)
4. T. Stepinski, K. Mattson, K, *Instrument for Rock Bolt Inspection by Means of Ultrasound*, in IEEE International Ultrasonics Symposium Proceedings, pp. 4–7 (2015)
5. M.D. Beard, *Guided Wave Inspection of Embedded Cylindrical Structures*, PhD Thesis, Imperial College (2002)
6. F. Seco, AR. Jiménez, *Modelling the Generation and Propagation of Ultrasonic Signals in Cylindrical Waveguides*. In Ultrasonic Waves, Dr Santos (Ed.), pp. 1–28 (2012)

Development and validation of a robust integrated thermal power plant model for load loss analysis and identification

Alton Marx[1*], *Pieter* Rousseau[1], and *Ryno* Laubscher[1]

[1]Department of Mechanical Engineering, Applied Thermofluid Process Modelling Research Unit, University of Cape Town, Library Rd, Rondebosch, Cape Town, 7701, South Africa.

Abstract. The development of deep learning methodologies for the analysis of thermal power plant load losses requires a combination of real plant data and data derived from fundamental physics-based process models. For this purpose, a robust integrated power plant thermofluid process model of a complete +600MW coal-fired power plant was developed within the Flownex Simulation Environment. It consists of standard and compound components, combined with specially developed scripts to ensure complete energy balance, specifically on the two-phase tank components. This enables simulation of the complete plant operation to determine power output as a function of any given set of internal and external operational variables, boundary conditions and component states. The model was validated against real plant design and acceptance test data. In order to demonstrate the ability of the model it was used to evaluate the plant performance related to three specific load loss inducing scenarios. The results show that a combination of mechanical faults, process anomalies and operational phenomena can be analysed. This provides the basis for generating model-based performance data that can be combined with real plant data to facilitate the development of deep learning analytics tools for load loss fault diagnosis and root cause analysis, as well as fault propagation and load loss forecasting.

1 Introduction

With the current shift towards renewable energy, conventional thermal power plants are forced to respond with agility to the fluctuation in demand. This requires operational flexibility while maintaining optimal plant performance. However, conventional designs focussed on efficiency at a fixed operating point, and therefore flexible operation often results in an increase in load loss events.

A load loss is defined by VGB [1] as the incapacity of a power plant to produce electricity at its Maximum Continuous Rating (MCR). A load loss may be caused by any number and combination of internal or external factors, but is commonly categorised as either planned or unplanned, and full or partial. Faults within the mechanical, chemical and electrical equipment are the primary causes of unplanned load losses in the power generation industry,

* Corresponding author.
E-mail address: altonmarx@gmail.com

followed by human error and external factors. Identifying what portion of load loss is associated with which plant area, system, or equipment plays a critical part in the formulation of operation and maintenance strategies.

Existing load loss classification systems are predominantly physics-based, utilising lumped zero-dimensional thermodynamic models. These models are developed to evaluate each power plant system based on its current performance against the original design [2]. A collective of systems or plant areas is typically integrated within a physics-based model grouping, for example thermal performance or critical electrical plant area trending models. These models provide a specific visualisation of plant performance. Power plant physics-based models are rarely integrated into one model capable of the evaluation of the load losses in relation to specific system faults. Moreover, many power plants still manually identify the causes of load losses.

Various integrated power plant process models have been developed in the past for different purposes. Alobaid et al.[3] provide a comprehensive review of recent dynamic thermal power plant modelling methodologies as well as their development and application. The most notable of these are:

- A 550MW anthracite coal-fired power plant model developed using the ThermoPower library [4] presented by Meinke et al.[5]. It consists of a predefined water/steam cycle, boiler combustion chamber and air passage power plant components that work in conjunction with control schemes. The study focussed on the dynamic behaviour at start-up and different operating modes.
- A 500MW bituminous coal-fired power plant model developed using the gPROMS [6] simulation software presented by Oko et al.[7]. The model consists of a complete power plant, including the flue gas cycle. The accuracy of the model was improved by incorporating first principles modelling equations into the predefined components. The study focussed on identifying optimal load change implementation methods.
- A 750MW anthracite coal-fired power plant developed using the APROS [8] process simulation software presented by Starkloff et al.[9]. The model consists of a complete power plant that includes different furnace firing levels. APROS includes all power plant components and their associated control schemas. The study focussed on power plant operational flexibility.

The dynamic coal-fired power plant models described above were validated against either steady-state or dynamic real plant data and good results were obtained. The models could be used to simulate different operating conditions and evaluate the performance of each plant component under dynamic simulation scenarios. However, the focus was not on load loss analysis or classification.

Deep learning has the potential to provide a reliable approach to the analysis of thermal power plant load losses. However, it requires large training datasets. In the case of load loss analysis, a combination of real plant data and load loss classification data is required. Due to the sparseness of load loss classification data, this study proposes the development of an Integrated Power Plant Process Model (I3PM) for the controlled generation of load loss classification datasets. The objective is therefore the development of a model that enables simulation of the complete power plant operation to determine power output as a function of any given set of internal and external operational variables, boundary conditions and component states. This will enable the generation of model-based load loss classification data that can be combined with real plant data for the training of deep neural networks.

The integrated model development employs an operational plant mirroring framework based on the original plant design, combined with component level fundamental thermofluid physics models in the Flownex Simulation Environment [10].

2 Model development

A South African +600MW lignite coal-fired power plant was used as the design base and for validation of the I3PM. The real power plant design specifications, construction drawings, acceptance tests, c-schedules, control schemes and condition monitoring data were used in the development of each plant component. Flownex was selected for the development of the model based on its collective features and fundamental component-level analysis capabilities that are ideally suited for the development of power plant thermofluid components, as illustrated via studies performed by the Applied Thermofluid Process Modelling Research Unit [11].

Flownex is a simulation software package based on a one-dimensional thermofluid network solver that can simulate various steady-state and dynamic thermofluid process scenarios. Flownex solves the fundamental governing equations using an Implicit Pressure Correction Method set inside a segregated solution algorithm. It consecutively solves the governing equations along with additional component specific equations, enabling the user to manipulate various component and solver parameters and/or criteria. Numerous additional physical domain (mechanical, electrical, C&I, chemical, nuclear, etc.) solvers have been stacked on top of the thermofluid network solver, expanding Flownex's capabilities and general application. Additionally, Flownex enables the creation of custom compound components, has an internal C# script component and a Python API enabling the execution of any additional calculations, algorithms and/or logic that may be required. The C# and Python code executions occur at different computation stages [10].

Each Flownex power plant component was first independently developed and calibrated to its unique operational parameters and design specification before systematically being connected to its respective neighbouring component. The complete Flownex I3PM is presented in Figure 1, where power plant areas are grouped by coloured sections. The modelling methodology of each component will be discussed under its specific plant area.

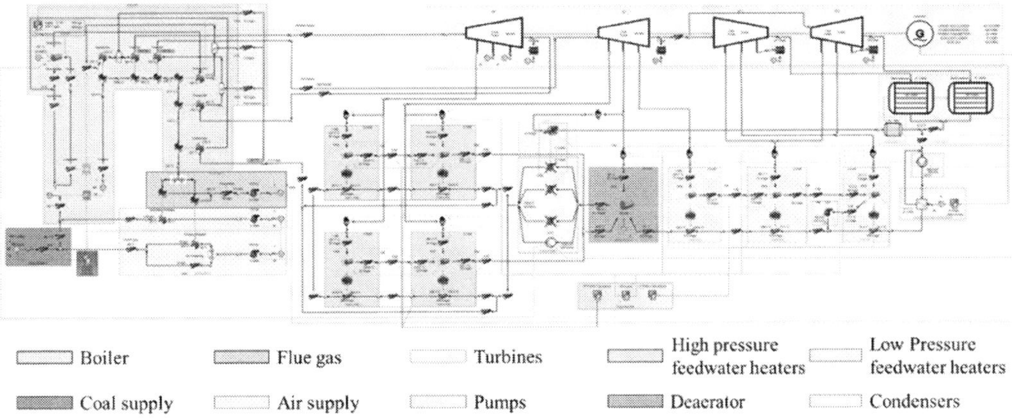

Figure 1: Flownex integrated power plant thermofluid process model.

Since the primary focus of this study is the generation of a load loss classification dataset, the focus was not on the simulation of power plant controls and response to disturbances, as required for dynamic modelling. On the contrary, the I3PM was designed to evaluate the impact of any given set of internal and external operational variables, boundary conditions and component states in steady-state operation before any corrective response could be initialised. This allows for the identification and allocation of load loss to plant areas, systems, or equipment.

2.1 Coal and air supply

The *Coal supply* plant area is comprised of a coal composition and a coal quantity section. The coal composition uses a fluid data reference and boundary condition component that enables the user to define the fuel constituents (coal: carbon, oxygen, ash, water, etc.) and state (temperature, pressure, etc.). In this case, lignite coal is used with a high ash content. The coal quantity is comprised of a coded algorithmic section within a C# script component, the Boiler Supply Control (BSC), which calculates the final coal quantity as a function of the number of active mills, the mass flow rate output of each mill and the performance of each mill. Additionally, the mill power consumption is calculated based on the real power plant mills design. The result of the coal supply plant area is the Pulverized Fuel (PF) condition.

The *Air supply* plant area is comprised of a Primary Air (PA) and a Secondary Air (SA) section. The PA section is comprised of two PA fan components, calibrated to the real PA fan performance curves. Air is fanned through the air side heat exchanger component of the Primary Airheater (PAH) and then through an Air Tempering (AT) path. The AT ensures that the final PA/PF mixture has a temperature below 100°C, which in practice reduces the probability of spontaneous combustion occurring. The ratio of AT to PA air is controlled by the BSC and a compound splitter valve component developed by Fuls [12]. Similarly, the SA section is comprised of two Forced Draught (FD) fans, calibrated to the real FD fan performance curves. Air is fanned through the Secondary Airheater's (SAH) air side heat exchanger component. Both the PAH and SAH air side heat exchanger components are induced with heat, transferred from their respective flue gas side heat exchanger components. The PA and FD fans draw air from a common ambient boundary condition. The fan power consumptions are calculated by their respective fan components.

The integration of the *Coal* and *Air supply* plant areas occur at two network nodes, the first mixes the PA and PF, the second mixes the SA with the PA/PF mixture. The final air/fuel ratio is either calculated by the BSC, based on the real power plant's combustion design, or can be set manually by the user.

2.2 Boiler and flue gas

The modelled power plant uses a natural circulated steam drum boiler design. The *Boiler* plant area is comprised of a combustion section, an evaporator loop section, and a boiler heat exchangers section. The combustion section is comprised of two custom C# script components, developed by the ATProM research unit [11], namely an Adiabatic Flame Temperature (AFT) component and a Furnace component. The AFT takes in the final air/fuel mixture and calculates the energy released and combustion products produced during the combustion process. The furnace component is based on the widely-used Gurvich method [13]. It takes the AFT results and distributes the energy released based on user defined input parameters. These include burner elevation and fuel flow, furnace geometry and characteristics, and the combustion side water wall elevations. Additionally, it assigns the combustion products and remaining energy to the furnace outlet gas. The combustion section allows the user to induce various known boiler furnace faults.

The evaporative loop section includes a downcomer, an evaporator, a header, and a steam drum. The downcomer is comprised of simple pipe components and the number and geometry of downcomer tubes, including their elevations, are user defined input parameters. The inlet to the downcomer receives a combination of saturated liquid from the bottom of the steam drum component and subcooled liquid from the economiser heat exchanger component. This approach ensures a physically realistic analysis where the feed water from the economiser is fed into the steam drum via a manifold situated below the baffle plate.

Similarly, the evaporator uses a pipe component with the number and geometry of furnace wall tubes, including their elevations, as input parameters. The evaporator is fed water from the downcomer component whilst heat is transferred from the Furnace component onto a set of pipes that are discretised along their length, producing a two-phase water/steam mixture at the outlet. This results in the density of the water within the evaporator being lower than that of the downcomer, which induces the natural circulation effect. The header uses a pipe component to merge the evaporator pipes before feeding the two-phase fluid into the steam drum. For the two-phase boiling heat transfer Flownex uses a combination of heat transfer correlations and lookup tables depending on the boiling regime [14].

The steam drum consists of a two-phase tank component that performs the phase separation of the water/steam mixture, based on the real steam drum geometry and elevation. The drum pressure and water level are specified as boundary values since in reality these are actively controlled via inter alia the feedwater flow rate. The drum water level may be specified either as a volume fraction or as a vapour mass fraction (quality). Since the drum pressure is specified the enthalpies of the saturated vapour main steam flow and saturated liquid leaving towards the downcomer are effectively fixed. The recirculation mass flow rate and enthalpy that enters the drum are determined by the performance of the evaporator heat exchanger. This means that if the drum is assumed to be perfectly insulated from the environment, there will be only one possible main steam mass flow rate that will ensure an energy balance over the drum. It is therefore necessary to determine the resultant main steam mass flow rate for each simulation case. This can be achieved with the aid of a built-in iterative “designer” feature that is available for steady-state simulation [15]. The iterative “designer” feature executes multiple steady state simulations while adjusting user defined parameters (control valve position, quality, pressure, etc.) until the desired outcome is achieved. Unfortunately, using this built-in feature was found to be computationally slow. An alternative approach was therefore developed which is discussed in section 2.8.

The boiler heat exchanger section includes a platen superheater, final and primary superheaters, final and primary reheaters, attemperators and an economiser, all consisting of a gas side heat exchanger component and a steam side heat exchanger component. The radiant superheaters absorb the direct radiation from the furnace or the preceding heat exchanger as well as the radiative and convective heat transfer from the flue gas directly surrounding the heat exchanger tubes. The gas side heat exchanger component transfers the heat absorbed from the gas path into the steam path through a connected steam side heat exchanger component. Each heat exchanger was calibrated to it respective design operating conditions.

In line with the real boiler design, attemperators are used to ensure that the calculated tube metal temperatures do not exceed the real plant boiler metal temperatures. The custom attemperator component [12] mixes the required spray water mass flow into the steam flow to achieve the specified steam outlet temperature. The spray water is extracted from the feedwater flow before it enters the economiser.

The economiser is supplied with feedwater from the final high-pressure Feedwater Heaters (FWH). Similar to the superheaters and reheaters the economiser uses a gas side heat exchanger component that transfers heat to the water side heat exchanger component, thereby heating the water to a temperature just below boiling point. The economiser outlet gas connects to the airheaters whilst the outlet water connects to the manifold at the base of the steam drum below the baffle plate.

The complete *Boiler* plant area was developed by first merging the evaporator and combustion sections while ensuring accurate natural circulation simulation under varying furnace operating conditions. This was followed by connecting the boiler heat exchanger section. Various evaporator loop and furnace operating conditions were tested ensuring accurate gas and steam path simulation. Finally, the attemperators were integrated into the model.

The *Flue gas* plant area consists of a gas side airheater section, Electrostatic Precipitator (ESP) component and an Induction (ID) fan component. As the gas is drawn out of the economiser it is split into PAH and SAH paths. The compound splitter valve component is used in combination with the BSC to assign the PAH to SAH gas ratio. The gas side PAH and SAH consists of heat exchanger components and transfer heat as previously described. The PAH and SAH outlet gas paths merge prior to the air being drawn into the ESP by the ID fans.

The ESP consists of a flow restriction component that enables the user to adjust the design calibration parameters allowing the introduction of component faults. The ID fans follow the same methodology as the PA and FD fans. Two ID fans are calibrated based on the actual fan performance curves. The boiler stack consists of a boundary condition component, set at the real boiler stack outlet elevation and ambient conditions.

The complete boiler model was created by first merging the *Flue gas* plant area with the integrated *Coal* and *Air supply* plant areas, before merging it with the *Boiler* plant area.

2.3 Turbines

The modelled power plant has a High-pressure (HP), Intermediate-pressure (IP) and two Low-pressure (LP) turbines that drive the Generator as well as a low-pressure Boiler Feed Pump Turbine (BFPT) that drives the boiler feedwater pump. The HP, IP and LP's were modelled using a compound steam turbine component [12]. The compound turbine component consists of four default simple turbine components that are optimised with a flow correction and an Ellipse-law loss C# script. The rotational speed and process design parameters of the real steam turbines, including the steam extraction designs, are used to develop the turbine train model. Additionally, a default labyrinth seal component is used to model the turbine gland seals. A default LP turbine component was used to model the BFPT. The complete turbine train model was obtained by integrating the various turbine and gland seal components.

The generator was modelled using a C# script component that determines the gross and net power production. The generator component collects the power production and consumption from each component, via data transfer links, thus capturing the holistic power distribution critical to the accuracy of the generator component calculations.

2.4 Pumps

Numerous pumps and booster pumps exist on a real power plant and for simplification only four pump components are included in the I3PM. These are the Condensate Extraction Pump (CEP), the low-pressure FWH Distillate Pump (FWHDP), the Boiler Feed Pump (BFP) that is driven by the BFPT and the two 50% Electrical Feed Pumps (EFP). The CEP consists of a compound pump component [12] that ensures accurate pressure balance through all the low-pressure FWH's and that the required pressure is obtained in the deaerator.

The FWHDP ensures that the distillate, which cascades down through all the low-pressure FWHs, is pumped into the main feedwater path. A simple pump component was used for the development of the FWHDP. A default variable speed pump component was used to develop the BFP and EFPs. The power consumption of the BFP is matched with the power output of the BFPT, including mechanical losses, while the power consumption of the EFPs is determined by its performance at the specified set point. All pumps were calibrated against their real power plant process design parameters and pump curves.

2.5 Low- and high-pressure feedwater heaters

The modelled power plant has six FWH stages. There are three low-pressure FWHs and four high-pressure FWHs with the high-pressure FWH split into two banks, namely paths A and B. The development of the FWH components expands on the Flownex feedwater heater modelling methodology presented by le Grange [16]. Each FWH consists of a two-phase tank component that represents the shell side, a pipe component that represents the tube bundle, and a heat transfer component that represents the convection-conduction-convection heat transfer path between the turbine bled-steam mixed with cascaded distillate, and the feedwater.

The tube bundle is modelled in a similar manner as the evaporator component, whilst the FWH shell follows the two-phase tank methodology described for the boiler steam drum. An added capability that the heat transfer component brings is the ability to model the level of fouling inside the FWH. Additionally, a low- and high -pressure FWH performance C# script component is used to determine the FWH's terminal temperature difference (TTD) and drain cooler approach (DCA).

2.6 Deaerator

The modelled power plant deaerator has a stork spray feedwater tank configuration that uses the mixing process of turbine bled-steam, high-pressure cascaded distillate, and feedwater to remove the non-condensable gasses from the water/steam cycle. The deaerator tank was modelled by using the two-phase tank methodology described for the boiler steam drum. The ventilation of the non-condensable gases is not included.

The *Low-pressure feedwater heaters* and *Deaerator* plant areas, along with the FWH distillate pump, were merged into a low-pressure feedwater train. The *High-pressure feedwater heaters* and the Boiler/Electrical Feed Pumps were merged into a high-pressure feedwater train. Both low- and high-pressure feedwater trains were tested under various operating conditions before being integrated into a complete feedwater train model.

2.7 Condensers

The *Condenser* plant area consists of two main condensers that extract steam from the LP turbine, an auxiliary condenser that extracts steam from the BFPT, and a gland steam condenser (GSC) that extracts the steam from all the turbine gland seals. A compound condenser component [12] was used to model the main and auxiliary condensers. The condensers are calibrated against the real condenser process design and enables the user to manipulate the back pressure with ease. The compound condenser component enforces a saturated liquid outlet flow, ensuring a complete phase change of the extracted turbine steam. As C# script component is used to ensure the mass and energy balance of the component.

A GSC is more accurately described as a low-pressure noncontact feedwater heater that discharges distillate to the condenser via a flash box. Therefore, the GSC model consists of a compound gland steam condenser component [12] that discharges distillate from the steam side into the condensate path through a common condenser flash box modelled with a flow restriction component.

The main condensers merge into a single condensate path before being connected to the condenser flash box. The flash box connects the outlet at the top of the auxiliary condensers to the discharge at the base of the GSC. The condensate is then pumped from the condenser flash box into the GSC condensate side by the CEP, which completes the condensate train model.

2.8 Operational plant mirroring framework

An operational plant mirroring framework was used in the development of the I3PM which consists of a systematic model development methodology that approximates the construction and commissioning of a power plant. The various component models are developed (manufactured) before being merged (assembled) into plant area sections. Plant area sections are tested (inspected) before being integrated into a complete plant area model. What remains is the integration of all plant area models. The complete boiler model is merged with the feedwater train model simulating the feedwater supply to the boiler, whilst the turbine train model is merged with the condensate train model enabling the simulation of the power production. This is followed by connecting the boiler to the turbines, the turbines' bled-steam extractions to their respective FWH's and connecting the condensate to the feedwater. The final model development step involves connecting the generator to all the relevant power producing and consuming components.

The mirroring framework next imposes a systematic control methodology that approximates the control of a real power plant. An I3PM control methodology is required to ensures complete mass, energy, and momentum balance throughout the cycle for any given set of steady-state input parameters, specifically on the two-phase tank components. The control methodology development was informed by the hierarchical power plant control approaches described by Busa et al. [17]. A custom-built iterative C# – Python control method was developed.

Two C# script components (drum control and feedwater control) were developed to balance the mass, energy, and momentum of the two-phase tank components used for the modelling of the steam drum, deaerator and FWH's. The C# scripts evaluate any imbalance that may occur within a two-phase tank and corrects by adjusting a corresponding control valve. The size of the correction is based on a scaled thermofluid process balancing calculation. Like the default "designer" feature described earlier, multiple iterations are required to reach each balance point with a sufficiently small residual value.

The Python API runs multiple steady-state simulations while allowing the C# script components to balance the model using a linear programming optimization algorithm. The computational time that already outperformed the "designer" feature was further improved by implementing a hierarchical power plant control approach, using a multilevel optimization algorithm that first balances the steam drum, then the deaerator, followed by the FWH's.

2.9 Data generation

The Python API is used for the controlled generation of load loss classification datasets. The API enables the altering of any I3PM parameter, which allows for the simulation of a wide range of operational anomalies and different states of component degradation. A list of load loss inducing I3PM parameters was derived from mechanical and process anomalies identified in historic load loss event manual root cause analysis data.

A Monte-Carlo (MC) algorithm was developed for the induction of I3PM component level faults that will result in a load loss. The MC algorithm assigns new randomly generated values, that lie within the operational and component design limits, to selected component parameters. This means that the values will not force a scenario onto the I3PM that would cause the real plant to trip or result in a severe component failure.

After the MC algorithm induces faults, and a simulation has converged, the results are stored and the I3PM is reset to a default state. The stored results are comprised of I3PM parameters that correspond to existing plant condition monitoring measurements as well as the specific MC algorithm configuration, enabling the classification of the generated data.

3 Model validation

To evaluate the accuracy and appropriateness of the results produced by the I3PM different steady state validation simulations were conducted. The boundary conditions were configured to simulate 60%, 80% and 100% MCR in accordance with the real power plant acceptance test data. Additionally, 69% and 97% MCR were simulated to specifically evaluate the boiler gas path against the manufacturers' design data. The validation temperature results are summarised in Figure 2.

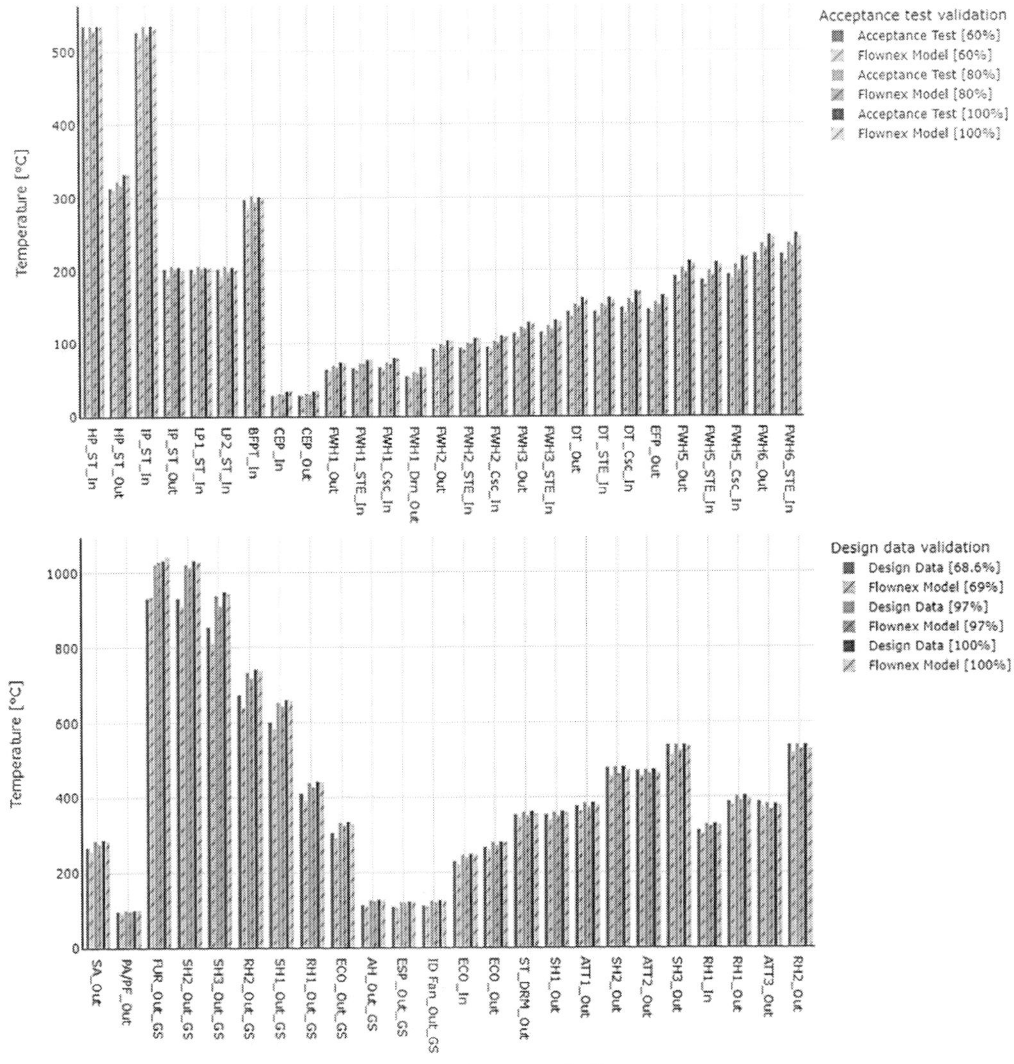

Figure 2: Integrated power plant thermal process model validation results

Only measurable parameters (temperature, pressure, and mass) were used for the validation of the model to align it more closely to the real power plant. As shown in Figure 2 the results from the validation simulations are all within 5% compared to the acceptance test data and a relative error of less than 8% when compared to the boiler manufacturer's design data.

The validation results established confidence in the ability of the I3PM to predict the performance of a real power plant with sufficient accuracy based on a component level configuration.

4 Data generation results

To demonstrate the ability of the I3PM to capture the effect of different combinations of anomalies it was used to evaluate the plant performance related to three specific load loss inducing scenarios. These are the use of low-grade coal, poor condenser performance and the plugging of feedwater tubes. The specific plant parameters used to represent these scenarios are a reduction in the Gross Calorific Value of the coal, an increase in the back pressure in the main condenser and an increase in the number of plugged tubes in the second low-pressure FWH, respectively. First, each of the three parameters were perturbed independently via the MC algorithm and for each case the I3PM simulation was run to determine the impact on the overall power output and cycle thermal efficiency. Following this, random combinations of any two of the three parameters were tested, and finally random combinations of all three of the parameters.

4.1 Single parameter scenarios

The results of the single parameter scenarios are presented in Figure 3. The figure shows each of the three parameter values on the y-axes together with the resultant cycle thermal efficiency on the x-axis, while the colour coding indicates the resultant load loss compared to the original design specification. The results show that the plugging of tubes has the least severe impact on plant performance, whilst poor coal quality has the greatest. Poor condenser performance primarily affects the plant efficiency. The results are aligned with expectations.

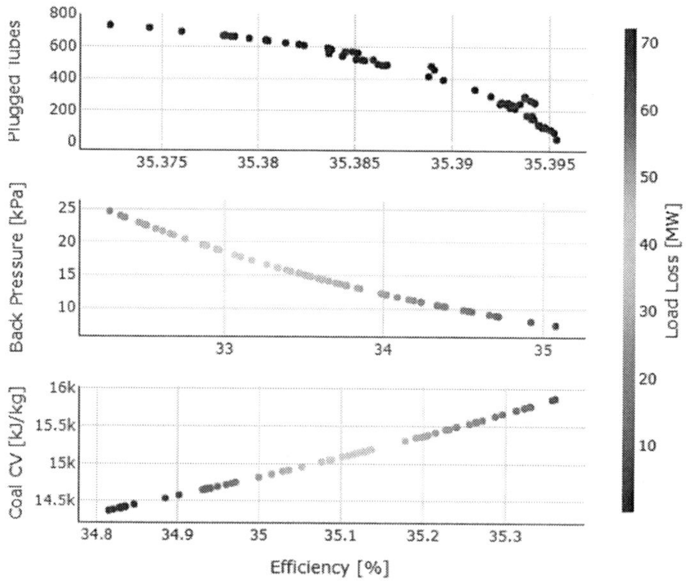

Figure 3: Single parameter scenarios

4.2 Dual parameter scenarios

The results of the dual parameter scenarios are presented in Figure 4 with parameter 1 on the x-axis and parameter 2 on the y-axis, while the colour coding indicates the magnitude of the resultant load loss. The combination of poor coal quality and poor condenser performance has the greatest impact on plant performance. The relative insignificance of the number of plugged tubes is visible on both the top and bottom graphs, since there is very little variation in the load loss in the y-axis direction. All results are again aligned with expectations.

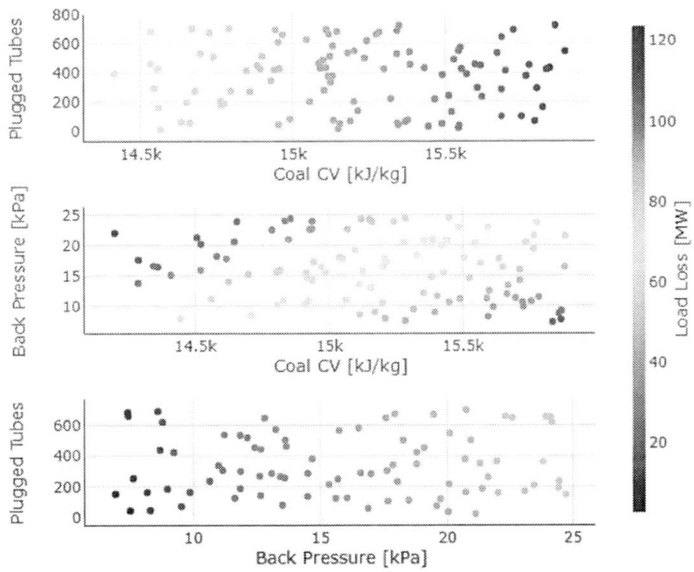

Figure 4: Dual parameter scenarios

4.3 Triple parameter scenarios

The results of the triple parameter scenarios are presented in Figure 5 with the three parameters represented on the x, y and z-axes respectively, while the colour coding indicates the magnitude of the resultant load loss. As expected, the results show that the combination of poor coal quality and poor condenser performance still has the greatest impact on plant performance relative to the number of plugged tubes. However, it also shows that the simultaneous impact of a combination of parameters can be determined, which will be required in order to emulate physically realistic load loss scenarios.

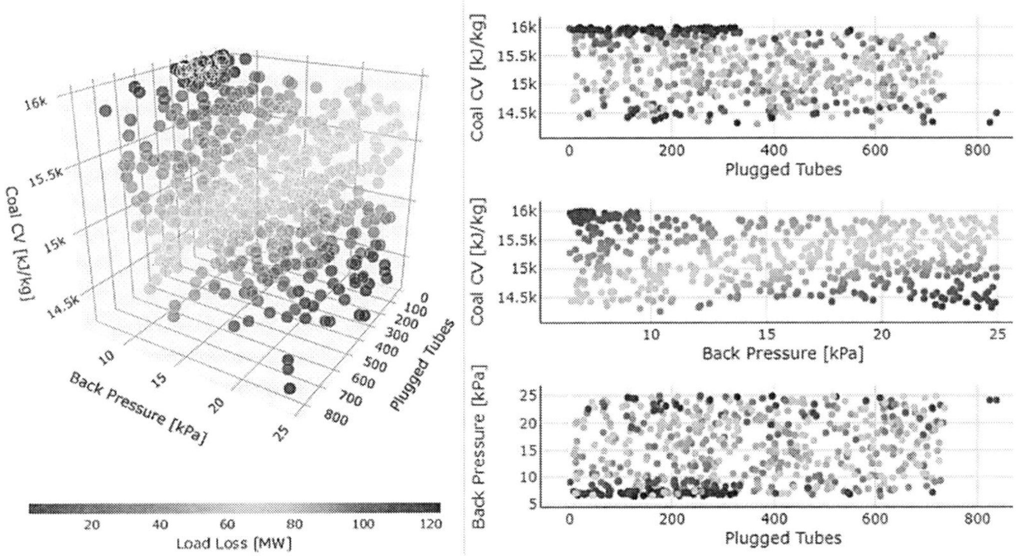

Figure 5: Triple parameter scenarios

5 Conclusion

In this study an Integrated Power Plant Thermofluid Process Model (I3PM) was developed for the controlled generation of load loss classification datasets. The validation showed that the model can successfully emulate real power plant operating conditions. The results from the single and multiple parameter load loss scenarios show that the simultaneous impact of a combination of parameters can be determined, which will be required in order to emulate physically realistic load loss scenarios. The model further establishes its suitability and capability to generate load loss classification data.

References

[1] VGB PowerTech Members, "Fundamentals and systematics of availability determination for Thermal Power Plants," in *VGB PowerTech - Terms of utility industry*, 7th ed., vol. Part B-P, VGB PowerTech e.V., 2008, pp. 0–131.

[2] G. F. M. de Souza, Ed., *Thermal Power Plant Performance Analysis. Springer series in reliability engineering*. Springer International Publishing, 2012.

[3] F. Alobaid, N. Mertens, R. Starkloff, T. Lanz, C. Heinze, and B. Epple, "Progress in dynamic simulation of thermal power plants," *Prog. Energy Combust. Sci.*, vol. 59, pp. 79–162, 2016.

[4] "ThermoPower | An open-source Modelica library." https://build.openmodelica.org/Documentation/ThermoPower.html.

[5] S. Meinke, F. Gottelt, M. Müller, E. Hassel, X. R. G. Simulation, and H. Schlossstr, "Modeling of Coal-Fired Power Units with ThermoPower Focussing on Start-Up Process," *Proc. 8th interna- tional Model. Conf.*, pp. 353–364, 2011.

[6] "gPROMS | A Siemens Process Systems Engineering product." https://www.psenterprise.com/products/gproms.

[7] E. Oko and M. Wang, "Dynamic modelling, validation and analysis of coal-fired subcritical power plant," *Fuel*, vol. 135, pp. 292–300, 2014.

[8] "APROS | Dynamic Process Simulation Software for Nuclear and Thermal Power Plant Applications." https://www.apros.fi/.

[9] R. Starkloff, F. Alobaid, K. Karner, B. Epple, M. Schmitz, and F. Boehm, "Development and validation of a dynamic simulation model for a large coal-fired power plant," *Appl. Therm. Eng.*, vol. 91, pp. 496–506, 2015.

[10] "Flownex Simulation Environment | Precise 1D simulation." https://flownex.com/.

[11] "Applied Themofluid Process Modelling Research Unit," *University of Cape Town, Department of Mechanical Engineering*. http://www.atprom.uct.ac.za/.

[12] F. Wim, "Simple Rankine cycle components," *ATProM Flownex Database*, no. July. University of Cape Town, Department of Mechanical Engineering, 2017.

[13] A. G. Blokh, *Heat transfer in steam boiler furnaces*. Berlin: Springer-Verlag, 1988.

[14] Flownex SE, "Flownex Theory Manual," no. December, 2020.

[15] Flownex, "Flownex Library Theory Manual," no. December, 2020, [Online]. Available: www.flownex.com.

[16] W. Le Grange, "Component development for a high fidelity transient simulation of a coal-fired power plant using Flownex SE," University of Cape Town, 2018.

[17] S. Basu and A. K. Debnath, *Power Plant Instrumentation and Control Handbook | A Guide to Thermal Power Plants*, Second. Elsevier Academic Press, 2019.

Analytical first derivatives of the direct Hermite-Simpson collocation formulation of optimal control problems

Aravind Arunakirinathar[1,*], *Jean-Francois de la* Beaujardiere[2], and *Michael* Brooks[2]

[1]Graduate Student,Mechanical Engineering, University of KwaZulu-Natal, 4041,
[2]Mechanical Engineering, University of KwaZulu-Natal, 4041

Abstract. In order to assess the capabilities of South Africa as a launch site for commercial satellites, an optimal control solver was developed. The developed solver makes use of direct Hermite-Simpson collocation methods, and can be applied to a general optimal control problem. Analytical first derivative information was obtained for direct Hermite-Simpson collocation methods. Typically, a numerical estimate of the derivative information is used. This paper will present the solver algorithm, and the formulation and derivation of the analytical first derivative information for this approach. A linear and non-linear sample problem are provided as validation of the solver.

1 Introduction

An optimal control solver making use of the direct Hermite-Simpson collocation formulation of optimal control problems was developed. Formulae for determining the analytical first derivative information of the defect constraints of the direct Hermite-Simpson collocation formulation of optimal control problems have been formulated. The control variables were represented by linear mean splines. Irrespective of this, the first derivative formulae obtained can be extended to other representations of the control variables. The formulae obtained can also be extended to solving ordinary two-point boundary value problems with Hermite-Simpson collocation.

Typically, these first derivatives are estimated using sparse finite-differencing, alternatively they can be obtained with automatic differentiation or symbolic differentiation. The formula provided allows for an easy and compact representation and calculation of analytical first derivatives. Provision of analytical derivative information, versus estimating this information, typically increases the robustness and accuracy of solvers using Newton-iterative techniques [1]. Symbolic differentiation algorithms tend to not present derivatives in compact forms, which has the potential to significantly increase computation time. Forward automatic differentiation is easy to use, but inefficient for large problems, whereas reverse automatic differentiation requires significant effort to implement, but is efficient for large problems [2].

Optimal control is a field of study with various application in fields such as aerospace, robotics and chemical engineering. It is concerned with the determination of trajectories which optimise some cost function of a given dynamic system. The techniques of optimal control have been used, successfully, to optimise a host of dynamic system trajectories. These

*This project was supported by the National Research Foundation (NRF).This is much appreciated.
e-mail: aravind519@yahoo.com,

trajectories include: minimum fuel orbital insertion trajectories, re-entry trajectories and minimum energy robot trajectories.

This paper first outlines the general transcription process of an optimal control problem to an Nonlinear Programming (NLP) problem using Hermite-Simpson collocation with linear mean control splines. The transcription process presented will assume the period of any phase of the state variables trajectory is a free variable. The derivation of the first derivative information of this formulation of the optimal control problem is then presented. Following which the first derivatives will be applied with example problems, using Matlab's Sequential Quadratic Programming (SQP) algorithm.

2 Problem Formulation

2.1 Transcription

Optimal control is the field of study concerned with determining the control history of a dynamic system which optimises a performance index or cost function. The cost can be a function of terminal values of the state variables or an integral function of the system, or both. The following summarises the optimal control problem, with terminal constraints to the state variables, and some of the state variables initial values defined. The systems analysed will be autonomous. It should be noted that other constraints such as path constraints can also be considered.

$$\min_{\mathbf{u}} \phi(\mathbf{x}_f, t_f) + \int_{t_0}^{t_f} L(\mathbf{x}(t), \mathbf{u}(t))dt \tag{1}$$

subject to

$$\begin{aligned} (\mathbf{x}_f, t_f) &= 0 \quad \text{(end conditions)} \\ \dot{\mathbf{x}} &= g(\mathbf{x}(t), \mathbf{u}(t)) \end{aligned}$$

Where $\mathbf{x}_f$ is the terminal state vector, t_f is the unknown final time of the phase of the trajectory. $\mathbf{x} \in \mathbb{R}^{m\times 1}$ and $\mathbf{u} \in \mathbb{R}^{n\times 1}$ are the state and control vector respectively. The functions $g : \mathbb{R}^{(m+n)\times 1} \rightarrow \mathbb{R}^{m\times 1}$, outputs the first time derivative of the state vector, $\phi : \mathbb{R}^{(m+1)\times 1} \rightarrow \mathbb{R}^{m\times 1}$, outputs the component of the cost function which is dependent on terminal values of the system, and the function $L : \mathbb{R}^{(m+n)\times 1} \rightarrow \mathbb{R}^{m\times 1}$, outputs the integrand of the integral component of the cost function. $\psi : \mathbb{R}^{(m+n)\times 1} \rightarrow \mathbb{R}^{m\times 1}$, outputs the component of terminal boundary constraints, which is dependent on terminal values of the system.

In the event the final time is unknown, the following augmentation to the problem may be performed in order to allow for numerical solutions to be obtained.

Let $t = t_f\tau$ where $\tau \in [0, 1]$ then $\frac{dt}{d\tau} = t_f$.
Let there be a function $f : \mathbb{R}^{(m+n+1)\times 1} \rightarrow \mathbb{R}^{m\times 1}$, which outputs the first derivative of the state vector with respect to the new independent variable τ. That is

$$f(\mathbf{x}(\tau), \mathbf{u}(\tau), t_f) = \frac{d\mathbf{x}}{d\tau} \tag{2}$$

Then by the chain rule $f(\mathbf{x}(\tau), \mathbf{u}(\tau), t_f) = t_f g(\mathbf{x}(\tau), \mathbf{u}(\tau))$

The cost (or objective) function could, for example, be formulated in a manner so as to minimise the fuel usage of a particular rocket mission. The dynamic equations, along with any path constraints, essentially constrain the feasible solutions of the optimal objective function. In order to solve this problem using an NLP, the problem is converted, or transcribed, by Hermite-Simpson collocation into an NLP problem. This is achieved by discretising the independent variable of the system (typically time), in order to create a grid, or mesh. Each grid point of this grid represents discrete values of the dependent variables, in the case of optimal control these variables are a state and control vector. The differential equations which represent the system are replaced with defect constraints for each independent variable interval, which relate the dependent variables of adjacent grid points by some suitable quadrature. In the case of this paper, Hermite-Simpson is the quadrature used.

C. Rosmann et al. [3] presented time-optimal solutions using direct Hermite-Simpson collocation. The solutions which used control splines with a free midpoint led to oscillations in the time-optimal solutions. The solutions which used linear mean control splines had no such oscillations. Since such oscillations are undesirable in practical application, and the ultimate goal of this solver is to deal with phases with time as a free variable, the control variables were represented using linear mean control splines.

Let $\mathbf{c}_k \in \mathbb{R}^{m\times 1}$ represent the collocation constraint vector at a grid point $k+1$. Let there be N grid points. Then at a particular grid-point k: for $\mathbf{x}_k \in \mathbb{R}^{m\times 1}, \mathbf{u}_k \in \mathbb{R}^{(m+n)\times 1}$

$$\bar{\mathbf{z}}_k = \mathbf{I}_{m+n}\mathbf{z}_{k+1} + \frac{h_k}{8}\begin{bmatrix} \mathbf{f}_k - \mathbf{f}_{k+1} \\ \mathbf{0}_{n\times 1} \end{bmatrix} \tag{3}$$

where

$$\mathbf{z}_k = \begin{bmatrix} \mathbf{x}_k \\ \mathbf{u}_k \end{bmatrix} \in \mathbb{R}^{(m+n)\times 1} \quad \text{and} \quad \mathbf{f}_k = \mathbf{f}(\mathbf{z}_k, t_f).$$

The transcribed form of the optimal control problem, with terminal constraints to the state variables is now presented as follows:

$$\min_{\mathbf{u}_k, \forall k} \phi(\mathbf{x}_f, t_f) + \sum_{k=1}^{N-1} \frac{h_k}{6}\Big(L(\mathbf{x}_k, \mathbf{u}_k, t_f) + 4L(\bar{\mathbf{x}}_k, \bar{\mathbf{u}}_k, t_f) + L(\mathbf{x}_{k+1}, \mathbf{u}_{k+1}, t_f)\Big) \tag{4}$$

subject to

$$\begin{aligned} (\mathbf{x}_f, t_f) &= 0 \quad \text{(end conditions)} \\ \mathbf{c}_k &= \mathbf{I}_m\mathbf{x}_k - \mathbf{I}_m\mathbf{x}_{k+1} - \tfrac{h_k}{6}(\mathbf{f}_k + 4\bar{\mathbf{f}}_k + \mathbf{f}_{k+1}) = \mathbf{0}, \quad \text{for all } k \end{aligned}$$

Here,$\mathbf{c}_k$ is referred to as a defect constraint. In the event that there are multiple trajectory phases, the following constraints must be imposed:

$$\mathbf{x}_f^i = \mathbf{x}_0^{i+1} \tag{5}$$

where the superscript i indicates the phase, and $\mathbf{x}_0^i$ indicates the initial state vector of phase i.

This new problem can be posed to an NLP solver in order for a solution to be obtained. It is often of importance to scale variables in order to improve robustness of an NLP solver [4]. In this work, state variables were scaled manually as the author saw fit.

2.2 Warm Starts

The developed optimal control solver will make use of warm starts. That is, obtaining the solution of simpler versions of the actual problem, in order to initialise the solver when attempting to obtain the solution of the actual problem. This provides solvers with good initial guesses. Provision of good initial guesses to Newton-iterative solvers increases speed and robustness of the solver, hence it is advantageous to make use of warm starts [5].

In this work, warm starts will be taken advantage of by obtaining a solution for a coarse grid. An estimate of this solution's discretisation error will be tested, if this estimate of the discretisation error is above a defined tolerance, the solver will make use of the coarse grids solution to determine an initial guess for a finer grid. The solver will then obtain a solution for this new grid, and the process will repeat.

2.3 Choice of Non-Linear Programming

The two popular NLP algorithms for solving direct optimal control problems are SQP and Interior point methods. SQP algorithms have the capacity to make use of warm starts. Thus the SQP algorithm can take advantage of a re-meshing strategy.

Interior point methods tend to increase the non-linearity of functions, which can present difficulties for already highly non-linear systems. Each iteration of the Interior point method must be strictly feasible, this prevents it from using warm starts. Hence interior point methods are unsuitable if a re-meshing strategy is to be used [6].

The class of problems the developed optimal control solver is intended to solve are highly non-linear. In addition to this, it is very desirable to take advantage of warm starts, hence a SQP algorithm was chosen.

2.4 First Derivative Information

The following section presents the analytical first derivative information of the defect constraints of the direct Hermite-Simpson collocation method, with linear mean control splines. In addition to this, first derivative information of the integral component of the cost function is provided.

For a particular phase of the trajectory:

Let

$$\bar{\mathbf{f}}_k = \mathbf{f}(\bar{\mathbf{z}}_k, t_k), \quad \text{and} \quad \mathbf{J}_k = \mathbf{J}_{\mathbf{f}_k}(\mathbf{z}_k), \bar{\mathbf{J}}_k = \mathbf{J}_{\bar{\mathbf{f}}_k}(\bar{\mathbf{z}}_k)$$

Then the first derivative information of the defect constraint of a particular grid point is represented as follows:

$$\begin{bmatrix} \left(\frac{\partial \mathbf{c}_k}{\partial \mathbf{z}_k}\right)^T \\ \left(\frac{\partial \mathbf{c}_k}{\partial \mathbf{z}_{k+1}}\right)^T \end{bmatrix} = \begin{bmatrix} \mathbf{I}_m \\ \mathbf{O}_{n\times m} \\ \mathbf{I}_m \\ \mathbf{O}_{n\times m} \end{bmatrix} - \frac{h_k}{6}\begin{bmatrix} \left(\frac{\partial \mathbf{f}_k}{\partial \mathbf{z}_k}\right)^T \\ \left(\frac{\partial \mathbf{f}_{k+1}}{\partial \mathbf{z}_{k+1}}\right)^T \end{bmatrix} - \frac{2h_k}{3}\left(\frac{\partial \bar{\mathbf{z}}_k}{\partial \mathbf{p}_k}\right)^T\left(\frac{\partial \bar{\mathbf{f}}_k}{\partial \bar{\mathbf{z}}_k}\right)^T \tag{6}$$

$$= \begin{bmatrix} \mathbf{I}_m \\ \mathbf{O}_{n\times m} \\ \mathbf{I}_m \\ \mathbf{O}_{n\times m} \end{bmatrix} - \frac{h_k}{6}\begin{bmatrix} \left(\mathbf{J}_k\right)^T \\ \left(\mathbf{J}_{k+1}\right)^T \end{bmatrix} - \frac{2h_k}{3}\left\{\begin{bmatrix} \mathbf{I}_{m+n} \\ \mathbf{I}_{m+n} \end{bmatrix} + \begin{bmatrix} \left(\mathbf{J}_k\right)^T & \mathbf{O}_{n\times(m+n)} \\ \left(\mathbf{J}_{k+1}\right)^T & \mathbf{O}_{n\times(m+n)} \end{bmatrix}\right\}\left(\bar{\mathbf{J}}_k\right)^T$$

$$\begin{aligned} \left(\frac{\partial \mathbf{c}_k}{\partial t_f}\right)^T &= \frac{-h_k}{6}\left(\left(\frac{\partial(\mathbf{f}_k + 4\bar{\mathbf{f}}_k + \mathbf{f}_{k+1})}{\partial t_f}\right)^T + 4\left(\frac{\partial \bar{\mathbf{z}}_k}{\partial \mathbf{t}_k}\right)^T\left(\frac{\partial \bar{\mathbf{f}}_k}{\partial \bar{\mathbf{z}}_k}\right)^T\right) \\ &= \frac{-h_k}{6}\left(\left(\frac{\partial(\mathbf{f}_k + 4\bar{\mathbf{f}}_k + \mathbf{f}_{k+1})}{\partial t_f}\right)^T + \frac{h_k}{2}\begin{bmatrix} \left(\frac{\partial(\mathbf{f}_k - \mathbf{f}_{k+1})}{\partial t_f}\right)^T \\ \mathbf{0}_{n\times m} \end{bmatrix}^T \left(\bar{\mathbf{J}}_k\right)^T\right) \end{aligned} \tag{7}$$

If the system is autonomous (not explicitly dependent on the independent variable):

$$\left(\frac{\partial \mathbf{c}_k}{\partial \mathbf{z}_k}\right)^T = \frac{-h_k}{6}\left(\left(\frac{\partial(\mathbf{f}_k + 4\bar{\mathbf{f}}_k + \mathbf{f}_{k+1})}{\partial t_f}\right)^T + \frac{h_k}{2}\begin{bmatrix} \left(\frac{\partial(\mathbf{f}_k - \mathbf{f}_{k+1})}{\partial t_f}\right)^T \\ \mathbf{0}_{n\times 1} \end{bmatrix}^T \left(\bar{\mathbf{J}}_k\right)^T\right) \tag{8}$$

Let

$$\mathbf{y}_1 = [\mathbf{z}_1^T, \mathbf{z}_2^T, \ldots, \mathbf{z}_N^T, t_f]^T \in \mathbb{R}^{(N(m+n)+1)\times 1}, \quad \text{and } \mathbf{c} = [\mathbf{c}_1^T, \mathbf{c}_2^T, \ldots, \mathbf{c}_{N-1}^T]^T \in \mathbb{R}^{(N-1)m\times 1}$$

The transpose of the Jacobian of the nonlinear constraints is then

$$\left(\mathbf{J}_{\mathbf{y}_1}(\mathbf{c})\right)^T = \left|\begin{array}{ccccc} \left(\frac{\partial \mathbf{c}_1}{\partial \mathbf{z}_1}\right)^T & \mathbf{0}_{(m+n)\times m} & \mathbf{0}_{(m+n)\times m} & \cdots & \mathbf{0}_{(m+n)\times m} \\ \left(\frac{\partial \mathbf{c}_1}{\partial \mathbf{z}_2}\right)^T & \left(\frac{\partial \mathbf{c}_2}{\partial \mathbf{z}_2}\right)^T & \mathbf{0}_{(m+n)\times m} & \cdots & \mathbf{0}_{(m+n)\times m} \\ \mathbf{0}_{(m+n)\times m} & \left(\frac{\partial \mathbf{c}_2}{\partial \mathbf{z}_3}\right)^T & \left(\frac{\partial \mathbf{c}_3}{\partial \mathbf{z}_3}\right)^T & \cdots & \mathbf{0}_{(m+n)\times m} \\ \mathbf{0}_{(m+n)\times m} & \mathbf{0}_{(m+n)\times m} & \left(\frac{\partial \mathbf{c}_3}{\partial \mathbf{z}_4}\right)^T & \ddots & \vdots \\ \vdots & \vdots & \vdots & \ddots & \left(\frac{\partial \mathbf{c}_{N-1}}{\partial \mathbf{z}_{N-1}}\right)^T \\ \mathbf{0}_{(m+n)\times m} & \mathbf{0}_{(m+n)\times m} & \mathbf{0}_{(m+n)\times m} & \cdots & \left(\frac{\partial \mathbf{c}_{N-1}}{\partial \mathbf{z}_N}\right)^T \\ & & \sum_{k=1}^{N-1}\left(\frac{\partial \mathbf{c}_k}{\partial t_f}\right)^T & & \end{array}\right| \tag{9}$$

The gradient of the integral component of the cost function for a particular phase is as follows:
Let

$$L_k = L(\mathbf{z}_k, t_f), \text{ and } \bar{L}_k = L(\bar{\mathbf{z}}_k, t_f)$$

and let

$$\mathbf{G}_k = \left(\frac{\partial L_k}{\partial \mathbf{z}_k}\right)^T; \bar{\mathbf{G}}_k = \left(\frac{\partial \bar{L}_k}{\partial \bar{\mathbf{z}}_k}\right)^T$$

Then define $C_k = \frac{h_k}{6}\left(L(\mathbf{x}_k, \mathbf{u}_k, t_f) + 4L(\bar{\mathbf{x}}_k, \bar{\mathbf{u}}_k, t_f) + L(\mathbf{x}_{k+1}, \mathbf{u}_{k+1}, t_f)\right)$, and thus

$$\nabla_{\left[\begin{array}{c}\mathbf{z}_k \\ \mathbf{z}_{k+1}\end{array}\right]} C_k = \frac{h_k}{6}\left(\left[\begin{array}{c}\mathbf{G}_k \\ \mathbf{G}_{k+1}\end{array}\right] + 4\left(\frac{\partial \bar{\mathbf{z}}_k}{\partial \mathbf{t}_k}\right)^T\left(\frac{\partial \bar{\mathbf{f}}_k}{\partial \bar{\mathbf{z}}_k}\right)^T\right) \tag{10}$$

$$= \frac{h_k}{6}\left(\left[\begin{array}{c}\mathbf{G}_k \\ \mathbf{G}_{k+1}\end{array}\right] + 4\left(\left[\begin{array}{c}\mathbf{I}_{m+n} \\ \mathbf{I}_{m+n}\end{array}\right] + \left[\begin{array}{cc}(\mathbf{J}_k)^T & \mathbf{O}_{n\times(m+n)} \\ (\mathbf{J}_{k+1})^T & \mathbf{O}_{n\times(m+n)}\end{array}\right]\right)\bar{\mathbf{G}}_k\right) \tag{11}$$

Now, we define

$$\mathbf{H}_k = \begin{cases} \left[\begin{array}{c}\nabla_{\left[\begin{array}{c}\mathbf{z}_1 \\ \mathbf{z}_2\end{array}\right]} C_1 \\ \mathbf{0}_{N-2m+1}\end{array}\right] & \text{if } k = 1 \\ \left[\begin{array}{c}\mathbf{0}_{(m+n)(k-1)} \\ \nabla_{\left[\begin{array}{c}\mathbf{z}_k \\ \mathbf{z}_{k+1}\end{array}\right]} C_k \\ \mathbf{0}_{(m+n)(N-k-1)+1}\end{array}\right] & \text{if } 1 < k \le N-1 \end{cases} \tag{12}$$

Then:

$$\frac{\partial C_k}{\partial t_f} = \sum_{k=1}^{N-1}\frac{h_k}{6}\left(\frac{\partial(L_k + 4\bar{L}_k + L_{k+1})}{\partial t_f} + \frac{h_k}{2}\left(\frac{\partial \bar{\mathbf{z}}_k}{\partial t_f}\right)^T\left(\frac{\partial \bar{L}_k}{\partial \bar{\mathbf{z}}_k}\right)^T\right)$$

$$= \sum_{k=1}^{N-1}\frac{h_k}{6}\left(\frac{\partial(L_k + 4\bar{L}_k + L_{k+1})}{\partial t_f} + \frac{h_k}{2}\left[\begin{array}{c}\left(\frac{\partial(\mathbf{f}_k - \mathbf{f}_{k+1})}{\partial t_f}\right)^T \\ \mathbf{0}_{n\times m}\end{array}\right]^T\bar{\mathbf{G}}_k\right)$$

If the system is autonomous, then $\frac{\partial C_k}{\partial t_f}$ becomes:

$$\frac{\partial \mathbf{C}_k}{\partial t_f} = \sum_{k=1}^{N-1}\frac{h_k}{6}\left(\frac{\partial(L_k + 4\bar{L}_k + L_{k+1})}{\partial t_f} + \frac{h_k}{2}\left[\begin{array}{c}\frac{(\mathbf{f}_k - \mathbf{f}_{k+1})}{t_f} \\ \mathbf{0}_{n\times m}\end{array}\right]^T\bar{\mathbf{G}}_k\right) \tag{13}$$

The gradient of C is thus

$$\nabla_{\mathbf{y}_1} C = \sum_{k=1}^{N-1} \mathbf{H}_k + \begin{bmatrix} \mathbf{0}_{(m+n)N} \\ \frac{\partial C}{\partial t_f} \end{bmatrix}$$

In the event that multiple trajectory phases exist, the derivative information is as follows: If t_f^i represents the final time of phase i, $\mathbf{y}_k$ represents a vector of all the unknown variables of phase k, and $\mathbf{c}^i$ represents a vector of the collocation constraints of phase i, and P be the number of phases. Let

$$\mathbf{y} = [\mathbf{y}_1^T, \mathbf{y}_2^T, \ldots, \mathbf{y}_p^T]^T \quad \text{and } \mathbf{cc} = [\mathbf{c}^1, \mathbf{c}^2, \ldots, \mathbf{c}^p]^T$$

Then

$$\left(\mathbf{J}_y(\mathbf{cc})\right)^T = \begin{bmatrix} \left(\mathbf{J}_{y_1}(\mathbf{c}^1)\right)^T & 0 & 0 & 0 \\ 0 & \left(\mathbf{J}_{y_2}(\mathbf{c}^2)\right)^T & 0 & 0 \\ 0 & 0 & \ddots & 0 \\ 0 & 0 & 0 & \left(\mathbf{J}_{y_p}(\mathbf{c}^p)\right)^T \end{bmatrix} \tag{14}$$

and $\nabla_{\mathbf{y}} C = [(\nabla_{\mathbf{y}_1} C)^T, (\nabla_{\mathbf{y}_2} C)^T, \ldots, (\nabla_{\mathbf{y}_p} C)^T]^T$.

2.5 Interpolation of Solution and Error Estimation

In order to make use of coarse grids when using direct collocation, interpolation of the solution is used to both represent the solution, and estimate discretisation error. The interpolation method used must be consistent with the quadrature method used, in this case Hermite-Simpson.

The interpolation function of the state and control variables for a particular grid interval is as follows: First, if $p = t - t_k$, then

$$\tilde{\mathbf{x}}(t) = \mathbf{x}_k + \left(\frac{p}{h_k}\right)\mathbf{f}_k + \frac{1}{2}\left(\frac{p}{h_k}\right)^2(-3\mathbf{f}_k + 4\bar{\mathbf{f}}_k - \mathbf{f}_{k+1}) + \frac{1}{3}\left(\frac{p}{h_k}\right)^3(2\mathbf{f}_k - 4\bar{\mathbf{f}}_k + 2\mathbf{f}_{k+1}) \tag{15}$$

and

$$\frac{d\tilde{\mathbf{x}}(t)}{dt} = \mathbf{f}_k + \left(\frac{p}{h_k}\right)(-3\mathbf{f}_k + 4\bar{\mathbf{f}}_k - \mathbf{f}_{k+1}) + \left(\frac{p}{h_k}\right)^2(2\mathbf{f}_k - 4\bar{\mathbf{f}}_k + 2\mathbf{f}_{k+1}) \tag{16}$$

with $\tilde{\mathbf{u}}(t) = \mathbf{u}_k + \frac{p}{h_k}(\mathbf{u}_{k+1} - \mathbf{u}_k)$.

The relative discretisation error for each mesh or grid interval is calculated as follows [6]:

$$\eta_k = \int_{t_k}^{t_{k+1}} \left| \frac{d\tilde{\mathbf{x}}(t)}{ds} - \mathbf{f}(\tilde{\mathbf{x}}(s), \tilde{\mathbf{u}}(s), s) \right| ds \tag{17}$$

$$w_j = \max_k \left\{ \left| \frac{d\tilde{x}_{j,k}}{dt} \right|, |\tilde{x}_{j,k}| \right\} \tag{18}$$

$$\epsilon = \max_{j,k} \frac{\eta_{j,k}}{w_j + 1} \tag{19}$$

Where j indicates the jth state variable, k indicates the kth grid interval, and ϵ is the maximum relative local error. The integral which is equivalent to η_k is solved with Romberg quadrature [6].

3 Examples

3.1 Linear case

The validation example presented is an exercise taken from page 60 of Lawden [7]. This problem was chosen as it has an analytical solution. The problem is as follows:

The dynamic system subject to the following state equation

$$\dot{x} = -x + u \tag{20}$$

Where x is the state variable, and u is the control variable. The system starts at a time of 0 and ends at a time of 1 second. The initial values for the state variable is 1. Then the optimization problem is set as follows:

Optimise the cost function, C,

$$C = \int_0^1 \left(x^2 + \tfrac{1}{3}u^2\right) dt$$

Then the Hamiltonian of the problem is

$$H = x^2 + \tfrac{1}{3}u^2 + \lambda(u - x)$$

and thus

$$\frac{\partial H}{\partial u} = \tfrac{2}{3}u + \lambda = 0 \tag{21}$$

$$\dot{\lambda} = -\frac{\partial H}{\partial x} = -2x + \lambda \tag{22}$$

Substituting the value for λ from the equation (21) leads to the following system of ordinary differential equations

$$\dot{x} = u - x \tag{23}$$

$$\dot{u} = u + 3x \tag{24}$$

Since the final value of x is unconstrained, the value of λ, at the final time, must be 0. Therefore the value of u must be zero at the final time. This system of differential equations (24) has been solved for using the Laplace transform, providing the following solution for the optimal control and cost

$$u(t) = -\frac{3\sinh 2(1-t)}{2\cosh 2 + \sinh 2}$$

$$C = \frac{1}{1 + 2\tanh 2} = 0.3252424$$

The cost, C, and control are both consistent with the solution provided in [7].

The following matrices were defined in order to solve the problem using the developed solver.

$$J_k = [-1, 1], \quad \text{and} \quad G_k = \begin{bmatrix} 2x_k \\ \tfrac{2}{3}u_k \end{bmatrix}$$

The problem was solved in 0.17 seconds with re-meshing, using the developed optimal control solver. Without the provision of analytical first derivative information, the solver required 3.41 seconds to obtain the same solution. This solver time is the sum of the time taken by the NLP to determine a solution for each mesh. The relative discretisation error of the presented solution is 5.2845×10^{-10}. The maximum difference between the analytical and numerical solution of the control variable is 1.8201×10^{-4}. The cost found by the solver is 0.3252424, implying a 0% objective function error. The variables were unscaled. Figure 1 below depicts the time history of the numerical solution of the state and control variables, the control error, and the analytical optimal control. The control error is the difference between the analytical and numerical solution of the control history, at discrete points in time. It should be noted that only first order conditions were used to determine the analytical optimal control. Figure 2 depicts the phase trajectory.

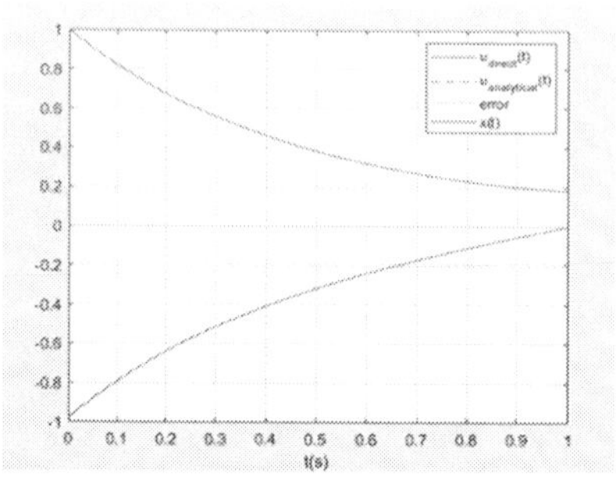

Figure 1. Numerical results of the linear example

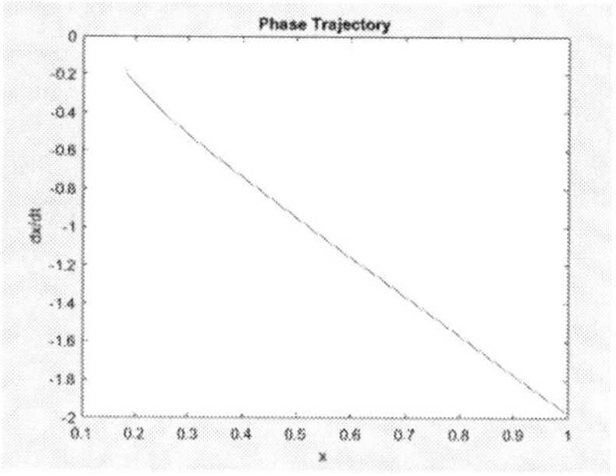

Figure 2. Phase trajectory of the linear example

3.2 Nonlinear case

The validation example presented is a minimum time co-planar circular orbital transfer. The system dynamics was augmented from Moyer & Pinkham [8], such that the problem was autonomous. The specific version of this problem solved for in Moyer & Pinkham will be solved. This problem attempts to find the minimum time taken for a spacecraft to transfer from the average circular orbit of the Earth around the Sun, to the average circular orbit of Mars around the Sun. The system of units used in Moyer & Pinkham were normalised. The time unit is equivalent to 58.18 days. The distance unit used is AU (astronomical unit). The mass unit is equivalent to 46.58 slugs, or 14.594 kg.

The following assumptions were used:

- Constant Thrust
- Gravity and Thrust are the only Forces acting on the rocket

- Earth and Mars have circular, co-planar orbits
- Only the gravitational attraction of the Sun is considered.

The system dynamics are as follows:

$$\begin{aligned}
\dot{r} &= v_r \\
\dot{\theta} &= \frac{v_r}{r} = \omega \\
\dot{v}_r &= \frac{v_\theta^2}{r} - \frac{\mu}{r^2} + \frac{T u_1}{m} \\
\dot{v}_\theta &= -\frac{v_r v_\theta}{r} + \frac{T u_2}{m} \\
\dot{m} &= -|\dot{m}| \\
\text{subject to} \quad u_1^2 + u_2^2 &= 1
\end{aligned}$$

Where r is the radial distance of the spacecraft from the sun, θ is the angular position of the spacecraft, v_r is the radial speed, v_θ is the circumferential speed, ω is the angular speed, T is the thrust constant, m is the mass, μ is the standard gravitational parameter of the sun, and u_1 and u_2 are the components of the unit vector which represent the thrust direction. In this example, the aforementioned unit vector components are the control variables.

The normalised boundary constraints are:

$$r(0) = 1; \qquad m(0) = 1; \qquad r(t_f) = 1.525; \qquad v_r(0) = 0; \qquad v_r(t_f) = 0$$

$$v_\theta(0) - \sqrt{\frac{\mu}{r(0)}} = 0; \qquad v_\theta(t_f) - \sqrt{\frac{\mu}{r(r_f)}} = 0$$

The following constants are used:

$$T = 0.8098; \qquad |\dot{m}| = 0.07487; \qquad \mu = 1$$

The solution to this problem, found in [8] by using an indirect Newton-Raphson optimal control method, is a minimum time of 3.3207 time units or 193.2 days. This solution was found in 36 seconds. The following figure, obtained from [9], is a representation of the optimal trajectory, and the corresponding optimal control history. The arrows represent the thrust direction unit vectors.

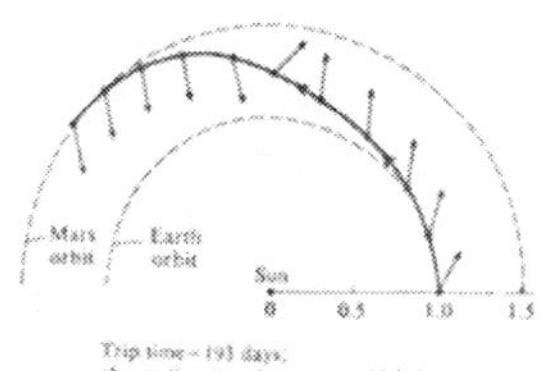

Figure 3. Bryson & Ho Trajectory solution [9]

The matrix below was defined in order to solve the problem using the developed solver:

$$J = \begin{bmatrix}
0 & 0 & 1 & 0 & 0 & 0 & 0 \\
-\frac{v_\theta}{r^2} & 0 & 0 & \frac{1}{r} & 0 & 0 & 0 \\
-\frac{v_\theta}{r^2} + \frac{2\mu}{r^3} & 0 & 0 & \frac{2v_\theta}{r} & -\frac{Tu_1}{m^2} & \frac{T}{m} & 0 \\
\frac{v_\theta v_r}{r^2} & 0 & -\frac{v_\theta}{r} & -\frac{v_r}{r} & -\frac{Tu_2}{m^2} & 0 & \frac{T}{m} \\
0 & 0 & 0 & 0 & 0 & 0 & 0
\end{bmatrix}$$

The problem was solved in 1.34 seconds with re-meshing, using the developed optimal control solver. Without the provision of analytical first derivative information, the solver required 14.82 seconds to obtain the same solution. This solver time is the sum of the time taken by the NLP to determine a solution for each mesh. The relative discretisation error of the presented solution is 7.659×10^{-7}. The cost found by the solver is 3.206 time units, implying a 0.00301 % objective function error. The variables were unscaled. The figure below is the optimal trajectory solution found by the optimal control solver.

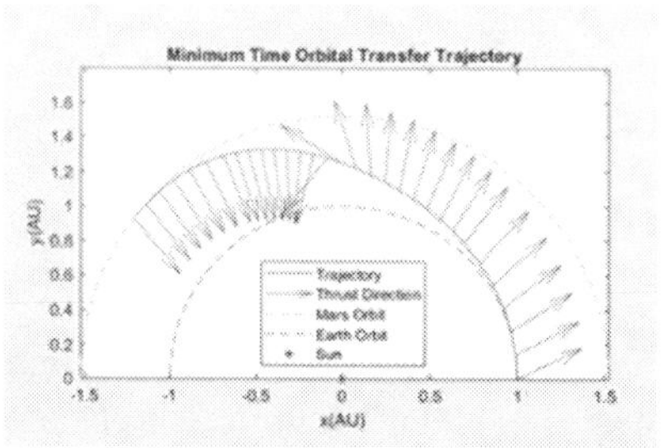

Figure 4. Numerical results of the example

As stated in [9], the optimal radial thrust component is outward for roughly the first half of the trajectory, and inward for the rest. The results obtained are consistent with this observation. Figure 5 below compares the thrust direction history determined by Moyer & Pinkham [8], to the history determined by the developed optimal control solver. This figure shows that the optimal thrust direction history solved for is consistent with the optimal thrust direction history (represented by θ^{*}) found by Moyer & Pinkham. In this figure, the thrust direction is represented by the angle measured from the local horizontal to the thrust direction vector.

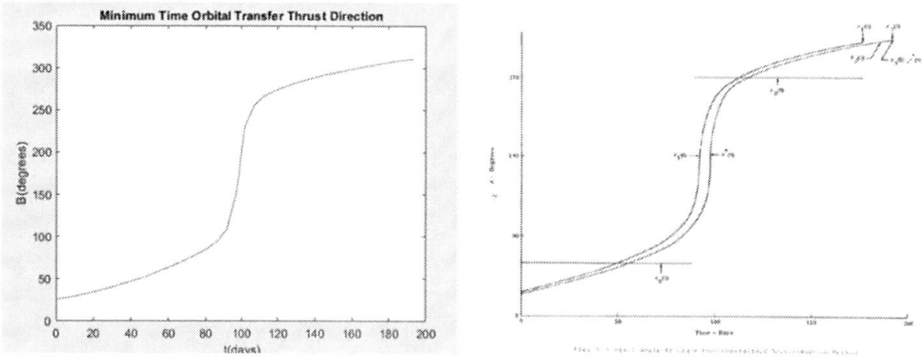

Figure 5. Control history found by solver (left) and Control history found in [8] (right)

The phase trajectory plots, are as follows:

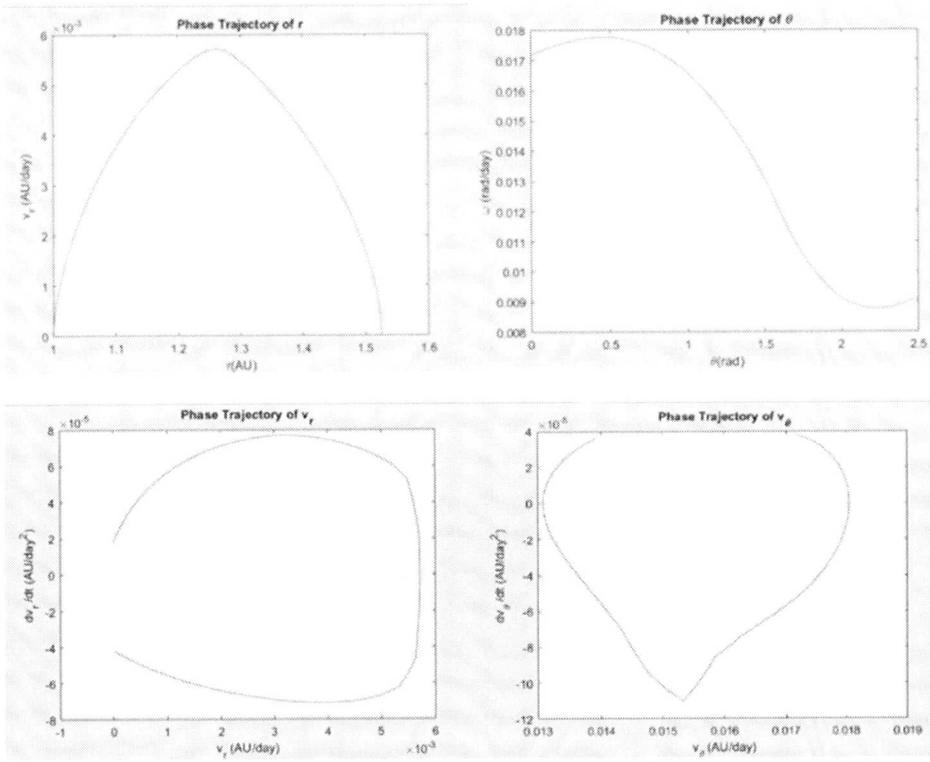

Figure 6. Phase trajectory of state variables

4 Conclusion

Hermite-Simpson direct collocation methods have been effectively used to solve various optimal control problems. However typically the derivative information found using these methods have been estimated with finite difference methods. This paper presented a relatively simple formulae for determining first order analytical derivative information, for the Hermite-Simpson direct collocation method. This information allows for robust and faster solving of optimal control problems. The derivatives were tested in a developed solver, and have been proven to function in various validation examples, one of which was provided in this paper.

The first derivative information of the initial mesh was made to undergo a derivative information check by Matlab. Matlab tests the derivative information by comparing the analytically determine derivatives with numerical estimates (using finite differencing methods). The first derivative information passed Matlab's test, further validating the obtained results. The provision of first order analytical derivative information had successfully reduced the solve time of the solver by a significant margin.

This work, was constrained by Matlab's SQP algorithm, which estimates second derivative information using dense finite differencing methods. In order to improve the overall solver performance, and to compare it fairly to other optimal control solvers, the solver should be used in conjunction with an SQP algorithm capable of exploiting the sparsity of the second derivative information.

References

[1] Betts J.T., *A Survey of Numerical Methods for Trajectory Optimization,* Journal of Guidance, Control, and Dynamics, **21** No. 2, 193-207 (1998)

[2] Boudjemaa R., Cox M.G., Forbes A.B., and Harris P.M., *Automatic Differentiation Techniques and their Application in Metrology,* March 26,(2003) http://www.npl.co.uk/ssfm/download/documents/cmsc26 03.pdf. (Accessed 30 April 2021).

[3] Rosmann C., Makarow A., and Bertram T., *Time-optimal control with direct collocation and variable discretization,* arXiv preprint arXiv:2005.12136, 2020).

[4] Nocedal J. and Wright S.J., *Numerical Optimization* (Springer, New York City, 1999)

[5] Forsgen A.,Ceragioli F., Dontchev A., Futura K., Marti K., and Pandolfi L., *System Modeling and Optimization* (Springer-Verlag, Boston, pp 51-66, 2006)

[6] Betts J.T., *Practical Methods for Optimal Control and Estimation Using Nonlinear Programming (2nd ed.,* (SIAM, Philadelphia, 2010)

[7] Lawden D.F., *Analytical Methods of Optimisation,* (Scottish academic press Ltd. Edinburgh, 1975)

[8] Moyer, H.G. and Pinkham, G., *Several Trajectory Optimization Techniques Part II: Application,*(in Computing Methods in Optimization Problems, New York, 1964)

[9] Bryson, A.E. and Ho, Y.C., *Applied Optimal Control* ,(Washington D.C.: Hemisphere Publishing Corporation, 1975)

A Ffowcs Williams-Hawkings numerical aeroacoustic study of varied and fixed-pitch blades of an H-Rotor vertical axis wind turbine

Errol Molatudi[1,*], *Thokozani Justin* Kunene[1], and *Lagouge Kwanda* Tartibu[1]

[1]University of Johannesburg, Mechanical & Industrial Engineering Technology Department, Doornfontein Campus, Johannesburg, South Africa

Abstract. The effects of sound pressure level at two observation positions of a fixed and varied blade pitch angle at Low-Mach unsteady incompressible Reynolds-Average Navier-Stokes flow approach, on an H-rotor Vertical Axis Wind Turbine. The objective of the study is to compare the noise dissipation and output power/energy of the airfoil blades design of the vertical axis wind turbine in residential zones. The Ffowcs Williams -Hawkings (FHWH) techniques were applied to validate the output noise and vortex shedding of the different angles of attacks (AoA). The study postulated that the time history of the calculated sound pressure level at two observers positions: the aeroacoustic, blade vortex interaction noise, flow separations, dynamic stall experience from varied angled of attacks are found to produces less noise and vortex shedding compared to the fixed angle of attack.

Keywords:
Aeroacoustic, Angle-of-Attack, Ffowcs-Hawkings, Sound Pressure level (SPL)

1. Introduction

There is a need for renewable energy globally, and wind energy is another alternative that can provide clean energy to residential areas. Wind energy has shown great potential since 2006 in China. China is currently leading in wind energy development, with a cumulative installed capacity of wind power increase from 23,900 MW in 2001 to 651,000 MW in 2019 [1]. A wind turbine is an attractive form of electricity production for residential households. However, the wind harnessing energy process comes with excellent noise pollution. The noise generated by the operating wind turbine significantly impacts surrounding residential areas, particularly humans and animals, respectively. Mechanical noise and aerodynamic noise are two types of noise produced by wind turbines. Moving machinery parts, gearboxes, and generators are common sources of mechanical noise [2]. Aerodynamic noise can be narrow-band or broadband, depending on the rotor geometry, blade geometry, and flow environment [3]. Because there are two types of wind turbines that can create electricity for residents,

* Errol Molatudi molatudierrol@gmail.com

A vertical axis wind turbine is preferred over a horizontal axis wind turbine because it offers greater benefits. The noise pollution produced by a vertical axis wind turbine is lower than that produced by a horizontal axis wind turbine [4]. To obtain less aerodynamic, the turbine must run at a low-speed ratio range and a higher angle of attack[5]. Reduced noise pollution is dependent not only on the angle of attack, but also on how the wind interacts with the blades, known as vorticity; research has shown that minimizing vorticity with the blade can result in improved turbine performance [6] The different trailing edges, tip forms, and airfoil profiles have all been studied in order to reduce aerodynamic noise from wind turbines. The airflow speed around the blades, which is related to the rotor tip-speed ratio, is an important contributor to noise emission. As a result of these discoveries, several blade angles of attack have been developed, including fixed blades and swivel blades with a fixed angle of attack. [7].

A 5kW semi-scale three-blade H-rotor vertical axis wind turbine with a fixed and variable angle of attack will be investigated in this study. The flow dynamics of a Darrieus H-rotor three-blade vertical axis turbine are significantly influenced. The airfoils are controlled by the leading edge slats and flaps, which increase cambering and improve flight or flow via lifting characteristics. The impacts of blade form, speed ratio, solidity, and distance between noise sources and receivers would be determined using an Ffowc William-Hawking model. Because the Ffowcs William-Hawking is concerned with rotating zones and sliding mesh models. In ANSYS-FLUENT, Ffowcs Willam - Hawking models have been implemented. As a result, all of the findings were obtained using the commercial CFD program ANSYS-FLUENT 2019 R1. The time-accurate solutions are obtained using an Unsteady Reynolds-averaged Navier-Stokes model [8]. H-rotor vertical axis wind turbines, also known as Darrieus vertical axis wind turbines, have been the subject of a lot of research since they produce a lot of electricity with little noise pollution [9].

2. Noise pollution

A wind turbine produces two sorts of noise when it is in operation: mechanical and aerodynamic noise. Machinery parts such as gearboxes and generators generate mechanical noise. It could be effectively reduced using a well-known engineering design [2]. Aerodynamic noise, on the other hand, remains a challenging task to predict. VAWT noise sources can be categorised. The high-frequency noise created by an airfoil's interaction with flow disturbances produced in its boundary layer and near wake is known as airfoil self-noise. Airfoil self-noise occurs in a vertical axis wind turbine when excess turbulence interacts with the blade surface, causing Stall/Separation noise. Interaction causes tip vortex noise. Theoretical concepts are typically utilized to forecast an airfoil's noise, whereas experimentally generated components are employed to better agree with what is observed in practice. Serrated trailing edges, various trailing edge and tip forms, and various airfoil shapes have all been researched to reduce wind turbine aerodynamic noise.

.

3. Acoustics model theory

3.1 The Ffows Williams Hawkings model

The same continuity and Navier stokes equations are used to derive the Ffows Williams Hawkings equation. The Ffowcs Williams technique, which assumes that the detailed pressure and velocity fluctuations are known, is frequently used to enhance turbine noise prediction. Ffowcs William and Farassat derived the genesis of this wave equation, propeller and helicopter rotor noise. The Ffowcs William and Hawkings equation is an inhomogeneous wave equation that includes the effects of moving surfaces. When the surface is in the near field of a solid body, the FW-H technique filters out the part of the solution that does not radiate as sound. [10].The key to determining any rotor noise applying Ffowcs William Hawking method, there are three terms of importance [11] The sound radiated by turbulence and flow distortion. A dipole is controlled by blade surface loading. The motion of the source compresses the time history as the source moves towards the observer, thus an increase in radiated sound.

The Ffowcs Williams Hawking equation helps to calculate noise signals. There have been improvements in determining the varying flow of practical interest. The only limit would be the size of the computer to generate unsteady flow and acoustics inside a computational grid, provided the Mach number is not too high. The linear wave equation can model the wave propagation outside the flow, so increasing the size of the computational domain is wasteful. The Ffowcs William Hawking will provide solutions to wave equations, given that the numerical calculations are accurate on a surface that bounds the source and that compressible flow fluctuation and the acoustic waves are captured accurately. [12] The frequency-domain version is totally dependant on the equation developed by Lockard. The Ffowcs William Hawking uses the equation s for two-dimensional problems, the derivation of this equation a Cartesian tensor notation was employed which is equally valid in three-dimensions. At the location of each receiver is predicted from small amplitude acoustic pressure fluctuations time-accurate solutions of the flow field variable around wind turbine are required.

Ffowcs Williams Hawkings' acoustic analogy is developed for simulating the moving surface acoustic radiation in the far-field. The Ffowcs Williams Hawkings is developed from the Navier-Stokes equations. Mohamed et al [8] represent the FHWH equations as:

$$\frac{1}{a_0^2}\frac{\partial^2 \acute{p}}{\partial t^2} - \nabla^2 \acute{p} = \frac{\partial^2}{\partial x_i \partial x_j}\{T_{ij}H(f)\} - \frac{\partial}{\partial x_i}\{[P_{ij}n_j + \rho u_i(u_n - v_n)] \delta(f)\} + \frac{\partial}{\partial t}\{[\rho_o v_n + \rho(u_n - v_n)]\delta(f)\} \quad (1)$$

Where :

u_i = Fluid velocity component in the x_i direction
u_n = Surface velocity component normal to the surface $f = 0$
v_i = Surface velocity components in the x_i direction
v_n = Surface velocity component normal to the surface
$\partial(f)$ = Dirac delta function
$H(f)$ = Heaviside function
$\acute{p}$ = The sound pressure at the far-field, $\acute{p} = (P - P_0)$
T_{ij} = The Lighthill's stress tensor, $T_{ij} = \rho u_i u_j + P_{ij} - a_o^2(\rho - \rho_o)\delta_{ij}$
a_0^2 = The speed of sound
P_{ij} = The compressive stress tensor

Where the compressive stress tensor can be further broken:

$$P_{ij} = p\delta_{ij} - \mu\left[\frac{\partial u_i}{\partial x_j} + \frac{\partial u_j}{\partial x_i} - \frac{2}{3}\frac{\partial u_k}{\partial x_k}\delta_{ij}\right] \quad (2)$$

A green function is used to compute the sound pressure at the receiver location. The subscript o denotes Free-stream quantities. The free-space Green function ($\delta(g)/4\pi r$) is used to find the solution to Eq. (1). The entire solution is made up of surface and volume integrals. The surface integrals explain the contributions from monopole and dipole acoustic sources and specific contributions from quadrupole sources, whereas the volume integrals describe quadrupole (volume) sources outside the source surface. When the flow is below subsonic, as in the Darrieus turbine instance, the contribution of the volume integrals becomes minor. Volume integrals are not used in ANSYS-FLUENT 2019 R1 [8]. As a result:

$$\acute{p}(\vec{x},t) = \acute{p}_T(\vec{x},t) + \acute{p}_L(\vec{x},t) \tag{3}$$

Where:

$$4\pi\acute{p}_T(\vec{x},t) = \int_{f=0}\left[\frac{\rho_o(\dot{U}_n+U_{\dot{n}})}{r(1-M_r)^2}\right]ds + \int_{f=0}\left[\frac{\rho_o U_n\{r\dot{M}_r+a_o(M_r-M^2)\}}{r^2(1-M_r)^3}\right]ds \tag{4}$$

$$4\pi\acute{p}_L(\vec{x},t) = \frac{1}{a_0}\int_{f=0}\left[\frac{\dot{L}_r}{r(1-M_r)^2}\right]ds + \int_{f=0}\left[\frac{L_r-L_M}{r^2(1-M_r)^2}\right]ds + \frac{1}{a_0}\int_{f=0}\left[\frac{L_r\{r\dot{M}_r+a_o(M_r-M^2)\}}{r^2(1-M_r)^3}\right]ds \tag{5}$$

Where:

$$U_i = v_i + \frac{\rho}{\rho_o}(u_i - v_i) \tag{6}$$

$$L_i = P_{ij}\hat{n}_j + \rho u_i(u_n - v_n) \tag{7}$$

With the integration surface synchronizing with an impenetrable wall, the two components on the right in Eq. (3), $\acute{p}_T(\vec{x},t)$ and $p_L(\vec{x},t)$ are sometimes referred to as thickness and loading terms, respectively, due to their physical definitions. In Eqs. (4) and (5), the square brackets denote that the kernels of the integrals are computed at the corresponding retarded times, which are defined as follows, given the observer time, t, and distance to the observer, r.

$$\tau = t - \frac{r}{a_o} \tag{8}$$

The different subscripted numbers in Eqs. (6) and (7) are the inner products of a vector and a unit vector, which are made tacit by the subscript. For example, $L_r = \vec{L}\cdot\vec{r} = L_i r_i$ and $L_r = \vec{U}\cdot\vec{n} = U_i n_i$, where $\vec{r}$, and $\vec{n}$ represent the unit vectors in the radiation and wall normal directions, respectively. The dot over a variable indicates that variable's source-time differentiation [8]. There are a few comments about the applicability of this integral solution:

a. The FW-H formulation can be used to treat both revolving and stationary surfaces.
b. It is not necessary for the surface $f = 0$ to coincide with body surfaces or walls. Because the formulation makes source surfaces permeable, they can be positioned in the flow's interior.
c. When a permeable source surface (either an interior or non-conformal sliding interface) is positioned at a specific distance from the body surface, the integral solutions obtained by Eqs. (5) and (6) include contributions from quadrupole sources within the source surface's region. The mesh resolution must be fine enough to resolve the

transient flow structures inside the volume contained by the permeable source surface when utilizing a permeable source surface.

4. Numerical model

4.1 Geometry and topology of VAW

The three-blade configuration is on the assumption that the three blades are more stable in comparison to the two or four-blade configuration(see Fig: 1) [13]., furthermore research is conducted on the aspect on a vertical axis wind turbine, that a higher aspect ratio has a lower power coefficient and vice versa [14] [15].

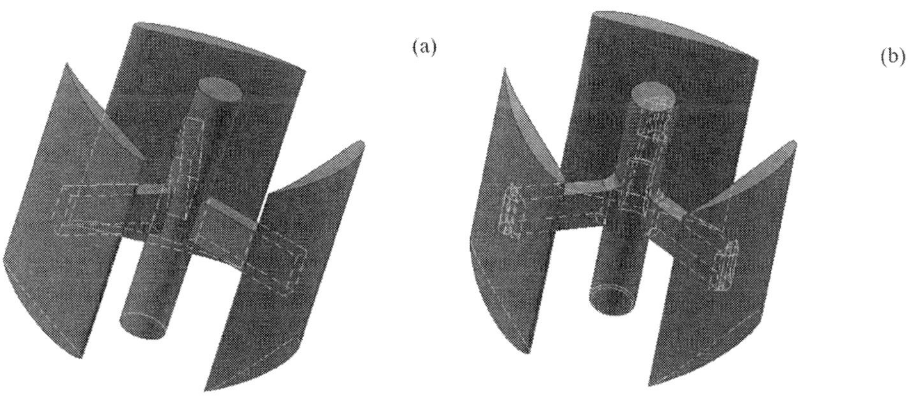

Fig. 1. Three-blade H-type VAWT, rigid (fixed AOA) and (b) swivel (varied AoA, sinusoidal profiled),

4.2 Mesh and grid sensitivity

The mesh was generated using the Body-influence method (see Fig: 2) to refine local bodies, i.e., the rotating core that houses the blades. The stationary body around is coarsened, but the rotating core is refined. The interface between the rotating core and the stationary body was included in the refinement of the mesh. The first layer of $1\text{x}10^{-5}$ m from 12 prismatic layers leading up to the wall plus value of, $y^+ = 5$. The limit of dynamic stall of blade Reynolds number, $Re_b = 771\ 707$ at an azimuth angle, $\theta = 1.5^0$ of the blades (or rotating core in this case) when the rotation reaches 1 second.

- Re_b is found on a fixed tip speed ratio (TSR), of 0.56.
- This TSR places the blades out of the static stall [15]. The modelled blades have a maximum stall angle of 0.9620.

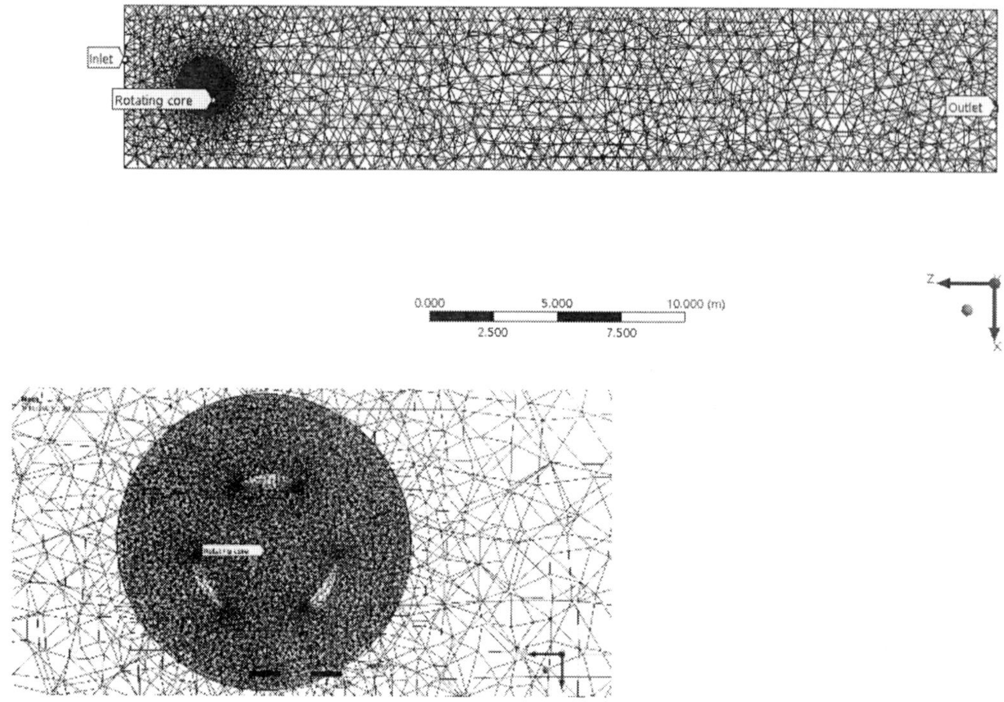

Fig. 2. Mesh density of a far-field computational domain, tetrahedral elements

4.3 Computational domain and boundary conditions

As published, the velocity inlet of 15 m/s is a wind speed around residential areas [16], [17]. We modelled a 10% turbulence intensity at the inlet (upstream) of the rotating core. The rotating core rotates the blades at an angular speed of, $\Omega = 15$ rad/sec, based on the Tip Speed Ratio (TSR) of 0.56.

4.4 Solver and turbulence model

Unsteady incompressible Reynolds-averaged The commercial CFD code ANSYS Fluent 2019 R1 was used to solve Navier-Stokes (URANS) simulations. Allmaras, Spalart (S-A model). Kostic [18] articulates the SA one-dimensional equations effectively. The impacts of blade form, speed ratio, solidity, and distance between noise sources and receivers would be determined using an Ffowc William-Hawking model. Since the Ffowcs William-Hawking is concerned with rotating zones and a sliding mesh model. As this Ffowcs Willam -Hawking models have been built in ANSYS-FLUENT, all results are calculated using a commercial CFD program ANSYS-FLUENT. The time accurate solutions are obtained using an Unsteady Reynolds-averaged Navier-Stokes model [8]. Many studies on the H-rotor vertical axis wind turbine, also known as the Darrieus Vertical axis wind turbine, have revealed high energy production with reduced noise pollution. [9]. For pressure-velocity coupling, a SIMPLE system is utilized. The spatial discretization utilized was a second-order accurate Finite Volume Method (FVM) [15]. To capture sources that are time-dependent turbulent quantities, such as the vortex expanding behind the aerofoils, a transient solution was used.

The convergence criterion for all residuals was set to 1x10^{9} for all governing equations. The temporal discretization of this unstable flow problem necessitated a solution in which a dynamic condition rotated blades and extracted energy from the incoming stream [19]. A Dynamic Mesh and an adapted mesh to two cells around each blade were employed for that actuation. It was anticipated that some mesh cells will exhibit negative meshing. The parameters that were affected included a time step (t) that was set to the time required to complete an azimuthal increase ($d\theta = \ = 1$ x10^{-5}) degrees, blade rotation, Δt = 1x10^{-5} s with 1x10^{+5} iterations [15]. For the interpreted six degree freedom function of all three blades rotating and a compiled UDF (centre of gravity, C G Motion for rotation) of sinusoidal pitching of the blades, two different User-Defined-Functions (UDF) were utilized (see Appendices A and B) (ANSYS Release 14, 2011). The shifting pitch angles (β, Angles-of-Attack) of the blades, which are a bearing between the rotor arm and the blade, are assimilated by this UDF.

4.5 Model validation

For all the available and probable investigations around VAWT and aerofoils in general, there are still few 3D simulations. Even without the rotor arms, the 3D models provide a degree of freedom that occurs in reality, accounting for the flow moving in an axial vertical orientation. Figure 3 shows the distribution of the three observer positions (observer time positions) investigated for this study. At this time of reporting, the lack of experimental data for the current study about acoustic noise is only validated using similar CFD aeroacoustics studies found in the literature. Figure 4 validates this current study with the results obtained by Su *et al.* [4] for the three-bladed H-rotor VAWT turbine. The sound pressure level the observer positions from our study have a similar trend and magnitudes of the harmonics of the literature findings of Su *et al.* [4]. This work validates the zero angle-of-attack model. We continued to find results of the varied angle-of-angle, β.

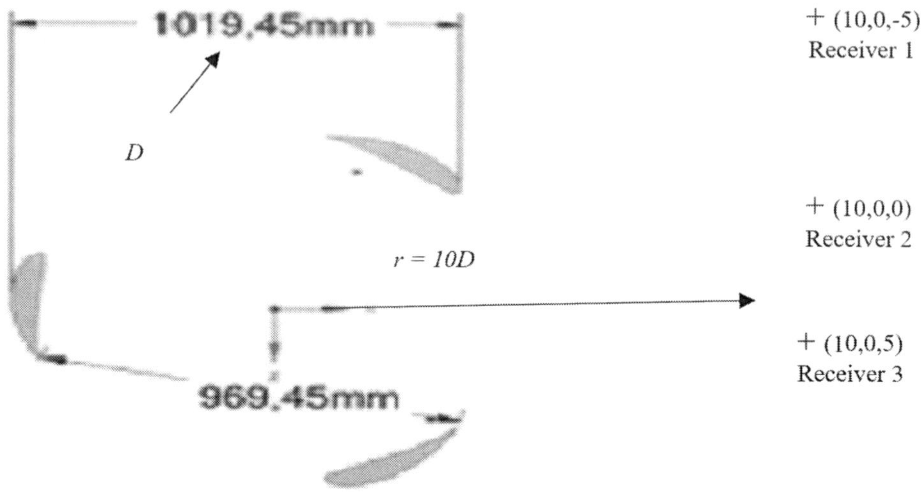

Fig. 3. The three prescribed observer positions at Cartisin coordinates (unit: m)

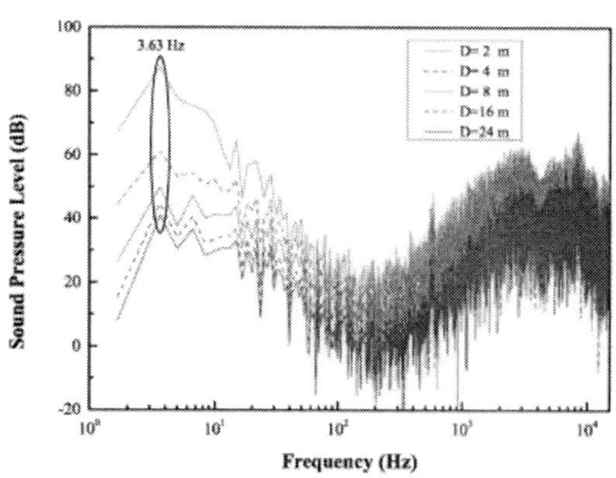

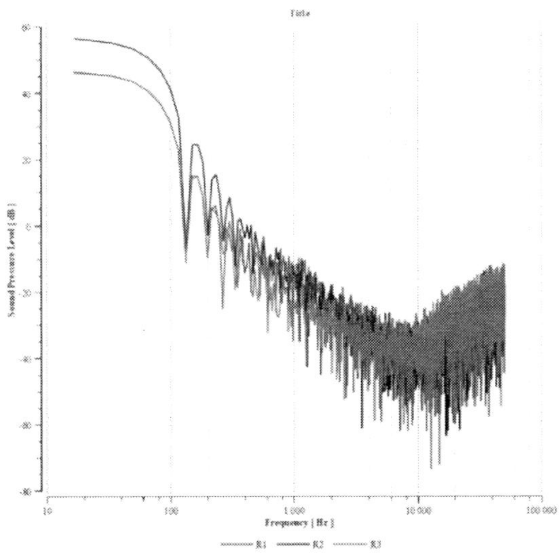

Fig. 4. Sound pressure level spectra at the prescribed observer positions, (a) Three-bladed H-rotor Darrieus turbine by Su *et al* [4] $\beta = 0$, AoA, $\lambda = 0.95$ at fcousing on D = 8 m for the recivers (b) current CFD study at $\beta = 0$, AoA, $\lambda = 0.56$ at D = 10 m for the recivers.

5. Results and discussion

Figure 5a shows that the blades are more potent in vortex sheets forming downstream and travelling further on a fixed angle of attack (AoA). However, the varied AoA to lower the stall separation as since with fewer vortex sheets being formed (Figure 5b). Therefore, the vortex sheet from varied AoA is quieter too. The varied AoA reduces flow separation and weakening the instability that periodically generates separation vortices, which leads to the vibration of rotating machinery and noise generation. We found the power coefficient (C_p) in VAWT is maximum at low Tip-Speed-Ratio value ($\lambda = 0.56$). The results demonstrate that an SG6043 aerofoil blade with a chord length of $c = 0.5$m captures the most energy in the presence of varying or fixed angle-of-attack.

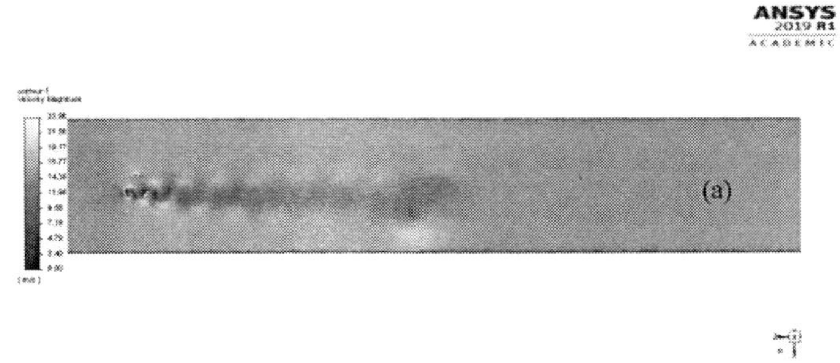

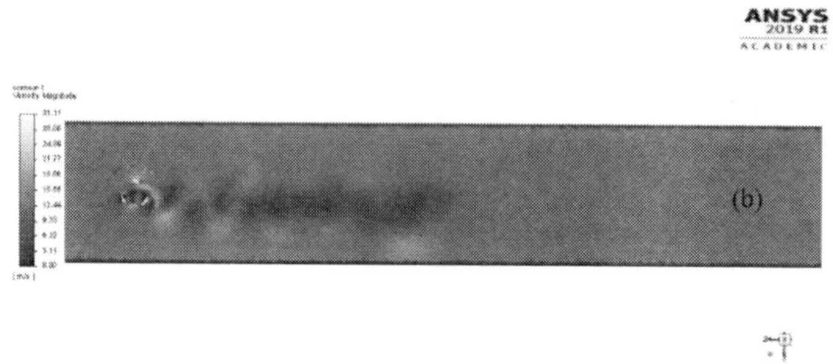

Fig. 5. Vortex sheet formation downstream the blades at a symmetry plane and time, 4 sec, for an AoA of (a) $\beta = 0$; (b) $-30^0 < \beta < 30^0$.

Figure 6(a) shows the sound pressure level spectra with the tonal peak in the sound spectrum of about, SPL= 30 dB at f= 150 Hz. Figure 6(b) that the agreement between the predictions and measurements is quite reasonable. Su *et al.* [4] reported a tonal frequency ranging from 3.83 to 4.11 Hz and the power spectral density amplitudes decreased from 5.2×10^{-3} to 4.4×10^{-4} W/Hz varying the angle of attack from 0^0 and 2^0 for a varied angle of attack.

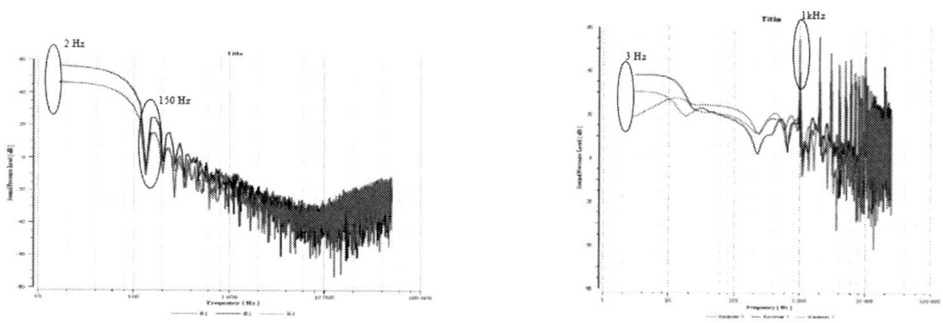

Fig. 6. Sound pressure level spectra AoA of (a) $\beta = 0$; (b) $-30^0 < \beta < 30^0$ at three receiver positions

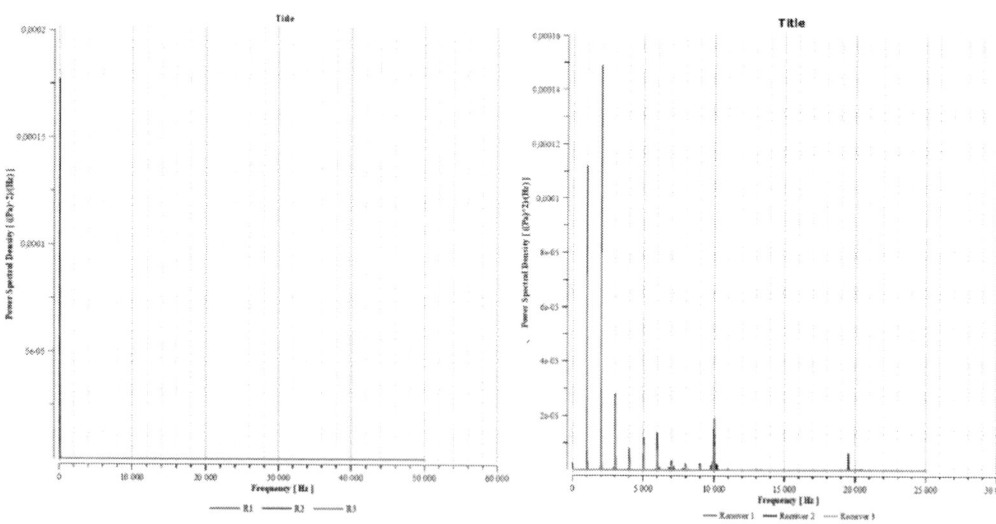

Fig. 7. Power spectra density AoA of (a) $\beta = 0$; (b) $-30^0 < \beta < 30^0$ at three receiver positions

Figure 7 is the power spectra density (PSD) of the current study as measured by the sound pressure level at observer positions. The sound emitted by the source is referred to as sound emission. The most essential quality, however, is noise immission [8], which relates to the sound perceived by an observer. There is a similar obversion with Su *et al.*[4] relating to the increase in tonal frequency and decrease in the power spectra density. Our results show a increase in the SPL at frequencies of 150Hz and 1kHz at delayed higer frequencies for amplitudes. Amounting to a decrease in the PSD amplitudes from 1.75×10^{-4} to 1.5×10^{-4} W/Hz (overall from all the receivers), respectively for fixed and varied AoA.

6. Conclusion

The aeroacoustic spectra of the fixed and varied angle of attack were simulated and analyzed: for a vertical-axis wind turbine with blades derived from a static tip speed ratio of 0.56. This investigation has aided in the understanding of noise sources and aerodynamics responsible for aeroacoustic and improved power generation in vertical-axis wind turbines for an urban area with blades that vary their angle of attack as the turbine rotates.

The key findings of aerodynamics and aeroacoustics are summarized as follows: (1) The numerical investigation of a varied angle of attack and fixed angle of attack on a vertical - axis wind turbine. The research postulates that the varied angle of attack is an ideal choice for a reduction in noise emission. The vortex sheet indicates less noise projection when comparing the varied angle of attack to that of the fixed angle of attack. When comparing the wind direction of the fixed angle of attack, the velocity is constant. As the angle of attack increases, the frequency of wake vortex shedding decreases, but the frequency of shear-layer instabilities starts to rise. The azimuth angle enhances streamwise wind, thus improving stall behaviour within the low tip speed ratio range. (2) The impact of the varied angle of attack on the tip speed ratio improves the turbine's power efficiency and torque output. However, the downstream and upstream noise intensifies as the angle of attack increases [21]. As a result, noise prediction at take-off must account for the effect of angles of attack. (3) The loudness and sharpness, which are the metrics that are important for the human perception of sound, are also observed to increase the noise pollution through a varied angle of attack [22]. However, discomfort is associated with the boundary layer tones viewed to be at amplitudes of 60 dB at frequencies above 1000 Hz [20]. As a result, the findings of this paper can contribute to a better understanding of noise-generating mechanisms and the design of quieter vertical axis wind turbines.

Furthermore, these findings can be used to establish the most efficient position for wind turbine installation so that the noise emitted is as inconvenient to nearby neighbours as possible. [2]

We want to improve the accuracy of our results with a computational domain that has a higher cell elements well above the million cell count. Then perform the numerical simulation using a computer cluster for parallel computing. We will also perform experiments using signal processing is performed by the beamforming technique to compare numerical results.

Acknowlegdement

The author would like to thank supervisor Thokozani Justin Kunene and co-supervisor Lagouge Kwanda Tartibu and for the support guidance patience during the preparation of

 https://doi.org/10.1051/matecconf/202134700013

this paper. The institution University of Johannesburg for financially committing to the conference cost.

Nomenclature

A	Sweep area [m^2]
c	Blade chord length [m]
C_L	Coefficient of Lift
C_D	Coefficient of Drag
C_P	Coefficient of Power
D	Turbine diameter [m]
$d\theta$	Azimuthal increment [^{0}]
F_L	Lift force [N]
F_D	Drag force [N]
L	Length [m]
H	Blade height [m]
P	Static Pressure [Pa]
Q	Torque [N.m]
R	Radius of rotor [m]
Re	Reynolds number over the blade [-]
U	Wind speed upstream [m/s]
v	Wind velocity [m/s]
V_∞	Wind speed upstream [m/s]

Greek

λ	Tip Speed Ratio (TSR) [-]
ρ	Density of air [kg/m^3]
υ	Kinematic Viscosity [m2/s]
μ	Dynamic Viscosity [N.s/m^2]
β	Angle-of-Attack [^{0}]
ω	Angular velocity [rad/sec]

Subscripts

geo	Geometry of the blade
p	Power

Appendix A

```
# include "udf.h"
# include "dynamesh_tools.h"
DEFINE_CG_MOTION(rotate,dt,vel,omega,time,dtime)
{
real freq_t;
omega [2] = freq_t; }
```

Appendix B:

```
#include "udf.h"
DEFINE_CG_MOTION(yaw, dt, vel, omega, time, dtime)
{
real a, w, pi;
pi = 3.1415;
/* define motion variables */
a = 0.05; /* 0.05m movement amplitude */
w = 2 * pi * 2; /* 2Hz frequency */
/* define object movement law */
omega [0] = -.5* sin(w*time);
}
```

References

1. M. Dyrholm, "Global Wind Energy councli, Global Wind Report 2019.2020.," 2019. [Online]. Available: eqmagpro.com/wp-content/upload/2021/03/GWEC-I-Global-Wind-Report_2021
2. J.N Pinder, "Mechanical noise from wind turbine," Wind Energy, pp. 158-168, 1992
3. Sally Wright, M.S;Anthony L, Rogers, "Renewable Energy Research Laboratory," Massachusetts Amherst, 2006.
4. J. Su et al., "Aerodynamic noise assessment for a vertical axis wind turbine using Improved Delayed Detached Eddy Simulation," Renew. Energy, vol. 141, pp. 559–569, 2019.
5. CS, Feirreira; G, Van Kuik; G, Van Bussel; F, Scrarano, "Visualization by piv of dynamic stall on vertical axis wind turbine," Exp Fluid, pp. 97-108, 2009.
6. LA, Danao; R, Eboibi OHowell, "An experimental investigation into the influence of unsteady wind on the performance of vertical axis wind turbine," Appl Energy, pp. 107:403-411, 2013.
7. Doolan,CJ;Moreau,DJ;Brooks,LA, "Wind turbine noise mechanism and some concept for its control," Acoust Aaust , pp. 40(1)7-13, 2012.
8. Mohamed, "Aero-acoustic noise evaluation of H-rotor Darius wind turbine," Energy 65, pp. 596-604, 2014.
9. Journal of Mechanical Engineering and Sciences, vol. 1, no. 122, pp. 3399-3409, 2018.
10. Ian Carlo ,M. Lositano; Louis Angelo M. Danao, "Steady wind performance of a 5KW threee blade H-rotor Darrieus Vertical Axis Wind Turbine with cambered tubercle leading edge blades," Energy 175, pp. 278-291, 2019.
11. K.Brenther and F.Farassat, "Modeling aerodynamically generated sound of helicopter rotors," Aeronaut.SCI 39, pp. 83-12, 2003.
12. J.Ffowcs William;D.Hawking, "Sound generation by turbulence and surfaces in arbitrary motion," Pilos Trans Soc, pp. 321-342, 1969.
13. M.Wang;J.B.Freund;S.K.Lele, "Computational prediction of flow-generated sound," Fluid Mech, pp. 483-512, 2006.
14. F. A.Sunyoto, "The effects of number of blades on the performance of H-Darrieus type wind turbine," QIR Quality in research.IEEE, 2013.
15. R. S.Brusca, "Design of a vertical axis wind turbine:how the aspect ratio effects the turbine's performance," Int J Energy Environ Eng, pp. 5(4):333-40, 2014.
16. A.R.Jha,, Wind turbine techonology., New York: CRC Press, 2011.
17. Al-Shemmer, " Wind turbines.," Wind turbines., vol. 978, no. Al-Shemmeri & Ventus., pp. 978-87,681-692, 2010.
18. L.Kostic, "Review of the Spalart-Allmaras Turbulence Model and its Modifications to Three-Dimensional Supersonic Configurations," Scientific Technical Review,, vol. 1, no. LXV, pp. 43-49, 2015
19. W. S, B. R and G. G, Wind turbine noise, Berlin, 1996.
20. M.Mohamed,G. Janinga,Optimization of Savonius blade using an obstabcle shielding the returning blade. Energy convers management, 52(1)236-42. (2011)
21. Viqueira-Moreira M, Ferrer E. Insights into the Aeroacoustic Noise Generation for Vertical Axis Turbines in Close Proximity. Energies. 2020; 13(16):4148. https://doi.org/10.3390/en13164148

 https://doi.org/10.1051/matecconf/202134700014

Neural Network for Modeling the Mechanical Properties of Gelatin-Cellulose Nanocrystals Hydrogel Membrane for Heavy Metal ions Removal from Wastewater

John Kabuba[1*]

[*]Department of Chemical and Metallurgical Engineering, Vaal University of Technology, Private Bag X021, South Africa

Abstract. The mechanical properties of Gelatin-cellulose nanocrystals hydrogel membrane were investigated for the removal of heavy metal ions from wastewater. The membrane was characterized using Scanning Electron Microscopy (SEM) and Fourier Transform Infrared Spectroscopy (FTIR) analysis. Neural Network (NN) model was developed to predict the mechanical properties such as Young's modulus, tensile strength, and elongation. The NN predicted results are very close to the experimental results with $R^2 = 0.99315$. The predicted values were found to be in excellent agreement with the experimental data and the current model has a good learning precision and generalization. The results revealed that the developed model is very accurate.

1 Introduction

Heavy metals in wastewater are highly toxic, not biodegradable and effortlessly accumulated in living organisms [1]. They can be the origin of illnesses such as cancer, high blood pressure, nervous system diseases and kidney failure at elevated concentrations [2]. The removal of heavy metal ions from wastewater is mandatory to protect the population and the environment. Various wastewater treatment technologies including convention filtration, coagulation-flocculation, reverse osmosis and adsorption have been introduced using membranes [3]. According to Singh and Hankins [4], the use of membrane in the removal of heavy metal ions from wastewater is economic, environmentally friendly and easily accessible. The varying nature and complexity of wastewater requires the improvements, in terms of quality of permeate and efficiency to enhance the reduction in membrane fouling. Nowadays, various researchers are focused on solving this problem by obtaining a new material applicable in any wastewater treatment facilities. Proteins are natural polyelectrolytes and have attracted great interest due to their ability to form self-assembled structures [5]. Gelatin is a protein, commercially available, biodegradable and presents good processability and film-forming properties [6]. Gelatin films present moisture-sensitive and exhibit low water vapor barrier and mechanical resistance under high relative humidity due to the hydrophilicity of the polypeptide chains [7]. Cellulose nanocrystals (CNCs) were used

*Corresponding author: johnka@vut.ac.za

to overcome these limitations due to their hydrophilic, renewable and non-toxic characteristic [8]. Several researchers have described a joint synergistic effect between gelatin and CNCs toward the formation of percolated networks stabilized by hydrogen bonding. The work done by George and Siddaramaiah [9] showed that the use of CNCs, obtained from cellulose maximized the tensile strength, Young's modulus, and stiffness of gelatin films. Leite, Ferreira, Correa, Moreira, and Mattoso [10] demonstrated that adding CNCs increased the tensile strength and high elastic modulus and low thermal expansion coefficient. Tensile strength and elongation at break play an important role in determining the flexibility and strength of a film specially to overcome the external stress that occurs during the process of heavy metals removal in wastewater. The estimation of mechanical properties is a very promising method as there is a growing interest in minimizing the quality control and conducting tests with real-time prediction. Neural networks (NN) have been found to be one of the most promising numerical simulation techniques to make reliable predictions [11]. Recently, several studies have reported the successful implementation of NNs for data modelling for a wide range of engineering applications [12]. NN is a system of data processing established on the structure of a biological neural system [13]. Even though there are numerous studies focusing on the development of NN models for predicting the properties of membrane system, not much research has been reported on the application of the NN models to predict the mechanical properties of gelatin-cellulose nanocrystals hydrogel membranes.

2 Materials and methods

2.1 Preparation of the gelatin-cellulose nanocrystals hydrogen membrane and metal solution

All the chemicals and reagents used in this work were analytical reagent grade purchased from Sigma-Aldrich, South Africa. Adequate quantities of CNCs and gelatin (purity $\geq$ 98%, 0.5 wt% and 5.0 wt% on a dry gelatin mass basis) were supplied by Council for Scientific and Industrial Research (CSIR), Pretoria, South Africa and dispersed in 50 mL of water. The CNCs suspension was then homogenized using a homogenizer to ensure the CNCs suspended uniformly. Certain amount of gelatin was then added into CNCs suspension. The mixture was then stirred at 55°C until a homogeneous viscous mixture was obtained. The cross-linking agent (EDTA 1%) was then added drop wise. After 4 hours, the mixture was poured into a petri dish and placed in oven at 45°C until the mixture was dried. Hydrogels as films were removed from the petri dish and washed with distilled water to remove unreacted chemicals. The unreacted chemicals have been removed from hydrogel using acetone. Heavy metal ions were prepared by dissolving copper sulfate ($CuSO_4 \cdot 5H_2O$), nickel nitrate $(NiNO_3)_2 \cdot 6H_2O$ and lead chloride ($PbCl_2$) for feed solution for testing the membrane performance. The initial pH of the solution was adjusted with 0.1 M HCl or 0.1M NaOH. All experiments were carried out in triplicate and the results were reported as an average value at 95% confidence level.

2.2 Characterization of the membrane

2.2.1 Determination of Zeta-potential (pHz)

The Zeta-potentials (pHz) provide of the net charges on particle surfaces as a function of pH. Zeta-potentials of aqueous solution of CNCs dispersion and gelatin were obtained at pH 4, 6 and 8 (0.5 wt% CNCs on a dry gelatin mass basis). Zetasizer Malvern Nano ZS at 25°C was used for the charge on the membrane surface. The sample was mixed with a sonicator for 5 min with microtip before analysis. All the measurements were performed in triplicate.

2.2.2 Mechanical tests

The mechanical tests were carried out in an Instron Universal Testing Machine equipped with 100 N load cell at room temperature. The Gelatin-CNCs hydrogel membrane samples were stretched using crosshead speed of 10 mm/min with clamps initially separately by 100 mm. Tensile strength, Young's modulus, and elongation were calculated from the stress-strain curves. The tensile tests were performed with five replicates per membrane samples.

2.2.3 Characterization analysis

The characterization of gelatin-CNCs hydrogel membranes was done using Fourier transform Infra-Red (FT-IR) spectroscopy and scanning electron microscopy (SEM). Perkin Elmer Spectrum 100 spectrometer on CaF_2 plates was used for FT-IR spectroscopic analysis and recorded in the 500–4000 cm^{-1} range at a resolution of 4 cm^{-1}. SEM analysis using JEOL JSM-840 Field Emission Scanning Emission Microscope instrument was used for the morphology and composition of the materials were characterized, running at 2 kV accelerating voltage. The quantity of membrane used was between 0.5–1.0 g and backfilled with nitrogen gas (99 % High Purity) and subsequently outgassed at 25°C for 24 hrs to remove adsorbed impurities acquired during ambient exposure. The Thermogravimetric analysis (TGA) was done using a Q500 equipment. A mass of 10 mg of sample was taken in a standard Ptcrucible and heated from 25 to 600°C with heating rate of 10°C min^{-1}, under inert atmosphere of nitrogen (60 mL/min). Samples before the TGA tests were conditioned in a ventilated climatic chamber at 23 ± 1°C and 50 ± 5% RH for 48hrs.

2.2.4 Heavy metal ions removal

The heavy metal ions removal tests were made for Cu^{2+}, Ni^{2+} and Pb^{2+}. The removal tests were done with solutions containing 200 mg/L for each heavy metals. The pH of the solution was 6.5 and for a solution containing 30 mg/L of each heavy metal where the pH value was adjusted to 3. The samples were left on a shaking board at 25°C for 24 hrs after which the hydrogels were removed from the solution. The heavy metal mass concentration in each solution before and after removal was measured using Atomic Absorption Spectroscopy (AAS, thermos electrical corporation 3300). All heavy metal solutions were diluted prior to the measurements. The heavy metal ion removal percentage was calculated using Equation (1).

$$\text{Heavy metal ion removal } (\%) = \frac{C_o - C_e}{C_o} x100 \quad (1)$$

where C_e and C_o is the mass concentration (mg/L) of heavy metal remaining after and the initial in the solution, respectively.

2.3 Neural Network modelling development

The Neural Network Toolbox V4.0 MATLAB, 2020a mathematical software was used for modelling the mechanical properties of gelatin-cellulose nanocrystals hydrogel membrane. In this study, 120 data sets were used to develop the NN model. The input variables were Young's modulus, tensile strength and elongation. The corresponding heavy metal ions removal (%) was used as target. A three-layered backpropagation NN with tangent sigmoid transfer function (*tansig*) at hidden layer and a linear transfer function (*purelin*) at output layer were used. The backpropagation algorithm was used for the network training. The number of hidden neurons was fixed in order to study the influence of the input variables to optimize the selection of the wastewater treatment parameters which have to be taken into account for membrane prediction. The connection between the neurons is assigned with a weight to each neuron, is added to the sum of the weighted inputs. Moreover, an activation function is defined for each layer of the NN. The activation functions are specific to a layer and can be different for the hidden and output layers. In order to fit the demands of the activation functions, inputs and targets of the NN are normalized between 0 and 1. The normalization function is given in the Equation (2).

$$Z = \frac{X - X_{min}}{X_{max} - X_{min}} \quad (2)$$

where Z is the normalized input, X is the original input and X_{min} and X_{max} are the minimum and maximum values, respectively of the input variables over the whole range of input values of the dataset. The training phase of NN which is randomly defined to minimize its performance factor defined as the mean squared error (MSE). The learning procedure initialised the random initial weights and biases for the NN and applied the Levenberg-Marquart (LM) algorithm to minimise the MSE. The mean squared error is allows to compare the outputs given by the NN and the outputs measured during the experiment. The MSE is given in the Equation (3).

$$MSE = \frac{1}{N} \sum_{N}^{i=1} \left(Y^{(i)} - Y_t^{(i)}\right)^2 \quad (3)$$

where $Y^{(i)}$ and $Y_t^{(i)}$ are the ith predicted and target outputs, respectively over N data.
The validation dataset is used to avoid overfitting of the network and another part of the dataset is used to test the network after the training phase and evaluate its accuracy [12]. The performance of the NN can be evaluated using the root-MSE (RMSE) along with the coefficient of determination (R^2) and the relative error presented in Equations (4) and (5), respectively.

$$R^2 = \frac{1 - \sum_{N}^{i=1} \left(Y_t^{(i)} - Y^{(i)}\right)^2}{\sum_{N}^{i=1} \left(Y_t^{(i)} - \hat{Y}\right)^2} \quad (4)$$

$$Error\ (\%) = \left| \frac{Y_t^{(i)} - Y^{(i)}}{Y_t^{(i)}} \right| x100 \quad (5)$$

where $\hat{Y}$ is the average of the predicted outputs over the N data. In order to check the generality of the network prediction and to prevent the data overfitting, the data sets were divided into training, validation and test dataset with the percentage of 70, 15 and 15, respectively. Data were subject to analysis of variance (ANOVA). The mean values were tested at a confidence level of 95%.

3 Results and Discussion

3.1 Physical and rheological properties of Gelatin-Cellulose Nanocrystals Hydrogel Membrane

Figure 1 illustrates the evolution of the Zeta-potential of gelatin solution and CNCs suspension upon pH variation. The isoelectric point (pI) of gelatin solution occurred at around pH 6, at which a pHz = 0. Gelatin has the ability to present different charges depending on the pH [8]. At pH < pI, the gelatin solution showed a net positive charge, which was attributed to the protonation of the amino groups, while at pH > pI the gelatin solution showed a net negative charge, due to ionization of the carboxyl groups. For the CNCs suspension, the pHz value was found to be negative regardless of the pH. The pHz (< - 40 mV and > + 40 mV) could maintain a stable colloidal system due to the repelling forces between particles. In aqueous solutions, electrostatic charges are one of the main driving forces for interactions between charges biopolymers [7].

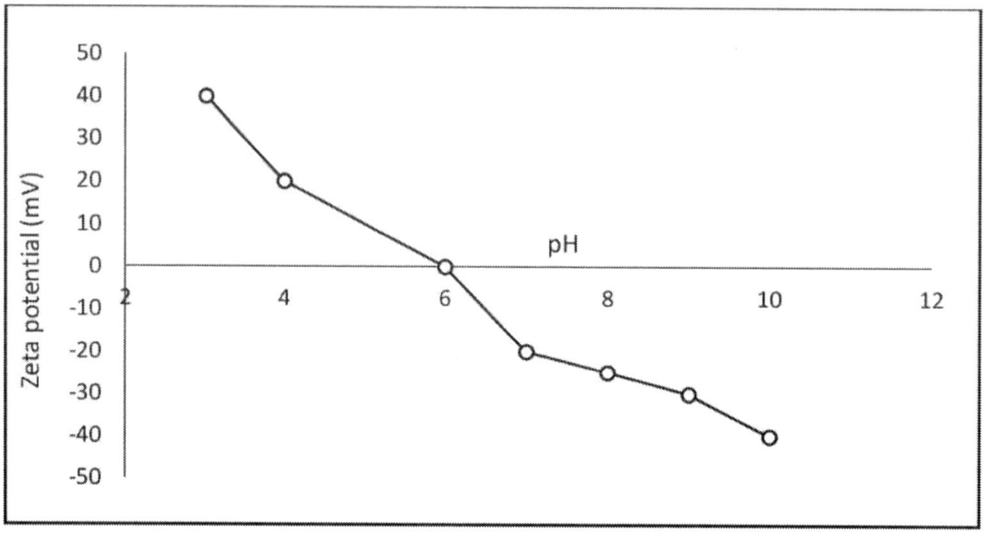

Fig. 1. Effect of pH on zeta potential

3.2 Characterization of Gelatin-Cellulose Nanocrystals Hydrogel Membrane

3.2.1 FTIR analysis

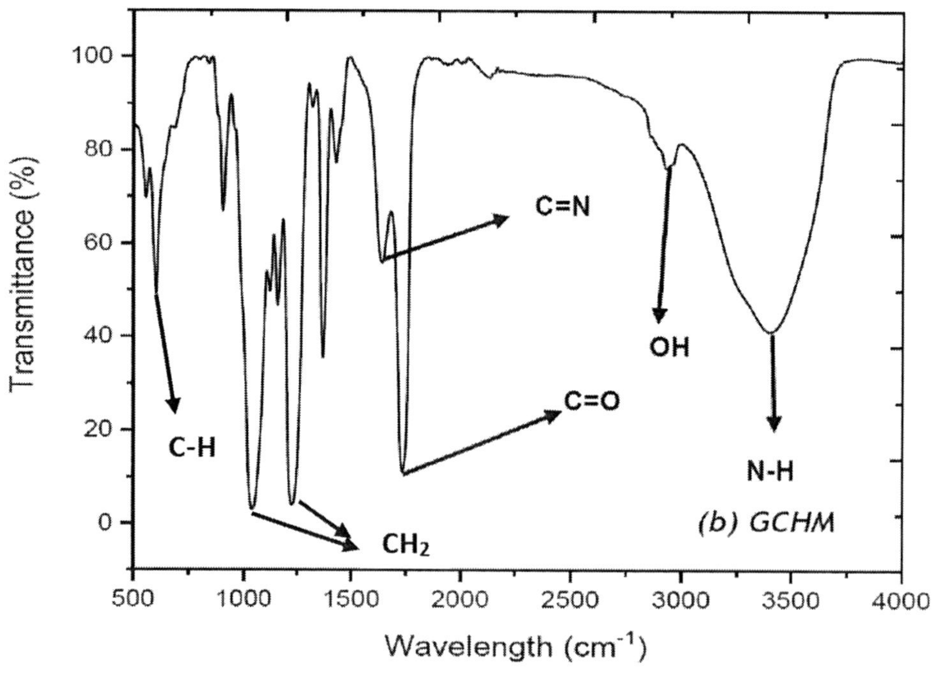

Fig. 2. FT-IR spectra of the Gelatin-cellulose Hydrogel Membrane

The FT-IR analysis of gelatin and CNCs membrane is presented in Figure 2. The FT-IR spectra were used to confirm the crosslinking of the gelatin chain and to study the changes that occurred in the functional groups of the gelatin following the reaction with bacterial cellulose. Gelatin-CNCs membrane has shown the characteristic bands around 2500 and 3000 cm^{-1} that represents the strong elongation vibration band of the OH function. The band around 1450 cm^{-1} is the vibration shear of CH_2 and it is regarded as crystallinity band in any cellulosic material [13]. The peak at 1700 cm^{-1} is characteristic of the elongation vibration band of the carboxyl (C=O) confirmed by the characteristic band around 890 cm^{-1} that corresponds to the C-H of aldehyde group [13]. The amide is represented by the deformation vibration peaks of N-H at 3500 cm^{-1}. Moreover, the amide is characterized by the elongation vibration of C=O to which is added a contribute band of elongation vibration of C-N. The peaks below 1000 cm^{-1} are characteristic of low molecular weight amides [14]. The incorporation of CNCs on the membrane is confirmed by the peak at 3500 cm^{-1}, which is attributed to the stretching vibration of hydroxyl groups from CNCs. In addition, a higher intensity of this peak is observed when increasing the CNCs content. This occurrence can contribute to the increased hydrophilicity of the membrane.

3.2.2 SEM analysis

Fig. 3. SEM image of the Gelatin-cellulose Hydrogel Membrane

Figure 3 indicates the SEM analysis in the morphological structure of the Gelatin-CNCs Membrane. Porosity characterization is based on the presence of open pores which are related to properties such as permeability and surface area of the porous structure. The SEM image indicates the presence of bigger particles with irregular shapes (Fig. 3). Figure 3 also shows the cavities of different shapes and sizes, and larger pores between the particles could be observed which will be helpful for the solution permeating through the membrane [15]. The high porosity provides a favourable removal of heavy metal ions [16]. The presence of such granules increases the surface area of the composite which is suitable for effective adsorption of heavy metal ions. Small openings and holes on the surface increase the contact of the adsorption and therefore lead to pore diffusion during the removal process [17]. A morphology that changes as the gelatin concentration increases (Fig. 3) and consequently, several pores have been observed.

3.2.3 TGA analysis

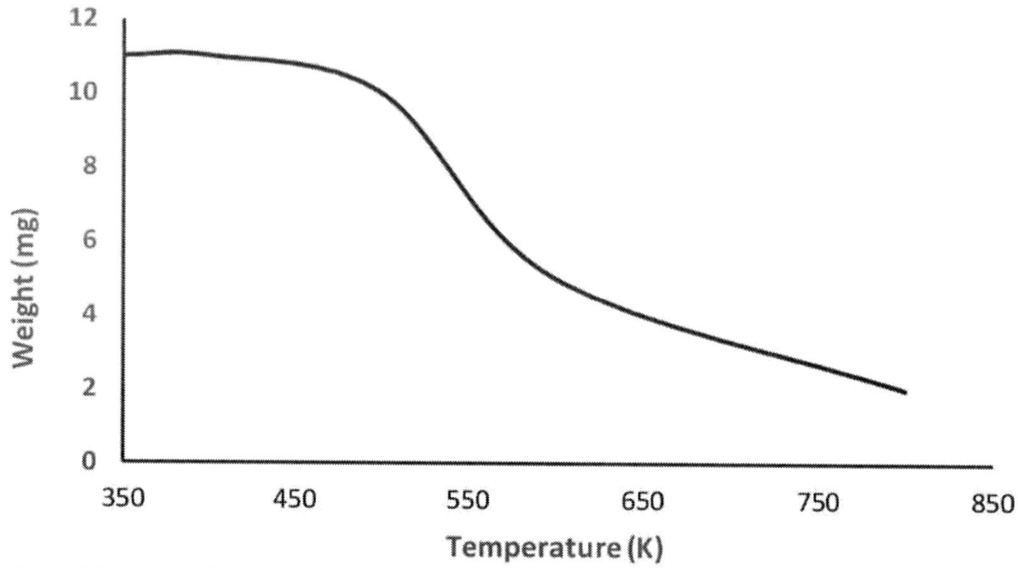

Fig. 4. TGA curve of membrane shows the relationship of weight and temperature

The characterization of the membrane was further studied by using TGA analysis. TGA analysis monitored the thermal degradation behaviour of the membrane, as a function of temperature in controlled surroundings. The TGA analysis result of membrane is shown in Figure 4. The first degradation of all films started from 350 to 800°C with the mass weight loss ranged from 12 to 2 mg. This observation might be due to the decomposition of highly connected amino acids group in the protein films [18]. Due to its crystalline structure, the incorporation of cellulose into the gelatin film forming solution is able to improve the thermal stability the films. The thermal decomposition of gelatin–cellulose blend films is shifted towards higher temperature as compared to gelatin films. The enhancement in thermal stability of gelatin–cellulose films can be attributed to the intermolecular and intramolecular interaction between the polymers. Overall, the results obtained from TGA analysis corroborate the earlier findings for FT-IR analysis, particularly the interaction behaviour between the polymers.

3.3 Selection of backpropagation training algorithm and transfer function

In this study, 10 backpropagation algorithms were compared to select the best. For all backpropagation algorithms, a 3-layer NN with *tansig* at hidden layer and a *purelin* at output layer were used. Levenberg-Marquardt (LM) backpropagation training algorithm (*trainlm*) with a minimum mean squared error (MSE) was found as the best. The results presented in Table 1 revealed that LM backpropagation has the smallest MSE value (MSE = 0.00029). So LM backpropagation was considered as the training algorithm. The next factor is the transfer function, different transfer functions were examined in each layer. Among the different transfer function, *log-sigmoid* function was selected for all neurons due do its better prediction performance than other transfer function [19].

Table 1. Comparison of 10 backpropagation algorithm

Backpropagation algorithm	Function	MSE	R^2
Batch gradient descent	*traingd*	0.44628	0.85
Batch gradient descent with momentum	*traingdm*	0.50724	0.56
BFGS quasi-Newton backpropagation	*trainbfg*	0.06100	0.88
Fletcher-Reeves conjugate gradient backpropagation	*traincgf*	0.05422	0.82
Levenberg-Marquardt backpropagation	*trainlm*	0.00029	0.94
One step secant backpropagation	trainoss	0.07461	0.82
Polak-Ribiere conjugate gradient backpropagation	*traincgp*	0.04871	0.84
Powell-Beale conjugate gradient backpropagation	*traincgb*	0.15762	0.88
Scaled conjugate gradient backpropagation	*trainscg*	0.14287	0.89
Variable learning rate backpropagation	*traingdx*	0.64783	0.45

3.4 Neural network modelling results

The NN model is designed to incorporate the mechanical properties of the Gelatin-CNCs membrane as input data and to predict the removal efficiency of heavy metals from wastewater.

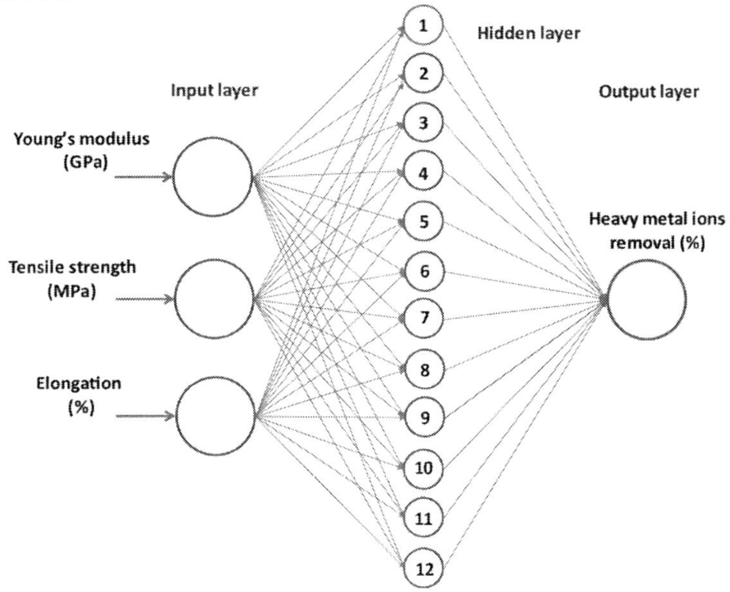

Fig. 5. Structure of Neural Network representing inputs and outputs of the model

Figure 5 shows the structural design of the NN model with input, hidden and output layers. The structure of the NN was obtained by trying different number of hidden layers and neurons. The NN structure model (3-12-1) with a tangent sigmoid transfer function (*tansig*) at hidden layer with 12 neurons and a linear transfer function (*purelin*) at output layer was obtained as presented in Figure 5.

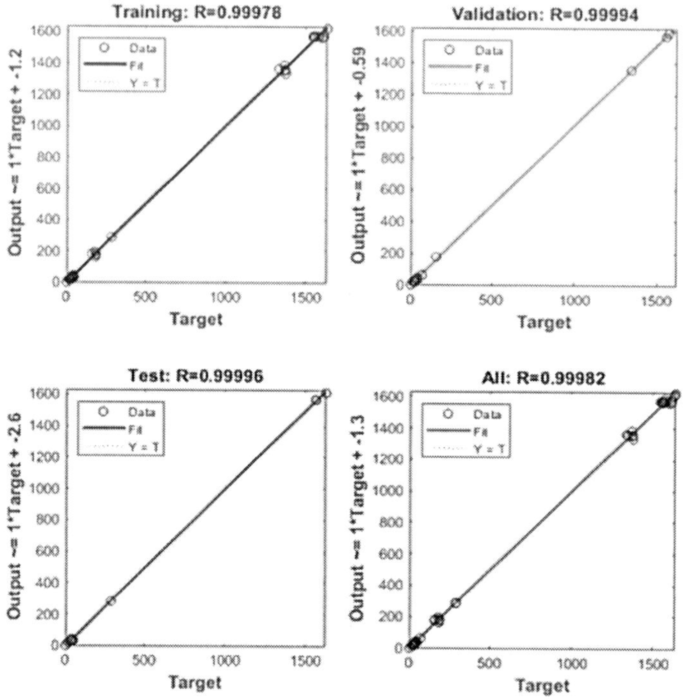

Fig. 6. Regression analysis of the network response between NN output and the corresponding targets

The scatter plots in Figure 6 illustrate a high efficiency of the neural network. Measurements of mechanical properties are based on the most probable predictions from the NN model for training, validation, testing, and overall data set. All R^2 have high values and are close to each other, which signifies the high performance of the model. For all set (R^2 = 0.99315), a significant part of predictions is in excellent correlation with the experimental data. Hence, it can note that the developed NN is appropriate and can predict tensile and impact values accurately. The testing results indicate a good generalization capability between input and output data. Figure 6 provides a comparison of the mean values of experimental and predicted data. At the output, the developed NN model successfully predicts mechanical properties of Gelatin-CNCs membrane for the removal efficiency of heavy metals.

3.5 Significance of the developed Neural network model

The objective of this study is to enable to predict accurately the mechanical properties of membrane as a function of the removal of heavy metals from wastewater. However, most of the NN models developed in previous research. In this case, the measurement data are scattered and the repetition rate is low. At the same time, modelling of mechanical properties is very difficult due to the complex interaction of several parameters. These interactions are too complex and poorly understood. In the current work, a well-trained model was elaborated, which could be an excellent candidate to predict the mechanical properties as a function of heavy metals removal efficiency. NN approach model can significantly reduce the need for experimental measurement of mechanical properties, which shall be greatly appreciated in membrane technology for wastewater treatment industries.

3.6 Mechanical Properties

The mechanical properties such as Young's modulus, tensile strength, and elongation of the Gelatin-cellulose hydrogel membrane were investigated with tensile tests and the results in the form of stress–strain curves shown in Figure 7. pH and CNCs content significantly regulate the mechanical properties of gelatin films. Gelatin-CNCs membrane for pH 4 presented significantly increased of stress between 0 and 5% strain. This may be attributed to the electrostatic attractive interaction between gelatin and CNCs reinforcing effect through stress transfer from the gelatin matrix to CNCs. When pH is 6, the increase of stress was observed to be almost half compared to pH 4 most likely due to the reduction in the electrostatic interactions above the gelatin. For the alkaline condition at pH 8, the stress increases at 15 MPa and remains constant due to the regeneration, which increases the overall crystallinity of the films. The mechanical properties behaviour of Young's modulus, tensile strength, and elongation can be attributed to the formation of agglomerates as revealed by the SEM result in Figure 3. The gelatin film with CNCs showed an elongation at break higher than that of the pure gelatin film, indicating that the gelatin-CNCs membranes were much more ductile and flexible. The interaction formed between the gelatin and CNCs weakened the protein-protein interactions, which were effective in stabilizing the gelatin network, as described in the work done by Zhuang et al. [20]. Therefore, it is clear that CNCs play an effective role in enhancing the mechanical properties of gelatin, with promising features for heavy metal ions removal from wastewater. The results agree with the works done by Leite et al., [6] for the CNCs nanocomposite films.

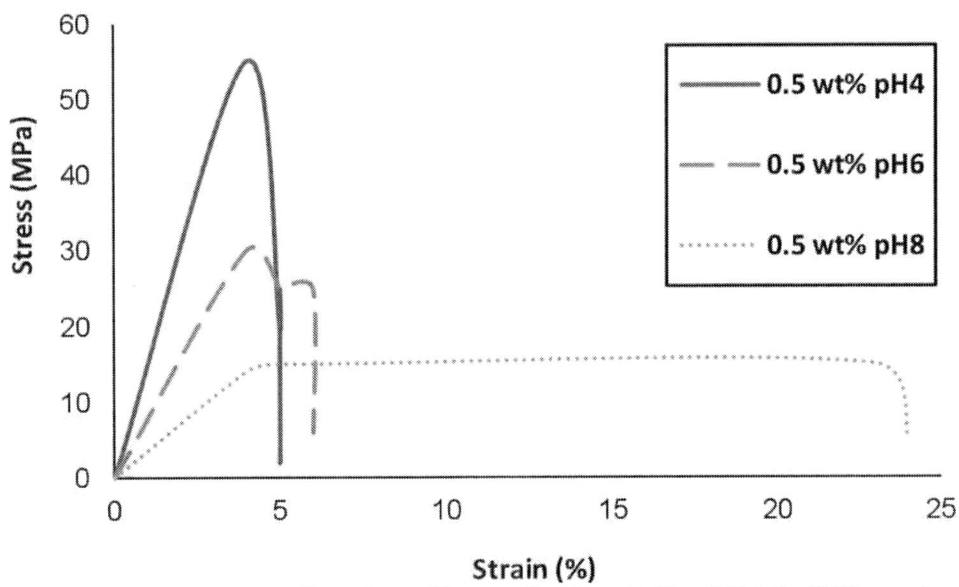

Fig. 7. Stress-strain curves of membrane film as a function of pH and Gelatin-CNCs membrane content

It can be seen from the Figures 8 that the maximum removal (72%) for the tensile strength was 50 MPa; in Figure 9, the elongation was 72% at the maximum removal (74%); and in Figure 10, the maximum removal (77%) for the Young's modulus was 19 GPa. Figures 8-10 provide a comparison of the mean values of experimental and predicted data. At the output, the developed NN model successfully predicts various mechanical characteristics. In Figures 8-10, the results also show a slight variation for the tensile strength and elongation at break except for the Young's modulus due to the high inherent scatter of the impact properties. In terms of the relation between the experimental data and the predicted values of heavy metals removal by the NN models. Figures 8-10 show that predicted values are in good agreement with the experimental data.

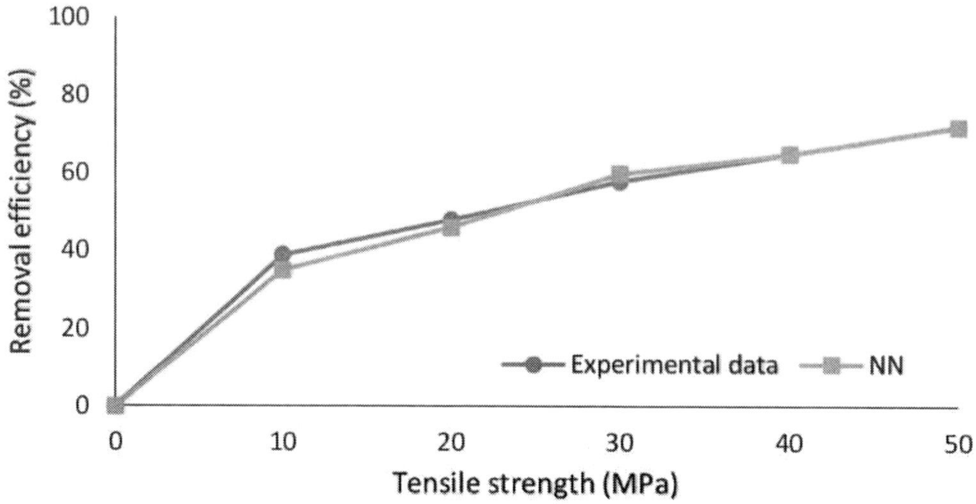

Fig. 8. Comparison between NN outputs and experimental data for Tensile strength at pH 4

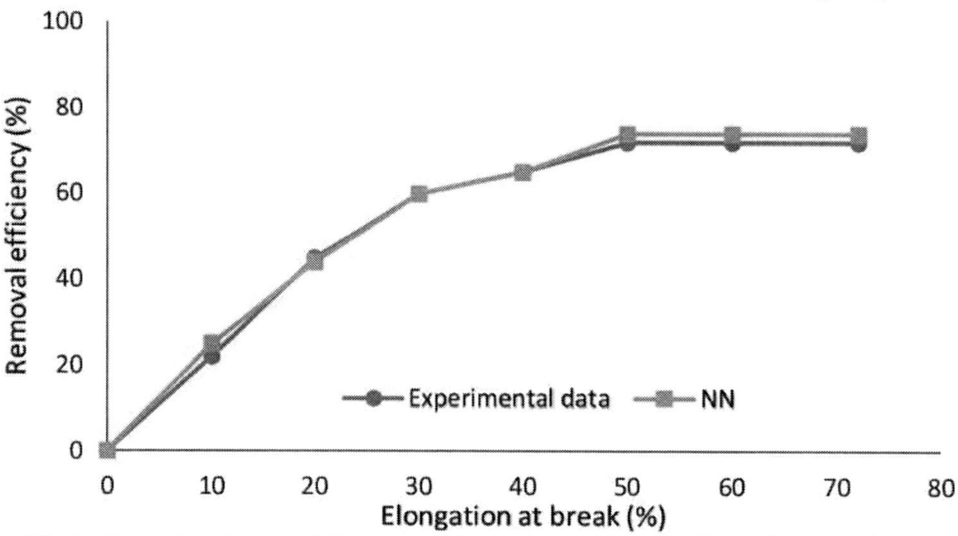

Fig. 9. Comparison between NN outputs and experimental data for Elongation at break at pH 4

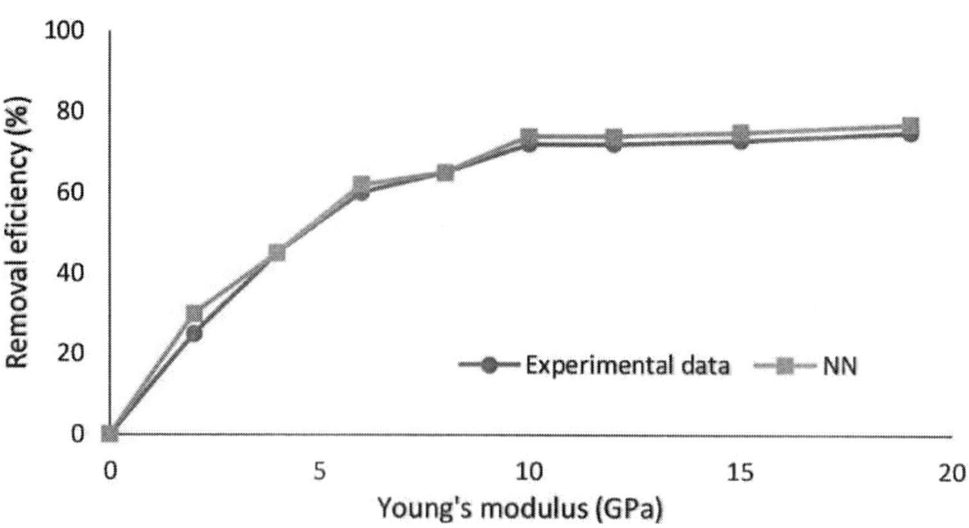

Fig. 10. Comparison between NN outputs and experimental data for Young's modulus at pH 4

4. Conclusion

Neural network model was designed to predict the mechanical properties of Gelatin-cellulose hydrogel membrane. The predicted and experimental values for the present model are in a good correlation and highlight the success of the usage of NN for easier prediction of mechanical properties such as tensile strength, elongation and Young's modulus for the removal of heavy metal ions from wastewater. The configuration of the BPNN giving the smallest MSE was LMA with a tangent sigmoid transfer function at hidden layer with 12 neurons and a linear transfer function at output layer. The NN predicted results are very close to the experimental results with R^2 = 0.99315. The developed model possesses superior predictive capabilities as compared to the results obtained from the regression analysis. Neural networks are powerful tool that can be used to minimize the cost of experimental research.

References

1. I. García-Díaz, F.A. López, F. J. Alguacil. Metals, **8**, 914 (2018)
2. J. Kabuba. Eng. Letters. **29**, 158 (2021)
3. E. O. Ezugbe, S. Rathilal. Membranes, **10**, 10050089 (2020)
4. R. Singh, N. Hankins. Elsevier: Amsterdam, The Netherlands (2016).
5. Z. Wang, J. Zhou, X. Wang, N. Zhang, X. Sun & Z. Ma. Food Hydrocolloids, **35**, 51 (2014).
6. L.S. F. Leite, F.K.V. Moreira, L.H.C. Mattoso, J. Bros. Food Hydrocolloids. **113**, 106424 (2021)
7. S. T. Hosseini & M. C. Gomez-Guillen. Trends in Food Sci. Technol. **79**, 125 (2018)
8. M. T. Hoang, T. D. Pham, D. Verheyen, M.K. Nguyen, T.T. Pham, J. Zhu, B. Van der Bruggen, Chem. Eng. Sci. **228**, 115998 (2020)
9. J. George & Siddaramaiah. Carbohydrate Polymers, **87**, 2031 (2012)
10. L. S. F. Leite, C.M. Ferreira, A. C. Correa, F.K. V. Moreira & L. H. C. Mattoso, Carbohydrate Polymers, **238**, 116198 (2020)
11. J. Kabuba, A. Mulaba Bafubiandi, K. Battle. Arab. J. Sci. Eng. **39**, 6793 (2014)
12. A. Alemdar, M. Sain. (2008). Bioresource Technol. **6**, 1664 – 1671 (2008)
13. P. Fernandes de Almeida, S. D. da Silva Lannes, F. Araújo Calarge, T. M. de Brito Farias, J. C. C. Santana. J Chem Chem Eng. **6**, 1029 – 1032 (2012)
14. R. M. Silvestein, F. X. Webster, D. J. Kiemle, D. L. Bryce. Spectroscometric identification of Organic Compound, 8th edition. Wiley. New York, 81 – 108 (2007)
15. X. Wang, P. Liu, F. Liu, X. Wang, M. Ji, L. Song. Adsorption Sci Technol. **36**, 1571 – 1594 (2018).
16. M. J. Hossan, M. A. Gafur, M. R. Kadir, M. M. Karim. Polymer Composites. **14**, 24 – 32 (2014)
17. J. Kabuba. Eng. Letters. **27**, 467-474 (2019)
18. M. S. Samsi, A. Kamari, S. M. Din, G. Lazar. J Food Sci Technol. **56**, 3099-3108 (2019)
19. J. Kabuba and K. Aziza, Artificial Intelligence Neural Network Techniques to Predict the Removal of Anions from Wastewater, in Proceedings of the 11th South African Conference on Computational and Applied Mechanics, SACAM 2018, 17-19 September 2018, Vanderbijlpark, South Africa (2018)
20. C. Zhuang, F. Tao, Y. Cui, J. Sci. Food Agric. **97**, 3384–3395 (2017)

A tri-phasic continuum model for the numerical analysis of biological tissue proliferation using the Theory of Porous Media.

Application to cardiac remodelling in rheumatic heart disease

Adam Mosam[1,2,*], *Sebastian* Skatulla[1,2], and *Ntobeko* Ntusi[3]

[1]Computational Continuum Mechanics Group, Department of Civil Engineering, University of Cape Town,South Africa

[2]Centre for Research in Computational and Applied Mechanics, University of Cape Town, South Africa

[3]Division of Cardiology, Department of Medicine, University of Cape Town, South Africa

Abstract. This research is part of an on-going project aimed at describing the mechanotransduction of rheumatic heart disease (RHD), in order to study long-term effects of new therapeutic concepts to treat inflammatory heart diseases and ultimately, estimate their effectiveness to prevent heart failure. RHD is a condition which is mostly common amongst low-income countries and accounts for approximately 250 000 deaths per annum. The Theory of Porous Media (TPM) can represent the proliferative growth and remodelling processes related to RHD within a thermodynamically consistent framework and is additionally advantageous with application to biological tissue due to the ability to couple multiple constituents.

The research presented will extend an existing biphasic TPM model for the solid cardiac tissue (solid phase) saturated in a blood and interstitial fluid (liquid phase) [1], to a triphasic model with the inclusion of a third nutrient phase towards growth. This inclusion is motivated by the reason to constrain the volume of the liquid phase within the system in response to the description of growth, which is modelled through a mass exchange between the solid phase and liquid phase within the biphasic model. Although the nutrient phase acts as a source for growth, the proposed mass supply function used to correlate the deposition of sarcomeres in relation to growth is predominantly mechanically driven and bears no connection to any biochemical constituent, which therefore renders the nutrient phase as a physiologically arbitrary quantity. However, the provision of the nutrient phase is a platform for the inclusion of known constituents which actively contribute towards growth, which may be explored in future research.

The triphasic model is applied to a full cardiac cycle of a left ventricle model, extracted from cardiovascular magnetic resonance (CMR) scans of patients diagnosed with RHD.

*e-mail: msmada002@myuct.ac.za

1 Introduction

1.1 Cardiovascular disease

Cardiovascular disease is the most prevalent cause of death globally, accounting 40% of all human mortality [2]. Even through significant medical advancement, approximately 25 million people suffer from heart failure, annually [3]. Unlike other types of tissue in the body, cardiac tissue is unique in the sense that it can rarely regenerate after being damaged [4].

Attention will be given to rheumatic heart disease (RHD), a chronic heart condition which is caused by a series of attacks of *rheumatic fever* [5]. Valvular dysfunction is the critical feature of RHD whereby the valves of the left ventricle are most commonly left in a state of permanent damage or complete failure [5, 6]. Hypertrophy of the muscle walls of the left ventricle is induced as a result of valvular dysfunction [6, 7]. Hypertrophy of the ventricle may either be in the form of concentric or eccentric growth, whereby cardiomyocytes are added in parallel (increase in thickness) or in series (elongation of the tissue), respectively [8]. Within this study, attention will be focused on the case of eccentric hypertrophy in relation to left ventricle dilation (LVD).

1.2 Biological and cardiac modelling

Cardiac modelling is viewed as the benchmark for soft tissue modelling, with much attention given to growth in arteries, with respect to cardiovascular disease [9, 10], and full scale heart modelling [11–14]. The characterisation of cardiac tissue growth in relation to biochemical stimulants, is not well understood. Hence, the studies which model growth proliferation are mechanically driven. A common example is found in the growth laws formulated to model eccentric and concentric hypertrophy through stress and strain based laws, respectively, proposed by Goktepe et al. [11]. Growth related modelling of cardiac tissue generally makes use of transversely isotropic growth laws, due to the composition of the muscle structure[12]. In the work of Werner et al. [15], transversely isotropic and isotropic growth was initiated on circular discs, defined as cardiac tissue material. This study employed a theory of porous media framework. The Theory of Porous Media (TPM) is advantageous in terms of soft tissue modelling as it is able to capture solid-fluid interaction [16]. In addition to the aforementioned citation, TPM has been utilised in describing soft tissue and multiphase systems in recent times, such as with the study of the liver in the work of Ricken et al. [16], and with a patient-specific left ventricle, conducted by Hopkins [1].

This research entails adding a third phase to an existing biphasic TPM framework, implemented in an in-house computational mechanics software SESKA. The biphasic constituents comprise of a solid and fluid phase, which poses a few problems in relation to growth proliferation [1]. Mitigating these problems associated with the biphasic model could be realised by creating a triphasic model through introduction of a third nutrient phase [1]. In this research, the triphasic model proposed is based on a model developed by Ricken et al. [17].

In terms of growth, a more efficient, and mathematically consistent method of expressing the growth evolution, which is derived from the balance equation of mass and represented through the growth Jacobian, is available in the work of Werner [18], as compared to that implemented in the current biphasic model. Coupling this growth evolution method with a strain driven, transversely isotropic growth law, for the case of LVD, will hence form the second major objective of this research.

2 Triphasic material model

In the framework of TPM, a solid porous skeleton is saturated with pore fluids, which constitute the κ constituents of the control space. The control space, at time t, is defined by B_S and bordered by the surface ∂B_S. The *volume fraction* n^α, may be defined as the local volume portion which belongs to the individual constituents φ^α, and relates the constituent volume elements dv^α to the bulk volume elements dv, by

$$n^\alpha(\mathbf{x}, t) = \frac{dv^\alpha}{dv}, \qquad \sum_\alpha^{S,L,N} n^\alpha(\mathbf{x}, t) = \sum_\alpha^{S,L,N} \frac{\rho^\alpha}{\rho^{\alpha R}} = 1 \tag{1}$$

with α being one of the κ constituents. For the triphasic model used in this research, solid, liquid and nutrient constituents are used, such that $\alpha \in \{S, L, N\}$. Furthermore, $\mathbf{x}$ is the spatial position at time t and ρ^α is the partial density, which relates to the real density $\rho^{\alpha R}$, by $\rho^\alpha = n^\alpha \rho^{\alpha R}$. Eq.$(1)_2$ describes what is termed the *saturation condition*, and should always be fulfilled.

2.1 Kinematics

The saturated porous solid will be idealized as a statistically distributed immiscible mixture comprising of constituents φ^α with particles $\mathbf{X}_\alpha$, that have independent motion functions $\mathbf{x} = \mathcal{X}_\alpha(\mathbf{X}_\alpha, t)$. This represents the *Lagrange* description of motion. With $\mathcal{X}_\alpha$ being unique and continuous, its inverse mapping leads to the *Eulerian* description of motion $\mathbf{X}_\alpha = \mathcal{X}_\alpha^{-1}(\mathbf{x}, t)$. This statement may only be true if the mathematical condition of a non-zero *Jacobian* $J_\alpha = \det\mathbf{F}_\alpha \neq 0$, is satisfied for each constituent, with $\mathbf{F}_\alpha$ being the *deformation gradient* $\mathbf{F}_\alpha = \frac{\partial \mathcal{X}_\alpha}{\partial \mathbf{X}_\alpha}$. Each constituent possesses its own velocity and acceleration field. In the Lagrangian configuration, these are stated as

$$\mathbf{x}'_\alpha = \frac{\partial \mathcal{X}_\alpha(\mathbf{X}_\alpha, t)}{\partial t}, \qquad \mathbf{x}''_\alpha = \frac{\partial^2 \mathcal{X}_\alpha(\mathbf{X}_\alpha, t)}{\partial t^2} \tag{2}$$

In the Eulerian configuration, the motion of a particle is governed by time t, and the spatial position $\mathbf{x}$. For an arbitrary scalar-value function $\Gamma(\mathbf{x}, t)$, the *material time derivative* is defined as $(\Gamma)'_\alpha = \frac{\partial \Gamma}{\partial t} + \operatorname{grad}\Gamma \cdot \mathbf{x}'_\alpha$, where "grad(·)" denotes the partial derivative of (·) with respect to the spatial position. Similarly, Grad(·) is defined as the partial derivative with respect to the material position.

For problems involving a multiphase coupling of a porous solid and $\kappa - 1$ fluids, one would generally continue from a Lagrangian description of the solid phase, with the solid displacement vector $\mathbf{u}_S$, being the unknown variable, determined by $\mathbf{u}_S = \mathbf{x} - \mathbf{X}_S$. However, with respect to the fluid phases, greater ease is found by adopting the Eulerian setting and by using the relative velocity between each respective fluid phase and the solid skeleton [19]. Consequently the *seepage velocity* is introduced as $\mathbf{w}_{\beta S} = \mathbf{x}'_\beta - \mathbf{x}'_S$.

The *spatial velocity gradient* is given by $\mathbf{L}_\alpha = \frac{\partial \mathbf{x}'_\alpha}{\partial \mathbf{x}}$, or alternatively as $\mathbf{L}_\alpha = (\mathbf{F}_\alpha)'_\alpha \mathbf{F}^{-1}{}_\alpha$. The spatial velocity gradient may be decomposed into its symmetric part $\mathbf{D}_\alpha = \frac{1}{2}\left(\mathbf{L}_\alpha + \mathbf{L}_\alpha^T\right)$, and skew-symmetric part $\mathbf{W}_\alpha = \frac{1}{2}\left(\mathbf{L}_\alpha - \mathbf{L}_\alpha^T\right)$.

2.2 Kinematics of Growth

The kinematic approach used employs a multiplicatively split deformation gradient $\mathbf{F}_S$ into an elastic part $\mathbf{F}_{Se}$, and a growth part $\mathbf{F}_{Sg}$, such that $\mathbf{F}_S = \mathbf{F}_{Se}\mathbf{F}_{Sg}$ [20]. This results in the formation of a stress free intermediate configuration, whereby the deformation resulting parts are captured within the elastic zone (between the intermediate and actual configuration).

Following the description of the multiplicative decomposition of $\mathbf{F}_S$, the same principle may be applied to the Jacobian $J_S = J_{Se}J_{Sg}$, whereby it's comprised of an elastic part J_{Se}, which translates into volume change due to deformation, and an irreversible growth part J_{Sg}, which causes a volume change due to deposited matter. [21]

With $\mathbf{n}_0$ being the unit normal to the surface in the reference configuration, the total stretch ϵ in the actual configuration, in the direction of $\mathbf{n}_0$, is given by

$$\epsilon = ||\mathbf{F}_S\,\mathbf{n}_0|| = \epsilon_e\epsilon_g \tag{3}$$

where ϵ_e and ϵ_g represent the stretch components, in similar analogy to the previous quantities. With respect to cardiac tissue growth, ϵ_g would hence represent the stretch caused by a generation of sarcomeres in relation to a maladaptive response to a pressure or volume overload [11]. It should be noted that, ϵ_g may be taken to be equivalent to J_{Sg}. Finally, for transversely isotropic growth, the following expression is used to determine the growth part of the deformation gradient

$$\mathbf{F}_{Sg} = \mathbf{I} + (\epsilon_g - 1)\mathbf{n}_0 \otimes \mathbf{n}_0 \tag{4}$$

2.3 Balance equations

The balance equations of the constituents φ^α are analogous to those found in mixture theory [16]. The balance equations for a porous body may be given in terms of the balance of mass, momentum and the moment of momentum

$$(\rho^\alpha)'_\alpha + \rho^\alpha \mathrm{div}(\mathbf{x}'_\alpha) = \hat{\rho}^\alpha\,, \quad \mathrm{div}\mathbf{T}^\alpha + \rho^\alpha(\mathbf{b}^\alpha - \mathbf{x}''_\alpha) = \hat{\rho}^\alpha\mathbf{x}'_\alpha - \hat{\mathbf{p}}^\alpha\,, \quad \mathbf{T}^\alpha = (\mathbf{T}^\alpha)^T \tag{5}$$

respectively, where "div" denotes the spatial divergence operator, $\mathbf{T}^\alpha$ is the Cauchy stress tensor, and $\rho^\alpha\mathbf{b}^\alpha$ is the volume force. The source terms, $\hat{\rho}^\alpha$ and $\hat{\mathbf{p}}^\alpha$, represent the mass supply and interaction forces of momentum between the constituents φ^α, respectively.

2.4 Assumptions

For the case of the left ventricle comprising of a porous solid skeleton saturated with blood, material incompressibility may be assumed for the entire system, such that $\left(\rho^{\alpha R}\right)'_\alpha = 0$. The mass supply function for the nutrient phase $\hat{\rho}^N$ accounts for the increase in mass of the solid phase and an equivalent reduction in mass of the nutrient phase, which leads to $\hat{\rho}^N = -\hat{\rho}^S$. The motion function of the nutrient phase $\mathcal{X}_N(\mathbf{X}_N, t)$ is taken to be equivalent to that of the liquid phase $\mathcal{X}_L(\mathbf{X}_L, t)$. As a result, the motion functions of the liquid and nutrient phases will hence be represented in terms of a general fluid phase $\mathcal{X}_F(\mathbf{X}_F, t)$. In addition, all body forces will be neglected and a quasi-static approach will be adopted, which results in all acceleration terms for the phases to be ignored $\mathbf{x}''_\alpha = \mathbf{0}$. Furthermore, the mass exchange term which contributes to momentum will be neglected due to the insignificant amount of momentum generated from the exchange of mass between the phases $\hat{\rho}^\alpha\mathbf{x}'_\alpha = \mathbf{0}$. The temperature difference between all the phases will be treated as negligible, as it is assumed that there is no differential temperature within the saturated tissue matrix.

2.5 Field equations

Considering the assumptions and restriction of mass supply terms, the balance of mass and momentum may be adapted to form the field equations required to model the proliferative process of growth and remodelling of cardiac tissue. Starting with the balance equation of mass, the local form may be rewritten in the form shown below for each constituent

$$\left(n^S\right)'_S + n^S\mathbf{D}_S : \mathbf{I} = \frac{\hat{\rho}^S}{\rho^{SR}}, \quad \left(n^L\right)'_L + n^L\mathbf{D}_L : \mathbf{I} = 0 \quad \left(n^N\right)'_N + n^N\mathbf{D}_N : \mathbf{I} = \frac{\hat{\rho}^N}{\rho^{NR}}, \tag{6}$$

such that the mass supply term for the liquid phase is neglected. The adapted forms of the balance equations of momentum for the mixture and for each constituent yields

$$\text{div}\mathbf{T} = \mathbf{0}, \quad \text{div}\mathbf{T}^{\text{S}} + \hat{\mathbf{p}}^{\text{S}} = \mathbf{0}, \quad \text{div}\mathbf{T}^{\text{L}} + \hat{\mathbf{p}}^{\text{L}} = \mathbf{0}, \quad \text{div}\mathbf{T}^{\text{N}} + \hat{\mathbf{p}}^{\text{N}} = \mathbf{0} \tag{7}$$

whereby the sum of the interaction forces $\hat{\mathbf{p}}^{\alpha}$, is zero and the quantity $\mathbf{T}$, represents the sum of the Cauchy stress tensor's of each constituent.

3 Constitutive modelling

Unlike the theory of mixtures, which is a closed framework that has an equivalent amount of unknowns and field equations, the theory of porous media differs in that it is constrained by the saturation condition, which stems from the volume fractions concept. This results in an additional field equation. Therefore, an additional scalar quantity needs to be introduced in the form of a *Lagrange multiplier* λ[16].

The Lagrange multiplier is introduced as a quantity by means of the entropy inequality, through the material time derivative of the saturation condition, shown in Eq.(1), and the adapted forms of the balance equations of mass Eq.(6). After evaluation of the entropy inequality (see [16]), expressions for the Cauchy stress tensor $\mathbf{T}^{\alpha}$ are yielded as

$$\mathbf{T}^{\text{S}} = 2\rho^{\text{S}}\mathbf{F}_{\text{S}}\frac{\partial\psi^{\text{S}}}{\partial\mathbf{C}_{\text{S}}}\mathbf{F}_{\text{S}}^{\text{T}} - \lambda \text{n}^{\text{S}}\mathbf{I}, \quad \mathbf{T}^{\text{F}} = \text{n}^{\text{F}}\lambda\mathbf{I}, \quad \text{n}^{\text{F}} = \text{n}^{\text{L}} + \text{n}^{\text{N}} \tag{8}$$

where the influence of the free Helmholtz free energy function of the fluid phases on the stress is neglected [16]. The following condition should be applied to the interaction force of the lumped fluid phase $\mathbf{p}^{\text{F}}$ and the solid mass supply function $\hat{\rho}^{\text{S}}$, in order to result in a valid entropy inequality:

$$\hat{\mathbf{p}}^{\text{F}} = \lambda \text{grad}\, \text{n}^{\text{F}} - \mathbf{S}^{\text{F}}\mathbf{w}_{\text{FS}}, \quad \hat{\rho}^{\text{S}} = \delta(\Psi^{\text{N}} - \Psi^{\text{S}}) \tag{9}$$

where Ψ^{α} is the chemical potential of the constituents φ^{α}, and $\mathbf{S}^{\text{F}}$ is a positive definite tensor which dictates the direction of fluid flow. The scaling parameter δ needs to be greater than or equal to zero in order to fulfil the entropy inequality conditions [16]. These expressions will be elaborated on in more detail in the following sections.

3.1 Stress

When the mitral valve reopens at the onset of the diastolic filling stage, blood is allowed to enter the ventricular cavity. It is assumed that, during this phase, the left ventricle does not exert any force but rather deforms and behaves in a passive manner. This results in a passive strain energy response, and may be represented using a strain energy function ψ^{S}. The last three phases of the cardiac cycle, namely; the isovolumetric contraction, ejection and isovolumetric relaxation, is termed the *active contraction* stage, and usually results in the muscle fibres contracting, which as a result causes the ventricular pressure within the cavity to increase due to the accumulated strain energy developed [7]. The total stress may hence be expressed as a superposition of the passive and active stresses

$$\mathbf{S} = \mathbf{S}_{\text{P}} + \mathbf{S}_{\text{A}} \tag{10}$$

where $\mathbf{S}_{\text{P}}$ is the passive stress and $\mathbf{S}_{\text{A}}$ is the active stress. $\mathbf{S}$ is related to the Cauchy stress tensor by $\mathbf{S} = \mathbf{F}_{\text{S}}^{-1}\mathbf{T}\mathbf{F}_{\text{S}}^{-\text{T}}$, where $\mathbf{T} = \mathbf{T}^{\text{S}} + \mathbf{T}^{\text{F}}$. A more descriptive overview of the constitutive formulation used to represent the passive and active material behaviour, and the subsequent integration with a TPM framework, may be viewed in the works of Mosam [22] and Hopkins [1].

3.2 Seepage velocity

From Eq.(9), a constitutive law for the seepage velocity $n^F \mathbf{w}_{FS}$ is proposed, for transversely isotropic flow [16]

$$n^F \mathbf{w}_{FS} = \frac{\left(n^F\right)^2}{\alpha_{F0}} \mathbf{M} \,\text{grad}\, \lambda \tag{11}$$

which is dependent on the spatial change in pore pressure and the permeability of the system (controlled by the first term in Eq.(11)). The direction of the flow is governed by the structural tensor $\mathbf{M}$. In order to characterize the parameter α_{F0} the following relationships which involve either, the initial Darcy permeability k^F_{0S}(m/s) and the specific weight γ^{FR}(N/m^3) or the intrinsic solid permeability k^F_{0S}(m^2) and the effective shear viscosity μ^{FR}(Ns/m^2), may be applied such that

$$\frac{\left(n^F\right)^2}{\alpha_{F0}} = \left(\frac{n^F}{1\text{-}n^S_{0S}}\right)^m \frac{k^F_{0S}}{\gamma^{FR}} = \left(\frac{n^F}{1\text{-}n^S_{0S}}\right)^m \frac{k^F_{0S}}{\mu^{FR}} \tag{12}$$

where m is a dimensionless constant, see also Eipper [23], Ricken [16], Ehlers et al. [24] and the citations therein.

3.3 Mass supply

The proposed method of representing growth proliferation in cardiac tissue is based on that presented in the work of Werner [18]. This growth evolution model will hence be combined with a strain driven growth law, developed by Goktepe et al. [11], in order to model a dilated left ventricle. Making use of the balance equation of mass for the solid phase, the Jacobian may be expressed in its decomposed form

$$J_S = \underbrace{\frac{\rho^S_{0S}}{\rho^S}}_{J_{Se}} \underbrace{\exp\left(\int \frac{\hat{\rho}^S}{\rho^S} dt\right)}_{J_{Sg}} \tag{13}$$

comprised of the elastic part J_{Se}, and growth part J_{Sg}, see Sec.(2.2). Using a finite difference approximation, and by discretising the equation above with respect to the nth and $(n+1)$ time step

$$J^{n+1}_{Sg} = \underbrace{\exp\left[\sum_{n=1}^{n} \left(\frac{\hat{\rho}^S}{\rho^S}\right)^n \Delta t^n\right]}_{J^n_{Sg}} \cdot \underbrace{\exp\left[\left(\frac{\hat{\rho}^S}{\rho^S}\right)^{n+1} \Delta t^{n+1}\right]}_{\Delta J^{n+1}_{Sg}} \tag{14}$$

results in a form where J_{Sg} may be updated every iteration. The mass supply function $\hat{\rho}^S$ is postulated by

$$\hat{\rho}^S = \hat{\rho}^{max}\hat{\rho}^{lim}, \quad \hat{\rho}^{lim} = \begin{cases} \left(\frac{J^{max}_{Sg} - J_{Sg}}{J^{max}_{Sg} - 1}\right)^{\gamma} (\epsilon_e - \epsilon_{max}) & \text{if } J_{Sg} < J^{max}_{Sg} \;\&\; \epsilon_e > \epsilon_{max} \\ 0 & \text{if } J_{Sg} \geq J^{max}_{Sg} \text{ or } \epsilon_e \leq \epsilon_{max} \end{cases} \tag{15}$$

making use of a strain driven growth law [11]. $\hat{\rho}^{max}$ (kg/(s m^3)) denotes the maximum mass exchange rate. J^{max}_{Sg} represents the maximum growth factor, which acts to limit the amount of growth as J_{Sg} tends towards it, which is additionally controlled by the parameter γ.

4 Numerical treatment

Considering the set of unknown quantities for the triphasic model, shown by Eq.(16)

$$\mathcal{U} = \left\{\mathbf{u}_S,\ n^S,\ n^N, \lambda\right\} \tag{16}$$

the governing equations may be solved in order to quantify these variables. The displacement $\mathbf{u}_S$ will be solved using the balance equation of momentum. The pore pressure λ will be calculated through the balance equation of mass of the mixture. Lastly, the solid and nutrient volume fractions n^S and n^N, will be determined using the respective balance equations of mass. Using the Galerkin approach, which transforms a system of equations into a discrete problem, allows the weak forms to be generated from the aforementioned balance equations.

The finite element method (FEM) is used as a numerical framework to approximate the differential equations present within the triphasic TPM model. The *Newton-Raphson* method is used to solve the system of non-linear equations, and the *Newmark-β* method is used as a temporal discretization scheme. These are implemented in the in-house software SESKA.

5 Numerical example using an RHD affected heart

The previous sections have established the mathematical framework needed for a triphasic TPM model, and additionally with application to cardiac systems which include growth and remodelling proliferation. This section will continue with an analysis of a patient specific heart, diagnosed with RHD. Investigation into the behavioural changes of the left ventricle will be conducted and will be furthermore used to validate the adequacy of the proposed TPM model.

The patient under study was diagnosed with RHD in 2014 after CMR (cardiovascular magnetic resonance) scanning, at the Groote Schuur Hospital. From the assessment, it was found that the patient had a severely stenosed mitral valve with moderate regurgitation. The patient was admitted for a valve replacement in 2015 whereby subsequent follow-up scans were taken in 2016. This investigation will involve the two scans taken in 2014 and 2016, which were modelled by Hopkins [1].

5.1 Applied boundary conditions

In order to build the boundary value problem needed for the geometry of the left ventricle, *Dirichlet* and *Neumann* boundary conditions need to be prescribed. In terms of the Dirichlet boundary conditions, the left ventricle is firstly fixed at the base, from moving in the vertical direction. During muscular contraction, the left ventricle experiences a degree of twist. In order to allow this torsional behaviour but still restrict the deformation to a level which is physiologically normal, a Dirichlet boundary condition is weakly imposed through application of an elastic boundary condition, with a prescribed stiffness k, around the epicardial base, as done in [25]. Lastly, in order to restrict motion in the xy axis, but still allow for torsional motion, zero displacement is enforced in the xy direction at the point of the apex of the left ventricle.

The Neumann boundary conditions are considered next. The inflow of blood into the ventricular cavity, during the diastolic filling stage of the cardiac cycle, imparts a hydrostatic pressure on the myocardial walls. To model this, an incremental surface pressure load is prescribed on the inner walls of the left ventricle. As specified in the work of Rama [25], and the citations therein, the surface pressure is assigned a value of 0 kPa and 1.5 kPa, at the start and end of diastolic filling, respectively. Due to blood not being able to seep out from the ventricles, it is assumed that the seepage flux for the entire domain of the left ventricle is zero.

5.2 Material parameters

With respect to cardiac modelling, a variety of parameters which influence the overall characteristics of the simulated model exist. Briefly noting, the parameters of concern include those associated with the strain energy function ψ^S during the diastolic filling stage, the active tension model during ejection, and the Windkessel model (see [26]). Incorporating the theory of porous media comes with an array of additional parameters linked to the various constituent phases and those needed to characterize the constitutive equations formulated for the seepage velocity and mass exchange function.

The rest fibre orientations θ_0 and rest sarcomere lengths l_0 of the epicardium are taken as -51° and 1.91μm, respectively. For the endocardium, these quantities are 66° and 1.78μm.

The input parameters relating to the solid tissue phase, and the fluid phases, which are contained within the blood saturating the solid skeleton, consists of the real densities $\rho^{\alpha R}_{0\alpha}$, and the reference volume fractions $n^{\alpha}_{0\alpha}$, when the left ventricle is in a state of rest. The parameters are given in Table (1)

Table 1. Prescribed TPM parameters associated with the constituents in the reference configuration, and the seepage velocity constitutive law.

Material parameter	Value	Unit	Source
Tissue density ρ^{SR}_{0S}	220	kg/m^3	[1, 27]
Blood density ρ^{LR}_{0L}, ρ^{NR}_{0N}	1060	kg/m^3	[28]
Solid r.v.f. n^S_{0S}	0.75	-	[29]
Liquid r.v.f. n^L_{0L}	0.1	-	-
Nutrient r.v.f. n^N_{0N}	0.15	-	-
Anisotropy parameter α_{F1}	0	-	[16]
Darcy permeability K^F	$2 \cdot 10^{-9}$	m^4/N $\cdot$ s	[30]

In order to calibrate the passive and active material parameters, to the 2014 and 2016 patient-specific cardiac models, a comparison of the end diastolic volume (EDV) and end systolic volume (ESV) was performed. The EDV and ESV values were measured from experimental data accessed from the CMR scans. The measured EDV and ESV values for the 2014 and 2016 models may be viewed in Table (2).

Table 2. Experimental values for the EDV and ESV, in relation to the 2014 and 2016 models. [1]

Quantity	2014	2016	Unit
EDV	146	147	ml
ESV	72	80	ml

The calibrated passive and active material parameters may be viewed in [1, 22].

Calibration of the cardiac models has so far been conducted with the condition of $\hat{\rho}^S$=0. The adopted method for calibrating the growth model was proposed by Hopkins [1], whereby the 2014 scan is used as an initial starting point. With the addition of growth, this model is then calibrated against the EDV and ESV values of the 2016 model. The parameter which will be calibrated comprises solely of the maximum mass exchange rate $\hat{\rho}^{max}$. Growth function parameters were used in accordance with Göktepe et al. [11], which were selected specifically to model a dilated heart. As such, the final calibrated value of $\hat{\rho}^{\mathrm{max}}$ was found to be $3.3 \cdot 10^{-5}\,\mathrm{kg/(mm^3 \cdot s)}$.

5.3 Analysis of the growth induced model

So far within this investigation, two model geometries have been simulated, corresponding to two different time periods, namely in the years of 2014 and 2016. The growth model discussed in the last section was implemented such that it would be able to predict the physiological changes that occurred between the interval, in terms of the mechanical and thermodynamic behaviour. This was done by calibrating the 2014 growth induced model ($\hat{\rho}^{\alpha} \neq 0$), to the 2016 ESV and EDV values. As stated, the 2016 CMR scans were taken after valve replacement surgery in 2015. While the EDV and ESV properties match those of left ventricle dilation characteristics, some do not. Therefore, further analysis of the 2016 data is too complex and lies outside the scope of this study. In addition, the triphasic growth model presented cannot distinguish between LVD conditions brought upon by valvular regurgitation, as this would involve a more in-depth study into the hemodynamic mechanics. As a starting point, the presented model is built to model a general case of LVD. This section will hence continue with a comparison of the initial state of the 2014 model, simulated with and without the inclusion of growth.

5.3.1 Analysis of the pressure-volume curves

After simulating the 2014 model with and without mass supply, over one heartbeat, the following results were produced, as shown by the pressure-volume curves in Fig.(3), and in Table (3).

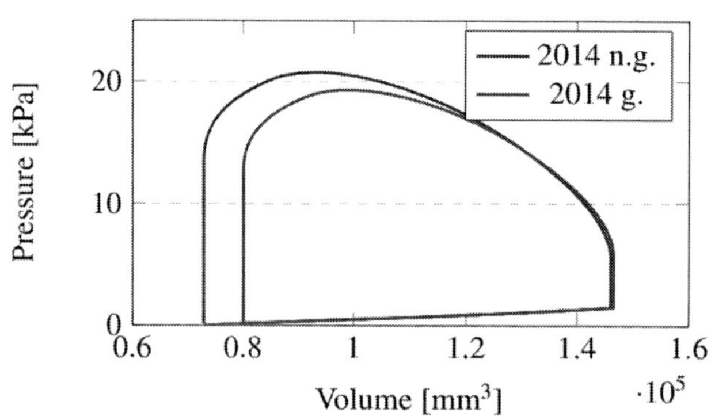

Figure 1. Pressure-volume curves of the 2014 model simulated with and without the inclusion of growth

From the data shown in the Table 3, it is apparent that deterioration in the condition of the heart is reflected by the growth model. The increase in EDV and ESV, with respect to the growth induced model, is an indication of the increase in volume which has occurred as a result of growth. In addition, the growth induced model appears to have reduced in peak pressure by 1.4 kPa, which denotes a reduced capacity. This result is analogous to the condition of LVD. The enlarged dilated left ventricle has an effect on the mechanical function, whereby weakening of the muscles is generally observed [31]. This deficiency may also be

Table 3. Cardiac results pertaining to the 2014 model, simulated with and without growth.

Parameter	2014 no growth	2014 growth induced	Unit
EDV	146	146.5	ml
ESV	72	80.0	ml
SV	74	66.5	ml
EDPVR gradient	28.0	26.4	Pa/ml
Peak pressure	20.8	19.4	kPa

noted by the reduction in SV, which correlates with the amount of blood ejected out the cavity into the various circulatory systems. Another feature which highlights the presence of LVD is the end diastolic pressure volume relationship EDPVR, given in terms of the slope at the point of EDV. In the case of LVD, the EDPVR gradient is generally observed to decrease, which signifies a decrease in vascular stiffness and hence a debilitation of the cardiac tissue [31]. This is in agreement to the results shown in Table (3), whereby the EDPVR is seen to reduce by 2.4 Pa/ml with the addition of growth. [31]

5.3.2 Changes in mechanical behaviour

Cardiac dilation results in mechanical and functional changes in the left ventricle. Understanding the abnormalities which occur may be found in analysing the stresses and strains that the cardiac tissue experiences during its activity. This section will comprise an observation of the fibre stresses and strains of the initial state of the left ventricle, simulated using the 2014 geometrical model with no mass supply, and compared against that using the growth induced model. For the purpose of illustrating this, the fibre stress and strain within the mid section of the ventricle, and along the outer surface of the epicardium, will be focused on. This data is shown in Fig.(2).

From the depiction of the pressure-volume curve in Fig.(1), which is represented quantitatively in Table (3), the growth induced model indicates that the ejection capacity had been compromised. It is expected that this reduction in mechanical load will coincide with the stresses and strains of the myocardium, which is confirmed through analysis of Fig.(2). From Fig.(2), it may be seen that during diastolic filling, at a point where minimal growth has developed (due to low states of stress), the fibre stress and strain of the two models, occur in tandem. However, as the stress, and hence the amount of growth increases due to stretching of the sarcomeres, the stress and strain of the growth induced model is seen to reduce below that of no-growth model. The pattern and values of the strains, shown in Fig.(2), is analogous to that presented in the work of Yu et al. [32], where a reduction of strains were documented in relation to an evaluation of left ventricular behaviour in patients with LVD.

In terms of the theory of porous media, the perfusion effects are greatly influenced by the mechanical action that the heart is subjected to. It may be understood that with a decrease in stress and strain, as a result of the inclusion of growth and remodelling, the pore pressure is expected to behave in a synchronous manner. This relationship may be viewed in the graph which denotes the evolution of the pore pressure against time over one cardiac cycle, as shown by Fig.(3). Commenting on the general characteristics of the pore pressure evolution of the two models, the pore pressure is observed to decrease during diastolic filling. This reduction may be attributed to the stretching of the muscle fibres, which would translate into expanded pore spaces. As the ejection phase is initiated, and the left ventricle contracts, the pore spaces

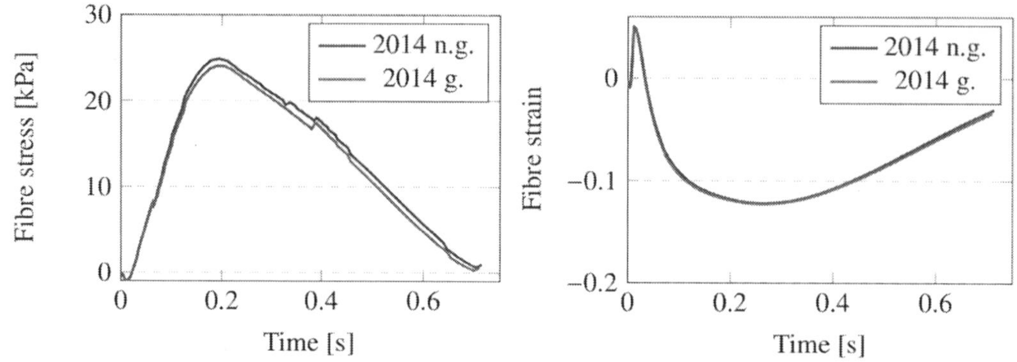

Figure 2. Fibre stress (left), and fibre strain (right), results generated from the growth induced and no-growth 2014 model.

would similarly shrink in size, which would therefore increase the pore pressure until the relaxation phase begins, whereby a stabilization of the pore pressure towards the initial state would then take place. The disturbance in the pore pressure distribution is a result which is inconclusive, and could however be a result of numerical instabilities.

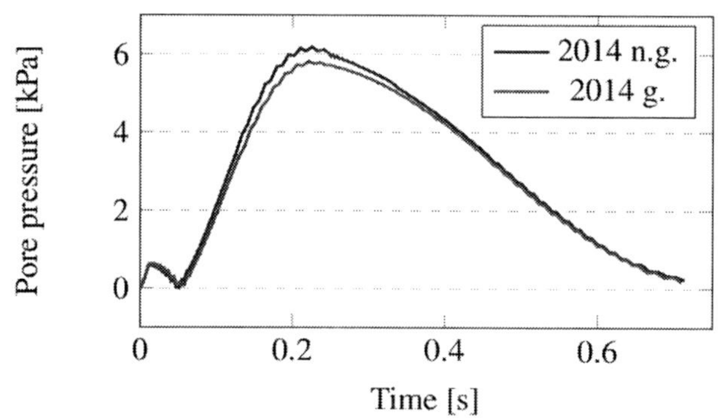

Figure 3. Pore pressure evolution, over a full cardiac cycle, which displays the relationship between the growth induced and no-growth 2014 model.

5.3.3 Changes in the volume fraction quantities

Predicting how the densities of the various constituents change within the heart in relation to growth proliferation is an important aspect. Degradation of the heart may correlate to abnormalities associated with a starvation of nutrients (such as oxygen supply)[7]. In addition, hypertrophy results in an increased solid tissue density, which results in a reduced relative blood tissue ratio [29, 33]. Consequently, a stiffening of the cardiac tissue and a reduction in nutrient provision may arise in these areas, which could lead to impaired ventricular function [7].

Fig.(4) illustrates the relationship between the sarcomere length, growth Jacobian J_g and the solid volume fraction n^S. For the case of LVD, a stretch induced growth function is employed. Hence, the degree of growth may be closely related to the sarcomere length, which also then has influence in relation to n^S. This relation is apparent through observation of Fig.(4), whereby high values across all three parameters are viewed to occur at the top portion of the myocardium, in particular, at the end of diastolic filling. With eccentric growth, an elongation of the muscle fibres occurs due to a deposition of sarcomeres. Consequently, the deposition of sarcomeres results in an increased solid tissue mass.

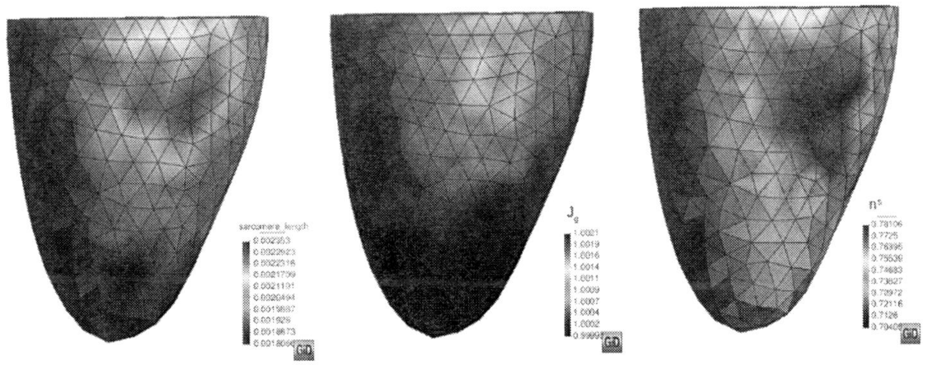

Figure 4. Illustration of the sarcomere length, growth Jacobian and solid volume fraction, at EDV.

To demonstrate the effect of the inclusion of growth on the solid and nutrient volume fractions, an evolution of these parameters over an entire cardiac cycle, measured at the point which displayed the greatest level of growth, was conducted and is presented graphically in Fig.(5). In addition, the point evolution of the growth Jacobian J_g, is shown in Fig.(6).

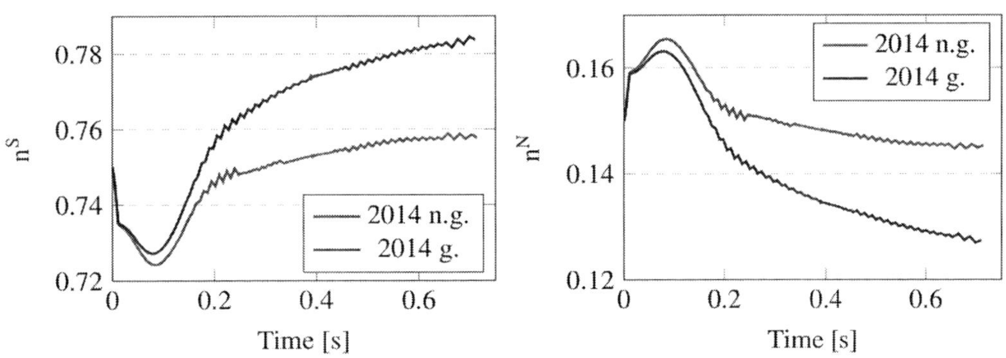

Figure 5. Evolution of the solid and nutrient volume fraction quantities, over a complete cardiac cycle, for the case of growth and no-growth.

From Fig.(5) and Fig.(6), it may be observed that with an increase in growth, an overall reduction in the blood tissue ratio occurs. During diastolic filling and early parts of the contraction stage, during which the amount of growth is still minimal, the blood tissue ratio is seen to increase. This feature is influenced by the mechanical loads exerted on the left ventricle within that phase of the cardiac cycle, whereby the left ventricle expands due to diastolic filling and an overall stretching of the muscle fibres occurs. The expansion of tissue would thus result in an increased pore space size, allowing the fluid to enter this region. A greater fluid density would therefore be incurred. As the pore spaces contract, the pore pressure would increase, causing the fluid to be expelled. This would hence result in an

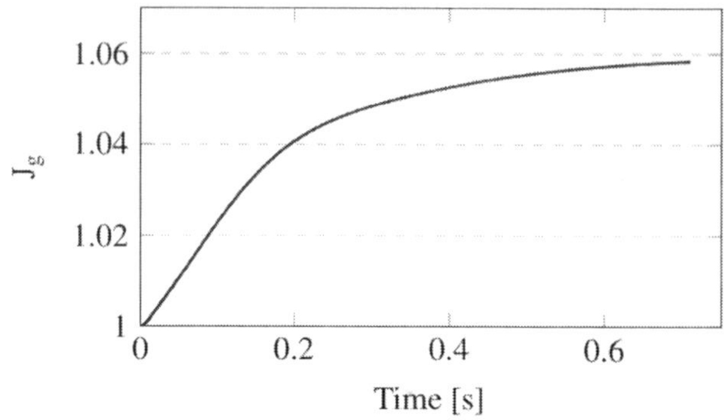

Figure 6. Progression of the growth Jacobian J_g, over a full cycle, at the point which displayed maximum growth.

increased relative tissue density. It may be observed however, that the increase in n^S is greater with respect to the growth induced model. This suggests that the tissue density is greatly influenced by the mass supply function. In addition, with the no-growth model, the volume fractions are observed to stabilize and approach the reference state ($n^{\alpha}_{0\alpha}$), once the heartbeat is complete. In contrast, with respect to the growth induced model, higher values of n^S are noted at the end of the heartbeat, compared to those at the beginning of the simulation, which signifies permanent tissue growth. This feature correlates well with the description of growth in relation to LVD.

Inclusion of the nutrient phase was motivated on the basis that more control over the depletion of the total fluid volume, as a result of growth, would be realised. This was done by coupling the mass supply function solely between the solid tissue phase and the nutrient phase. As such, during simulation of the growth induced model, it is expected that the liquid volume will be consumed less compared to the nutrient quantity. Table (4) displays the final volume fraction values, obtained at the end of the simulation, between the growth induced and no-growth models.

Table 4. Volume fraction values at the end of the cardiac cycle, using the no-growth and growth TPM models. Changes in the volume fraction quantities and the percent contribution of the fluid phases towards the solid volume are also presented

Parameter	n^S	n^N	n^L
No-growth model v.f.	0.758	0.146	0.096
Growth model v.f.	0.784	0.127	0.089
Change in v.f.	0.026	-0.019	-0.007
% fluid contribution	-	73%	27%

From Table (4), it may be observed that the contribution towards growth of the tissue phase, is greater with respect to the nutrient phase as compared to the liquid phase. This result agrees with the expected assumption that the nutrient phase will deplete more during growth. However, the reason that n^L decreases with growth may be found from the mathematical formulation of the mass supply terms shown in the field equations. A solution to making the growth fully dependent on the solid and nutrient volume fractions may lie in adapting the balance equations, and the term containing the mass supply function specifically. By replacing the respective real density $\rho^{\alpha R}$, shown in the final term of Eq.(6)$_1$ and Eq.(6)$_3$, with a scalar factor, the mass supplied would be fully coupled between the nutrient and solid phase. In this regard, it is hypothesised that the liquid phase would not be affected by growth.

6 Conclusions

This section will highlight the conclusions drawn in relation to the implementation of the triphasic TPM model, and its application to the example of a patient specific left ventricle model.

This research consisted of implementing a triphasic framework, based on the Theory of Porous Media. This was firstly achieved by extending the current biphasic model present within the in-house built solid mechanics software SESKA. An additional degree of freedom was included, in the form of a nutrient volume fraction, which allowed for mass exchange in context of growth and remodelling proliferation. Mass supply was induced through a stretch driven growth function and was implemented using an adapted method developed by Werner [18]. Furthermore, the triphasic model was coupled with cardiac mechanics framework in SESKA, and was hence used to simulate patient specific models in relation to RHD. This study was however limited to using the 2014 CMR scans, as the 2016 scans were taken after the patient was admitted for valve replacement. This rendered the analysis of the 2016 scans, as outside the scope of this study. In addition, the model presented could only describe the general case of LVD, and not that caused by different valvular regurgitation conditions. This could however be investigated further and incorporated into the triphasic model in future work.

The results produced by the 2014 model, simulated with growth, matched up closely to those expected in dilated left ventricles. This was seen by the increase in ESV and EDV results, which indicated an increase in cavity volume in states of rest and at maximum capacity, respectively. In addition, reductions in the SV, peak pressure, and the EDPVR gradient, were observed in the 2014 growth model, which conveyed a reduced pumping capacity and weakening of the heart tissue, and are typical characteristics related to LVD. Reductions in pore pressures, and fibre stresses and strains were additionally noted. This correlated with observations documented in the work of Yu et al. [32]. As such, it may hence be concluded that the triphasic growth induced model may be sufficiently used to model hearts associated with LVD.

One of the key objectives of this research was to implement the third phase, such that it would contribute to the mass exchange process and would therefore control the volume levels of the liquid constituent. It was found that with growth, a slight reduction in n^L occurred, which was found to stem from the mathematical construct of the balance equations. However, the reduction in n^L was observed to be minimal in comparison to the reduction in n^N. Hence, the overall objective was still satisfied.

Acknowledgement

This research has been supported by the National Research Foundation (NRF) of South Africa (Grant Nos. 104839 and 105858). Opinions expressed and conclusions arrived at are those of the author and are not necessarily to be attributed to the NRF.

References

[1] G. Hopkins, *Growth, Modelling and Remodelling of Cardiac Tissue: A Multiphase Approach* (University of Cape Town, 2017)

[2] W. Rosamond, K. Flegal, G. Friday, K. Furie, A. Go, K. Greenlund, N. Haase, M. Ho, V. Howard, B. Kissela et al., Circulation **115**, e69 (2007)
[3] D.P. Zipes, P. Libby, R.O. Bonow, D.L. Mann, G.F. Tomaselli, *Braunwald's Heart Disease: A Textbook of Cardiovascular Medicine* (Elsevier Health Sciences, 2018)
[4] H.D. Allen, D.J. Driscoll, R.E. Shaddy, T.F. Feltes, *Moss & Adams' Heart Disease in Infants, Children, and Adolescents: Including the Fetus and Young Adult* (Lippincott Williams & Wilkins, 2013)
[5] E. Marijon, M. Mirabel, D.S. Celermajer, X. Jouven, The Lancet **379**, 953 (2012)
[6] J.E. Hall, *Guyton and Hall textbook of medical physiology* (Saunders, 2015)
[7] A. Katz, *Physiology of the Heart, Fifth Edition* (Wolters Kluwer Health/Lippincott Williams & Wilkins Health, 2010)
[8] L.H. Opie, P.J. Commerford, B.J. Gersh, M.A. Pfeffer, The Lancet **367**, 356 (2006)
[9] G. Himpel, E. Kuhl, A. Menzel, P. Steinmann, Computer Modeling in Engineering and Sciences **8**, 119 (2005)
[10] E. Kuhl, R. Maas, G. Himpel, A. Menzel, Biomechanics and modeling in mechanobiology **6**, 321 (2007)
[11] S. Göktepe, O.J. Abilez, E. Kuhl, Journal of the Mechanics and Physics of Solids **58**, 1661 (2010)
[12] R.C.P. Kerckhoffs, J.H. Omens, A.D. Mcculloch, Mechanics Research Communications **42**, 40 (2011)
[13] S. Göktepe, O.J. Abilez, K.K. Parker, E. Kuhl, Journal of Theoretical Biology **265**, 433 (2010)
[14] M.K. Rausch, A. Dam, S. Göktepe, O.J. Abilez, E. Kuhl, Biomech Model Mechanobiol **10**, 799 (2011)
[15] D. Werner, T. Ricken, A. Ferreira Pfeiffer, Proceedings in Applied Mathematics and Mechanics **13**, 63 (2013)
[16] T. Ricken, J. Bluhm, Archive of Applied Mechanics **80**, 453 (2010)
[17] T. Ricken, A. Schwarz, J. Bluhm, Computational Materials Science **39**, 124 (2007)
[18] D. Werner, *Two Scale Multi-component and Multi-phase Model for the Numerical Simulation of Biological Growth Processes in Saturated Porous Media - at the Example of Fatty Liver in Human* (Technical University of Dortmund, 2017)
[19] W. Ehlers, J. Bluhm, *Porous media: theory, experiments and numerical applications* (Springer Science & Business Media, 2013)
[20] A. Menzel, E. Kuhl, Mechanics Research Communications **42**, 1 (2012)
[21] E. Kuhl, Journal of the Mechanical Behavior of Biomedical Materials **29**, 529 (2014)
[22] A. Mosam, *A tri-phasic continuum model for the numerical analysis of biological tissue proliferation using the Theory of Porous Media. Application to cardiac remodelling in rheumatic heart disease* (University of Cape Town, 2019)
[23] G. Eipper, *Theorie und Numerik finiter elastischer Deformationen in fluidgesättigten porösen Festkörpern* (Inst. für Mechanik (Bauwesen), 1998)
[24] W. Ehlers, N. Karajan, B. Markert, A. Acartürk, Proceedings in Applied Mathematics and Mechanics **5**, 27 (2005)
[25] R. Rama, *Proper Orthogonal Decomposition with Interpolation-based Real-time Modelling of the Heart* (University of Cape Town, 2017)
[26] M.A. Essack, *Material Parameter Identification for Modelling the Left Ventricle in the Healthy State* (University of Cape Town, 2014)
[27] K.C. Vinnakota, AJP: Heart and Circulatory Physiology **286**, H1742 (2004)

[28] J. Cutnell, K. Johnson, D. Young, S. Stadler, *Physics, 10th Edition* (Wiley, 2015)

[29] A. Indermühle, R. Vogel, P. Meier, S. Wirth, R. Stoop, M.G. Mohaupt, C. Seiler, European Heart Journal **27**, 1571 (2006)

[30] W.A. Wall, L. Wiechert, A. Comerford, S. Rausch, International Journal for Numerical Methods in Biomedical Engineering **26**, 807 (2010)

[31] R. Klabunde, *Cardiovascular Physiology Concepts, Second Edition* (Wolters Kluwer Health/Lippincott Williams & Wilkins, 2011)

[32] Y. Yu, S. Yu, X. Tang, H. Ren, S. Li, Q. Zou, F. Xiong, T. Zheng, L. Gong, Journal of International Medical Research **45**, 2092 (2017)

[33] J.P. Cleutjens, W.M. Blankesteijn, M.J. Daemen, J.F. Smits, Cardiovascular Research **44**, 232 (1999)

Modeling a Large Air-Cooled Condenser

D.L. Louw[1,*], *C.J.* Meyer[1,**], and *S.J.* van der Spuy[1,***]

[1]Department of Mechanical and Mechatronic Engineering
University of Stellenbosch
Stellenbosch, 7600

Abstract.
This study reports on a modeling strategy for large Air-Cooled Condensers (ACCs). A large 64-fan ACC was modeled under various crosswind conditions that investigated the ACC's specific axial flow fan configuration. The ACC model was developed in two parts, an axial flow fan model and a heat exchanger model. The axial flow fans were modeled using an Actuator Disk Model (ADM). The heat exchangers' pressure drop was modeled using the Darcy-Forchheimer porosity model, and the Effectiveness Number of Transfer Units (ε-NTU) method was used to determine the air heat transfer rate. The ACC was configured using two different axial flow fans, identified in this study as the L-fan and the N-fan. Comparatively the L-fan has a steeper fan static pressure characteristic curve than that of the N-fan, at the cost of a greater shaft power consumption. Under normal operating conditions the average heat-to-power ratios were calculated at 89.91 W/W for the L-fan and 102.48 W/W for the N-fan. Under crosswind conditions of 9 m/s the heat-to-power ratios of the leading edge fan-units decreased by 80.6% and 87.0% for the L-fan and N-fan respectively. However, at the fan-units immediately downstream of the leading edge the heat-to-power ratios only decreased by 34.1% for the L-fan and 64.2% for the N-fan.

1 Introduction

Thermal power generation requires the discharge of heat to complete the steam cycle [1]. Multiple coal-fired power stations operated by Eskom, such as Matimba, Medupi and Kusile, utilize air-cooled condensers (ACCs) to facilitate this heat discharge. The dependence of ACCs on airflow for heat transfer renders them sensitive to atmospheric wind conditions.

This study investigates a 64 fan-unit forced draft configuration ACC and its performance subject to various crosswind conditions. Various forced draft configuration ACCs under adverse crosswinds conditions have previously been investigated [2–5]. However, these ACCs faced limitations in terms of the nature of the axial flow fan models used [3, 4] and/or the size of the ACC investigated (max 30 fan-units) [2, 3, 5].

Simplified fan models are required due to the prohibitive numerical cost of full 3-dimensional fan models [6]. The actuator disk model (ADM) [7] used in this study provides

*e-mail: 17562252@sun.ac.za
**e-mail: cjmeyer@sun.ac.za
***e-mail: sjvdspuy@sun.ac.za

greater accuracy than a PJM at a fraction of the computational cost of fully explicit 3D fan modeling [6].

This study investigates a traditional forced draft ACC with 64 A-frame fan-units in a 8×8 configuration. The performance of the ACC is investigated as the system is subjected to various crosswinds, which are applied in two directions, i.e. parallel and perpendicular to the streets of fan-units, and are applied at multiple reference velocities. A specific fan configuration utilising a high power fan at the front and back of the ACC is also tested as a means to improve performance under adverse crosswind.

2 Numerical Models

2.1 Governing Equations

A fluid's motion in 3-dimensions is governed by 3 principles [8, 9], which are the conservation of mass, the conservation of momentum given, and the conservation of energy.

The *standard* k-ε turbulence model [10], with its canonical constants, was used in this study. The model provided greater numerical stability than the k-ε *SST* turbulence model.

The equations are discretized and solved numerically using the finite volume method (FVM) using the open source CFD code `OpenFOAM` [11]. The `OpenFOAM` CFD toolbox allowed for the development and integration of additional numerical models needed to model the ACC.

The `OpenFOAM` incompressible flow solver with the Boussinesq buoyancy model (`buoyantBoussinesqSimpleFoam`) is used in this study [5]. The solver uses an alternative pressure field, p', which is obtained by subtracting the hydrostatic component from the actual pressure:

$$p' = p_{ref} - \rho_k g h \tag{1}$$

where ρ_k is the Boussinesq density, g is the gravitational acceleration, h is the height above ground level.

The incompressible flow solver was used for its numerical stability [5]. The Boussinesq buoyancy model was used to reduce computational costs, and is valid when $\Delta\rho/\rho_k \ll 1$.

2.2 Axial Flow Fan Model

The axial flow fan is modeled using a custom actuator disk model (ADM) that uses three axially aligned disks to represent the fan being modeled [7]. The fan's inlet and outlet flow field velocities are obtained from the upstream and downstream disks. These velocities are used to calculate the unique relative flow velocities, U_R, and angles of attack, α, at each cell in the actuator disk.

The lift and drag forces, F_L and F_D, imparted by the fan blades are calculated for each radial segment of the fan blade using blade element theory:

$$F_L = \frac{1}{2}\rho|U_R|^2 C_L c \tag{2}$$

$$F_D = \frac{1}{2}\rho|U_R|^2 C_D c \tag{3}$$

where C_L and C_D are the lift and drag coefficients of the respectively, and c is the chord length of the fan blade.

The fan blade forces are scaled according to the solidity ratio, σ, of the fan:

$$\sigma = \frac{n_b c}{2\pi r} \tag{4}$$

where n_b is the number of fan blades and r is the radius.

The fan blade forces defined by eqs. (2) and (3) are calculated as volumetric forces, f, in a cylindrical reference frame with axial, tangential and radial components:

$$f_a = \frac{1}{2}\rho|U_R|^2\frac{\sigma}{\tau}(C_L\cos(\gamma-\alpha) - C_D\sin(\gamma-\alpha)) \tag{5}$$

$$f_t = \frac{1}{2}\rho|U_R|^2\frac{\sigma}{\tau}(C_L\sin(\gamma-\alpha) + C_D\cos(\gamma-\alpha)) \tag{6}$$

$$f_r = 0 \tag{7}$$

where τ is the thickness of the actuator disk's cell layer, and γ is the fan blade's setting angle relative to its plane of rotation. The forces are added as a source term in the momentum equation.

The lift and drag coefficients for each fan blade were obtained for various Reynolds numbers using `XFoil` [12]. The lift and drag coefficients are approximated using flat plate theory outside of the upper and lower stall points of the fan blade profiles.

The fan shaft power consumption, $\dot{W}$, is obtained by the following volume integral:

$$\dot{W} = \omega\int_V f_t\, r\, dV \tag{8}$$

where ω is the fan's rotational speed, and f_t is the tangential component of the ADM's source term.

2.3 Heat Exchanger Model

The heat exchanger is represented using two simplified numerical models that account for its heat transfer component and its pressure drop component. The heat transfer component is modeled using the ε-NTU (Effectiveness Number of Transfer Units) method [1], which adds a source term to the energy equation. The heat transfer rate, $\dot{Q}$, for a given effectiveness, ε_{NTU}, is given by:

$$\dot{Q} = \dot{m}_a c_{pa}(T_v - T_{ai})\varepsilon_{NTU} \tag{9}$$

where $\dot{m}_a$ is the air mass flow rate, c_{pa} is the specific heat of the air, T_v is the steam vapor temperature, and T_{ai} is the air inlet temperature.

The pressure drop component is modeled using the Darcy-Forchheimer porosity model, which adds a source term to the momentum equation. The pressure drop gradient defined by the Darcy-Forchheimer porosity model is given by:

$$\frac{dp}{dx} = \mu k_d U + \rho k_f\frac{U^2}{2} \tag{10}$$

where k_d and k_f are the Darcy and Forchheimer coefficients, and U is the flow velocity.

3 Computational Setup

3.1 Computational Domain & Mesh

The ACC comprised of an 8×8 array of A-frame forced draft fan-units. The fan-unit schematic is shown in Figure 1 (a). Windwalls are present at the periphery of the domain. The ACC itself is located in an extended atmospheric domain as shown in Figures 2 and 3.

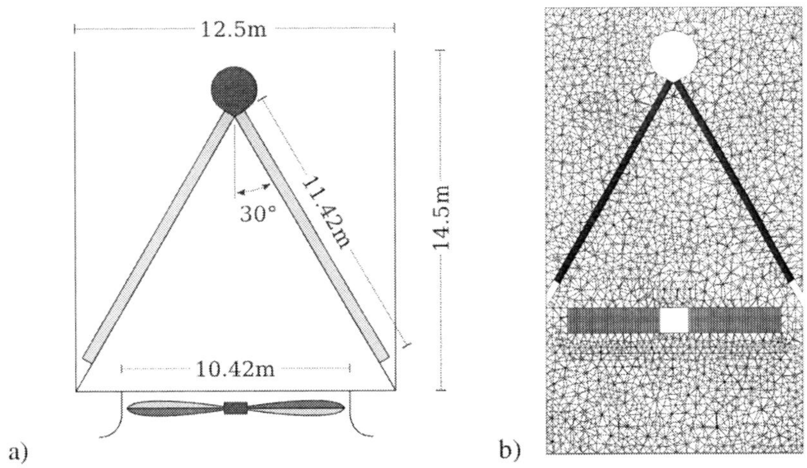

Figure 1. Single Fan-Unit: a) Domain Schematic, b) Mesh Slice

Table 1. Single Fan Forced Draft ACC Submesh Components

Submesh Region	Mesh Type	
Actuator Disk	Hexahedral	Structured
Heat Exchanger	Hexahedral	Structured
Fan-Unit Lower	Tetrahedral	Unstructured
Fan-Unit Plenum	Tetrahedral	Unstructured
Fan-Unit Upper	Tetrahedral	Unstructured
Atmosphere	Hexahedral	Structured

The domain and mesh for the numerical ACC model are created in a modular fashion, which reuses a copy of the same mesh for each fan-unit. The submesh components created to form the entire computational domain are listed in Table 1

The mesh for a single fan-unit is shown in Figure 1 (b). The model used structured hexahedral cells to mesh the actuator disk region, heat exchanger region and large atmospheric region due to the regular shape of these regions, while unstructured tetrahedral cells were used to accommodate the remaining complex geometries found around the fan-unit. The final mesh for the ACC model consists of a total of 45.97×10^6 cells, of which 20.98×10^6 are hexahedral cells and 24.81×10^6 are tetrahedral cells.

3.2 Axial Flow Fan Configuration

The 64 fan-units are identified using an (i, j)-indexing scheme, respectively identifying the x and y-coordinate of the fan-unit as shown in Figure 2. The i-index indicates which street the fan-unit belongs to, while the j-index where in the street the fan-unit is located. Fan-units of the same street (shared i-index) share a steam duct, but each fan-unit has a seperate plenum chamber.

The ACC is configured using two different axial flow fans, namely the L-fan and the N-fan:

1. The L-fan is located at the front, $j = 1$, and the back, $j = 8$, sides of the ACC, and is configured using a blade setting angle of 9.3°.

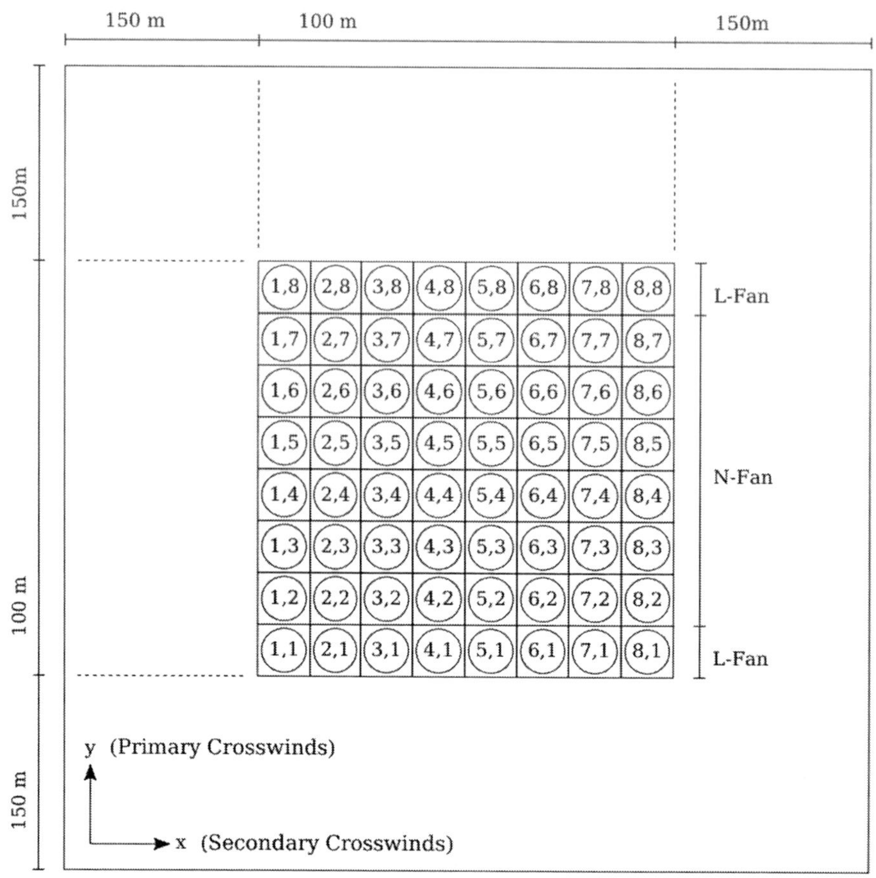

Figure 2. ACC Schematic: Top View

2. The N-fan is used for the remainder of the fan-units, $j \in [2, 7]$, and is configured using a blade setting angle of 11.6°.

Comparatively the L-fan has a steeper pressure characteristic curve and shaft power requirement, and the N-fan has a flatter pressure characteristic and shaft power requirement. The use of L-fan at the front and back of the ACC is an attempt to mitigate performance losses that occur under adverse crosswinds.

3.3 Ambient & Boundary Conditions

The ACC model's boundary conditions are set to match atmospheric conditions. The boundary conditions used are tabulated in Table 2. The ambient conditions for the simulation are calculated using a reference pressure, p_{ref}, of 101325 Pa and a reference temperature, T_{ref}, of 300 K.

A fixed value boundary condition is used to apply the crosswind's velocity profile at the appropriate atmospheric inlet boundaries of the domain using an atmospheric boundary layer (ABL) [13]. The velocity profile for a given reference velocity, U_{ref}, at a given reference

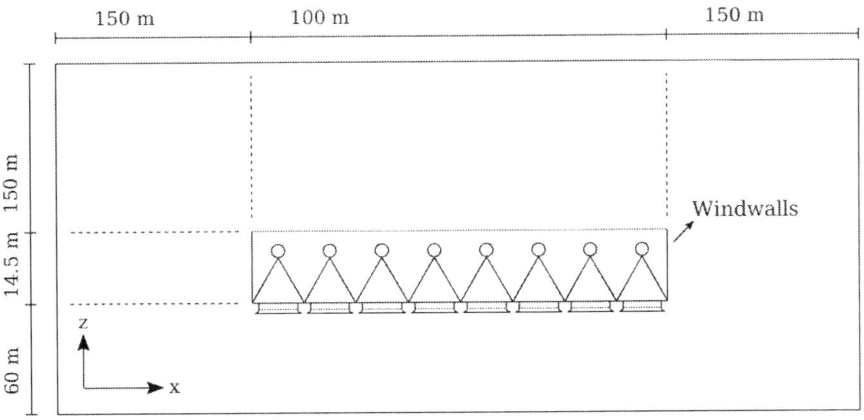

Figure 3. ACC Schematic: Front View

Table 2. Forced Draft ACC Boundary Conditions

Field	Boundary	Type	Value
p'	Sides / Top	Total-Pressure	p_{ref}
	Walls	Zero-Gradient	$\nabla p' = 0$
U	Inlet	Fixed-Values	Equation (11)
	Sides	Zero-Gradient	$\nabla U = 0$
	Walls	No-Slip	$U = 0$
T	Sides	Fixed-Value (Inlet) / Zero-Gradient (Outlet)	T_{ref}, $\nabla T = 0$
	Walls	Zero-Gradient	$\nabla T = 0$

height, z_{ref}, is specified as:

$$U = \frac{U_{ref}}{ln\left(\frac{Z_{ref}+z_0}{z_0}\right)} ln\left(\frac{z+z_0}{z_0}\right) \tag{11}$$

where z is the vertical coordinate, z_0 is the surface roughness (default of 0.1 m), and z_g is the minimum z-value.

Standard wall-functions are used for the *standard k-ε* turbulence model.

3.4 Performance Criteria

The following fan-unit performance criteria are defined to facilitate a comparison between the different axial flow fans used in the ACC. The volumetric effectiveness and heat transfer effectiveness respectively represent the normalised volumetric flow rate through a single fan-unit and the normalised heat transfer rate for a fan-units heat exchangers.

$$\varepsilon_{\dot{V}} = \frac{\dot{V}}{\dot{V}_{ideal}} \tag{12}$$

$$\varepsilon_{\dot{Q}} = \frac{\dot{Q}}{\dot{Q}_{ideal}} \tag{13}$$

where V_{ideal} is the volumetric flow rate of an idealized single fan-unit ACC axial flow fan and Q_{ideal} is the heat transfer rate of a idealized single fan-unit ACC's heat exchangers. The idealized volumetric flow rates and heat transfer rates were determined using an isolated single fan-unit ACC model with no crosswinds applied.

Finally, the heat-to-power ratio represents the ratio between a fan-unit's heat transfer rate and fan shaft power consumption:

$$\eta_{HTP} = \frac{\dot{Q}}{\dot{W}} \tag{14}$$

3.5 Scope

The numerical ACC model was solved for a total of seven different crosswind conditions. The first solution serves as the reference case for normal operation of the ACC with no crosswind applied, i.e. $U_{ref} = 0\ \mathrm{m/s}$. Three crosswinds with reference velocities, U_{ref}, of 3 m/s, 6 m/s and 9 m/s were applied in both the primary y-direction and secondary x-direction.

The crosswind directions chosen facilitate a comparison of the two different axial flow fans when they are used at the leading edge of the ACC. Under primary crosswind conditions the steeper pressure gradient L-fan is located at the leading edge of the ACC, while the flatter pressure gradient N-fan is found at the leading edge under secondary crosswind conditions (except for the corner fan-units).

4 Results & Discussion

4.1 Reference Conditions

The ACC model was evaluated without crosswind to provide a reference case. The volumetric effectiveness and heat transfer effectiveness of the ACC are shown in Figure 4

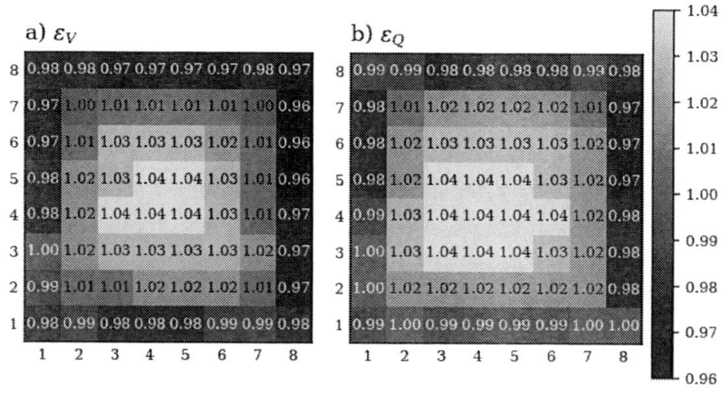

Figure 4. Reference Case: Effectiveness ($U_{ref} = 0\mathrm{m/s}$)

Under normal operating conditions the mean volumetric effectiveness of the ACC is 1.002 and the mean heat transfer effectiveness is 1.001. Both are shown to be slightly higher towards the center of the ACC, and slightly lower at the edges of the ACC. The increased performance at the center is attributed to airflow entering the axial flow fans at a more direct angle than at the sides of the ACC.

The mean heat-to-power ratio of the L-fan fan-units is 90.0 W/W and the mean heat-to-power ratio of the N-fan fan-units is 102.5 W/W. The overall mean heat-to-power ratio of the ACC is 99.4 W/W.

4.2 Primary Crosswind Conditions

The performance results of the ACC under the primary crosswinds of 3, 6 and 9 m/s are discussed in this section. The volumetric effectiveness is shown in Figure 5, the heat transfer effectiveness is shown in Figure 6, and the heat-to-power ratio is shown in Figure 7.

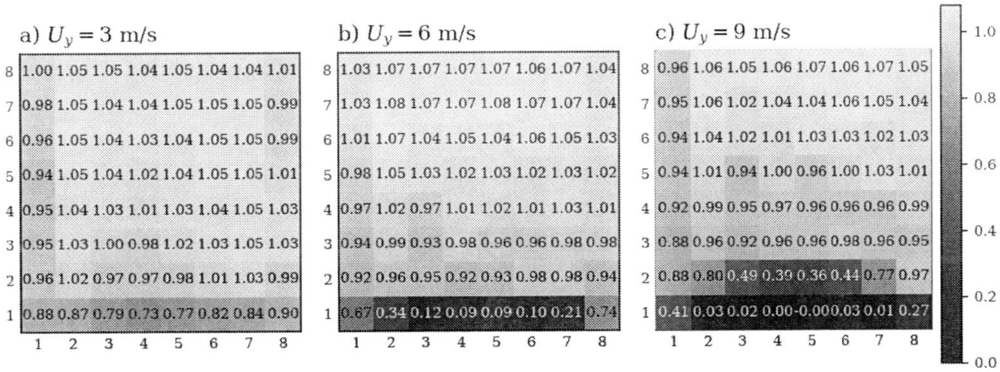

Figure 5. Volumetic Effectiveness: Primary Crosswinds (y-direction)

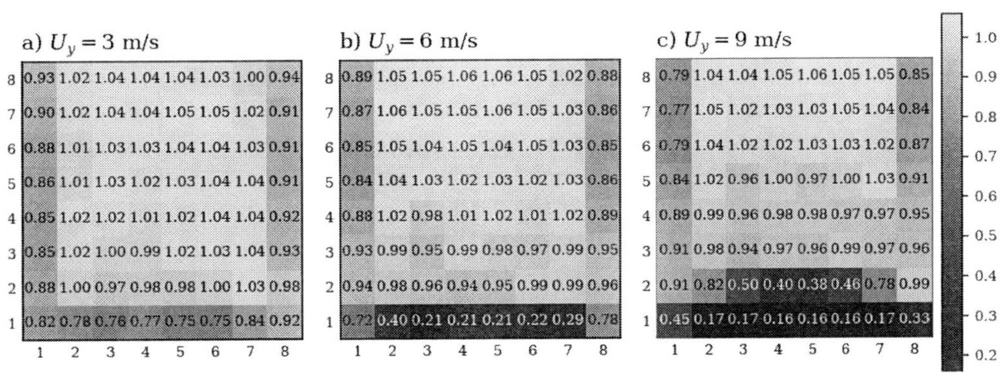

Figure 6. Heat Transfer Effectiveness: Primary Crosswinds (y-direction)

The decreases in the volumetric and heat transfer effectiveness are concentrated at the centre of the leading edge of the ACC, with smaller decreases towards the sides and downwind of the leading edge. Similar volumetric and heat transfer effectiveness were observed in previous studies of full sized ACCs [3–5].

The volumetric effectiveness of the leading edge fan-units, excluding the corner units, decreases from 0.99 under normal operating conditions to 0.80, 0.16 and 0.02 under crosswind velocities of 3, 6 and 9 m/s respectively, and comparatively the heat transfer effectiveness decreases from 0.99 to 0.78, 0.26 and 0.17.

The fan-units directly downwind of the leading edge also exhibit decreases in volumetric and heat transfer effectiveness. The effectiveness values of the above-mentioned fan-units decrease from the norm of 1.02 to 1.00, 0.95 and 0.54 under crosswind velocities of 3, 6 and 9 m/s respectively.

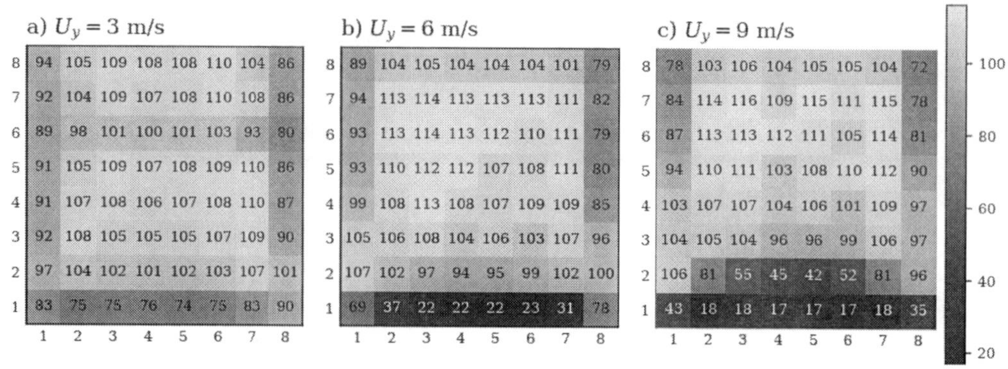

Figure 7. Heat-to-Power Ratio: Primary Crosswinds (y-direction)

Under the primary crosswinds the volumetric effectiveness and heat transfer effectiveness are shown to have a very close correlation. However, the differences between the two suggests that an additional factor contributes to the decrease in heat transfer rate, e.g. increased inlet air temperatures. At the leading edge the increased heat exchanger inlet tempreatures is attributed to reverse flow throught the fan-units which allows heated air to flow from the plenum chamber into the supply flow of the downwind fan-units. At the sides of the ACC aligned with the crosswinds the increase is attributed to hot-plume recirculation where hot air exiting the heat exchangers is captured in vortices that recirculate the air into the supply flow of downwind fan-units.

The heat-to-power ratio of the leading edge fan-units, excluding the corner units, respectively decreases from 90.7 W/W to 76.3, 26.2 and 17.5 W/W at crosswind velocities of 3, 6 and 9 m/s. Immediately downwind of the leading edge the heat-to-power ratio decreases from 103.2 W/W under normal operating conditions and 3 m/s crosswinds to 98.2 and 59.3 W/W under 6 and 9 m/s crosswinds respectively.

Some asymmetry is shown in the two effectiveness values and the heat-to-power ratios at the left and right hand sides of the ACC. The asymmetry is minimal at low crosswind velocities and is more prominent at high crosswind velocities. The asymmetry is attributed to the rotation of the fans as modeled by the ADM.

4.3 Secondary Crosswinds Conditions

The performance results of the ACC under the secondary crosswinds of 3, 6 and 9 m/s are discussed in this section. The volumetric effectiveness is shown in Figure 8, the heat transfer effectiveness is shown in Figure 9, and the heat-to-power ration is shown in Figure 10.

The volumetric effectiveness of the leading edge fan-units, excluding corner fan-units, decreases respectively from 0.98 under normal operating conditions to 0.37, −0.07 and −0.16 under crosswind velocities of 3, 6 and 9 m/s, and the heat transfer effectiveness decreased from 0.99 under normal operating conditions to 0.36, 0.14 and 0.10 respectively. The negative volumetric effectiveness values are indicative of reverse flow throught the fan-units.

The volumetric effectiveness of the fan-units immediately downwind of the leading edge decreases from 1.01 under normal operating conditions respectively to 1.01, 0.84 and 0.34 at crosswind velocities of 3, 6 and 9 m/s. Similar effectiveness values are seen for the heat transfer, as was the case for primary crosswind conditions.

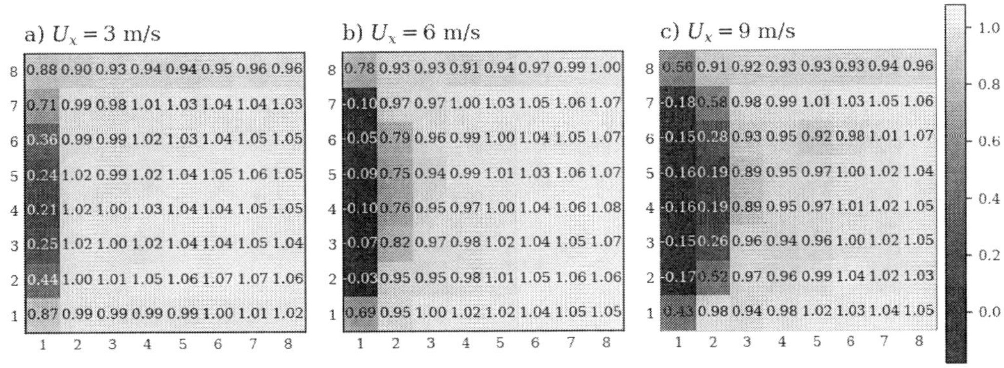

Figure 8. Volumetic Effectiveness: Secondary Crosswinds (y-direction)

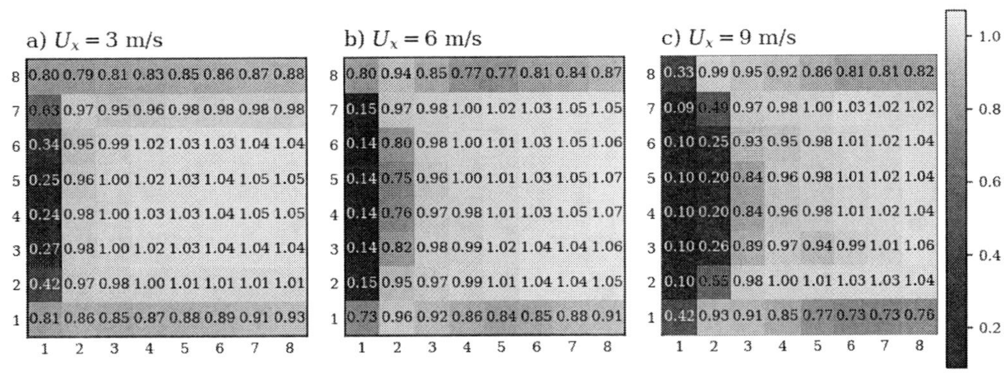

Figure 9. Heat Transfer Effectiveness: Secondary Crosswinds (y-direction)

The heat-to-power ratios of the leading edge fan-units, excluding the corner fan-units, decreases from 98.2 W/W under normal operating conditions respectively to 40.3, 18.7 and 13.3 W/W at crosswind velocities of 3, 6 and 9 m/s.

Similar to the primary crosswind cases, some asymmetry is shown in the two effectiveness values and the heat-to-power ratios at the sides of the ACC parallel to the crosswind direction.

4.4 Comparison Between Crosswind Conditions

A comparison of the ACC's volumetric effectiveness and heat transfer effectiveness and heat-to-power ratios between the two crosswind conditions is shown in Figure 11. The comparisons are shown for the ACC as a whole (overall mean), and for the leading edge fan-units (leading edge mean) excluding the corner fan-units. The corner fan-units are excluded due to their use of the L-fan and their ability to draw air in from multiple directions preventing a fair direct comparison of reduced fan performance at the leading edge.

The volumetric effectivenesso comparison, Figure 11 (a), shows that the fan performance of the entire ACC is marginally better under primary crosswind conditions for all crosswind velocities simulated. However the volumetric effectiveness at the leading edge fan-units is significantly better under primary crosswind conditions. This phenomena is directly

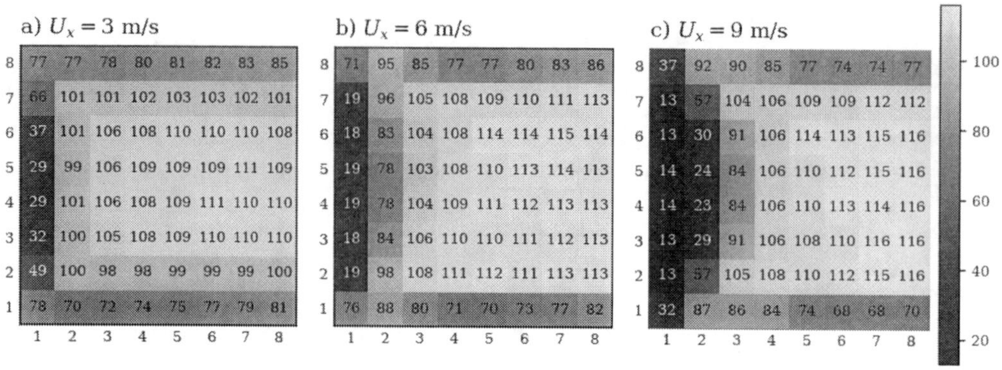

Figure 10. Heat-to-Power Ratio: Secondary Crosswinds (y-direction)

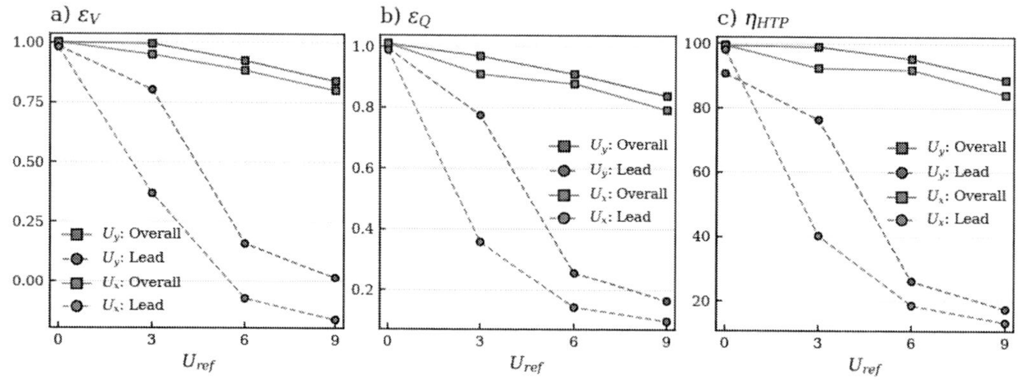

Figure 11. Comparison: a) Volumetric Effectiveness, b) Heat Transfer Effectiveness, c) Heat-to-Power Ratio

attributed to the use of the L-fan whose steeper pressure gradient is better able to mitigate the adverse effects of crosswinds at the leading edge. Similar performance trends regarding comparable fan located at the periphery of the ACC have previously been observed [5].

Similar to the volumetric effectiveness, the heat transfer effectiveness of the entire ACC, Figure 11 (b), is marginally better under all crosswind velocities, and also significantly better at the leading edge of the ACC. The volumetric effectiveness is also seen to be correlated with the heat transfer effectiveness with both effectiveness values showing similar overall trends and trends at the leading edge.

Finally the heat-to-power ratio comparison, Figure 11 (c), shows similar trends to the volumetric and heat transfer effectiveness with higher heat-to-power ratios for the entire ACC under primary crosswind conditions. It is also noted that under primary crosswinds, and despite a lower heat-to-power ratio at the leading edge, the L-fan is shown to be less susceptible to a decrease in the heat-to-power ratio as the crosswind velocity increases. This suggests that the use of the L-fan at the leading edge is an appropriate design choice to mitigate any expected adverse fan performance due to crosswinds.

5 Conclusions

The actuator disk model was used to model the two types of axial flow fans of a large ACC under adverse crosswinds conditions. The model was able to capture the asymmetrical flow field phenomenon under crosswind conditions and directly calculate the axial flow fan's power consumption.

The volumetric effectiveness of the ACC was found to correlate with the heat transfer effectiveness. This indicates that the primary cause of reduced ACC performance under crosswind conditions is a reduction in the axial flow fan's performance. Secondary causes such as increased heat exchanger inlet air temperatures were identified, but their effect on the ACC's performance is significantly less.

The fan-units affected by the crosswinds were found at the leading edge of the ACC. Consequently the ACC performed much better under primary crosswinds than under secondary crosswinds due to the use of the L-fan at the leading edge fan units. The L-fan's steeper pressure gradient able to mitigate the adverse fan performance at the leading edge much better than the N-fan.

References

[1] D.G. Kröger, *Air-cooled Heat Exchangers and Cooling Towers* (Departement of Mechanical Engineering University Stellenbosch, Stellenbosch, South Africa, 1998)
[2] J. Bredell, D. Kröger, G. Thiart, Applied Thermal Engineering **26**, 846 (2006)
[3] M.T.F. Owen, D.G. Kröger, Journal of Engineering for Gas Turbines and Power **133** (2011)
[4] F.G. Louw, Master's thesis, University of Stellenbosch (2011)
[5] R.A. Engelbrecht, J. van der Spuy, C.J. Meyer, A. Zapke, *Numerical Investigation of the Performance of a Forced Draft Air-Cooled Heat Exchanger* (2017), Vol. Volume 5A: Heat Transfer of *Turbo Expo: Power for Land, Sea, and Air*
[6] S.J. van der Spuy, T.W. von Backström, D.G. Kröger, R & D Journal of the South African Institute of Mechanical Engineers **26**, 12 (2009), revised 2010
[7] G.D. Thiart, T.W. von Backström, Journal of Wind Engineering and Industrial Aerodynamics **45**, 189 (1993)
[8] H.K. Versteeg, W. Malalasekera, *Introduction to Computational Fluid Dynamics: The Finite Volume Method*, 5th edn. (McGraw-Hill Education, New York, NY, 2015)
[9] M.T. Schobeiri, *Fluid Mechanics for Engineers* (Springer, Berlin, Heidelberg, 2010)
[10] B.E. Launder, D.B. Spalding, Computer Methods in Applied Mechanics and Engineering (1974)
[11] *The OpenFOAM FOUNDATION* (2021), `https://openfoam.org`
[12] M. Drela, Low Reynolds Number Aerodynamics (1989)
[13] D.M. Hargreaves, N.G. Wright, Journal of Wind Engineering and Industrial Aerodynamics **95**, 355 (2007)

Surface Response to Mechanics of Hardness and Wear Characteristics of Nanoscale Hydrophobic Film

Akinsanya D. Baruwa[1*], *Mamookho E. Makhatha*[1] *and Esther T. Akinlabi*[2,3]

[1]Department of Metallurgy, University of Johannesburg, South Africa
[2]Directorate, Pan African University for Life and Earth Sciences, Ibadan, Nigeria
[3]Department of Mechanical Engineering Science, University of Johannesburg, South Africa

Abstract. There is an increase in demand for durable and efficient organic nanoscale coatings for modern manufacturing systems or components for various applications. Due to this call, a new superhydrophobic silane compound was developed and deposited on mild steel, stainless steel and titanium substrates to determine its durability. The substrates are pretreated with alumina, and the coatings are deposited via the atomic layer deposition (ALD) method. The three substrates were considered for evolving properties comparison and determination of the substrates' response to the film's mechanical properties. The mechanical properties and the failure mechanics were investigated using nanoindentation and nanoscratch. The mechanical viability indicated that stainless steel showed the most robust properties compared with other substrates. Therefore, a nanoscale coating's mechanical strength can be influenced by the substrate's material compositions.

1.0 Introduction

The impact of chromium-based coatings on human health led to the clamour for alternative coating for surfaces of materials. Many coating materials were developed to eliminate the carcinogenic effect that most chromium-based exert on human health. Among the developed materials, using organic coating as a replacement was widely accepted by many researchers [1]. Afterwards, there are aggressive creations of diverse organic coatings for various applications such as aesthetic and protection [2], insulating and biomedical [3], microelectronics [4], among others. It found extensive use in the area of as anti-icing [5], medical [6], adhesion [7], semiconductors [8], and corrosion[9–11], to mention a view.
For a coating to function efficiently on a surface, the mechanical viability and mechanics have to be understood about substrate conformity and the coating process. The mechanical properties of a film are usually characterized by its deposition process, curing condition and chemical constituents [12]. The characteristics are determined through many methods; among them are the nanoindentation and scratch test due to their usefulness in measuring and understanding removal mechanism- scratch morphology-deformation relationship [13]. This relationship indicates that responses from nanoindentation and scratch tests can be transposed to wear properties.

*Corresponding author: darebaruwa@gmail.com

Zheng et al. [14] investigated the failure mode mechanism of TiSiCON and TiSiCN films deposited on cemented carbide tools and observed that both cohesive and adhesive failures were found at the beginning and end of the scratch. It was also established from the investigation that the composite that contains oxygen has a better adhesion when compared with TiSiCN coating. TiFeN, TiN and TiFeMoN films were deposited on silicon wafer in application areas where massive shearing force and strain exposures are required [15]. It was discovered that films with $H/E_r > 0.11$ exhibit delamination failure beyond scratch track while films with $H/E_r \leq 0.11$ exhibit combine hardness and toughness. Bartolomé et al. [16], also, deposited zinc dialkyldithiophosphate (ZDDP) tribofilms on diamond-like carbon (DLC) coatings and discovered that ZDDP tribofilm exhibited two mechanical responses. One response showed a linear elastic-plastic behaviour, while the other showed a compressed non-linear elastic behaviour.

A lot has been achieved concerning the mechanical behaviour of thin films. However, a stable mechanical strength has not been achieved with most of the organic thin films. Organic thin films found most of their applications in surface protection. The mechanical property usually found with the organic coating is rather adhesive strength than hardness or wear resistance. Nanoindentation and nanoscratch will be adopted for the investigation of newly developed organic compound (Repellix) to determine the behaviour of the films on different surfaces pretreated with alumina oxide

2.0 Experimental

2.1 Materials and deposition process

Superhydrophobic Repellix organic compound developed and prepared by Integrated Surface Technologies was used. Three substrates were considered for the experiment: AISI 304 stainless steel, mild steel, and Titanium grade 5. An atomic layer deposition (ALD) machine- RPX- 540 manufactured by IST was used for the deposition. The details of the deposition process and operating conditions have been provided elsewhere [17]. Before deposition of superhydrophobic Repellix, the surfaces of the substrates were first pretreated with alumina in the following steps via oxidation reactions:

$$2AlOH + 2Al(CH_3)_3 \rightarrow 2[Al - O - Al(CH_3)_2 + 2CH_4 \quad (1)$$

$$2[Al - O - (3H_3O)_2 \rightarrow Al_2O_3 + 2AlOH + 4CH_4 \quad (2)$$

2.2 Characterization

Mechanical stability was investigated through a Hystron TI 950 TriboIndenter (Hystron, Minneapolis, MN, USA) with Triboscan v.9 software for data acquisition of nanoindentations. A depth-controlled method was used to determine the actual hardness and Young Modulus of the films using a Berkovich diamond tip indent of radius 50 nm. A total of 15 indentations were made on every sample.

The nanoscratch was done using the Aton Paar micro/nano tester (USA) equipped with scratch 7.3.10 software and uses a Rockwell head with a radius of 100 μm. The scan was made at a progressive load from *3 mN* to *1000 mN* at a 50 Hz scan rate, *2 mm/min* of speed and scratch length of *1 mm*.

3.0 Results and Discussion

3.1 Nanoindentation

The mechanical strength or stability of the films on different substrates is dependent on the difference in the morphological evolutions of the film on each substrate. The mean indentation data obtained are presented in Figure 1 and the calculated values in Table 1. In order to characterize the properties of a thin solid, nanoindentation has proved to be of great advantage for mechanical properties [18]. Therefore, nanoindentation is used to determine the hardness and Young Modulus of the films through the depth-controlled method. The characteristics are measured through Oliver–Pharr method [19] by describing the unloading data from an empirical power-law in Equation 3.

$$P = \alpha(h - h_f)^m \tag{3}$$

where α and m are empirical data determined by the fitting parameters and h_f is the final displacement after complete unloading, and from curve fitting. $m = 1.5$ for Berkovich indenter.

The hardness and Young's modulus are determined from Equations 4 and 5,

$$H = \frac{P_{max}}{A} \tag{4}$$

and

$$S = \left(\frac{dP}{dh}\right)_{h=h_{max}} = \gamma\left(\frac{2}{\sqrt{\pi}}\right) E_r\sqrt{A} \tag{5}$$

where A equals the projected contact area; S is the unloading stiffness, h_{max} is the maximum penetration depth, γ depicts the correction factor due to the lack of axial symmetry and E_r is the reduced modulus as prescribed in Equation 6.

E_r can be expressed as

$$\frac{1}{E_r} = \frac{1+v^2}{E} + \frac{1-v_I^2}{E_I} \tag{6}$$

Where E and ν are Young's modulus and the Poisson's ratio of the test material, however, E_I and v_I are Young's modulus and the Poisson's ratio of the indenter, respectively. The area, however, can be calculated from Equation 7,

$$A = 24.5h_c^2 \tag{7}$$

where

$$h_c = h_{max} - \varepsilon\frac{P_{max}}{S} \tag{8}$$

The relationship between the stiffness and the $h_c$is presented as Equation 9

$$S = \frac{2}{\sqrt{\pi}} E_r\sqrt{24.5}(h_c + h_d) \tag{9}$$

Where h_d is the effective truncation length of the indenter tip- accommodated for the indenter-tip-rounding to determine the actual contact area.

Table 1. The summary of the data calculated from the averaged indentations

Sample	Young Modulus E (GPa)	Hardness H(GPa)	Wear Resistance H^3/E^2 (MPa)	Plasticity index H/Er
Stainless Steel	52.85±2.05	1.24±0.19	682.61±058.11	0.0235±0.0019
Mild Steel	17.90±1.71	0.25±0.08	487.66±025.07	0.0140±0.0031
Titanium	06.84±1.08	0.10±0.02	213.74±019.29	0.0146±0.0014

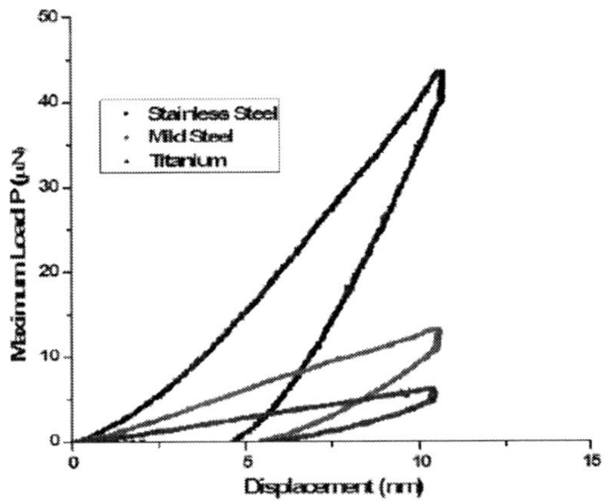

Figure 1: Depicts the load-displacement curve of stainless steel, mild steel and Titanium

From Figure 1, it could be seen that substrates have an essential role in the behavioural modalities of a deposited film. Difficult depth-dependent hardness values are observed on the coatings on different substrates. It takes the indenter to reach a maximum of *44µN* load to penetrate the film deposited on the stainless steel, while mild steel and Titanium required just *10µN* and *5µN* respectively. The low load requirement of Titanium could also be attributed to pores found in the microstructure of the Titanium (not shown). This means that the film deposited on stainless steel has better hardness than mild steel, and mild steel performs better than Titanium. Equations 3 to 9 were used to calculate the hardness, Young's modulus, wear-resistance and plasticity index. The summary of the characterization and calculations are presented in Table 1. The averaged indentations carried out in this experiment shows that coating deposited on the stainless steel has the highest Young's modulus and hardens of *1.24 GPa* and *52.5 GPa* with a standard deviation of 0.94 GPa and 45.67 GPa, respectively. These values are higher than Repellix films on both mild steel and Titanium. The low hardness and Young's modulus noticed in titanium substrate could be as a result of the vast pores observed in its structure (not shown). The least hardness and Young's modulus are found in the film deposited on the mild steel (MS) with values at *0.10 GPa* and *0.68 GPa* with a standard deviation of *0.10 GPa* and *6.84 GPa*. More so, the film's resistance to plastic deformation (wear) and plasticity index (the utmost) was also determined through nanoindentation [20]. Although the coating on mild steel has the lowest plastic index- a sign of good wear resistance, the high H^3/E^2 ratio, however, defeated its wear resistance. Stainless steel has both high plasticity index H/E and wear resistance (H^3/E^2) ratio. It was established that High H/E and H^3/E^2 ratios could also be beneficial in the rate of wear reduction [21]. The disparity in the set of data derived could be a result of the chemical compositions altered during preheating of the substrates. It could also be responsible for the difference in the microstructural evolution of the substrates due to heat. This analogy is in agreement with the work of Nunes and Piedade [22]. It is worthy of establishing that the high standard deviation observed from the data results from the hierarchical structures of the superhydrophobic coatings.

3.2 Nanoscratch

The mechanical strength of the Repellix deposit was also characterized through nanoscratch. Nanoscratch is a useful method for investigating the nature of wear, fracture, and friction in material remover mechanisms [23]. The coefficient of friction and the penetration details acquired from the tests are presented in Figure 2. The resultant coefficient of friction of the composite deposited on stainless steel is *0.033*, mild steel *0.029* and *0.032* on Titanium. These values are very close, and it is an indication that the film adheres strictly to the substrates irrespective of surface response to the formation of the structure. It also means that the hardness and Young's modulus on each substrate could differ, but it is not a factor determining the compound's adherence to the substrates.

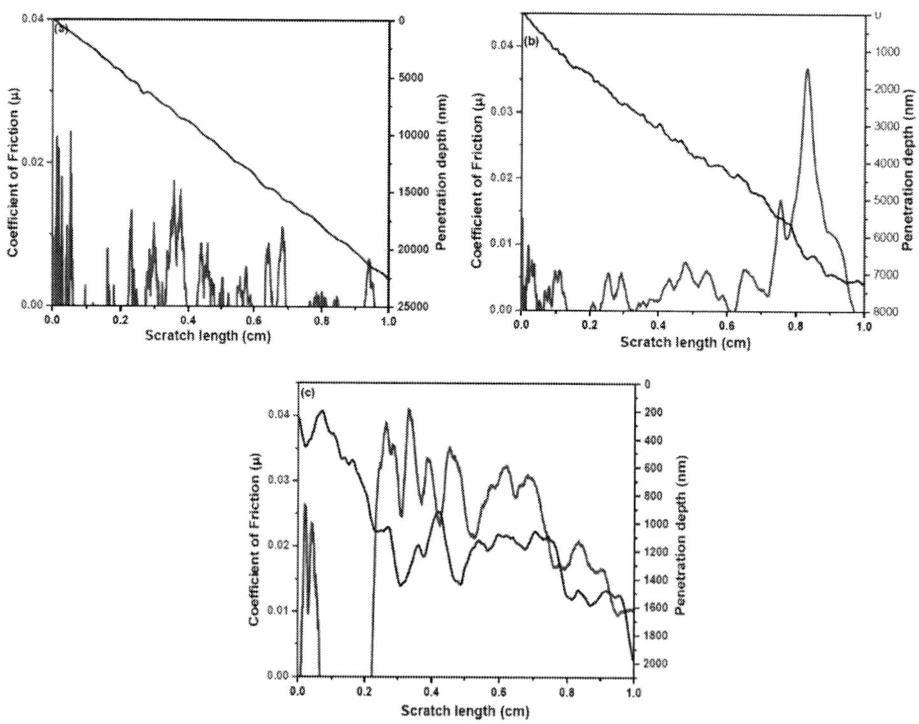

Figure 2: Penetration depth and coefficient of friction of the Repellix-alumina composite on a) Stainless steel (SS), b) Mild steel (MS), and c) Titanium (Ti)

From Figure 2, the effect of roughness is observed on the pattern of the coefficient of friction. Zero coefficients of friction are detected in stainless steel and titanium coated surfaces. This phenomenon is associated with air traps (gaps between particles) as a result of hierarchical structures in the films. The penetration depth, however, shows a different pattern. The scratch depth of stainless steel plunges smoothly while mild steel and Titanium plunge roughly into the composite. The rough plunges could result from roughness or wear chipping due to plasticity along the scratch track.

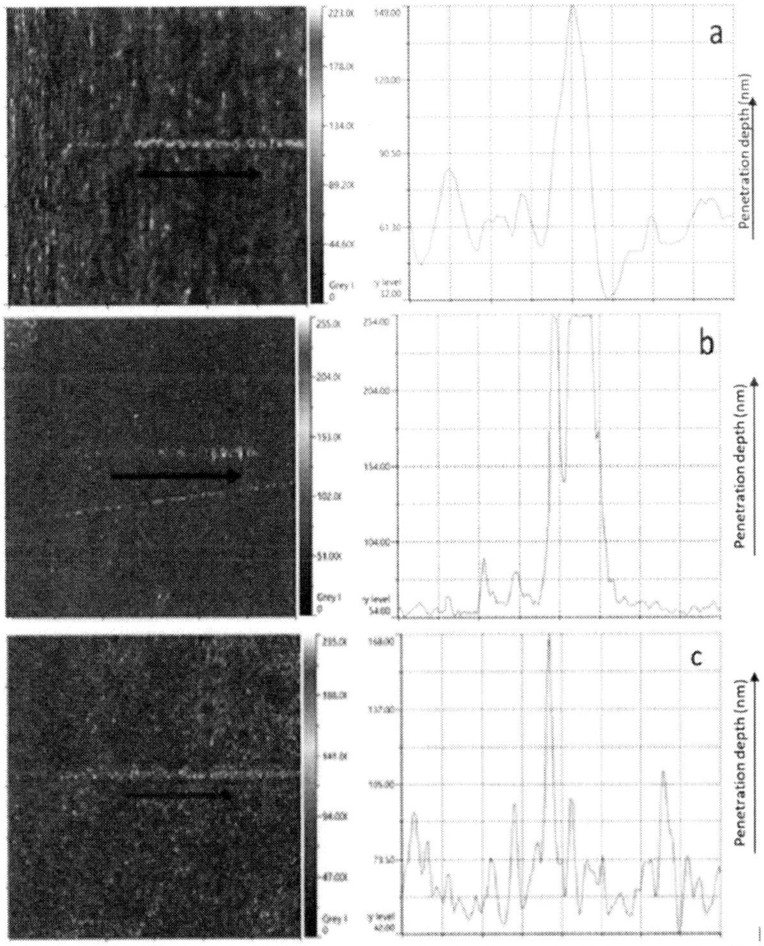

Figure 3: Penetration depth of measurement towards the end of the scratch of a) stainless steel, b) mild steel, and C) Titanium

Figure 3 shows the measurement of wear depth toward the end-tip of the scratch. Indenter encroached deeper into the composite of mild steel up to 254 nm, while stainless steel and titanium composites have 148 nm and 168 nm, respectively. High penetration could result from the smoothness of the surface, or the indenter had a smoother scratch without big particles obstructing the process. Consequently, Feng et al. [24] attributed the depth of penetration to the degree of crosslinked agent present in a compound's chemical matrix. The lower penetration depth observed in the films deposited on stainless steel and Titanium surfaces describes their toughness and resistance ability [25]. However, the coating on the surface of stainless steel does not show good adhesion.

4.0 Conclusion

A newly developed organic compound deposited on the surfaces of selected metals pretreated with alumina was characterized for mechanical properties. Mechanical tests proved that the coating deposited on stainless steel outperformed other substrates at both investigating methods. It also affirms that hardness cannot be the sole determinant factor in ascertaining the mechanical integrity of a coating or surface. Therefore, it can be said that the film deposited on the stainless steel could be durable more than the films deposited on mild steel and Titanium. In conclusion, an organic compound could be used for coating surfaces that require good mechanical properties when it forms a composite with inorganic structures as a surface pretreatment.

Acknowledgement

The authors would like to thank the management of the University of Johannesburg for the financial support for this project.

References

1. R. Figueira, I. Fontinha, C. Silva, and E. Pereira, Coatings **6**, 12 (2016)
2. F. Deflorian, S. Rossi, and L. Fedrizzi, Electrochim. Acta **51**, 6097 (2006)
3. K. Pacaphol and D. Aht-Ong, Surf. Coatings Technol. **320**, 70 (2017)
4. H. Kim, J. Vac. Sci. Technol. B Microelectron. Nanom. Struct. **21**, 2231 (2003)
5. P. Nguyen-Tri, H. N. Tran, C. O. Plamondon, L. Tuduri, D. V. N. Vo, S. Nanda, A. Mishra, H. P. Chao, and A. K. Bajpai, Prog. Org. Coatings **132**, 235 (2019)
6. G. Mani, M. D. Feldman, S. Oh, and C. M. Agrawal, Appl. Surf. Sci. **255**, 5961 (2009)
7. J. Liang, X. W. Wu, Y. Ling, S. Yu, and Z. Zhang, Surf. Coatings Technol. **339**, 65 (2018)
8. D. A. Tropin and A. V. Fedorov, J. Loss Prev. Process Ind. **49**, 762 (2017)
9. X. Cui, G. Zhu, Y. Pan, Q. Shao, C. (xinxin) Zhao, M. Dong, Y. Zhang, and Z. Guo, Polymer (Guildf). **138**, 203 (2018)
10. M. L. Zheludkevich, R. Serra, M. F. Montemor, K. A. Yasakau, I. M. M. Salvado, and M. G. S. Ferreira, Electrochim. Acta **51**, 208 (2005)
11. M. M. Khalaf and H. M. Abd El-Lateef, Mater. Chem. Phys. **177**, 250 (2016)
12. J. Gao, H. Lv, X. Zhang, and H. Zhao, Prog. Org. Coatings **76**, 1477 (2013)
13. P. Kardar, M. Ebrahimi, and S. Bastani, Prog. Org. Coatings **62**, 321 (2008)
14. Y. J. Zheng, Y. X. Leng, X. Xin, Z. Y. Xu, F. Q. Jiang, R. Wei, and N. Huang, Vacuum **90**, 50 (2013)
15. B. D. Beake, V. M. Vishnyakov, and A. J. Harris, Tribol. Int. **44**, 468 (2011)
16. L. Bartolomé, E. Oblak, and M. Kalin, Tribol. Int. **95**, 19 (2016)
17. E. T. Akinlabi, A. D. Baruwa, O. P. Oladijo, N. Maledi, and J. Chinn, J. Mater. Eng. Perform. 1 (2019)
18. G. A. Crawford, N. Chawla, M. Koopman, K. Carlisle, and K. K. Chawla, Adv. Eng. Mater. **11**, 45 (2009)
19. W. C. Oliver and G. M. Pharr, J. Mater. Res. **19**, 3 (2004)
20. A. Leyland and A. Matthews, Wear **246**, 1 (2000)
21. S. Hassani, M. Bielawski, W. Beres, L. Martinu, M. Balazinski, and J. E. Klemberg-Sapieha, Surf. Coatings Technol. **203**, 204 (2008)
22. J. Nunes and A. P. Piedade, Appl. Surf. Sci. **284**, 792 (2013)
23. H. Zhou, S. Qiu, X. Zhang, and C. Xu, Appl. Surf. Sci. **258**, 9756 (2012)
24. L. Feng, B. (Nadia) Benhamida, C. Y. Lu, L. P. Sung, P. Morel, A. T. Detwiler, J. M. Skelly, L. T. Baker, and D. Bhattacharya, Prog. Org. Coatings **125**, 339 (2018)
25. H. Li, J. Li, Z. Liu, J. Huang, J. Kong, and D. Xiong, Surf. Coatings Technol. **375**, 589 (2019)

Model-free Intelligent Control for Antilock Braking Systems on Rough Terrain

Ricardo de Abreu[1,*], *Theunis R.* Botha[1,**], and *Herman A.* Hamersma[1,***]

[1]Department of Mechanical and Aeronautical Engineering, University of Pretoria, South Africa

Abstract. Advancements have been made in the field of vehicle dynamics, improving the handling and safety of the vehicle through control systems such as the Antilock Braking System (ABS). An ABS enhances the braking performance and steerability of a vehicle under severe braking conditions by preventing wheel lockup. However, its performance degrades on rough terrain resulting in an increased wheel lockup and stopping distance compared to without. This is largely as a result of noisy measurements, and un-modelled dynamics that occur as a result of the vertical and torsional excitation experienced over rough terrain. Therefore, it is proposed that a model-free intelligent technique, which may adapt to these dynamics, be used to overcome this problem. The Double Deep Q-learning (DDQN) technique in conjunction with a Temporal Convolutional Network (TCN) is proposed as the intelligent control algorithm, and straight line braking simulations are performed using a single tyre model, with tyre characteristics approximated by the LuGre tyre model. The rough terrain is modelled after the measured Belgian paving with the normal forces at the tyre contact patch approximated using FTire in ADAMS. Comparisons are drawn against the Bosch algorithm, and results show that the intelligent control approach achieves lateral stability by preventing wheel lockup whilst braking over rough terrain, without deteriorating the stopping distance.

1 Background

Significant advances towards vehicle safety have been made with one such system being the development of the Antilock Braking System (ABS). Originally introduced for trains and later developed for passenger vehicles and light trucks, Kelsey-Hayes initiated the "automatic" braking system exploratory development program, which concluded that the braking system of a vehicle should not only reduce the vehicle's stopping distance but also prevent the loss of vehicle control [1]. ABS is a feedback control system that attempts to maintain controlled braking under all operating conditions. This is accomplished by controlling the slip at each wheel to ensure the optimum forces within the limits of the tyre-terrain interaction are achieved. Reports released by the National Highway Traffic Safety Administration (NHTSA) highlight notable decreases in multi-vehicle crashes and fatal pedestrian crashes with the inclusion of ABS [2], however, the performance of an ABS on rough terrain conditions is found to be suboptimal, resulting in wheel lockup, bad stopping distances and poor

*e-mail: ricardo.deabreu@tuks.co.za
**e-mail: theunis.botha@up.ac.za
***e-mail: herman.hamersma@up.ac.za

lateral control [3]. This has inspired the emerging field of study exploring the influence of several phenomena, such as tyre force generation, that impact the effectiveness of ABS on rough terrain [3].

Braking over rough terrain presents a tough challenge to the performance of an ABS by introducing effects that interfere with its normal operation, leading to subpar performances. These effects include noisy measurements, inconsistent terrain input excitations, vehicle body motions, and tyre oscillations. As a tyre travels along a rough terrain, the undulations in the terrain surface not only change the effective rolling radius of the tyre [3], but also cause changes in the vertical and torsional dynamics of the wheel that are un-modelled and often result in loss of contact between the tyre and terrain. The type of control approach used for controlling an ABS plays a crucial role in its performance, and forms the main focus of this study.

ABS control is a nonlinear problem and several control strategies exist, making use of modern control techniques such as sliding mode control, gain scheduling, and fuzzy logic control [4]. Majority of ABS control approaches rely on a set of rules (a defined model) and are robust on specific smooth terrains (dry, icy, or wet), however this does not hold over rough terrain where mismatches tend to occur between the control design model and the real process [5], and as a result, the un-modelled dynamics and parametric uncertainties lead to underwhelming performances of the ABS. This leads to an interest in model-free control designs. Model-free control techniques, commonly known as data-driven techniques, make use of data (usually input-output) in their control design. Reinforcement Learning (RL), also known as Adaptive Dynamic Programming (ADP), is representative of a model-free technique that exploits the process structure used when designing the control instead of using an identified process model. The field of model-free control techniques and RL in particular remains a relatively new concept, however application of this technique to non linear problems such as the control of ABS have been explored with promising results. In [6] a supervised actor-critic approach is proposed for adaptive cruise control, which approximates the optimal control policy for adaptive cruise control and can be adapted to the control of ABS. In [7] the use of ADP for the optimal control of an ABS over smooth terrain is proposed. This approach relies on penalizing the braking distance of the vehicle, and outperforms many existing solutions in literature. In [5] the design and implementation of a model-free slip control of ABS through Q-learning is implemented over smooth terrain, with the performance of the algorithm proving its capability over a wide operating range.

Majority of ABS controllers are simulated on smooth terrain with focus on improving their models stopping distance. However, excessive wheel lockup occurs as a result of the additional excitation in off-road conditions, this not only results in loss of directional control, but also one of the main focuses of ABS which is to prevent wheel-lockup. Thus developing a model-free ABS controller which prevents wheel-lockup on rough terrain without deteriorating the stopping distance forms the main objective of this study.

2 Implementation

2.1 Overview

The method of implementation is broken up into three sections: 1) modelling of an existing ABS algorithm for comparison, 2) modelling of the brake system dynamics and 3) modelling of the proposed ABS control method.

2.2 Bosch Algorithm

The best-documented ABS algorithm was published by Robert Bosch GmbH in 1999 [8], and is known as the Bosch algorithm. The Bosch algorithm is a bang bang type, rule-based control strategy that takes the wheel angular acceleration and longitudinal slip as inputs, and makes use of three different pressure control modes to control the brake pressure at each wheel: *pump* (increase pressure), *dump* (reduce pressure), and *hold* (maintain pressure). Dictated by two upper and one lower threshold for the angular acceleration and an upper longitudinal slip threshold, the Bosch algorithm combines these three pressure control modes in various configurations resulting in a repetitive ABS control cycle consisting of eight phases. Studies [10, 11] have shown that by fine tuning the parameters of the Bosch algorithm, reasonable results can be obtained when braking over rough terrain, thus it is a suitable algorithm to be used as a baseline.

2.3 Brake System Modelling

A mathematical model of the brake system dynamics is discussed and this includes the tyre model. This model provides a link between the tyre braking dynamics and the brake hydraulic dynamics, and although simple, retains the essential characteristics of the actual system.

2.3.1 Brake Model

In order to reduce solution times and determine all the initial aspects of the controller, a simplified model is used in the initial stages of this research, after which more complexity is added. A single tyre model is considered and demonstrated in Figure 1, where the vehicle velocity acts in the positive direction.

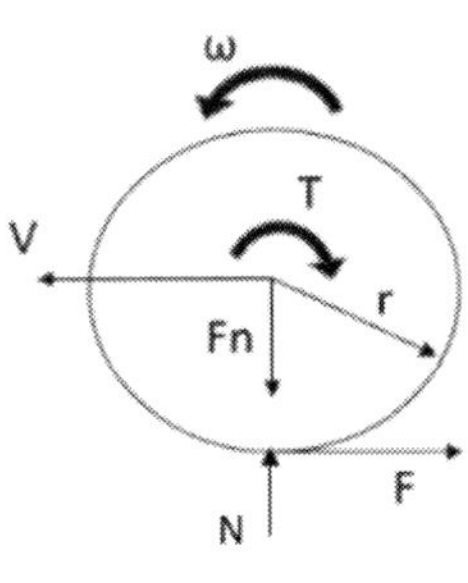

Figure 1: A single tyre free-body diagram with lumped friction

Applying first principles to the free-body diagram, with the sum of the forces acting positive in the left direction and the sum of moments acting positive in the anticlockwise direction, a system of the form is obtained:

$$m\dot{v} = -F, \tag{1}$$
$$I\dot{\omega} = rF - T_b, \tag{2}$$

where m is a 1/4 of the vehicle mass, I the moment of inertia around the centre of the wheel and r is the radius of the wheel. The vehicle velocity, angular velocity, brake torque and tyre-road friction force are represented by v, ω, T and F respectively. For the sake of simplicity, the vertical force is assumed to act in line with the centre of the wheel neglecting effects of rubber hysteresis, whilst effects such as the torsional deflection of the sidewall and rolling resistance are not considered. This simplified approach also neglects rolling distance or pneumatic trail as this component is small compared to the brake torque. The tyre-road friction force is described by the lumped LuGre model [9], defined in Equations (3)-(5).

$$\dot{z} = v_r - \frac{\sigma_0 |v_r|}{g(v_r)} z. \tag{3}$$

$$F = (\sigma_0 z + \sigma_1 \dot{z} + \sigma_2 v_r) F_n, \tag{4}$$

where

$$g(v_r) = \mu_c + (\mu_s - \mu_c) e^{-|v_r/v_s|^{\alpha}}. \tag{5}$$

The brake torque is the force applied at the brake disc to stop its motion, and is defined as a linear relationship with the brake pressure in Equation (6) [10].

$$T_b = 271 \times P, \tag{6}$$

$$\dot{P} = \frac{P_{set} - P}{\tau}. \tag{7}$$

The pressure as a function of time $\dot{P}$ is equal to the difference between the backpressure (P_{set}) and the current pressure in the valve, divided by a time delay constant τ. P_{set} and τ are set according to the specifications used in [10]. Therefore considering the four nonlinear, differential equations $[\dot{P}, \dot{z}, \dot{v}, \dot{\omega}]$, the relationship of the system dynamics and how they interact with each other is established.

2.4 Model of Controller

The purpose of the controller model is to actuate and perform the operation of the ABS, through regulating the brake pressure of the vehicle. Literature reveals several different processes can be used in the control of the ABS [4], however results from [5, 7] indicate the possibilities that exist in model-free control methods. In an attempt to overcome the unmodelled dynamics and parametric uncertainties that lead to underwhelming performances of the ABS over a rough terrain, an intelligent model-free control is proposed.

RL is a method that learns through the interaction between an autonomous active decision-making agent and its dynamic environment, with the objective of the agent to maximize a reward signal despite uncertainty about its environment [12]. Model-free RL methods make use of empirical evidence from past experiences hereby enabling the agent to learn from its environment through trial and error, despite no prior knowledge of the environment. Basic model-free approaches update according to Temporal Difference (TD) learning rule, and can be separated into two types: (1) Q-Learning and (2) Policy Optimization [12].

Q-Learning is a action-value method that makes use of Q-values Q(s,a) (action-value functions) that choose an optimal policy based on the values of the actions of the agent. This approach is almost exclusively achieved through off-policy methods, meaning that each update is not dependent on the agent, and data obtained at any stage during training can be used [12]. Q-learning is easy to implement, has a high sample efficiency and examples include the Deep Q-learning (DQN) and Double Deep Q-learning (DDQN) algorithms.

Policy Optimization methods learn a parameterized policy that can select actions without consulting a value function [12]. The value function may still be used to learn the policy parameter, however it is not required to select the action. These methods are referred to as policy gradient methods. Actor-critic methods are methods that learn approximations to both policy and value functions, where the policy structure is known as the actor because it is used to select actions, and the estimated value function is known as the critic which criticizes the actions made by the actor. Examples of the policy gradient methods include the Asynchronous Advantage Actor-Critic (A3C) and Advantage Actor-Critic (A2C), as well as the Soft Actor-Critic (SAC) algorithm.

The Q-learning and policy gradient methods present advantages and disadvantages for each separate task/problem, thus it is proposed that the different Q-learning (DQN and DDQN) and policy gradient methods (A2C, A3C and SAC) be compared as alternative ABS control algorithms, in order to find the best suitable method for this problem. Due to the availability of baseline implementations for RL algorithms, the GitHub repository of Deep-Reinforcement-Learning-Algorithms-with-PyTorch as provided by [13] is proposed as the main source of implementation and comparison of the above RL algorithms.

The explored RL algorithms make use of deep Neural Networks (NN) as their function approximators, however real-life applications generally involve time series data, as is the case with the control of ABS, hereby adding the complexity of sequence dependency amongst the input variables. Knowing this, it might be beneficial to store this information over a long period of time and learn from it, thus introducing an additional aspect of predictive modelling to the control problem to aid in its performance. Two types of NNs which have shown to be good at handling sequence dependency as well as the modelling of time series data are Long Short Term Memory (LSTM) [14] and Temporal Convolutional Networks (TCN) [15] architectures. It has been proven that the TCN not only exhibits longer memory but consistently outperforms the LSTM on a vast range of tasks [15] and is thus explored further.

In order to model the controller, a state space and action space are to be defined. Several combinations of states can be used as input, such as $[P, v, w, z, \lambda, F_n]$, however these inputs need to be able to be easily accessible in a physical system, and not limit the performance of the model controller by being redundant. Therefore taking this into account, internal investigations show that a simple state space consisting of $[\frac{v}{r}, w]$ yield the best results, with a discrete action setting of [pump, dump, hold].

2.4.1 Reward Function

The reward function determines the behaviour of the brake control, thus in order to ensure the reliability of the brake control, it is crucial to use a properly defined reward function. Different aspects to determining the ideal reward/penalty function exist, such as regulating the control slip. This approach involves regulating the slip by means of predetermined slip thresholds, and is the most commonly used penalty system for ABS systems, however it needs to be adapted according to the type of terrain and restricts the model - no longer representing a model-free approach. An alternative approach would be to monitor and prevent wheel lockup. This approach focuses solely on preventing wheel lockup and penalizes the model when the slip is 100% or the angular velocity is equal to 0. Lastly, the braking distance of the vehicle can be monitored and used as a penalty system. This approach is not ideal as it neglects the importance of lateral control of the vehicle and provides sparse rewards.

When considering the reward function for the ABS model, there is conflict between two intuitive objectives of the ABS: 1) the vehicle is encouraged to brake hard to achieve the shortest possible stopping distance and 2) the wheels are to not slip, which is common during hard braking. If it is unbalanced, the agent becomes either too conservative or reckless, leading to slow braking or wheels locked at 100% respectively. Taking this into consideration, the following reward function is proposed:

$$R = P - 10 - j(\lambda), \quad j(\lambda) = \begin{cases} 10\lambda & \lambda > 20\% \\ 0 & \lambda < 20\% \end{cases} \tag{8}$$

The function serves as a penalty function which always returns a negative reward. The agent tries to maximize the rewards which here it leads to minimizing the penalties received. The penalty function also penalizes the duration of the braking process, the longer the vehicle brakes - the more penalties received. This encourages the agent to additionally slow down the vehicle as quickly as possible. Knowing that the maximum pressure pumped from the brake value is 10 Mpa the negative rewards is achieved by vertically shifting the rewards to below 0 with the constant value -10. To further encourage the model to stop as quickly as possible, the braking force can be taken into consideration via two possible options: linear acceleration or pressure. By increasing either of these, the braking force is increased and the vehicle stops quicker. When travelling over rough terrain, the linear acceleration is sensitive to big changes in the normal force acting at the tyre contact patch, thus it is proposed that the pressure P be used. This ensures that the first intuitive objective of the ABS is met, to meet the second objective, the function $j(\lambda)$ is introduced. This function is dependant on the wheel slip and penalizes the model as the wheels begin to slip. Maximum braking forces tend to occur between the region of 10-30% slip [4], however the peak braking force is closer to the 10% region with slippage occurring almost instantly past 30%, thus slip greater than 20% is penalised. A constant weight of 10 is multiplied by the wheel slip to discourage wheel lockup.

3 Results and Discussion

In order to simulate braking over rough terrain, a varying normal force at the tyre contact patch is required. Previous studies have been conducted at the Gerotek Test Facilities' Belgian paving [3, 10, 11], and this is used as vertical excitation. Due to the difficulty in modelling a vertical tyre that can capture the tyre dynamics over rough terrain, an FFT of the vertical force over the Belgian paving during braking was obtained by simulating a validated FTire model in ADAMS at varying speeds. This model is used to artificially generate a vertical force during simulation, and captures the speed dependency effect of the model, allowing for a random excitation force to be created and better generalization. To further encourage generalization by the model several steps are taken:

1. The starting speeds of the vehicle are randomised between 15 and 19m/s
2. The moment of Inertia I, μ_s and μ_c of the tyre model and mass of the body are randomised for each episode
3. Different normal force profiles are randomly sampled every 5 episodes

It should be noted that all the simulation results are based on the assumption of straight-line braking in the brake system model, and neglect the effects of the transportation delay caused by the brake hydraulic systems and brake pad travel as in [16]. The simulation is terminated below a speed of v = 2m/s as ABS algorithm often stop working at very low speeds [3].

Figure 2 plots the mean rolling episode scores against the episode number for the five different RL algorithms using a deep NN as the function approximator. An average score -680 over 1000 episodes was achieved for the Bosch algorithm and is used as a benchmark for the intelligent algorithms.

It can be seen that none of the algorithms are capable of matching or exceeding the performance of the Bosch algorithm, however it is important to note that this is as a result of using the simplified single tyre model. The dynamics that cause the Bosch algorithm to fail namely: unmodeled wheel torsional dynamics as well as noisy wheel speed signal leading to problems in slip calculation and angular acceleration are non existent in these simulations i.e. the Bosch is performing non real world conditions. It is recommended that a more complex model be used to expose the Bosch to these real life conditions, however despite this, it is good practice to compare the proposed intelligent models to the Bosch algorithm. It can be seen that the DDQN, A3C and A2C algorithms demonstrate successful learning with an increase in the average number of rewards per each episode, and are used in further simulations.

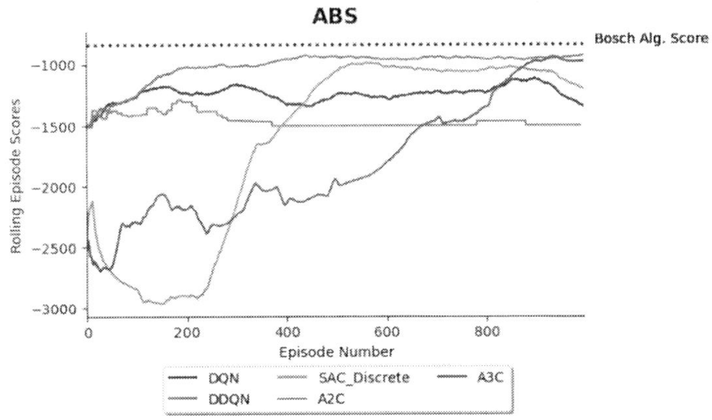

Figure 2: Comparison of rolling episode scores over 1000 episodes for each RL algorithm on rough terrain

As an alternative to the NN function approximator, the use of the intelligent models in conjunction with a TCN is investigated. By storing 100 previous step intervals, the benefit of using a history of data for time series data is seen in Figure 3, where despite the A3C and A2C models unsuccessfully learning, the DDQN algorithm which was able to better the Bosch algorithm over time.

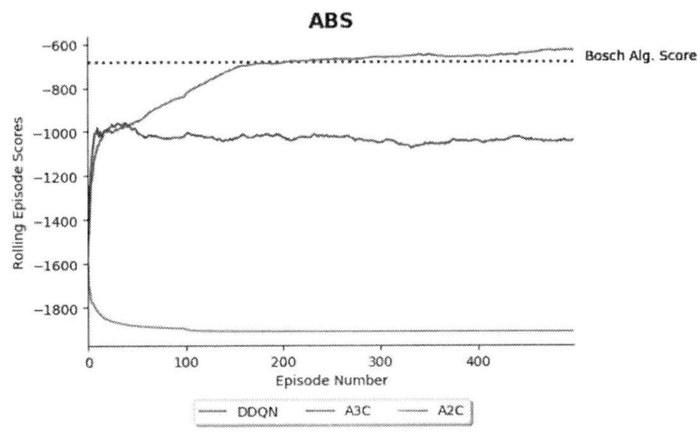

Figure 3: A comparison of the DDQN, A2C and A3C with the use of TCN as a function approximator, over 500 episodes on rough terrain

The DDQN algorithm has shown to be the most consistent and best performing of all the algorithms and is used for post processing. It is important to further validate the performance of this algorithm to ensure it is performing as expected. 10 episodes of the DDQN model are performed against 10 episodes of the Bosch algorithm at the same starting speed of 17m/s. Different performance measures that are conducted based on these runs include: the average stopping distance, the average longitudinal slip and its percentage time spent in certain slip ranges (this gives a good indication of the how much lockup occurs) as well as the vehicle and wheel speed profile, and the pressure, slip and reward profiles for a single episode. Table 1 summarises the results of the average stopping distances, as well as the different longitudinal slip performance measures of the Bosch, DDQN with and without the TCN (labelled DDQN-NN and DDQN-TCN).

Table 1: Summary of performance measures of the Bosch, DDQN-NN and DDQN-TCN

Stopping Distance (m)	**BOSCH**	**DDQN-NN**	**DDQN-TCN**
Mean	24.37	24.5	25.21
Std. dev	1.19	1.21	1.98
Best	23.45	23.74	23.17
Longitudinal slip (%)			
Mean	9.53	3.84	5.56
Std. dev	19.64	10.87	15.97
$\lambda < 10\%$	76.55	89.96	85.98
$10\% < \lambda < 20\%$	6.90	3.95	6.45
$\lambda > 20\%$	16.55	6.08	7.57

Analysing this summary, it can be seen that the DDQN models (with and without TCN) have a lower average slip percentage and produce a lower percentage of slip values greater than 20%, highlighting their ability to prevent wheel lockup on rough terrain. Along with ensuring the stability and lateral control of the model, the stopping distances are within 1m of the stopping distance achieved by the Bosch algorithm, and thus acceptable. Comparisons between the vehicle and wheel speed for the DDQN-TCN algorithm against the Bosch algorithm are presented in Figure 4.

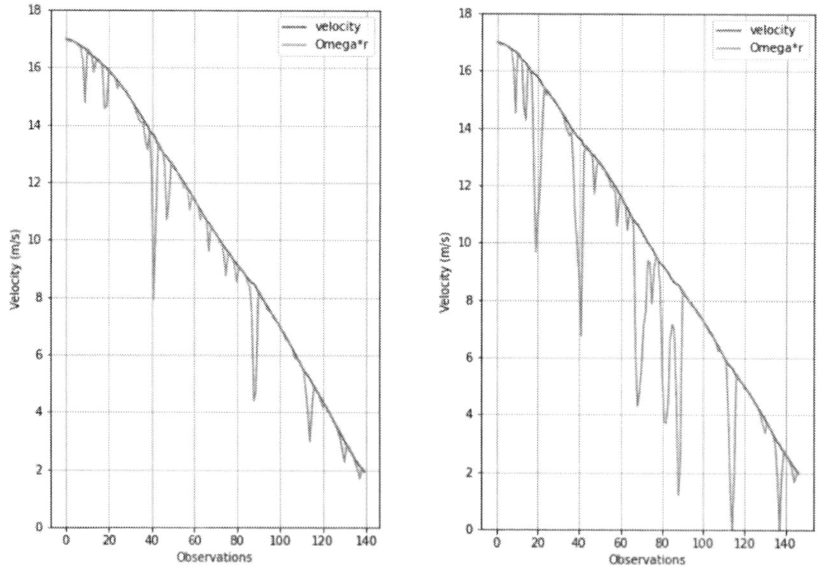

Figure 4: A comparison of the vehicle and wheel speed of the DDQN-TCN Algorithm (left) vs. the Bosch Algorithm (right)

It can be seen that the DDQN-TCN algorithm cycles the wheel speed better than the Bosch algorithm, within the same amount of step intervals (observations), hereby achieving small slippage as confirmed in Figure 5, where a maximum slip of 50% is achieved compared to 100% for the Bosch algorithm. This confirms the results from Table 1, where low percent of slippage is achieved by the DDQN-TCN model, resulting in better lateral control, as well preventing wheel lockup over rough terrain.

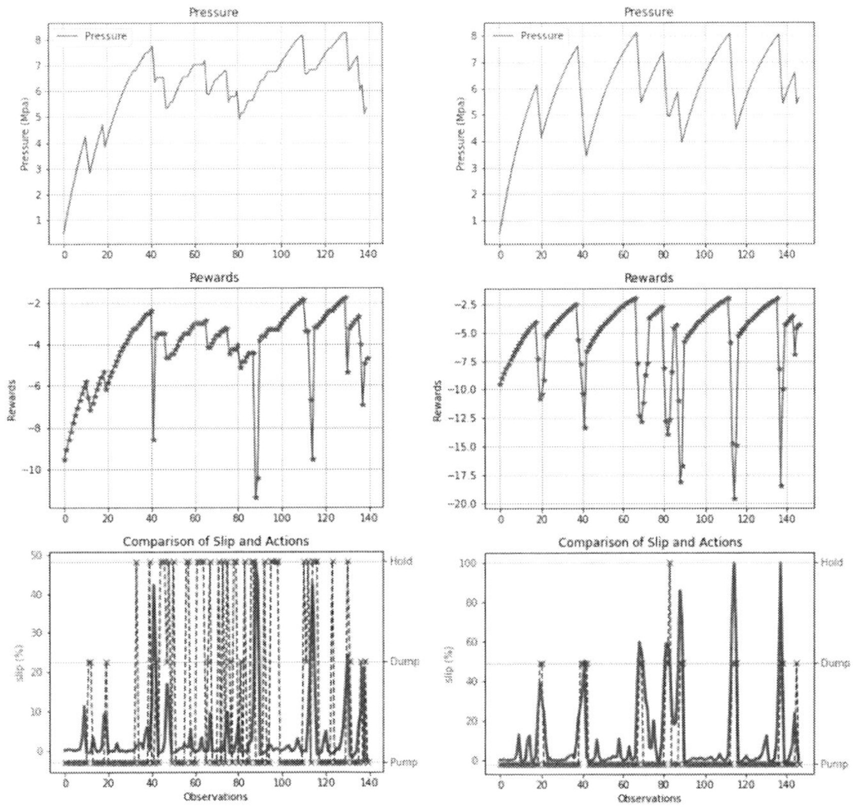

Figure 5: A comparison of the pressure, rewards and the slip and actions chosen of the DDQN-TCN Algorithm (left) vs. the Bosch Algorithm (right)

The following results prove that the proposed intelligent model-free ABS control algorithm is able to prevent wheel lockup over rough terrain, without deteriorating the stopping distance of the system. Future work includes increasing the complexity of the model from a single tyre to a validated full vehicle simulation model and training the intelligent algorithms on this vehicle model using FTire simulations in ADAMS. Additional tests to determine the success of the model are also to be performed according to the ABS performance metric proposed in [3]. These results should provide sufficient evidence on the application of a model-free approach as an ABS control approach.

4 Conclusion

Straight line braking simulations on rough terrain were performed using a single tyre system. The control of the ABS was modelled by multiple model-free RL techniques, and compared to the performance of a baseline Bosch algorithm. The DDQN algorithm with the use of TCN (labelled DDQN-TCN) demonstrated the most consistency of all the different algorithms considered, and was able to exceed the average rolling episode scores benchmark set by the Bosch algorithm. Through a series of ten straight line braking simulations, the DDQN-TCN model was able to achieve a lower average slip percentage as well as a lower percentage of slip values greater than 20%, highlighting its ability to prevent wheel lockup on rough terrain, whilst maintaining an average stopping distance within 1m of the Bosch algorithm. A more comprehensive model, along with the performance metrics defined in [3] are recommended for future work, in order to fully validate the performance of the model-free algorithm.

Bibliography

[1] SAE, Antilock Brake System Review, SAE International (2014)

[2] Garrott, W. R. & Mazzae, E. N., An overview of the national highway traffic safety administration's light vehicle antilock brake systems research program, Technical report SAE Technical Paper (1999)

[3] Hamersma, H. A, ABS braking on rough terrain, PhD thesis, University of Pretoria (2017)

[4] Aly, A. A., Zeidan, E.-S., Hamed, A., Salem, F. et al., 'An antilock-braking systems (abs) control: A technical review', Intelligent Control and Automation **2**, 186–195 (2011)

[5] Radac, M.-B. & Precup, R.-E , 'Data-driven model-free slip control of anti-lock braking systems using reinforcement q-learning', Neurocomputing **275**, 317–329 (2018)

[6] Zhao, D., Wang, B. & Liu, D, 'A supervised actor–critic approach for adaptive cruise control', Soft Computing **17**(11), 2089–2099, (2013)

[7] Sardarmehni, T. & Heydari, A, Optimal switching in anti-lock brake systems of ground vehicles based on approximate dynamic programming, in 'Dynamic Systems and Control Conference', American Society of Mechanical Engineers **57267**, p. V003T50A010, (2015)

[8] Bosch Gmbh, R. & Girling, P, Bosch Driving-safety systems, Society Of Automotive Engineers (1999)

[9] Canudas-de Wit, C., Tsiotras, P., Velenis, E., Basset, M. & Gissinger, G, 'Dynamic friction models for road/tire longitudinal interaction', Vehicle System Dynamics **39**(3), 189–226, (2003)

[10] Penny, W. C. W, 'Anti-lock braking system performance on rough terrain', Master's thesis, University of Pretoria, (2016)

[11] van der Merwe, N. A., Els, P. S. & Žuraulis, V, 'Abs braking on rough terrain', Journal of Terramechanics **80**, 49–57, (2018)

[12] Sutton, R. S. & Barto, A. G, Reinforcement learning: An introduction, MIT press, (2018)

[13] Christodoulou, P. Deep-reinforcement-learning-algorithms-with-pytorch, (2017). URL: Available at: https://github.com/p-christ/Deep-Reinforcement-Learning-Algorithms-with-PyTorch

[14] Hochreiter, S. & Schmidhuber, J, 'Long short-term memory', Neural computation **9**(8), 1735–1780, (1997)

[15] Bai, S., Kolter, J. Z. & Koltun, V, 'An empirical evaluation of generic convolutional and recurrent networks for sequence modeling', arXiv:1803.01271 (2018)
[16] Keshmiri, R. & Shahri, A. M, 'Intelligent abs fuzzy controller for diverse road surfaces', World Academy of Science, Engineering and Technology **2(2)**, 62–67, (2007)
[17] Stallmann, M. J. & Els, P. S, 'Parameterization and modelling of large off-road tyres for ride analyses: Part 2–parameterization and validation of tyre models', Journal of Terramechanics **55**, 85–94, (2014)

Prediction of boiler gas side effective heat transfer coefficients using mixture density networks and historic plant data

Renita Raidoo[1,*] and *Ryno* Laubscher[2,**]

[1]University of Cape Town, Rondebosch, Cape Town, 7700

[2]University of Stellenbosch, Stellenbosch Central, Stellenbosch, 7599

Abstract. Machine learning has received increased recognition for applications in engineering such as the thermal engineering discipline due to its abilities to circumvent thermodynamic simulation approaches and capture complex inter-dependencies. There have been recent headways to couple deep learning models to process simulations, given the deeper insight they can provide. The present study entails the development of a mixture density network (MDN) capable of predicting effective heat transfer coefficients for the various heat exchanger components of a utility scale boiler. Large boilers are susceptible to dead zones and other anomalous phenomena that influence performance and manifest as multimodalities in the measured data, which system-level 1D process models struggle to capture. The overall water-side heat load calculation, as well as mass and energy balances around the components were done to determine the heat transfer coefficients at each stage of the boiler using historic sensor data. The measured data was then used to train a deep learning model capable of outputting predicted heat transfer coefficients and local model uncertainty. The predictive model can be coupled to a water circuit process model which can be used to study aspects such as metal temperatures and operating philosophies at the different operating loads of the plant.

1 Introduction

Monitoring the thermal performance of boiler heat exchangers is crucial to ensure safe and efficient operation of power plants. Thermal monitoring is used in estimating residual life of boiler tubes and heat exchanger effectiveness. Boiler tubes are subjected to extreme conditions in coal-fired power plants which can lead to mechanical failures. These failures can be caused by a number of problems such as, slagging, fouling, caustic embrittlement, fire-side corrosion and thermal fatigue failure with the latter being one of the main reasons for unwanted plant downtime [1],[2]. Managing boiler tube failures is important as it can help in reducing forced outages and, in turn, improve plant availability and reliability [1]. Heat exchangers can experience rating problems whereby the actual heat absorption is lower when

*e-mail: rdxren002@myuct.ac.za
**e-mail: rlaubscher@sun.ac.za

compared to the intended design heating duty [3]. Heat transfer from the flue gas to the heating surfaces is largely reduced by the effects of ash deposition and non-uniform sweeping of flue gases. Increased levels of deposition and non-uniform flow leads to higher heat exchanger outlet temperatures which results in higher levels of slagging on downstream heat exchangers, lower boiler thermal efficiency and higher fuel demand [4]. Therefore, a need exists to estimate plant heat exchanger effectiveness and tube metal temperatures using actual plant measurement data, seeing as conventional empirical process models struggle to capture the various effects in the boiler. A model capable of capturing actual plant thermal performance would enable studying historic heat exchanger metal temperatures to estimate residual life and quantify effects of boiler input parameters on heat exchanger performance.

The present study proposes a data-driven method of determining effective heat transfer coefficients with minimal plant data using a mixture density network (MDN) approach. The proposed methodology will be applied to a 618 MWe utility-scale pulverised fuel boiler.

Practical machine learning problems can often have significantly non-Gaussian distributions due to the presence of unmeasured phenomena and measurement uncertainty [6]. The MDN model is useful as it has the ability to make better predictions on multimodal data since it is able to predict multiple Gaussian distributions for single output parameters. Bishop [6] proposes a framework for modelling multiple probability distributions such that a range of predicted values can be obtained with an inherent uncertainty. The benefit of this is that the operator can make an informed decision based on the probability and standard deviation given with an associated heat transfer coefficient prediction.

Artificial neural networks (ANNs) have been used in a broad range of applications including: pattern classification, pattern recognition, optimisation, prediction and automatic control [7]. Many recent studies in the field of thermal engineering and plant performance monitoring have incorporated the use of machine learning as an alternative to classic methods. Hu et al [8] make use of a short-term memory-based autoencoder network for early detection of anomalies. The study showed excellent results which indicate that the framework can be used as a legitimate tool for plant performance monitoring. Another study from Herawan et al [9] shows the success of using an ANN to predict the power generation of a waste heat recovery system. Their work highlights the method's ability to perform in highly fluctuating environments with multiple inputs affecting the output, as is common with thermal combustion systems. Other uses of machine learning in power plants include system enhancement and optimisation [10], health monitoring [11] and complex thermal management [5].

In the present work, heat transfer coefficients are predicted using plant data from an actual subcritical boiler located in Southern Africa over a period of 336 days. The heat absorbed by the heat exchangers is obtained by calculating the increase in enthalpy of the water/steam in each circuit using measured pressures and temperatures. The data in this study utilises 70 inputs for the calculations used to estimate heat coefficients. These heat exchanger calculations entail the use of combustion calculations and mass and energy balances. The required heat transfer rates are determined, and subsequently the flame temperature, mass flow rate of fuel and heat transfer coefficients at the various heat exchangers of the boiler system.

Following this, an ANN is developed and coupled with a probabilistic model to create a MDN which is capable of outputting multiple conditional probability distributions. It should be noted that the input to the model has only 13 features which are direct inputs to the boilers operation. The inputs include steam flow rate demand, ambient temperature, excess air ratio and final steam thermal conditions among other integral pressure and temperature readings. These inputs are user controlled and are not products of the boiler's operation. The output parameters of this model are the standard deviations, mixing coefficients and predicted means (the predicted heat coefficient value). The model is validated using the mean average error and statistical inference methods, namely prediction intervals and confidence intervals.

2 Materials and methods

2.1 Natural convection boilers

The boiler in the present study is of the radiant water-tube type, operating on a natural circulation principle. The boiler consists of 7 heat exchangers, namely the furnace, platen superheater, final superheater, secondary reheater, primary superheater, primary reheater and economiser. The layout of the flue gas and water circuits can be seen in Figure 1a and 1b.

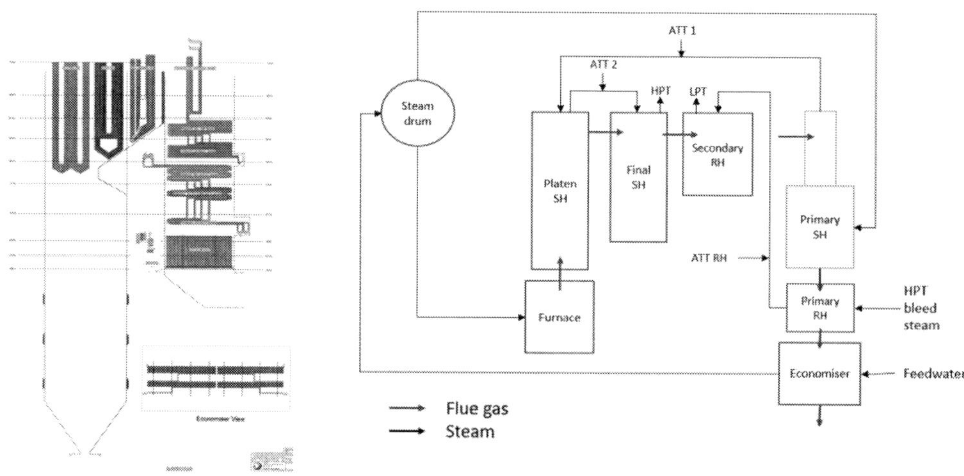

(a) Boiler Layout (b) Steam Flow

Figure 1: Schematic of boiler

2.2 Calculation of heat transfer coefficients from measurement data

Heat transfer coefficients represent the proportionality between the heat flux and the temperature gradient between two fluid control volumes. In order to determine the heat transfer coefficient θ_i for the ith heat exchanger, measured data is used for mass and energy balance calculations. These need to be performed to determine the average control volume temperatures as well as the heat transfer rate from the hot to the cold fluid. For the below heat transfer calculations a quasi-steady flow assumption was made. The heat transfer rate per heat exchanger is calculated using the difference between the water inlet and outlet enthalpies as seen in equation 1.

$$\dot{Q}_{HX,i} = \dot{m}_{s,i}(h_{s,out,i} - h_{s,in,i}) \quad (1)$$

The enthalpies are evaluated at the inlet and outlet positions of the heat exchanger using the measured temperatures and pressures in the water circuit. The mass flow of steam is evaluated at each stage of the boiler since there are added flows between the superheaters and reheaters from the attemperation. The furnace heat transfer calculation differs slightly for the subcritical boiler. The furnace heat load is determined using the feedwater flow into the steam drum and the latent heat of evaporation of water evaluated at the steam drum pressure. The total water heat load per measured observation is calculated by summing the heat transfer rates of each boiler heat exchanger, $\dot{Q}_{steam} = \Sigma \dot{Q}_{HX,i}$.

The gas composition of the flue gas flowing through the boiler is determined using the combustion products from a global step complete combustion calculation which takes the fuel ultimate analysis and the excess air ratio α as inputs. The ultimate analysis of the fuel is taken from the design data, C: 0.4156, H: 0.0222, N: 0.0097, O: 0.079, S: 0.0094, H_2O: 0.055 and Ash: 0.409 [12]. The quasi-steady fuel flow rate per measurement is found by balancing the total boiler energy inputs and energy outflows, as seen in equation 2. For the energy inputs; $h_{fuel,sensible}$ denotes the sensible enthalpy of the fuel and HHV denotes the higher heating value of the fuel. Furthermore, $\dot{m}_{fuel}$ is the mass flow rate of fuel, $m_{dry,air}$ is the dry air supplied per kilogram of fuel, $m_{moist,air}$ is the moisture entering the boiler along with the air per kilogram fuel and $h_{air,ambient}$ is the ambient enthalpy of the air. For the energy outflows, the heat transfer plus all the losses in the system need to considered. These losses include the total heat loss due to the unburnt carbon, radiation, bottom ash, unaccounted losses and losses caused by ash at the exit. $Y_{ar,C,loss}$ is the as-received fraction of unburnt carbon per kilogram of fuel, $f_{rad,loss}$ is the percentage of chemical energy lost to the environment, $Y_{ar,ash}$ is the as-received ash mass fraction, $h_{bottomash} = c_{P,ash}(1073K - 298K)$ is the bottom ash heat loss, $C_{unaccounted}$ is the unaccounted heat loss, $\dot{m}_{fg} = \dot{m}_{fuel} \cdot \left(1 - Y_{ar,ash} + m_{dry,air} + m_{moist,air}\right)$ is the flue gas mass flow rate and $h_{fg,out}$ is the boiler exit flue gas enthalpy calculated using a real gas mixture property database.

$$\begin{aligned} \dot{Q}_{sensible,fuel} + \dot{Q}_{combustion,fuel} + \dot{Q}_{combustion,air} = \dot{Q}_{steam} + \dot{Q}_{carbonloss} + \dot{Q}_{radiationloss} + \\ \dot{Q}_{bottomashloss} + \dot{Q}_{unaccountedloss} + \dot{m}_{fg,out} h_{fg,out} + \dot{Q}_{sensibleloss} \end{aligned} \tag{2}$$

$$\begin{aligned} \dot{Q}_{sensible,fuel} &= \dot{m}_{fuel} h_{fuel,sensible} \\ \dot{Q}_{combustion,fuel} &= \dot{m}_{fuel} HHV_{fuel} \\ \dot{Q}_{combustion,air} &= \dot{m}_{fuel}(m_{dry,air} + m_{moist,air}) h_{air,ambient} \\ \dot{Q}_{carbonloss} &= \dot{m}_{fuel} Y_{ar,C,loss} HHV_{carbon} \\ \dot{Q}_{radiationloss} &= \dot{m}_{fuel} HHV_{fuel} f_{rad,loss} \\ \dot{Q}_{bottomashloss} &= \dot{m}_{fuel} Y_{ar,ash}(1 - Y_{flyash}) h_{bottomash} \\ \dot{Q}_{unaccountedloss} &= \dot{m}_{fuel} HHV_{fuel} C_{unaccountedloss} \\ \dot{Q}_{sensibleloss} &= \dot{m}_{fuel} Y_{ar,ash} Y_{flyash} h_{flyash,exit} \end{aligned}$$

Once the fuel flow has been determined, the flame temperature (furnace inlet gas temperature) can be found using an adiabatic combustion control volume calculation. The flame temperature and furnace heat transfer rate, calculated using the measured data, is then used to determine the furnace exit gas temperature. Radiation leaving the furnace exit plane is ignored. Using the inlet gas temperatures and calculated heat transfer rates per heat exchanger, as for the furnace, the exit gas temperatures for each heat exchanger can be found using a simple energy balance calculation.

To account for the air flow that contributes to the flue gas composition, two modelling factors are included in the air flow calculation namely the excess air ratio α and the air leakage factor

α_{leak}. The percentage of O_2 is measured at the air-heater outlet and used to find the leak air flow at a 100% load. Using CFD simulation data [12] of the current boiler at 100% load case, α was found and in turn the α_{leak} using an iterative root-finding solver. For the remainder of the data observations α_{leak} was fixed and α recalculated to find the input required to match the measured O_2.

As mentioned, the average gas- and water-side inlet and outlet temperatures are required to determine the temperature gradient between the gas and water control volumes, as shown in equation 3. The temperature gradient is required to calculate the heat transfer coefficients. The water temperatures are measured and gas temperatures calculated as explained previously. The inlet gas temperature to a given heat exchanger is assumed to be the outlet gas temperature of the previous heat exchanger. Using the known heat transfer rate and inlet gas temperature the outlet gas enthalpy can be found as: $h_{fg,out}(T_{fg,out}) = h_{fg,in}(T_{fg,in}) - \dot{Q}_i/\dot{m}_{fg}$. The outlet enthalpy is then used to determine the outlet temperature of the heat exchanger.

$$\Delta T_i = \bar{T}_{fg,i} - \bar{T}_{s,i} \tag{3}$$

Once the average temperatures are found, the heat transfer coefficient can be determined using the effective heat transfer area of the heat exchanger as seen in equation 5.

$$\bar{T}_{s,i} = \frac{T_{s,in,i} + T_{s,out,i}}{2} \quad \bar{T}_{fg,i} = \frac{T_{fg,in,i} + T_{fg,out,i}}{2} \tag{4}$$

$$\theta_i = \frac{\dot{Q}_i}{A_i \Delta T} \tag{5}$$

Figure 2 shows the heat transfer coefficients of the boiler heat exchangers at various boiler loads, represented by $\dot{Q}_{steam}$. It can be seen that the heat transfer decreases along the flue gas path, the highest rates are found at the furnace and the lowest at the economiser. The clustering around each of the bands is indicative of the multi-modality of the data which suggests external forces have significant effects on the distribution of the coefficients. The heat transfer to the furnace walls, platen superheater and final superheater for the case study boiler is mainly due to radiation from the hot combustion gases [12]. The unsteadiness of the combustion processes leads to a large variation in estimated heat transfer coefficients per given load for these radiant heat exchangers. It is seen as the heat exchanger location moves further away from the furnace the band of data scatter reduces.

3 Deep Learning Method

3.1 Data

The input data to the machine learning model is taken from pressure, temperature and mass flow sensors on the plant and are listed in table 1. Reheater components have been abbreviated to RH. It should be noted that the inputs to the MLP are all direct inputs to the boiler operation and not results of the boilers operation; so essentially these parameters are operator controlled settings. The main steam temperature, which is the inlet steam temperature to the high pressure turbine, as well as the reheater 2 outlet temperature are in effect controlled parameters since they are directly affected by the amount of attemperation per boiler load.

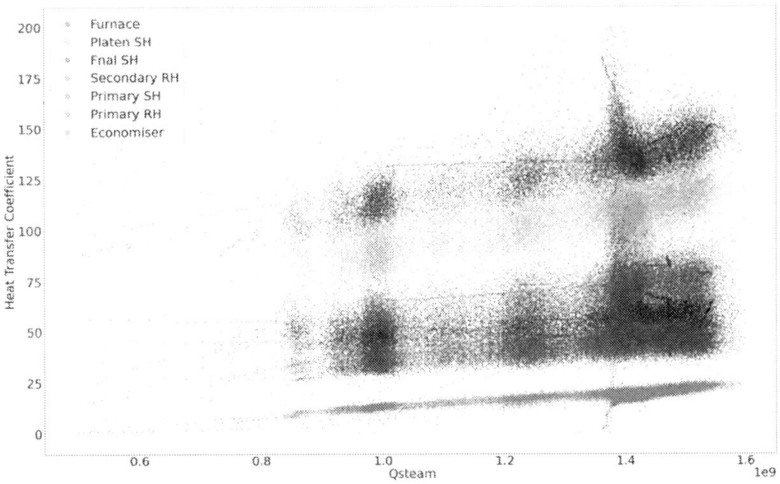

Figure 2: Heat Exchanger θ at various boiler loads

Table 1: Measured inputs to the ANN

	Feedwater	Steam drum	Main steam	RH 1 inlet	RH 2 outlet	Attemperator 1	Attemperator 2	RH attemperator
Temperature	x		x	x	x			
Pressure	x	x			x			
Mass flow rate	x			x		x	x	x

The output data matrix has 7 columns which contain the calculated heat transfer coefficient values for the various heat exchangers. The rows of the matrix correspond to the different observations.

The data is split into 85% training set and 15% validation set. The purpose of the validation set is to test the inference ability of the model. It is used to determine the model accuracy in a 'real-world' deployment. The model is trained using the training set, so the accuracy is not a true representation of the model's performance due to built in training bias. The model has technically never seen the validation set before, therefore its predictions are objective and impervious to prior information.

3.2 Multilayer perceptron neural networks

A multilayer perceptron (MLP) neural network is a fully connected layered feedforward network [13]. This means that every node in a given layer is connected to every node in the adjacent layer, however there are no connections between nodes of the same layer. The network is parameterised by weight W and bias b matrices. Different layers extract different representations of features from the input data, each layer learns to extract more complex and deeper features from its previous layer [14].

The function of a neural network can be seen as two phases, firstly training and secondly generalising i.e. making predictions.

Forward propagation can be seen as the forward pass of information from layer to layer. Each layer accepts inputs, applies some activation function and then passes it to the successive layer until the final layer outputs the prediction $\hat{Y}$. The activation function used for this model is a ReLu function which outputs a value directly proportional to the input. If the input is positive, or it outputs zero if the input is negative. In the first line of equation 6, the weights are applied to the output of the previous layer $A^{[l-1]}$. It should be noted that the first layer of the network is a matrix of inputs such that $A^{[0]} = X$. Once the activation function is applied, the pattern continues until the final activation $g(z)$ is applied to produce the predicted value at the output layer L.

$$\begin{aligned} Z^{[l]} &= W^{[l]}A^{[l-1]} + b^{[l]} \\ A^{[l]} &= g^{[l]}(Z^{[l]}) \\ A^{[L]} &= \hat{Y} \end{aligned} \tag{6}$$

Training is performed by the process of forward- and backward-propagation which aims to adjust the trainable parameters, W, b. The purpose of this is to minimise the cost function shown in equation 7. Once a complete forward propagation cycle has been completed, the model uses the predicted output to estimate the cost function value which indicates how close the prediction is to the actual value.

$$J_{MSE}(\bar{y}^n, \hat{y}^n) = \frac{1}{d_{target}} \sum_{j=1}^{d_{target}} \frac{1}{2}(\bar{y}^n - \hat{y}^n)^2 \tag{7}$$

Backward propagation is the automatic differentiation algorithm used to calculate the gradients of the weights and biases in a neural network graph structure with respect to the cost function value. These parameter gradients are used to update the trainable parameters using a gradient descent algorithm. Gradient descent is an optimisation algorithm which minimises the error of a predictive model with regard to a training dataset and the cost function. In the current work, the model is optimised using the ADAM algorithm shown in equations 8 and 9 [15]. It is known that the default configuration parameters do well on most problems and were therefore used in the present work.

$$\begin{aligned} V_{dW} &= \beta_1 V_{dW} + (1-\beta_1)dW & V_{db} &= \beta_1 V_{db} + (1-\beta_1)db \\ S_{dW} &= \beta_2 S_{dW} + (1-\beta_2)dW^2 & S_{db} &= \beta_2 S_{db} + (1-\beta_2)db^2 \end{aligned} \tag{8}$$

$$W = W - \alpha\frac{V_{dW}}{\sqrt{S_{dW}} + \varepsilon} \qquad b = b - \alpha\frac{V_{db}}{\sqrt{S_{db}} + \varepsilon} \tag{9}$$

Where V_{dW}, V_{db}, S_{dW} and S_{db} are initialised at 0. The process of forward and back propagation and gradient descent continue until optimal weights and biases have been found.

3.3 Mixture Density Networks

MDNs developed in the present study are standard feedforward artificial neural networks with a probabilistic extension module which outputs the mixture parameters, μ, π, σ. These parameters define κ probability distributions within the multi-modal data. The ANN, which incorporates stacked densely-connected neural layers has 7 neurons for the corresponding output parameters. The resulting MDN output for the current work, therefore, has shape K x 7 x 3. For any given input value, the mixture model provides the formal mathematical description for modelling an arbitrary conditional density function as seen in equation 10.

$$P(\bar{Y}|\bar{X}) = \sum_{k=1}^{k}(\bar{X})\mathbb{N}(\bar{Y}|\bar{\mu}_k(\bar{X}), \bar{\sigma}_k{}^2(\bar{X})) \tag{10}$$

The schematic below in figure 3 shows the general architecture of the MDN model. An arbitrary number of input nodes and hidden layers are shown.

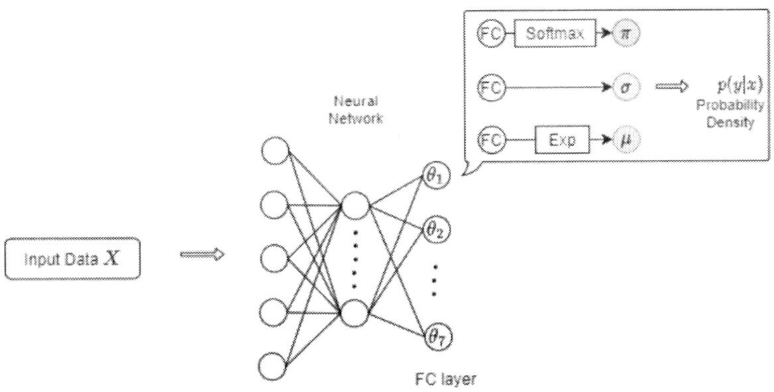

Figure 3: ANN-MDN Schematic

The mixing coefficients, denoted by π, must satisfy the constraints: $\sum_{k=1}^{k} \pi_k(x) = 1$ and $0 \leq \pi_k(x) \leq 1$ [6]. This is achieved by passing the activation from the fully connected layer through a softmax function as seen in equation 11.

$$\pi_k(x) = \frac{exp(a_k^{\pi,n})}{\sum_{l=1}^{k} exp(a_k^{\pi,n})} \tag{11}$$

The variance must satisfy $\sigma^2 \geq 0$, it is therefore represented in terms of the exponential of the corresponding activation such that $\sigma_k(x) = exp(a_k^{\sigma})$ [6]. The error function shown in equation 12 is defined as the negative logarithm of likelihood [6]. The derivatives with respect to w (vector of weights and biases) of the error function are determined in order to minimise the error.

$$J(\bar{Y}, \bar{\pi}, \bar{\sigma}, \bar{\mu}) = -\sum_{n=1}^{d_{target}} ln\left\{\sum_{k=1}^{k} \pi_k(\bar{x}^n, \bar{W}) \cdot N(\bar{y}^n|\bar{\mu}_k(\bar{x}^n, \bar{W}), \bar{\sigma}_k(\bar{x}^n, \bar{W}))\right\} \tag{12}$$

3.4 Model Development

A course grid search was performed to determine the best architecture for the network. The present study tested models with one and two layers, with varying numbers of neurons. The batch size was kept constant at 128 and a low learning rate of 0.0001 was used to ensure optimal convergence. The performance of the tested architectures were compared using the error metrics, MAE and MAPE, shown in equation 13.

Once the best-performing architecture has been identified using the standard feedforward ANN, the MDN module was added. Similar to the ANN various MDN hyperparameters were tested and the best-performing values used.

Once the final model had been developed and completely trained, its accuracy was tested using the validation dataset.

$$MAE = \frac{1}{n}\sum_{i=1}^{n}|y-\hat{y}| \quad MAPE = \frac{100\%}{n}\sum_{i=1}^{n}|\frac{y-\hat{y}}{y}| \tag{13}$$

4 Results and Discussion

As can be seen in table 2, the model with 2 hidden layers and 15 neurons performs the best since it had the lowest MAE. The model was tested using 2, 3 and 5 distribution components and it was decided that 2 components captured the modes in the data sufficiently well as there was no significant increase in prediction accuracy with more components.

Table 2: Hyperparameter Search Result: MAE on Development Dataset

Layers	25	15	10
1	21.96	9.83	10.155
2	10.86	3.22	20.77

The main requirement of the model was to be able to predict the heat transfer coefficients of all 7 heat exchangers in the boiler. The output signals are shown in figure 4. Specific points are indistinguishable however it can be seen that the model does capture the general trend of the data since the observations and predictions predominantly overlap. Another interesting feature of the model is its capacity to interpret noisy or disrupted signals. This is seen in the sections of straight lines. The model is able to fit a strongly correlated approximation which mimics this sensor failure. A detailed sample which includes the confidence and predictive interval bands is shown below in figure 6a, 6b and 6c. The final model's results are shown in table 3.

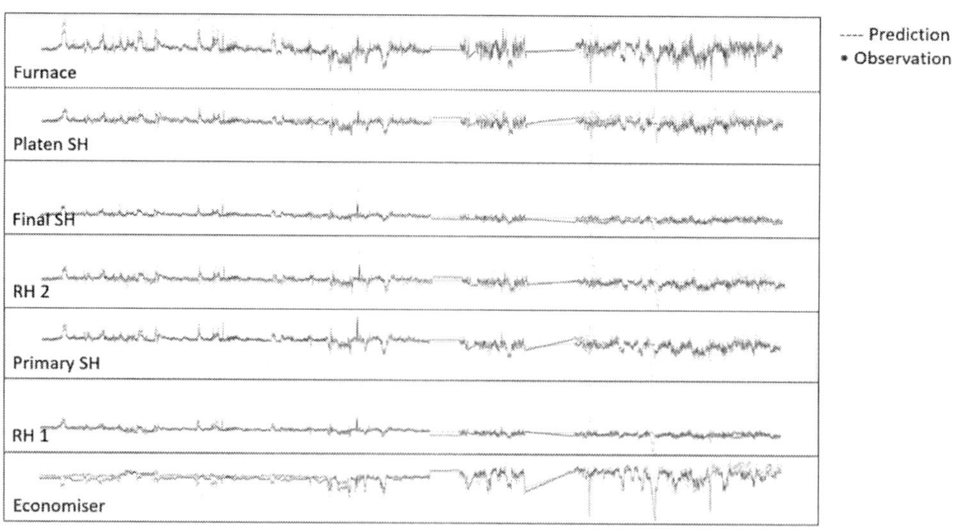

Figure 4: Heat transfer coefficient predictions for all heat exchangers in the boiler system

Table 3: Model Performance Results

	Training	Validation
MAE [$\frac{W}{m^2K}$]	3.35	3.11
MAPE	7.78%	7%

Figure 5 shows the error distribution graphs for the training and validation set. It can be seen that 57% of the predictions have less than a 10% error. The validation set has approximately 60% of the predictions within the 10% error margin.

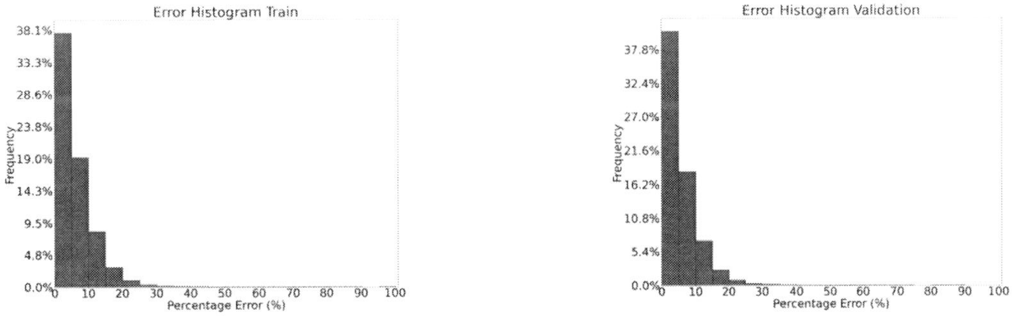

(a) Training Set (b) Validation Set

Figure 5: Error Distributions

Figure 6 shows the observed data, the predicted mean, the confidence interval band and the prediction interval band. Figure 6a shows a sample of the validation dataset of the furnace. The furnace, as was mentioned, is the stage of the boiler that is most susceptible to fluctuations in the boiler's operating parameters.

Figure 6c shows a sample of the development dataset of the economiser which is the final stage in the flue gas path of the boiler. A constant offset is noticed for the economiser heat transfer coefficient predictions, but the magnitude of the error is relatively low. Furthermore, the fluctuations in the observed data is significantly lower when compared to the furnace values.

The boiler tubes which typically fail due to localised overheating are the superheater elements, more specifically the final superheater seeing as it has the highest steam temperatures. It is, therefore, encouraging to see that the model is capable of predicting the actual heat transfer coefficients with relative accuracy. Although there are some outliers, the confidence interval captures the vast majority of these peaks.

The model's accuracy and precision were analysed using statistical inference methodologies. The main focus of this study is to determine the efficacy of this model on making predictions, therefor prediction intervals and confidence intervals are used as error metrics. Although prediction intervals and confidence intervals have underlying similarities, they represent different properties of the data. The confidence interval is the range which has a 95% chance of containing the true population of the data. It is based off the statistical parameters, standard deviation and mean of the estimated predictions generated by the model. It can be interpreted as the likelihood of the observed data falling within 2 standard deviations of the predicted mean. The predictive interval stipulates what range the predicted mean can fall between

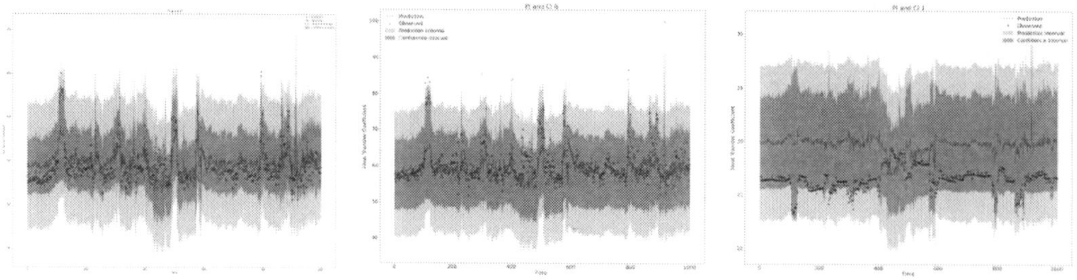

(a) Furnace (b) Final Superheater (c) Economiser

Figure 6: A sample of the actual and predicted outputs using the MDN on the development set

based on the range of the true data. This can be interpreted as the estimated range in which a future prediction will fall given the true observations that have already been observed within some level of confidence, also 95% in this case.

Table 4: Statistical Results

	% In-band	Band Width
Statistical Prediction Interval	91.98	22.33
Prediction Confidence Interval	76.83	12.49

It is interesting to note the performance of the model on each heat exchanger along the path of the flue gas. The economiser prediction, including the predicted standard deviation is less precise than the furnace. It is also noteworthy that the statistical prediction interval is substantially wider than the model's predicted confidence interval.

5 Conclusion

In conclusion, a methodology to use an MDN to predict the heat transfer coefficients of various components in a natural convection boiler was presented. The model had an overall accuracy of 93% , with 91.98% of the measured data falling within the confidence interval and 76.83% of the mean prediction falling within the prediction interval. Given that the confidence interval is wider than the prediction interval, it can be concluded that the model requires longer training and finer tuning to improve the precision of the standard deviation parameter. Overall the model was able to capture plant behaviour and showed distinct trends in the data.

References

[1] Akash Singh, Vivek Sharma, Siddhant Mittal, Gopesh Pandey, Deepa Mudgal, Pallav Gupta, International Journal of Industrial Chemistry **9**, (2018), 1 – 15

[2] Wu Haibo, Zhang Man, Lu Qinggang, Sun Yunkai, Journal of Thermal Science **21, No.4** (2012), 368 - 376

[3] Serth, Robert W., Lestina, Thomas, *Process heat transfer : principles and applications* (Oxford, England ; Waltham, Massachusetts : Academic Press, 2014) 67

[4] Zhang Yanguo, Li Qinghai and Zhou Hui, *Theory and Calculation of Heat Transfer in Furnaces* (San Diego: Elsevier Science Technology, 2016) 186

[5] Yang Liu,Nam Dinh, Yohei Sato and Bojan Niceno, Applied Thermal Engineering **144**, 308-320 (2018)

[6] Christopher M. Bishop, *Pattern Recognition and Machine Learning* (Springer Science+Business Media, LLC, 2006) Ch. 5.6

[7] Preeti Manke, Sharad Tembhurne, International Journal of Computer Science Issues **9**, 520 - 526 (2012)

[8] Di Hu, Chen Zhang, Tao Yang and Gang Chen, Sensors **20**, (2020)

[9] Safarudin Gazali Herawan et al, Soft Computing **22**, 5955–5964 (2018)

[10] Hosham Salim, Khalid Faisal and Raheel Jawad, Applied Computational Intelligence and Soft Computing **2018**, (2018)

[11] Prabhas Hundi, Rouzbeh Shahsavari, Applied Energy **256**,(2020)

[12] Ryno Laubscher, Pieter Rousseau, Applied Thermal Engineering 114057 **160**, (2019)

[13] Ke-Lin DuM. N. S. Swamy, *Neural Networks and Statistical Learning* (Springer, 2019)

[14] Ling Liu, Yanzhao Wu, Wenqi Wei, Wenqi Cao, Semih Sahin, and Qi Zhang, IEEE 38th International Conference on Distributed Computing Systems **38** (2018)

[15] Ian Goodfellow, *Deep Learning* (The MIT Press, Massachusetts, 2015) 308

Modelling of dew point operating conditions in a rotary regenerative air heater

Warren Brandt[1]*, *Walter* Schmitz[2], *Thomas John* Sheer[2], and *Mike* Lander[3]

[1]Eskom Power Plant Engineering Institute; Boiler Combustion; School of Mechanical, Industrial and Aeronautical Engineering, Private Bag 3, WITS 2050 Johannesburg, South Africa
[2] School of Mechanical, Industrial and Aeronautical Engineering, University of the Witwatersrand, Private Bag 3, WITS 2050 Johannesburg, South Africa
[3]Eskom; Research Testing and Development; PEI Draught Plant Specialist, Johannesburg, South Africa

Abstract: Dew point related fouling of regenerative rotary air heaters contributes to efficiency reduction of coal-fired boilers. This fouling commonly causes high pressure differences over the air heater steel matrix, along with elevated leakage rates, resulting in increased loading of draught group fans. These increases also conduce to excessive erosion rates, introducing premature failure of element packs and draught group components. The research problem aimed to develop, and validate, a modelling tool to predict the onset of dew point related fouling for regenerative rotary air heaters (VBA RAH model). A modelling tool equip system engineers with the knowledge to enhance operating conditions and maintenance strategies, to prevent the onset of dew point related fouling. The results yielded that the cold end of the air heater packs experience sulphuric acid dew point related fouling for at 670MW(90% MCR), at 544MW(80% MCR) and at 468MW(68% MCR) of the exposure time to the flue gas stream. Different operating and design change configurations were simulated to find potential solutions to minimize dew point related fouling and improving boiler efficiency. The tool provides a platform for identifying methods to reduce maintenance costs and production losses, which will contribute to improved and sustainable operating of coal fired power generation plants.

1 Introduction

In coal fired power generating plants, air heaters are used to transfer the thermal energy in the flue gas to the combustion air entering the boiler, as stated in Figure 1. A regenerative air heater comprises a steel matrix of stacked element plates. An induced draught fan extracts hot flue gas from the furnace which passes through the steel matrix, transferring heat from the flue gas to the element plates. A forced draught fan blows cold air through the hot element plates, which heat the combustion air to improve the boiler performance. If the thermal performance of an air heater deteriorates, the boiler performance reduce accordingly. Therefore, air heaters are usually one of the major contributors to reduction in energy

* Warren Brandt: brandtwa@eskom.co.za

availability from coal fired power generating plants. A common problem for regenerative air heaters is dew-point related cold end fouling, which leads to insufficient heat transfer.

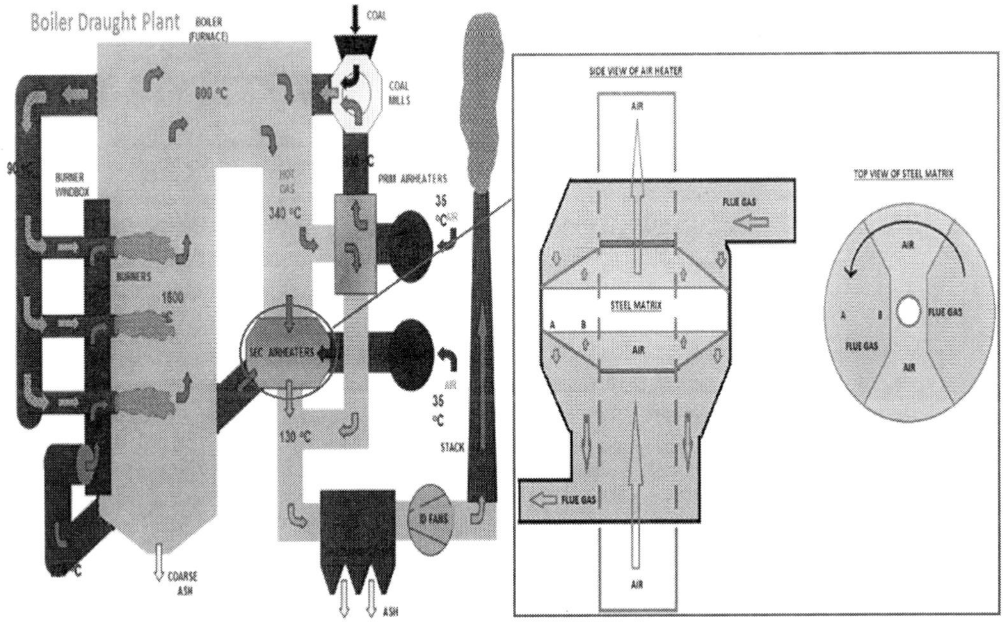

Fig. 1. Boiler process flow schematic.

The following operating conditions are indicators of regenerative air heater dew-point related fouling [1]:

- Occurrences of high pressure differences across the air heater steel matrix, resulting in high FD (Forced Draught) and ID (Induced Draught) fan loading (FD and ID fans operating at maximum capacity).
- Increased fly ash erosion rates occur on localised areas of the air heater steel matrix air heater hoods, dampers and ducts up to the Electrostatic Precipitators (ESP) or Fabric Filter Plant (FFP) inlet-casing.
- Accelerated wear and premature failure of element packs on the steel matrix periphery.
- Exaggerated dust emission levels occur because of augmented volumes of flue gas extracted from the furnace. The fouled element packs restrict flow, which then requires the radial vanes to open more from the FD and ID fans. This introduces increased volumetric flow, which also elevates the flue gas velocity to a point that the ESP becomes less efficient. Dust handling plants with filter bags experience excessive erosion, which reduces the life expectancy of the filter bags. The particle count for the emissions will increase, resulting in additional load losses. Second, the increased velocities also cause higher erosion rates in the air heaters and flue gas ducting, producing increased levels of air ingress. If the air leakage to the flue gas stream increases, the flue gas volume becomes diluted, introducing an increase in toxic gas emissions levels. These conditions are symptoms of excessive fouling of air heater element packs, which can only be rectified by either applying frequent high pressure washing or replacement of element packs. Both of these are costly activities, which can be avoided by applying the correct operating conditions and maintenance strategies. Through evaluating air heater performance, and finding critical parameters that create rotary regenerative air heater fouling conditions, a simulation model was developed to predict the onset of dew point related fouling.

The above conditions are symptoms of excessive fouling of air heater element packs, which require frequent high pressure washing or replacement of element packs. Dew point related fouling exists when the flue gas temperature and surrounding materials experience a temperature below the flue gas dew point temperatures. The fly ash particles in the flue gas adhere to each other and element pack surfaces when these conditions occur. Sulphuric acid, sulphurous acid, and moisture in the flue gas produce dew point related fouling. When precipitation inside the small spaces between the elements occur, the packs experience excessive clogging of ash deposits [3]. This process reduces the flow passages through the element packs. During normal operation, continuous exposure to high temperatures and drying (from the conditions internal to the air heater) cures and hardens the deposits between the elements. If the soot blowing system cannot remove the deposits, complete blocking of element packs will be the result. The application of correct operating conditions and maintenance strategies will mitigate the risk of incurring these costly activities. This research aimed to evaluate air heater performance through identification of critical parameters causing rotary regenerative air heater fouling conditions. These parameters were used to develop a simulation model to predict the onset of dew point related fouling. The research included plant experiments to validate the model.

2 Literature Overview

2.1 Regenerative Air heater Model

Caby and Gruen used a physical model to define correlations used to evaluate the heat transfer and pressure drop correlations of a thermal test facility used for developmental research of a regenerative air heater model [4, 5]. These correlations were later used by Habbitts to develop a mathematical model for rotary regenerative air heaters, known as the RAH model. A finite difference technique, to analyse the heat transfer for transient tests, by Mullison and Loerhke, was used by Caby and Habbitts to calculate the performance characteristics of the air heaters [6, 7]. This model was further developed by De Klerk, to estimate the fluid temperatures, element surface metal temperatures and differential pressure across the steel matrix for alternating thicknesses of element plates [8]. The following assumptions were made by Habbitts, and later reviewed by De Klerk, which was also considered during this research project [9]:

- Steady flow conditions are experienced flowing into and out of the air heater.
- The width and number of plates are sufficient to essentially reduce the problem to a one dimensional setup, with one directional flow in each block.
- Finite thermal conduction occurs in the solid parallel to the direction of flow, with no transverse temperature gradients. Thus the boundary conditions for the side walls of the elements are adiabatic. (As indicated in Figure 2 from section 4, below along the top and bottom plate centre lines.)
- Convection heat transfer occurs across the solid surface interfaces within the element. The magnitude of which is determined by a heat transfer coefficient h (temperature dependant). A single h value is assumed to apply throughout a particular block of plates at a particular instant in time. This is a space averaged convective heat transfer coefficient assumed to be applied across a series of parallel plates for a particular flow rate.
- All plates are made from the same material.
- The fluid thermal capacity is negligible in comparison with the solid thermal capacity.
- The effect of heat transfer by radiation is neglected due to its small contribution to the temperature difference between plate regions. The symmetrical gas molecules do not

participate in radiation, and the concentration of water vapour in air is too small. Therefore radiation from the air stream can be neglected. Radiation from the flue gas to the plates can be significant owing to temperatures differences of up to 40°C between gas and the plates. This effect is ignored due to the temperature difference occurring over small regions at the hot end of the matrix. Once a temperature drop is experienced, the effect of radiation decreases rapidly.

- The fluid flowing into and out of the control volume is considered to be at a steady state. This assumption is used for the formulation of the energy balance of the volume in Figure 2 in section 4. The energy exchanges across the boundaries of each region are assumed to be represented by conduction in the solid parallel to the direction of flow (this effect is also known as longitudinal conduction), convection between the solid and fluid region, and enthalpy flux into and out of the fluid region. Energy is stored within the solid regions and is proportional to thermal capacity of the solid.
- The mass flowrate for each block of plates for specific fluid stream is constant, but the temperature along with densities and velocities do change with time.

2.2 Dew point related fouling

Mathebula used a Quantitative Evaluation of Minerals by Scanning Electron Microscopy (QEMSCAN) surficial and cross-sectional analysis, to show the phases of fouling of ash deposits from an Eskom Power Station. These phases were anhydrite, AlSi-sulphate, Si (Al)-sulphate, iron sulphate and sulphur. The phases occur because of a reaction between sulphuric acid and fly ash components, and a reaction between sulphuric acid and fuel oil catalysts [10]. Kunze proposed the acid dew point concept based on the gas-liquid equilibrium. Therefore, the acid dew point was defined as the temperature of gas-liquid equilibrium interface when the apparent partial pressures of sulphuric acid vapour and water vapour at the interface equal to those in the flue gas [Kunze]. Sulphuric acid dew point temperatures vary according to changes in partial pressure for moisture (H_2O) and sulphur-trioxide (SO_3). Mathebula also concluded that fly ash adhere to the sulphuric acid that condenses onto the air heater plates, along with the fly ash particles adhering to each other when the gas temperature reduces below the sulphuric acid dew point temperature. The formation from the phases also shows that the sulphuric acid reacts with fly ash components and fuel oil char [10].

Wei compared the experimental data from previous acid dew point prediction methods using empirical formulas and the iterative calculation model. This comparison showed small deviations between the calculation results and experimental results, except for the studies of Haase & Borgmann and experimental constants-1 method. The results of Haase & Borgmann were abnormally low, while the results of experimental constants-1 showed a large-span increase, and the greatest deviation was around 41 °C [12]. Müller ignored the influence of water vapour on acid dew point, which resulted in under estimation of acid dew point temperatures. Halstead took the content of water vapour into consideration, but the results did not concur with the experiment data. The predictions of Verhoff & Branchero agreed well with the experimental data, and was found to be the preferred option. Verhoff & Branchero were the lowest compared with other deviations between other formula results and experimental data. Therefore, the prediction of Verhoff & Branchero is of great accuracy and has a wide range of application [12]. Ganapathy and Mathebula used Verhoff & Branchero's equation to calculate dew point temperatures (water, sulphuric acid and sulphurous acid) of flue gas [12, 11, 10].

3 Project Definition

The aim of the project was to develop a rotary regenerative air heater simulation model to predict the onset of dew point related fouling. This model aims to equip system engineers with the knowledge to enhance operating conditions and maintenance strategies, in order to prevent the onset of dew point related fouling. The study contributes to the possibility of increasing the life expectancy of air heater element packs and draught group components. It also contributes to improved boiler efficiency, reduced maintenance costs and production losses, which will conduce to improved and sustainable operating of coal fired power generation plants. The current maintenance approach incorporates a more reactive approach, and therefore creating preventative maintenance measures will be beneficial to Eskom. The following objectives were set in place to establish whether steel matrix metal temperatures can be controlled to prevent dew point related regenerative air heater fouling:

1. Identified dew point related fouling parameters.
2. Developed a modelling tool that predicts dew point operating conditions in a rotary regenerative air heater (VBA RAH model).
3. Verified the fouling condition by plant experiments to prove the accuracy of the VBA RAH model. The plant experiments comprised:
   a. Measurements of fluid (Air and flue gas) temperatures for each layer of the steel matrix.
   b. Measurements of metal surface temperatures for each layer of the steel matrix.
   c. Air heater leakage and operating parameters.
   d. Coal analysis which established coal composition.
   e. Flue gas analysis which established flue gas composition.

4 Theoretical Analysis

4.1 Regenerative air heater simulation model theory

For this research project the following theoretical basis was used to develop a rotary regenerative air heater model (Visual Basic for Application Regenerative Air Heater – VBA RAH) which incorporated a prediction of the onset of dew point related fouling. Figure 2 expresses the enhanced model for energy balance for the fluid region. This model forms the basis for estimating the temperature profiles of both the fluid and the solid regions. Equation 4.1 illustrates the effect of the occurrence of convection, where equation 4.2 and 4.3 consider the solid regions where conduction occurs because of the convective effect of the fluid region [8].

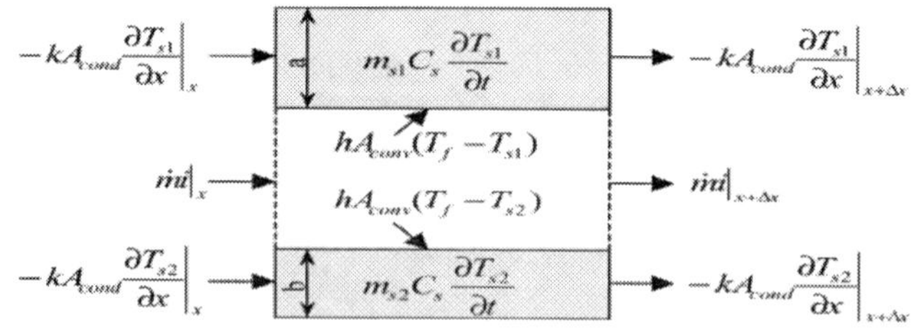

Fig. 2. Basic plate/channel element [9].

Fluid region:

$$\frac{\partial T_f}{\partial x} + \frac{h \cdot W}{\dot{m}_{fluid} \cdot C_f}\left(2T_f - T_{s1} - T_{s2}\right) = 0 \qquad (4.1)$$

Each solid region temperature profile will be formulated using:
Solid 1:

$$\frac{\partial T_{s1}}{\partial t} = \frac{k}{\rho_s C_s}\frac{\partial^2 T_{s1}}{\partial x^2} + \frac{h}{\rho_s C_s a}\left(T_f - T_{s1}\right) \qquad (4.2)$$

Solid 2:

$$\frac{\partial T_{s2}}{\partial t} = \frac{k}{\rho_s C_s}\frac{\partial^2 T_{s2}}{\partial x^2} + \frac{h}{\rho_s C_s a}\left(T_f - T_{s2}\right) \qquad (4.3)$$

Dimensionless parameters are used as part of the numerical method. T* is the dimensionless temperature and can be calculated using the following equation:

$$\frac{\partial T_{s2}}{\partial t} = \frac{k}{\rho_s C_s}\frac{\partial^2 T_{s2}}{\partial x^2} + \frac{h}{\rho_s C_s a}\left(T_f - T_{s2}\right) \qquad (4.4)$$

The following equations are used to convert the heat transfer parameters to non-dimensional parameters:

1. Dimensionless time:

$$\theta = \frac{\dot{m} C_f}{M_s C_s} \qquad (4.5)$$

2. Conduction Parameter:

$$\lambda = \frac{k A_{cond}}{\dot{m} C_f L} \qquad (4.6)$$

3. Dimensionless length:

$$z = \frac{x}{L} \qquad (4.7)$$

4. Number of transfer units:

$$NTU = \frac{h A_{conv}}{\dot{m} C_f} \qquad (4.8)$$

5. Thickness Ratio:

$$\beta = \frac{a}{b} \qquad (4.9)$$

Dimensionless temperatures must be calculated for both solid regions and the fluid region:

$$T_f^* = (T_f - T_{cold\ inlet})/(T_{hot\ inlet} - T_{cold\ inlet}) \qquad (4.10)$$

$$T_{s1}^* = (T_{s1} - T_{cold\ inlet})/(T_{hot\ inlet} - T_{cold\ inlet}) \qquad (4.11)$$

$$T_{s2}^* = (T_{s2} - T_{cold\ inlet})/(T_{hot\ inlet} - T_{cold\ inlet}) \qquad (4.12)$$

Therefore rewriting the equation 4.1 to 4.3 by substituting in the non-dimensional parameters:

Fluid region:

$$\frac{\partial T_f^*}{\partial z} + NTU\left(T_f^* - \frac{T_{s1}^* - T_{s2}^*}{2}\right) = 0 \qquad (4.13)$$

Solid 1:

$$\frac{\partial T_{s1}^*}{\partial \theta} = \lambda\frac{\partial^2 T_{s1}^*}{\partial z^2} + \frac{NTU}{2}\left(1 + \frac{1}{\beta}\right)\left(T_f^* - T_{s1}^*\right) \qquad (4.14)$$

Solid 2:

$$\frac{\partial T_{s2}^*}{\partial \theta} = \lambda\frac{\partial^2 T_{s2}^*}{\partial z^2} + \frac{NTU}{2}(1 + \beta)\left(T_f^* - T_{s2}^*\right) \qquad (4.15)$$

The same assumptions from the research conducted by Habbitts and De Klerk mentioned in section 2.1 were used [9]. The solution for the above mentioned equation uses the standard difference approximations and Crank- Nicholson implicit method. These three equations (equations 4.13 to 4.15) provide a method to calculate the temperatures for both solid regions and the fluid region along the elements at any point in time. Figure 3 is an illustration of the finite difference elements. A system of equations can now be planned in a matrix form. The non-dimensional parameters are substituted into equation 4.13 to 4.15, and then used in a finite difference scheme illustrated in equations 4.16 to 4.22:

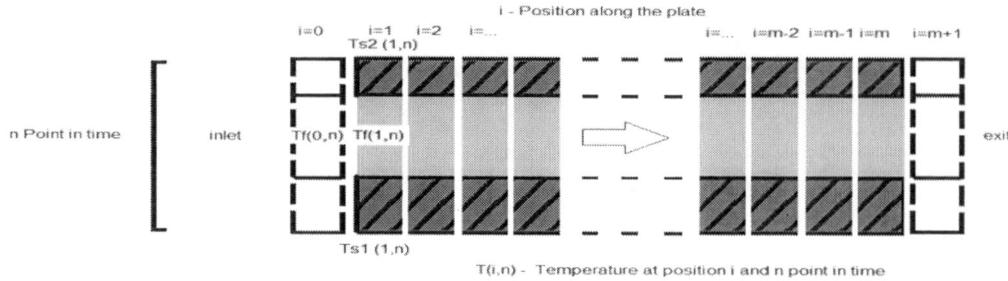

Fig. 3. Discretization of a flat double plated region into finite difference elements.

Fluid region:

$$\frac{T_f^*(i,n)-T_f^*(i,n)}{\Delta z} + NTU\left(T_f^*(i,n) - \frac{T_{s1}^*(i,n)-T_{s2}^*(i,n)}{2}\right) = 0 \qquad (4.16)$$

Solid 1:

$$\frac{T_{s1}^*(i,n+1)-T_{s1}^*(i,n)}{\Delta\theta} = \lambda(X+Y) + \frac{NTU}{2}\left(1+\frac{1}{\beta}\right)\left(T_f^*(i,n) - T_{s1}^*(i,n)\right) \qquad (4.17)$$

Solid 2:

$$\frac{T_{s2}^*(i,n+1)-T_{s2}^*(i,n)}{\Delta\theta} = \lambda(X+Y) + \frac{NTU}{2}\left(1+\frac{1}{\beta}\right)\left(T_f^*(i,n) - T_{s2}^*(i,n)\right) \qquad (4.18)$$

X and Y can be substituted with:

Solid 1:

$$X = \frac{T_{s1}^*(i+1,n+1)-2T_{s1}^*(i,n+1)+T_{s1}^*(i-1,n+1)}{2\Delta z^2} \qquad (4.19)$$

$$Y = \frac{T_{s1}^*(i+1,n)-2T_{s1}^*(i,n)+T_{s1}^*(i-1,n)}{2\Delta z^2} \qquad (4.20)$$

Solid 2:

$$X = \frac{T_{s2}^*(i+1,n+1)-2T_{s2}^*(i,n+1)+T_{s2}^*(i-1,n+1)}{2\Delta z^2} \qquad (4.21)$$

$$Y = \frac{T_{s2}^*(i+1,n)-2T_{s2}^*(i,n)+T_{s2}^*(i-1,n)}{2\Delta z^2} \qquad (4.22)$$

The above mentioned equations are numerically solved to find the metal surface temperature along with the fluid temperature profiles for a rotary regenerative air heater.

4.2 Dew point related fouling theory

Sulphurous- and sulphuric acid formation occurs during condensation of water vapour in the flue gas. This further reacts with the ash alkaline elements, resulting in air heater fouling. The dew point temperature is therefore one of the critical parameters requiring governance to operate above of the metal temperature. This process will avoid sulphuric acid deposition. The three critical dew point temperatures to be taken into consideration are the dew point temperature of water vapour, sulphurous acid and sulphuric acid. Equation 4.23 is used to calculate the partial pressure $Þ_i$ of the flue gas constituents [10]. The partial pressure for sulphur trioxide considers the percentage conversion initially assumed for the formation of sulphur trioxide (ranges from 1% to 5% of SO_2 to SO_3).

$$Þ_i = \left(\frac{\%\, mole\, fraction}{100}\right) \times P \qquad (4.23)$$

The partial pressure for water vapour and steams tables are used to find the dew point temperature of water vapour.

$$Þ_{H_2O} = \left(\frac{\%H_2O}{100}\right) \times P \quad (4.24) \quad ; \textit{Find } T_{H_2O} \textit{ from the steam tables}$$

The sulphurous acid dew point temperature estimation uses the partial pressures for water vapour and sulphur dioxide in equation 4.26 and 4.27 for the variable a, and b in equation 4.25 [11, 12].

$$\frac{1000}{T_{dp}} = 3.9526 - (0.1863 \times a) + (0.000867 \times B) - (0.000913 \times a \times b) \quad (4.25)$$

$$a = \ln(Þ_{H_2O}) \quad (4.26)$$

$$b = \ln(Þ_{SO_2}) \quad (4.27)$$

The sulphuric acid dew point temperature estimation uses the partial pressures for water vapour and sulphur trioxide as shown in equation 4.29 and 4.30 for the variable c, and d in equation 4.28.

$$\frac{1000}{T_{dp}} = 2.276 - (0.0294 \times c) - (0.0858 \times d) + (0.0062 \times c \times d) \quad (4.28)$$

$$c = \ln(Þ_{H_2O}) \quad (4.29)$$

$$d = \ln(Þ_{SO_3}) \quad (4.30)$$

5. Results

5.1 Sulphuric acid dew point temperature ranges

Figure 4 is an illustration of the sulphuric acid dew point temperatures for the three conditions of testing. From these results it is clear that the cold end layer mostly operate in the critical temperature ranges. With insufficient soot blowing the deposits harden and cause excessive fouling of air heaters. Sulphurous acid and moisture dew point temperatures are not reached during normal operation, only during a light up or shut down of a unit.

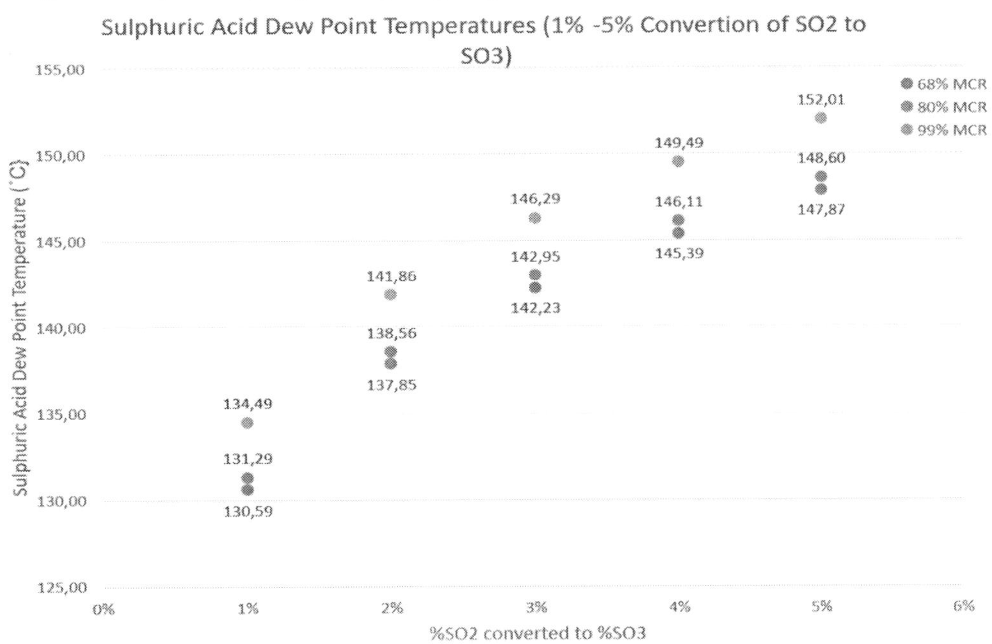

Fig. 4. Sulphuric acid dew point temperatures for 1% to 5% SO_2 to SO_3 conversion

This test considered three different load conditions (68%MCR - Maximum Continuous Rating, 80%MCR and 99%MCR). Measurement of the inlet and outlet metal- and fluid (Flue gas and air) temperatures for each layer of element packs, along with air heater leakage, were taken. The test also considered coal composition and flue gas compositions measured by an Eskom Research Testing & Development team. This measurement included SO_2% (volumetric) and moisture to estimate the relevant dew point temperatures. Coal composition provided input parameters to a mass balance for products of combustion. This allowed validation of the measured flue gas composition. A 1% to 5% range of SO_2 to SO_3 conversion were used for all three cases. The following results were obtained:
99% MCR: For 1% - 5% conversion of SO_2 to SO_3, 90% - 100% of the simulated metal temperatures at the cold end layer were exposed to sulphuric acid dew point conditions (25 seconds of the total 27 seconds of flue gas exposure). The measured metal temperatures were exposed completely (As shown in Figure 5).

80% MCR: For 1% - 5% conversion of SO_2 to SO_3, 81% - 100% of the simulated metal temperatures at the cold end layer were exposed to sulphuric acid dew point conditions (22 seconds of the total 27 seconds of flue gas exposure). The measured metal temperatures were exposed completely.

68% MCR: For 1% - 5% conversion of SO_2 to SO_3, 37% - 100% of the simulated metal temperatures at the cold end layer were exposed to sulphuric acid dew point conditions (10 seconds of the total 27 seconds of flue gas exposure). The measured metal temperatures experienced 46% - 100% exposure.

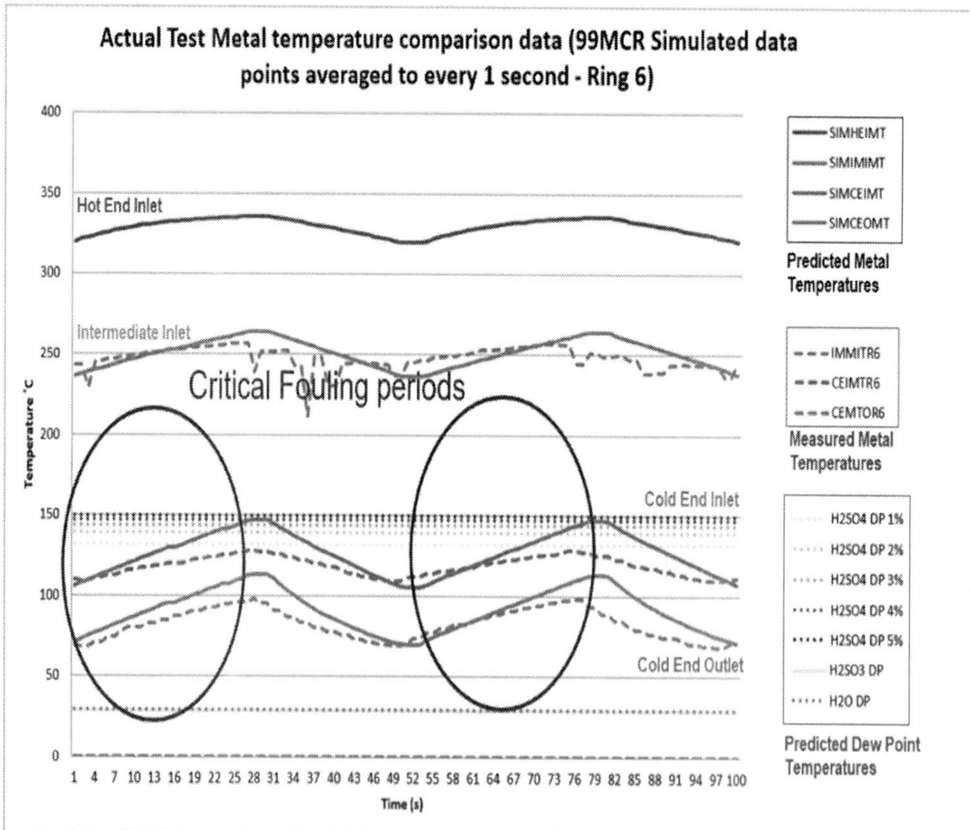

Fig. 5. Operating metal temperature for one full rotation per layer. (Simulated temperatures versus measured temperatures at 99%MCR)

Further validation of the VBA RAH model was done by comparing results from research conducted by Habbitts (1998) and De Klerk (2001). The major contributor to errors in this research, related to the omission of a K factor for the 99%MCR condition velocity measurement calculations. A factors of 0.85 were included which improved the correlations between the experimental and theoretical results. The RAH (De Klerk, 2001) model and recently developed VBA RAH model showed good correlation. This validation process proved that the VBA RAH model can be used as a tool to identify methods to optimize air heater performance. Figure 6 is an illustration of the correlation between the measured values from the research conducted by Habbitts and the simulation results for the RAH and the VBA RAH model.

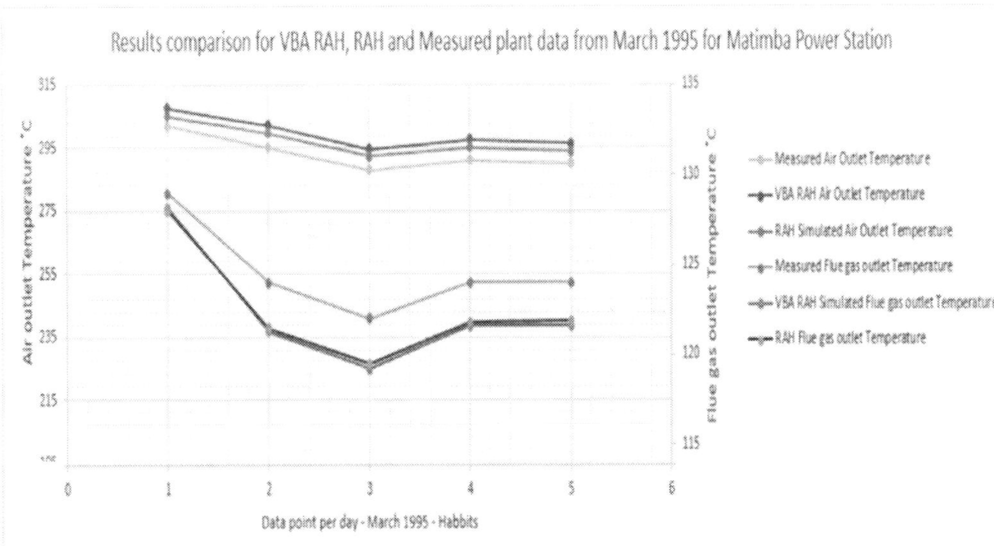

Fig. 6. Result comparison for the VBA RAH, RAH and measured plant data from tests conducted in March 1995 at Matimba Power Station.

Comparing the measured values to the simulated values, the correlation was fair but still not ideal, and some form of error existed. It was a challenging environment to measure temperate because of high levels of ash in the flue gas, and the element pack configuration in the steel matrix. Unaccounted for thermal inertia (affected by ash and wear protection), contact resistance, ash accumulation in perforated areas and incorrect flue gas flow measurements contributed to this error. An average deviation of 7.29% were found for the three load conditions. For the comparison of measured and simulated data, the 68% MCR condition showed the best correlations. The largest deviation existed for the 99% MCR condition, which was caused by an omission of a K factor during velocity measurements. The flow kinetics manometer used has a specified K factor of 0.85. This factor was included and the correlation between measured and simulated data improved, and the average deviation for the three loads reduced to 2.17%. Because of non-ideal allocation of measuring ports, the K factor was further reduced to 0.80 to find an ideal flow rate to match for the 99% MCR measured operating parameters. The average deviation between loads reduced to a 4.23% after this change, which proved that the 0.85 K factor is the preferred option.

The model also provides the ability to amend operating parameters to estimate changes in boiler performance. The inlet temperature was raised to 70 degrees which resulted in a 1.6 MW reduction in energy input to the boiler. The coal flow reduced by 0.1 kg/s and the boiler efficiency improved by 0.2%. The cost saving with regards to a reduction in coal flow is R1

127 033.57 per year. Changing the operating conditions in this way can be used as a method to reduce dew point related fouling, but this option is limited by the capacity of the pre-steam air heater and the fan design temperature. For the FD fans, the design maximum temperature is 50˚C. If the air temperature is increased beyond this point the fan will operate with air densities outside of its design range. Due to the fact that the fan is a constant volume machine, the volumetric flow rate will be maintained. With this in mind the density will decrease with an increase in temperature. This process reduces the pressure difference across the fan, but requires increased radial vane positions to maintain the required operating pressure after the air heaters. Other option were evaluated and the option of increasing the air temperature to 50˚C and removal of the third layer of element packs showed promising results. This estimated a reduction in the onset of dew point related fouling, and improving the boiler performance. Therefore this was a recommendation for further research with experimental validation.

6. Conclusion

The VBA RAH model was successfully developed to predict the onset of dew point related fouling and provides a platform to equip system engineers with the knowledge to enhance air heater performance. The research showed that the VBA RAH model can help improve life expectancy of air heater element packs and draught group components such as ducting, dampers and supple joints. It also proves to reduce the risk of the occurrence of dew point related fouling, and improve cleaning capacity of the element packs. Air heater operating and design amendments were simulated to evaluate the impact on overall boiler performance. The following solutions to reduce dew point related fouling were the suggested:

1. Increasing the air inlet temperatures from 35˚C to 70˚C.
2. Removing the third layer of element packs.
3. Removing the third layer and increase the air inlet temperatures to 50 ˚C.

The third option is a promising solution to reduce the onset of dew point related fouling and is recommended for further research through plant experiments. Although the simulation results did not show ideal correlations to the measured performance at all times, the impact assessment of boiler performance because of changes in air heater design and operating parameters proves to be valuable. This platform provides methods to improve boiler efficiency, reduce maintenance costs and production losses, which will conduce to improved and sustainable operating of Eskom coal fired power generation plants.

Nomenclature

$\dot{m}$ – Mass flow of fluid
C_f – Heat capacity of the fluid
M_s – Mass of the solid material of the element pack
C_s – Heat capacity of the solid material of the element pack
k – Conductivity of the solid material of the element pack
L – Length element pack (height of the element pack)
x – Length of the pack at the given time interval
a – Half of the thickness of the corrugated solid region
b – Half of the thickness of the undulated solid region
A_{conv} – Area of convection at the given time interval
A_{cond} – Area of conduction at the given time interval
h – Average heat transfer coefficient
T_f - Fluid temperature

T_{s1} - Corrugated solid temperature
T_{s2} - Undulated solid temperature
T_f^* - Dimensionless fluid temperature
T_{s1}^* - Dimensionless corrugated solid temperature
T_{s2}^* - Dimensionless undulated solid temperature

References

1. H. Kleynhans, J. Eganza, ID Fans at Maximum Capacity. Guideline-Possible Causes and Contributors Guideline, ESKOM Technology Engineering Guidelines, Johannesburg (2016).
2. W. Brandt, Unit 2 BMFT due to inadequate secondary air flow (<20% BMCR), ESKOM Matimba Power Station, South Africa, Lephalale (2015).
3. W. Brandt, Engineering Change Root Cause Analysis for Air heater Ash fouling Report, Eskom Matimba Power Station Engineering ECM –RCA documentation, South Africa, Lephalale (2018).
4. K. Gruen, Heat transfer and pressure drop of regenerative air pre-heaters elements, MSc (Eng.) dissertation, University of the Witwatersrand, Johannesburg (1998).
5. M. J. Caby, Experimental Investigation of Heat Transfer in Regenerative Heat Exchangers, MSc (Eng.) dissertation, University of the Witwatersrand, Johannesburg (1996).
6. S. Habbitts, "Computer Simulation of Power Station Regenerative heat exchangers," MSc (Eng.) dissertation, University of the Witwatersrand, Johannesburg (1998).
7. R. I. Loehrke R. S. Mullisen, A transient heat exchanger evaluation test for arbitrary fluid inlet temperature variation and longitudinal core conduction, J. of H. T., 108, May 1986.
8. G. De Klerk, An Improved Simulation Model for Rotary Regenerative Air Heaters," MSc (Eng.) dissertation, University of the Witwatersrand, Johannesburg (2001).
9. G. B. De Klerk, T. J. Sheer, H. H. Jawurek, M Lander., A Versatile Computer Simulation Model for Rotary Regenerative Heat Exchangers, H. T. Eng., 1-13, 3 April (2013).
10. N. V. Mathebula, Observation and the classification of the fouling mechanisms in rotary air heaters, MSc (Eng.) dissertation, University of the Witwatersrand, Johannesburg (2014).
11. V. Ganapathy, Cold end corrosion: Causes and cures, Abeline: ABCO Ind., 57-59, USA, Texas (1989).
12. W. Wei, Theoretical prediction of acid dew point and safe operating temperature of heat exchangers for coal-fired power plants, Apl. Therm. Eng., 782-790, 123 (2017)

Computational Investigation of Multiple Oscillating Hydrofoils and its Effect on Thrust Augmentation

Uma Krige[1], *Kim Boon* Lua[2], and *Wei Hua* Ho[1*]

[1] University of the Witwatersrand, School of Mechanical, Industrial and Aeronautical Engineering, South Africa
[2] National Yang Ming Chiao Tung University, Department of Mechanical Engineering, Taiwan
[*] Corresponding author. Email address: weihua.ho@wits.ac.za

Abstract. Previous investigations of flapping hydrofoils for the purpose of thrust production have been limited to one or two in tandem. Tandem foils were found to have superior performance because the performance of the aft foil was augmented by the vortices shed from the fore foil. It is however not clear if increasing the number of foils will continue to have increased performance or if there exist an optimal number after which the overall performance either stagnates or reduces. A 2D computational study was conducted to investigate the effect of increasing the number of hydrofoils to four at a Reynolds number of 8000 flapping in-phase and out-of-phase. Optimal and sub-optimal conditions found previously with tandem hydrofoils were found also be applicable to three and four hydrofoils.

1 Introduction

Engineers have often looked to mother nature in order to get inspiration for improved solutions that are both simple and robust [1]. Bio-inspired studies and designs include flapping flight of insects and birds for micro air vehicle (MAV) applications [2]–[5] and the swimming of marine animals for undersea vehicles [5]–[9]. Traditionally aquatic vehicles such as submarines and ships have used propeller-driven propulsion utilising the rotation of propeller blades to create thrust. This form of propulsion although simple in design and efficient, has certain disadvantages such as cavitation when the angular velocity is sufficiently high, and are unsuitable for stealth operations. Unlike propeller driven propulsion, the dominant source of propulsion and manoeuvring forces found in marine animals is generated by flapping foils. For instance fish, dolphins and whales utilise the flapping caudal fins to efficiently produce large propulsive forces and swim gracefully and without any chaotic wake generated [10].

Important fluid dynamic features of swimmers and flyers in the animal kingdom are large motion amplitude and unsteady flow features. The fluid structure interaction between fluid and foil is highly complex as it includes the formation of large-scale vortex structures, onset of shear layer formation and subsequent reattachment and corresponding flow field. Previous studies [11], [12] have shown that flapping parameters such as: amplitudes of motion, phase difference between various components, and the Strouhal number (St) strongly affect propulsive effectiveness and efficiencies. Full 3D flapping motion is a combination of heaving, pitching and twisting motion however [2] suggests that significant thrust can be provided by pure oscillating (sometimes termed as pitching) 2D motion. Studies in oscillation of tandem foils (hydro or aero) have shown that under certain conditions, such as spacing and phase difference in the motion, the overall thrust performance can be improved beyond that of a single foil [13]–[15]. To the author's knowledge, there has not been studies done beyond two foils to investigate if such performance augmentation continues and if they will stagnate or diminish beyond a certain number. The aim of this study is to investigate this using symmetric NACA0012 aerofoils.

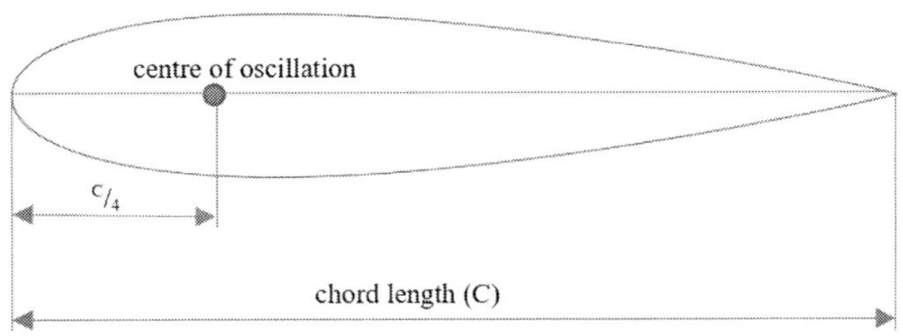

Figure 1: NACA0012 airfoil with quarter-chord located at the origin

2 Methodology

Rigid two-dimensional foils undergoing oscillating motion was modelled using ANSYS Fluent to simulate the flow-field and to calculate the thrust (axially aligned with drag but in the opposite direction, hence negative of drag) subject to a constant horizontal flow of Reynolds number (Re) of 8000 based on the chord length of the airfoil and upstream flow.

The method utilised for the motion is the overset mesh which overcomes the limitations of using the simpler sliding mesh method whilst not requiring the large computational resources of the dynamic mesh method. Overset mesh method works by having two layers of mesh located one above the other. The layer which contains the foil (overset mesh) is smaller and is situated "above" the larger overall domain (background mesh). Information is passed between the meshes during solving. In scenarios with multiple foils, each foil is located in its own "overset" mesh and in certain locations, there are three "sets of mesh" overlapping in the same location. The flow domain consist of a rectangular background mesh with circular overset meshes as shown in Figure 2.

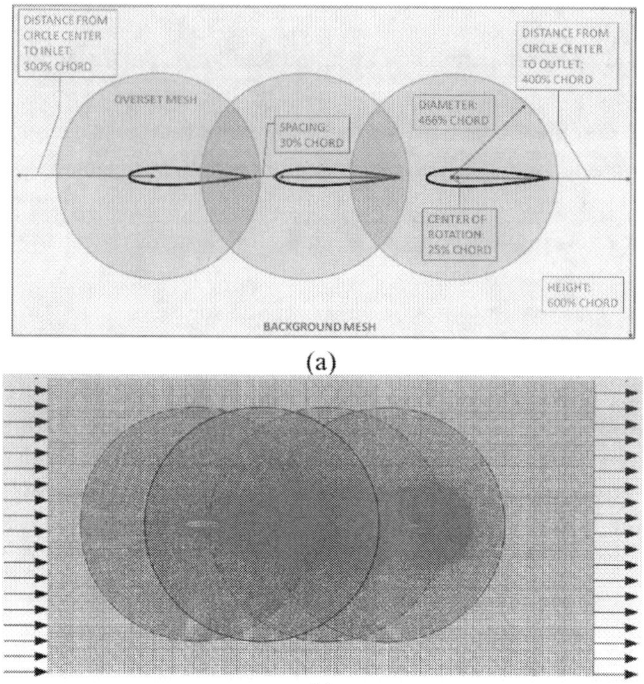

(a)

(b)

Figure 2:(a) Diagrammatic mesh setup (not to scale) (b)Representative mesh used in the simulations.

The solver settings used were double precision, pressure based, with the spacing between subsequent foils to be 30% of the chord length (0.3C). PRESTO! was used as the pressure discretisation scheme. The solution was first solved in steady-state without motion using the SST-kW turbulence model before switching to transient simulation using the Scaled-Adaptive Simulation (SAS) turbulence model. Simple harmonic pitching motion to the foil was prescribed using a user-defined function (UDF). Pseudo steady state was found to have been achieved after 5 cycles and all results presented are of the 6th cycle.

Models with 1 to 4 foils were solved both in-phase and out-of-phase at a reduced frequency (k) of 1.26 Hz and oscillating amplitude of 20° giving a total of 6 sets of simulations. When multiple foils are out-of-phase, alternate foils will be in-phase, i.e. when there are 3 foils, the first and the third foils will be in-phase with the second foil out-of-phase by 180°. Similarly, when there are 4 foils, the first and third foils will be in-phase with each other while out of phase with the second and fourth foils. These two phase choices were made based on previous research which found in-phase to perform in an optimal manner while 180° out of phase performed in a sub-optimal manner for a spacing of 0.3C [15].

Several validation exercises were performed before a suitable mesh was selected. The first is the shape of the overset mesh. Validation was performed by replicating the settings and foil shape used by Flint et al [2] (via communication with the corresponding author of that paper) using a single foil due to availability of data. Although only thrust (negative drag) is required for this study, both lift and drag were used for the validation exercise.

First, an overset mesh using a circle the same size as that study was compared to an oval and ellipse was performed, and the result is shown in Figure 3. The circular overset mesh was eventually selected as it closely corresponded to the results by Flint et al [2].

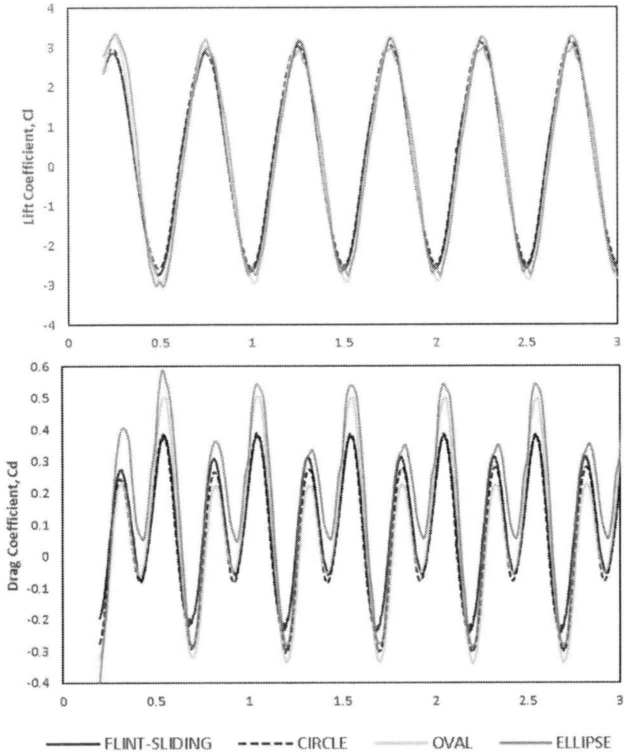

Figure 3: Comparison of overset mesh shapes.

The mesh sizing for the background mesh was refined to establish an appropriate mesh. The overset mesh utilised the settings by Flint et al [2] and is deemed to be suitable. Three background meshes termed "coarse", "medium" and "fine" were used and the results are presented in Tables 1 and 2 and Figure 3.

Table 1: Description of various background mesh tested.

Background Mesh	Description	Number of Nodes
Coarse	0.01m element size + face meshing, quadrilateral	7352
Medium	0.007m element size + face meshing, quadrilateral	14679
Fine	0.005m element size + face meshing, quadrilateral	28741

Table 2: Comparison of results from different meshes.

	Cl Difference			Cd Difference		
	max	average	average %	max	average	average %
Coarse	0.062	-0.008	0.606	0.017	0.00167	0.287
Medium	0.127	-0.009	-0.875	0.023	0.00221	0.423
Fine	0.061	-0.007	0.542	0.017	0.00122	-0.175

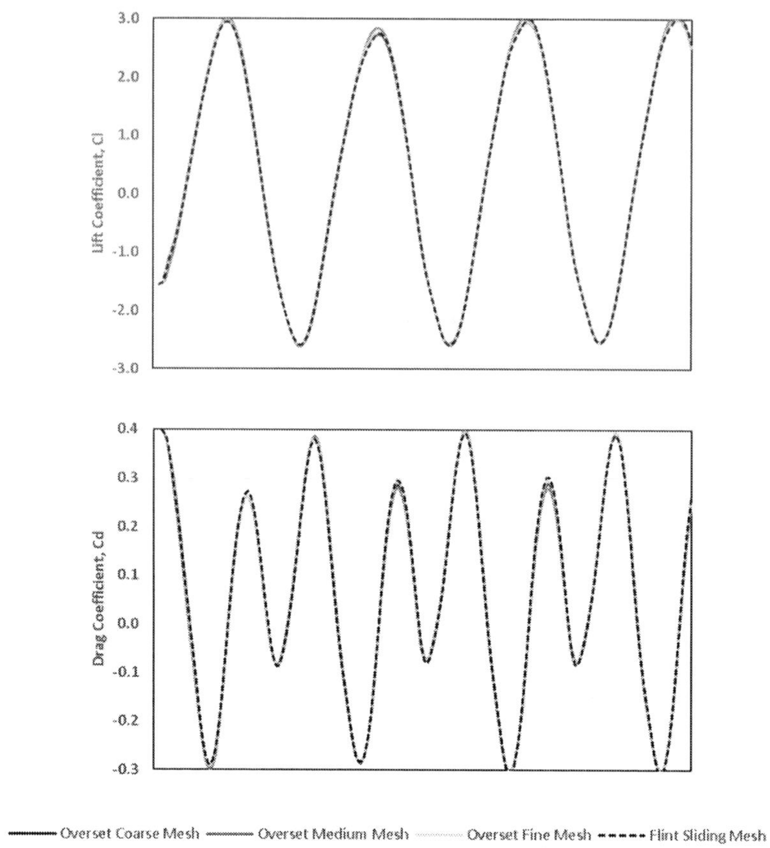

Figure 4: Graphical comparison of results from different meshes with respect to time

The solver settings used were double precision, pressure based, with the spacing between subsequent foils to be 30% of the chord length (0.3C). PRESTO! was used as the pressure discretisation scheme. The solution was first solved in steady-state without motion using the SST-kW turbulence model before switching to transient simulation using the Scaled-Adaptive Simulation (SAS) turbulence model. Motion was prescribed using a user-defined function (UDF). Pseudo steady state was found to have been achieved after 5 cycles and all results presented are of the 6th cycle.

Models with 1 to 4 foils were solved both in-phase and out-of-phase at a reduced frequency (k) of 1.26 Hz and oscillating amplitude of 20° giving a total of 6 sets of simulations. When multiple foils are out-of-phase, alternate foils will be in-phase, i.e. when there are 3 foils, the first and the third foils will be in-phase with the second foil out-of-phase by 180°. Similarly, when there are 4 foils, the first and third foils will be in-phase with each other while out of phase with the second and fourth foils. Figure 5 shows an illustration of the definition of foil motion being in-phase and out-of-phase.

(a) (b)

Figure 5: Illustration of (a) foil motions being in-phase and (b) foil motions being out-of-phase

3 Results and discussion

First it can be observed from Figure 6 that for all the scenarios, the drag coefficients against time and oscillation pitch angle show the expected trends. What is presented is only the net thrust produced by the entire system, but the individual foils exhibit similar behaviour and are omitted for the sake of brevity.

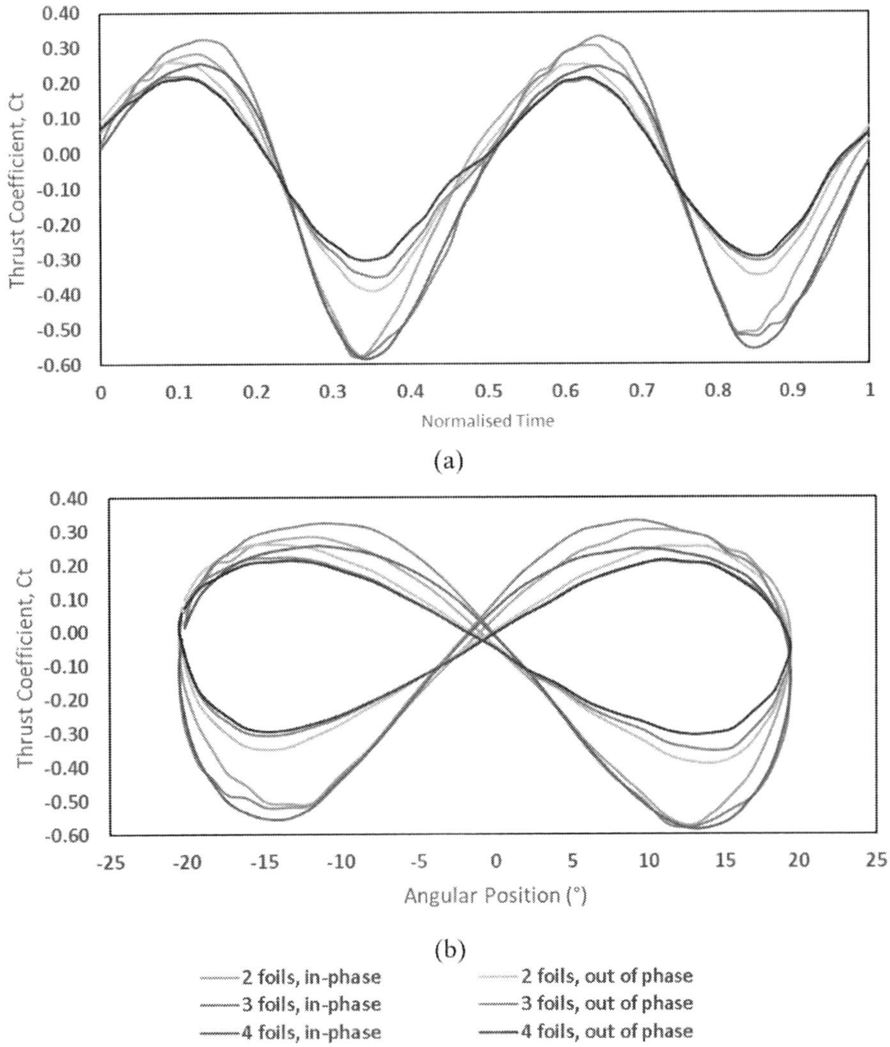

Figure 6: Thrust coefficient (a) against time and pitch angle (b) for systems with multiple foils.

Net thrust coefficient generated by the foil systems is given in Figure 7. An immediate observation of note is that within the conditions simulated, there was no net thrust generated. St in the current study is 0.21. The comparison between the current study and previous results are given in Figure 8. Since the Strouhal number is inversely proportional to the speed of the oncoming flow, a decrease in flow speed would increase the Strouhal number, potentially enhancing the net thrust. This is further supported by the fact that the thrust was unable to overcome the drag produced by the system. A system with slower oncoming flow would also experience less drag.

It is interesting to note that the out-of-phase scenario generated more thrust for the two and four foil system but there is a switch-over at three foils. For foils oscillating out of phase, there is drop in net thrust with three foils, but it increases again for four foils. It is not clear if the increase will continue beyond four foils.

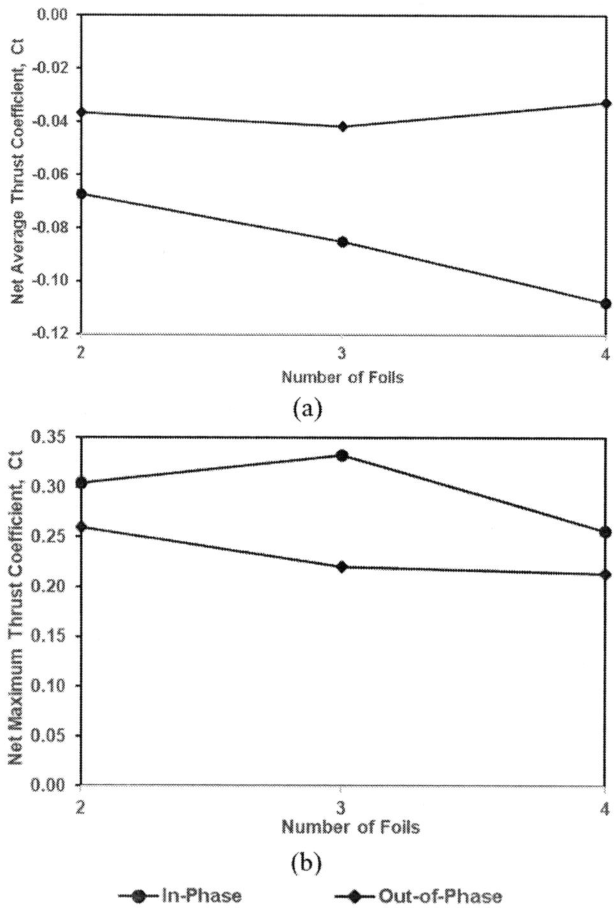

Figure 7: (a) Net average thrust coefficients and (b) net peak thrust-against number of foils

When two or more foils are pitching in tandem, the wake of the fore foil affects the flow around and hence the performance of the aft foil. The vortices shed by the fore foil interact with those being formed on the aft foil and, for this spacing, for a 0° phase difference these interact constructively and are hence rotating in the same direction while for a 180° phase difference these act destructively and can be observed to be rotating in opposite directions. The observation of a phase difference of 0° being optimal and 180° being suboptimal for thrust performance for two tandem foils is found to hold true for systems of three and four foils when considering the net maximum thrust coefficient of the system. The destructive effects of the suboptimal cases became less devastating to the last foil as more foils were placed between the first and last foils. This is evident in the flattening of the trend between the three and four foil system net maximum thrust coefficient as well as the upward trend for the net average thrust coefficient for out of phase cases seen in Figure 5. In such cases, it has been observed that the effects of vortex interaction on the last foil decreases with the number of foils. When vortices are formed and shed, energy is extracted from the flow to be converted into a rotational velocity of the fluid elements. As more energy is extracted from the flow, less energy is available for additional vortices to be formed and shed hence reducing the significance of vortex interaction by reducing the strength of the vortices involved.

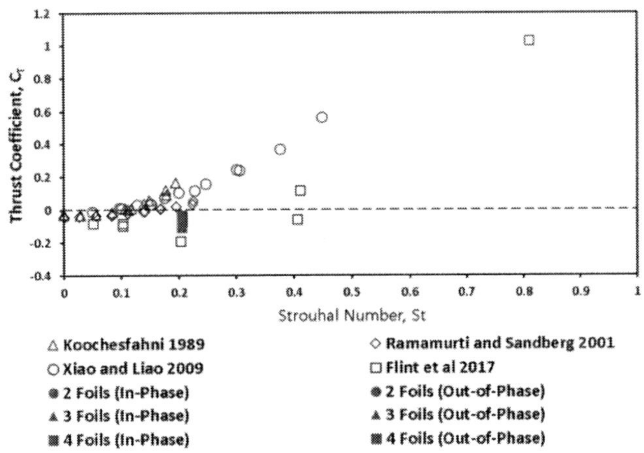

Figure 8: Comparison of current results against literature

To elucidate the results mentioned above, one needs to study the shedding of vortices during the motion. First, we compare the results between the case of two, three and four foils for the 20° amplitude motion.

During in-phase motion, the vortex shed off the second foil in a two-foil system (A) in Figure 7 is what develops into the strong leading-edge vortex (LEV) on the third foil of a three-foil system (C). The interaction between the first and second foils is similar for both two-foil and three-foil systems. This is, however, not the case for the four-foil system whose only similarity is in the vortex structures formed on the fore foil. This system has a more sheet-like interaction pattern between the second, third and fourth foils (H). It is interesting to note that there is far more roll-up of the shed sheet vortex at the midway amplitude of the two-foil system (B) than of the third foil of a three-foil system (D). This progressively becomes more sheet-like in the four-foil system (H).

An interesting thrust performance enhancing mechanism was discovered on the drag performance of the third foil resulting in a 68% increase in peak thrust (peak) without a significant increase in peak drag (trough) in Figure 6. This foil experienced a "double hump" peak which were also observed in the two previous studies [2], [16] but was not extensively explained. This increase could be attributed to the LEV formation being enhanced by the constructive interference of the vortex sheet rollup of the foil fore of it as shown in Figure 7(C). This phenomenon was not present on the third foil in the equivalent case of four foils wherein the sheet shed off the second foil experiences little to no rollup in its interaction with the leading edge of the fourth foil. A possible reason for the "double hump" could be linked to the double vortices being shed (E) which differ from those shed from the two-foil system (A). Another visual indicator in the difference between thrust performance of the three and four foil systems is the size and strength of the wake aft of the system: the wake of the three-foil system (F) occupies a wider area and has a higher vorticity than that of the four-foil system (I).

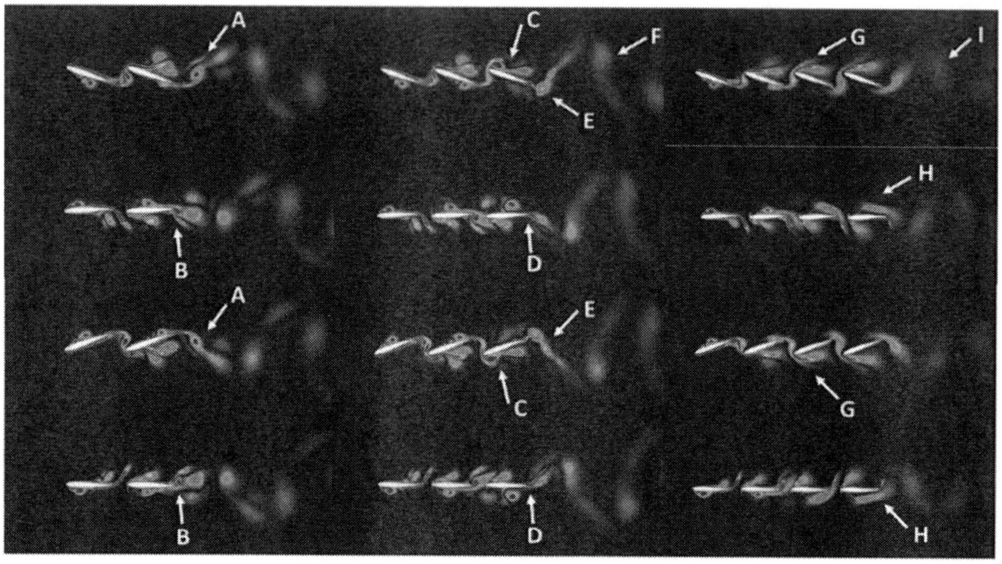

Two foil system Three foil system Four foil system

Figure 9: Vorticity contours for in-phase

During the out of phase motion, the vortex sheet shed off the fore foil of a two-foil system destructively interferes with the leading-edge vortex being formed on the aft foil (J) to produce an overall smaller leading-edge vortex (K). Contrasted to the in-phase motion, the vortex structure on the second foil (L) does not differ between two and four-foil systems. The interaction between the first and second foil (J) is also unchanged (N). For the four-foil system, the trailing edge vortex rollup of the third and fourth foils (M) has lower vorticity than the that of the second foil (N). The wake produced in the out of phase motion (O) occupies a smaller area and has a lower vorticity than that of the in-phase motion (I) and is shaped in more discrete clusters (O) rather than a continuous swirl (I). The difference of the wake is once again a visual indicator of difference in the thrust being extracted from the flow.

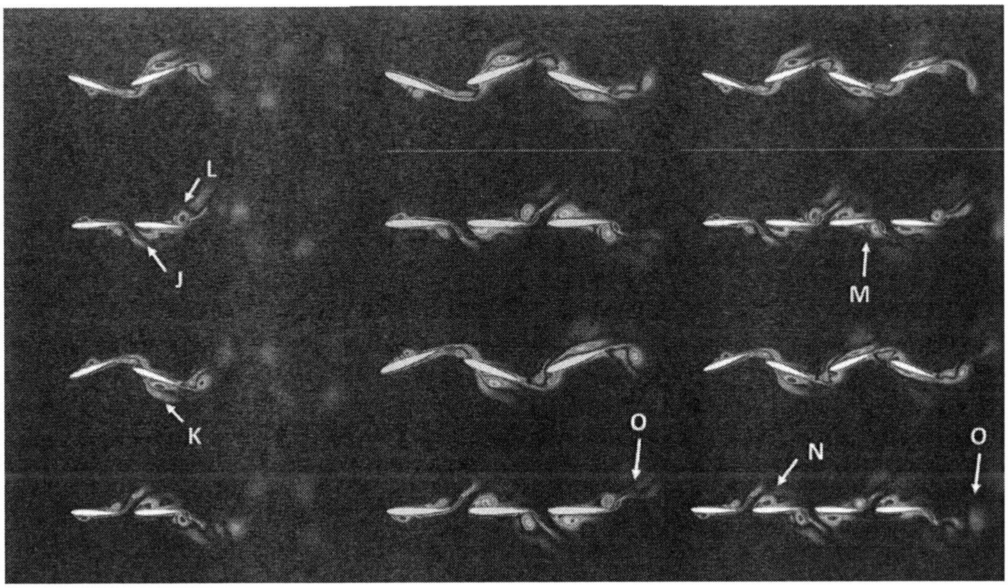

Two foil system Three foil system Four foil system

Figure 10: Vorticity contours for out of phase

4 Conclusion

A study has been performed on the thrust capabilities of symmetrical hydrofoils undergoing harmonic oscillating motion under different operating conditions. The Reynolds number, Strouhal number and reduced frequency of the simulations were 8000, 0.21 and 1.26 respectively. Known optimal and suboptimal conditions for tandem flapping double foil cases were tested for triple and quadruple tandem foil sets. These were found to have a similar effect on the enhancement or diminution of performance of triple and quadruple foil cases. There was no further system performance enhancement found after three foils and the addition of a fifth foil was not studied on the basis of this trend. The system was unable to produce a net positive thrust for all cases tested likely attributed to the incoming flow velocity being too great and resultantly inducing too much drag. Future work could investigate the effects of increasing spacing or the possibility of combined pitching-heaving motion to improve thrust performance by means of introducing an additional velocity component.

References

[1] J. C. Nawroth et al., Nat. Biotechnol., vol. 30, no. 8, pp. 792–797, (2012)

[2] T. J. Flint, M. Jermy, T. H. New, and W. H. Ho, Int. J. Heat Fluid Flow, vol. 65, pp. 328–341, (2017)

[3] N. M. Shumway, M. Gabryszuk, and S. J. Laurence, AIAA Aerospace Sciences Meeting, 2018, no. 210059 (2018)

[4] Y. H. Chen and M. Skote, Phys. Fluids, vol. 27, no. 3, (2015)

[5] H. Hu, L. Clemons, and H. Igarashi, Exp. Fluids, no. 51, pp. 347–359, (2011)

[6] A. H. Techet, J. Exp. Biol., vol. 211, no. Pt 2, pp. 274–279, (2008)

[7] F. E. Fish, P. W. Weber, M. M. Murray, and L. E. Howle, Integr. Comp. Biol., vol. 51, no. 1, pp. 203–213, (2011)

[8] F. E. Fish, Appl. Bionics Biomech., vol. 3, no. 1, pp. 49–60, Jan. (2006)

[9] H. Liu, Appl. Mech. Rev., vol. 58, no. 4, pp. 269–282, (2005)

[10] C. L. Yu, Y. H. Hsu, and J. T. Yang, Ocean Eng., vol. 70, pp. 51–60, (2013)

[11] D. A. Read and M. S. Triantafyllou, p. 153, (2001)

[12] F. S. Hover, Haugsdal, and M. S. Triantafyllou, J. Fluids Struct., vol. 19, no. 1, pp. 37–47, (2004)

[13] K. B. Lua, H. Lu, X. H. Zhang, T. T. Lim, and K. S. Yeo, Phys. Fluids, vol. 28, no. 12, (2016)

[14] L. E. Muscutt, G. D. Weymouth, and B. Ganapathisubramani, J. Fluid Mech., vol. 827, pp. 484–505, (2017)

[15] T. M. Broering and Y.-S. Lian, Acta Mech. Sin., vol. 28, no. 6, pp. 1557–1571, (2012)

[16] Y. Shen and H. Cai, J. Vibroengineering, vol. 18, no. 3, pp. 1839–1848, (2016)

Experimental Testing on an Air-Cooled Condenser Motor, Gearbox and Fan

C.J. van Rensburg[1,*], *M.P.* Venter[1,**], and *D.N.J.* Els[1,***]

[1]Department of Mechanical and Mechatronic Engineering, Stellenbosch University

Abstract. The majority of South Africa's electricity is generated from coal-fired power stations. After expansion through the turbine stage the heated steam is condensed by air-cooled condensors (ACCs). In water scarce areas air-cooled condensers are used to reject heat by means of convection through an array of fans that force air over the heated steam tubes positioned in a A-frame structure above the fan. The ACC system consists of an induction motor, two-stage gear reducer and an axial fan. For this research, the rotational velocity of the gearbox's output shaft is measured as well as the bending and torsional strain experienced during different stages of operation. A proximity sensor and various sets of strain gauges are used to record the measurements respectively. Testing was conducted at three different sites that utilize different control methods for the induction motors. The purpose of testing at different sites was to evaluate the effectiveness of each start-up control system and its application inside an ACC. The results obtained in each of the three cases are quite clear that any motor controller is beneficial to the overall system simply due to the reduced inrush current and torque peaks.

1 Introduction

South Africa has large coal reserves situated in the North-Eastern regions of the country. It is because of this reason that the majority of the country's electricity is generated by coal-fired power stations in those areas. Unfortunately, these regions are considered water scarce and cannot sustain the traditional wet cooling tower power plants. To avoid this problem, the latest power plants in the region use forced air to reject heat into the atmosphere through air-cooled condensers.

The induction motors used in ACCs typically have a rated output of ~250 kW. Usually within the first 5 s of start-up, the torque peaks exerted on the gearbox by the induction motor can reach up to four times the rated torque of the gearbox, which over a prolonged period of frequent start-ups can damage the gearbox and lead to subsequent failure thereof.

An ACC system consists of the induction motor, a gearbox and axial fan supported by a walk bridge. The method of starting an induction motor is important to consider as this can significantly reduce the start-up torque peak [7]. The three starting methods that are investigated here are: direct-online (DOL), variable frequency drive (VFD) and a soft-starter.

*e-mail: 19840918@sun.ac.za
**e-mail: mpventer@sun.ac.za
***e-mail: dnjels@sun.ac.za

In the case of a DOL connection, the motor is directly connected to the power grid and is also the fastest starting method, since no motor controller is used. A VFD reduces the start-up torque by altering the magnitude and frequency of the voltage supplied to the motor. The soft-starter reduces the voltage supplied to the motor and subsequently limiting the peak current and torque.

Figure 1: ACC cooling tower motor and gearbox installation

The most common areas of gearbox failure are found to be the bearings and gear tooth fatigue damage [4]. Measurements on both shafts of the gearbox was taken by [1] to relate the input torsion and bending strain to the output side of the gearbox.

Further contributions to shaft vibration and bearing wear is caused by the aerodynamic loading experienced by the fan. The most probable causes for unbalanced aerodynamic loads are: dynamically unbalanced blades (1); variances in attack angle of fan blades (2); and distorted inlet airflow conditions (3). The close proximity of the support bridge to the fan also affects airflow conditions downstream of the fan [2].

The aim for this research was to determine the most suitable motor controller by comparing the difference in start-up torque peaks and start-up time. The same type of measurements were conducted at three different test sites. Each test site is installed with one of the aforementioned motor starting controllers, a two-stage gearbox ($\sim$ 10 : 1) and a axial fan with a diameter of $\sim$10.36 m.

2 Induction Motor Control Methods

2.1 Direct on Line

For this starting method, no controller is implemented and the motor terminals are connected to a three-phase power supply. The motor terminals can be connected in either the star or delta configurations shown in Figure 2 with the differences between the two connections being the line and phase voltages and currents.

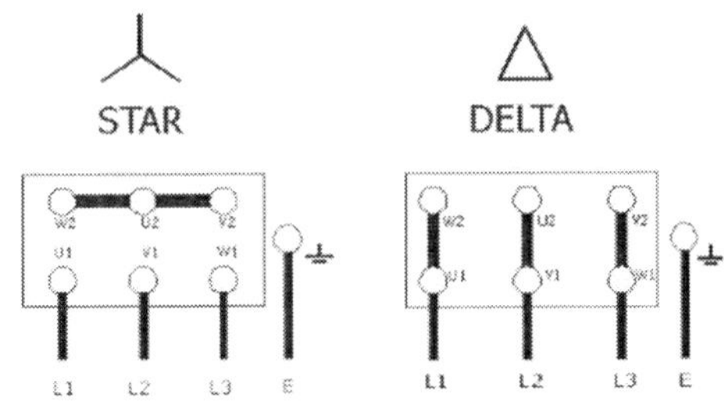

Figure 2: Three phase star & delta connections for induction motor terminals

2.2 Variable Frequency Drive

The MinWaterCSP facility constructed at Stellenbosch University has a VFD installed that works in conjunction with a programmable logic controller (PLC) to determine the start-up characteristics of the motor. For any testing purposes, the motor speed is set on the VFD and the motor is accelerated to the set speed.

The advantage of this VFD is that the motor speed can be changed while the motor is running. The time accelerating the motor from standstill to full speed varies depending on the power grid the VFD is connected to. This is illustrated in Section 4.1 where the start-up algorithm uses the voltage available to start the motor and the set speed to determine the start-up time. The size of the axial fan installed and power grid voltage available at the MinWaterCSP site makes it difficult to compare the startup time of the VFD against the other test sites. Figure 3 below illustrates the VFD's internal circuit diagram installed at [3].

The VFD receives three-phase alternating current (AC) power from the grid and rectifies this into a direct current (DC) component. The DC component of the power supply contains ripples which are smoothed in the DC bus circuit of the VFD. Finally, a series of insulated-gate bipolar transistor (IGBT) switches convert the DC power back to AC before reaching the motor terminals. The VFD used at the MinWaterCSP plant is the CFW701 model from WEG.

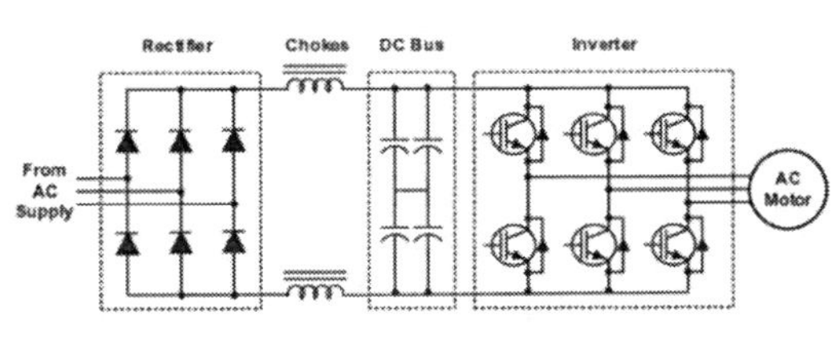

Figure 3: Internal circuit diagram of a VFD

2.3 Soft-Starter

A soft-starter motor controller is the cheaper alternative to install compared to the VFD, but only offers control of the start and stop phases of the induction motor. Once the motor reaches its rated speed, the soft-starter is bypassed (internally or externally) through a series of circuit breakers. There are two types of soft-starters available: IGBT- or thyristor-based. The soft-starter installed at power station II is thyristor-based. Referring to Figure 4, there are three pairs of thyristors connected in the anti-parallel configuration; one for the positive and negative half of each cycleof the supply voltage. The soft-starter controls the voltage supplied to the motor by only conducting for a short period of time based on gate pulses supplied to the thyristors or IGBT switches.

Figure 5 depicts an example of how the voltage supplied to the motor is increased during start-up and controlled by the firing angle of the thyristors. At start-up the angle is at a maximum to allow minimum voltage to the motor and gradually increase over time, allowing a larger part of the sinusoidal voltage to pass through.

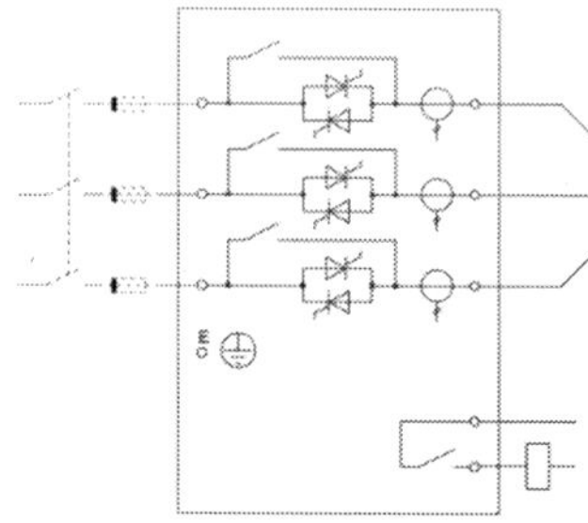

Figure 4: Soft-Starter internal circuit diagram

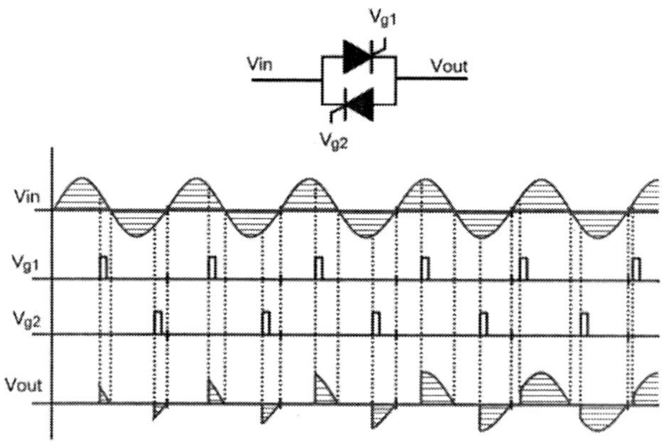

Figure 5: Start-up voltage increase of soft-starter

3 Experimental Design

There are three types of measurements of interest on the ACC motor & gearbox for this research: speed measurement of gearbox output shaft (1); bending strain on the gearbox output shaft(2); and the torsional strain experienced by the gearbox output shaft (3). All measurements are synchronized by a QuantumX MX1601B module.

3.1 Speed Measurement

The gearbox's low-speed shaft angular velocity can be determined using an optical or proximity sensor. To eliminate the effect of changing light conditions, an inductive proximity sensor was selected for this particular application. This type of sensor detects the presence of metallic objects close to the sensing surface. A wheel with 60 teeth was designed and fitted around the gearbox shaft for the proximity sensor to detect.

Figure 6 illustrates the final solution implemented for measuring the gearbox's shaft speed at all the test sites. This was the most effective solution given the physical constraints presented by the support platform of the motor and gearbox. Surface imperfections on the shaft's surface and manufacturing defects of the toothed wheel's parts, made it difficult to accurately align the two half discs. The vertically aligned sensor is used for a single pickup point whereas the horizontal sensor measures the frequency of the passing teeth. The two sensors are used to confirm the shaft speed.

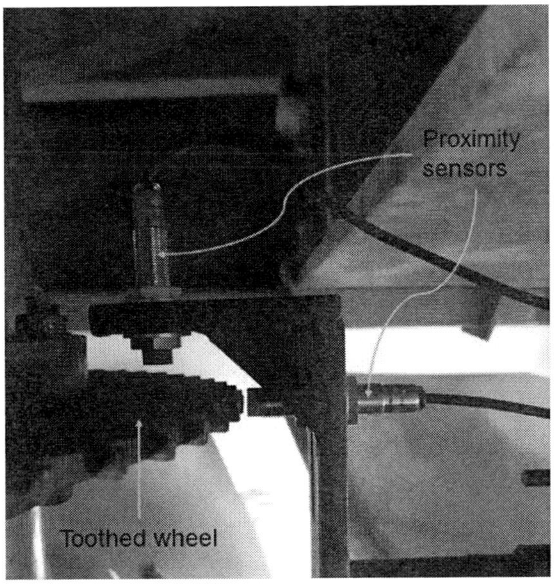

Figure 6: Toothed wheel installation with proximity sensors

3.2 Bending Strain Measurement

At start-up, the gearbox output shaft undergoes a transient bending phase whereafter it oscillates around a constant value. The circumferential positioning of the strain gauges doesn't affect the readings, but the distance away from the gearbox bearing along the length of the shaft does. To measure the bending strain, two sets of strain gauges (each gauge of 350Ω) are positioned 180° opposite each other.

Figure 7b illustrates a half-bridge configuration foil. Two half-bridges are then connected to form a full-bridge configuration and complete the Wheatstone bridge for maximum sensitivity. On one side of the shaft, Figure 7b would represent R_1 and R_3 in Figure 7a on the left. The two full-bridge configurations are positioned 90° apart from each other and from Figure 8 it is possible to calculate the absolute bending moment of the shaft.

The strain gauges are connected to a wireless V-Link 200 node from LORD MicroStrain that transmits data to a base station receiver (WSDA-Base-101-LXRS). A shunt calibration was performed from within the software before any readings were taken. The bending moment experienced in the shaft at the position of the strain gauges is then calculated using Equation 1:

$$\sigma_{shaft} = \frac{\pi}{4} \cdot E \cdot \epsilon \cdot R^3 \tag{1}$$

where,

- R is the shaft radius in [m]
- ϵ is the strain measured in $[\frac{m}{m}]$
- E is the shaft material's Young's modulus in [Pa]

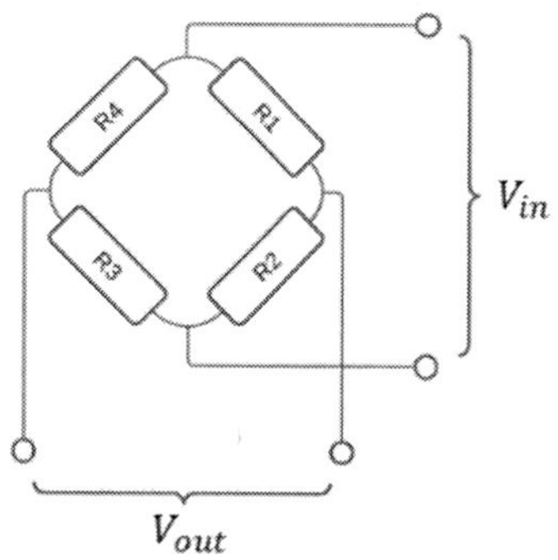

(a) Wheatstone bridge

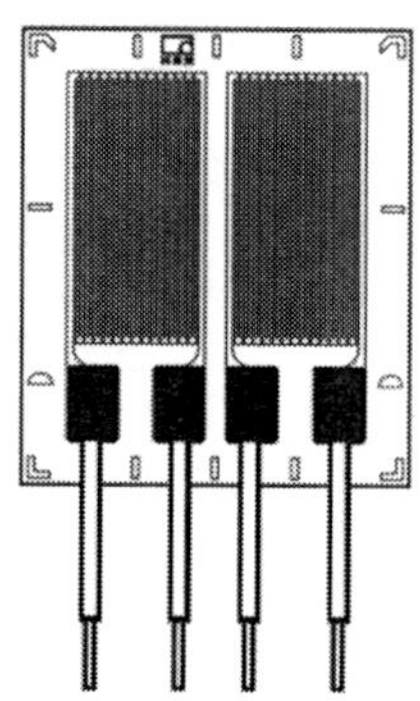

(b) Half-bridge strain gauge configuration

Figure 7: Bending strain gauge setup

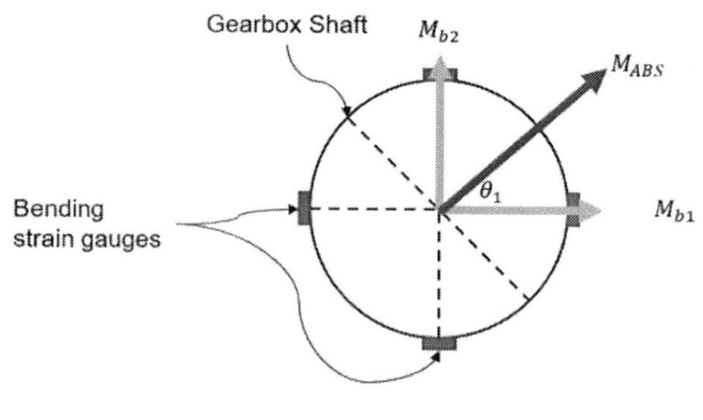

Figure 8: Bending strain gauges shaft positioning

3.3 Torsional Strain Measurement

Shaft torsion is measured at the same position on the shaft as the bending strain. Figure 9 illustrates how the strain guage foil accomodates two 350Ω strain gauges that are 90° apart from each other. Similar to the bending strain gauges, a shunt calibration is also performed before readings were taken. In the shunt calibration process, the wireless node uses an internal precision resistor to load part of the sensor bridge while the sensor remains unloaded. The setup shown in Figure 9 represents again a half-bridge configuration.

Once more, two half-bridges are wired together 180° around the shaft to form a full-bridge configuration. Another advantage of using this configuration is it does not require internal circuit completion which is not available with the QuantumX MX1601B module used. From the measured torsional strain, the torque experienced by the shaft can be calculated using Equation 2.

$$\tau_{shaft} = \frac{\pi E R^3}{2(1 + \upsilon)} \cdot \epsilon \tag{2}$$

where,

- R is the shaft radius in [m]
- ϵ is the strain measured in $[\frac{m}{m}]$
- E is the shaft material's Young's modulus in [Pa]
- υ is the Poisson's ratio of the shaft material

Figure 9: Torsional half-bridge strain gauge foil

4 Results

4.1 MinWaterCSP Facility

During the first 5 s of start-up, there are five clear peaks in the recorded shaft torsion as the VFD accelerates the motor (see Figure 10a). A possible cause for this is the control algorithm implemented in the PLC that controls the VFD according to the load on the motor. Figure 10b illustrates the sudden decrease of shaft torsion during the stoppage period of the induction motor. Figures 10a and 10b illustrates the angular velocity of the gearbox output shaft just before it reaches the set speed and deceleration of the shaft after motor shutdown. The measured angular velocity for both periods of opertation are fitted with a smoothing spline function implemented in MATLAB R2020b. The reason for the values of the smoothing parameters are to fit the measured data as accurately as possible. Figure 12 illustrates the bending moments calculated for the gearbox low-speed shaft for a 90 s period at steady-state. These circumferential position of the strain gauges measuring bending strain is 45° rather than the preferred 90° discussed in Section 3.2. The circumferential offset of the strain gauge sets causes the two channels to record different values for the bending strain. M_{B1} and M_{B2} refer to the two bending channels recording data.

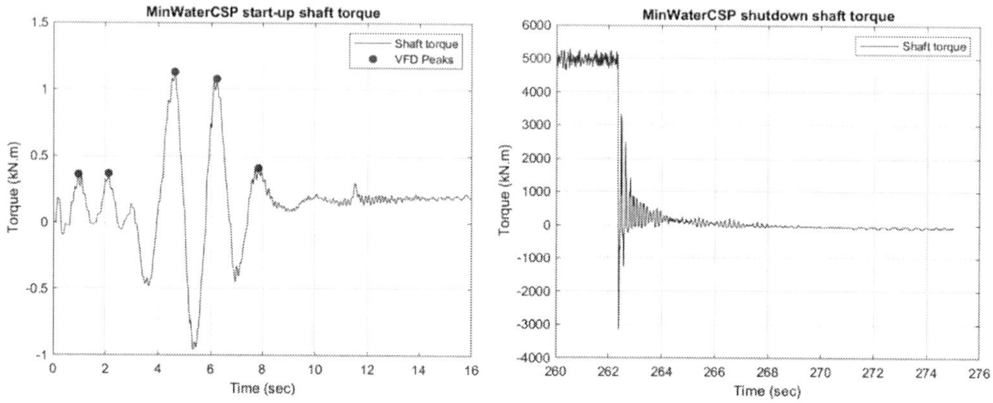

(a) (b)

Figure 10: MinWaterCSP gearbox different shaft torque periods

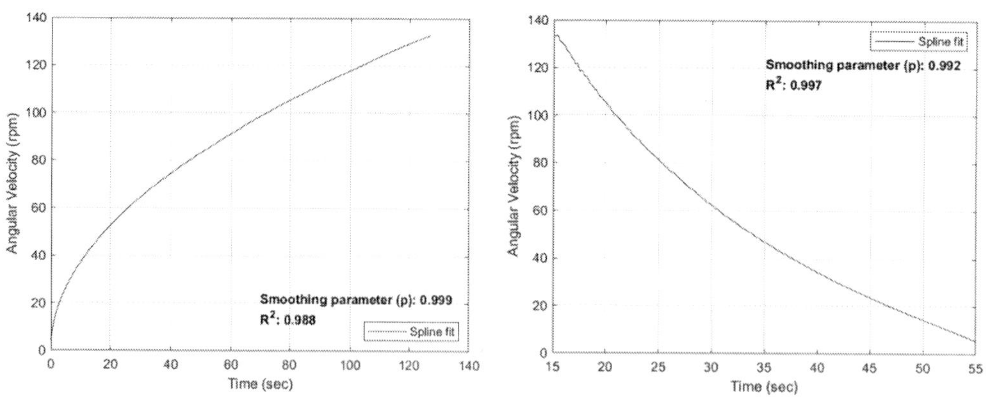

(a) Start-up shaft velocity profile (b) Shutdown shaft velocity profile

Figure 11: MinWaterCSP gearbox different velocity profiles

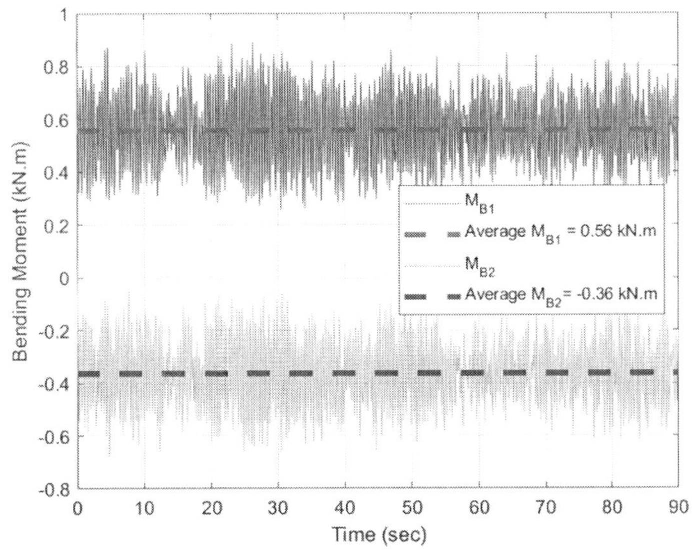

Figure 12: Average bending moments for 90 second period - MinWaterCSP

4.2 Thermal Power Station Test Sites

The power station with the DOL starting method is termed power station I and the soft-starter controlled induction motor, power station II. The different starting methods are directly compared against each other using the data mentioned in Section 3. From Figure 13, it is possible to compare the different velocity profiles of the DOL and the soft-starter. The DOL takes ~4 s to reach full speed whereas with this soft-starter it takes approximately ~13 s.

The same procedure implemented in Section 4.1 for the shaft velocity is used here. Smoothing splines are again fitted with the smoothing parameter chosen so to fit the data as best as possible without jeopardising the validity thereof. Both fans were rotating at the start of measurement, hence the angular velocity graphs not starting from zero.

In Figure 14 the low-speed shaft torques are directly compared for the start-up period. The high starting voltage of power station I's induction motor causes the gearbox shaft to experience a torque peak of 62.45 kN·m which is approximately four times the steady-state value. The soft-starter is able to start the motor at a reduced voltage and the initial torque peak is 14% of that experienced for a induction motor with a direct-on-line connection.

At power station II, the motor is started with a lower voltage (approximately 30 %) of the supply voltage and the torque is gradually increased as a result. The excitation period seen between 10 s and 14 s is caused by the soft-starter introducing harmonic excitations during start-up [7] as a result of the ratio between the instantaneous supply voltage and the rated supply voltage. In Figure 16a, the calculated bending moment for both channels are shown during steady-state operation for 2 minutes as well as a polar plot in the rotating coordinate system of the shaft. The offset in the data indicates that there is a continuous unbalance force on the shaft caused by the fan during steady-state operation.

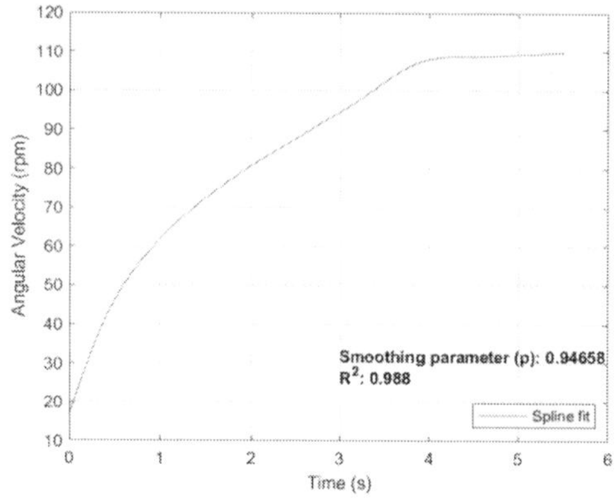

(a) Power station I start-up velocity profile

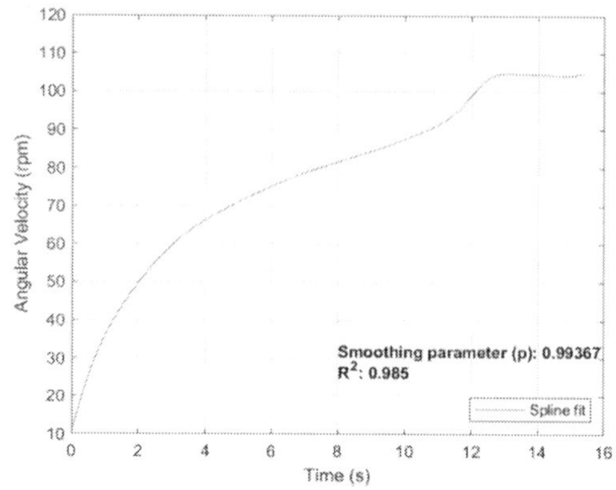

(b) Power station II start-up velocity profile

Figure 13: Comparison of gearbox shaft velocity profiles

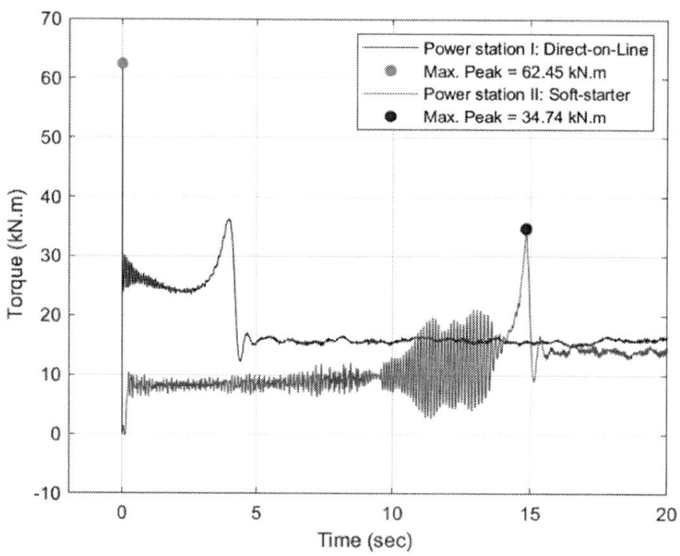

Figure 14: Power station I & II start-up gearbox shaft torque comparisons

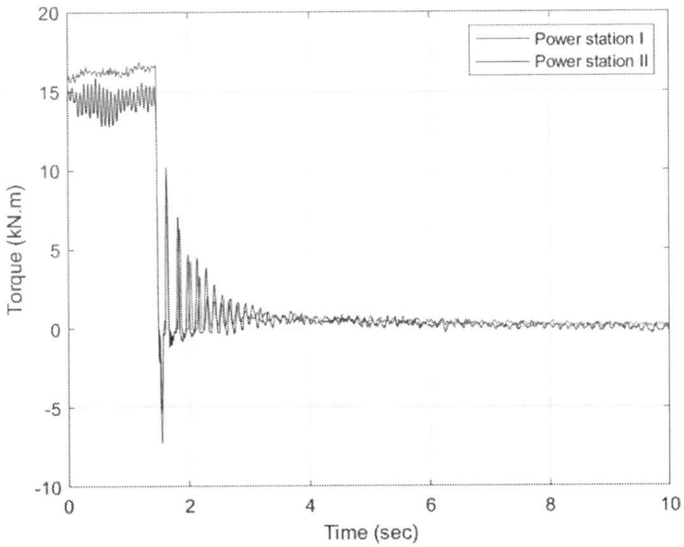

Figure 15: Power station I & II shutdown gearbox shaft torque comparison

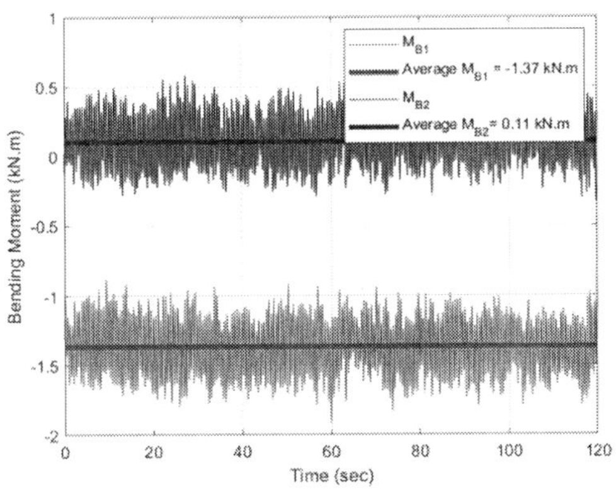

(a) Average bending moment for 2 minutes

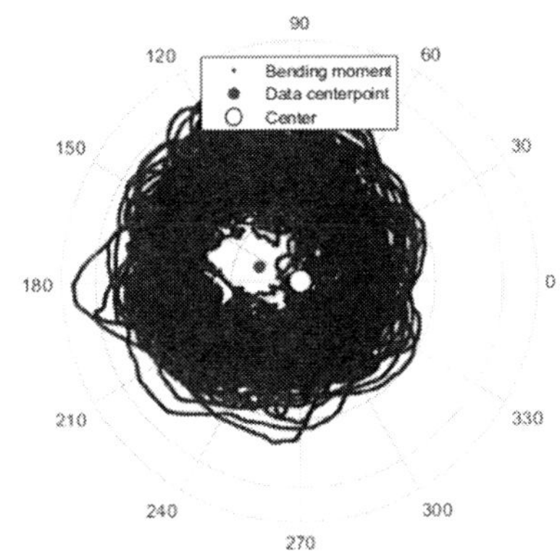

(b) Polar plot of absolute bending moment calculated

Figure 16: Power station I bending moments

5 Conclusion

This research set out to take various measurements on three different test sites of an air-cooled condenser. The measurements set out in Section 3 were carried out with repeatable results. The data acquisition (DAQ) system was developed by performing a system analysis at the MinWaterCSP facility to determine the measurements that are possible. Thereafter, the correct sensors and equipment were chosen and following test runs.

The average calculated absolute bending moment for power station I is ~2.52 kN·m. This coincides well with that measured by [1] at ~2.8 kN·m at the same power station. This confirms that the experimental setup used and calibration of the sensors were correctly done. From Sections 4.1 and 4.2, it can be seen that the velocity profile of power station I has a steaper slope compared to power station II and the MinWaterCSP facility.

Every bending moment recorded follows a particular trend, where the initial value is close to zero and after the start-up phase is over, it oscillates around a constant value. Another difference noticed in Section 4 is the different start-up torque profiles for each test site. The five peaks found in the torque profile of the MinWaterCSP facility cannot be accounted for until the PLC's start-up algorithm is understood. At the time of motor shut down, there is a sudden decrease in shaft torque of ~8 kN·m as the fan decelerates.

Figure 14 directly compare the two different torque profiles between power stations I and II. Power station I's torque profile is typical for an induction motor with a direct-on-line start. In each case, the shaft torque is gradually increased after the initial peak and settles at the stead-state value after the breakdown torque is reached.

The axial fans tested on at power stations I and II have an inertia of 9501 kg·m^2 and 7962 kg·m^2 respectively. This explains the near exact steady-state torque value for both sites. From Figure 15, one can see a similar profile for the shaft torque to the MinWaterCSP motor, although here a difference of ~20 kN·m is noted. Another advantage of the soft-starter controller, is the ability to control the stoppage of the induction motor which is evident from the small upward peaks. In the case of power station I, power is cut from the motor and it then decelerates freely.

The lower shaft torque peak and relatively fast starting time of power station II's motor provide a convincing argument for the soft-starter. Since these induction motors at the power stations always run at rated speed, there is little need to install a VFD, which is more expensive. Considering the results found in Section 4 and the cost of installation of each starting method, the soft-starter can be concluded as the best motor controller amongst the different methods evaluated.

References

[1] C.H.O. Lombard, *Analysis Of Gearbox Loads In Large Air Cooled Condensers* (2017)

[2] Dr J. Muiyser, *Analysis of vibration inducing sources of a large-scale cooling system fan blade* (2016)

[3] MinWaterCSP, https://www.minwatercsp.eu/, (2018)

[4] J. Muiyser, *Simultaneous measurement of air flow conditions and the resultant blade and gearbox loading at large-scale cooling system fans* (2012)

[5] H. Goldschagg, *ACC Fan Gearboxes: Eskom's Experience in the Selection and Maintenance of ACC gearboxes* (ACCUG Nevada Conference, 2013)

[6] R.J. Drago, *The effect of start-up load conditions on gearbox performance and life - Failure analysis and case study* (Meeting of the American Gear Manufacturers Association, 2007)

[7] ABB, *Softstarter Handbook* (2021)

Performance analysis of a two-stage travelling-wave thermo-acoustic engine using Artificial Neural Network

Miniyenkosi Ngcukayitobi[1,*], *Sphumelele* Sibutha[1,**], *Lagouge K.* Tartibu[1,***], and *Flávio C.* Bannwart[2,****]

[1] Department of Mechanical and Industrial Engineering Technology, University of Johannesburg, Johannesburg, South Africa
[2] School of Mechanical Engineering, University of Campinas, UNICAMP, Campinas, Brazil

Abstract. Thermo-acoustic systems can convert thermal energy into acoustic waves and vice-versa. This conversion is due to the thermo-viscous interaction between the acoustically oscillating gas fluid within a porous medium, referred to as a regenerator, and the pore internal walls. The thermo-acoustic approach is proposed in this study as an alternative sustainable solution for addressing the issue of electricity in remote areas of developing countries. This approach is environmentally friendly as it utilises air as the working medium and therefore does not generate harmful emissions. In this study, a two-stage travelling-wave thermo-acoustic engine has been modelled using DeltaEC. The simulation was performed by considering various input heat for both of the engine stages. The heat input for the first stage was set within the range of 359.48 to 455.75W, while in the second stage was within the range of 1307.99 to 1656.35W. Hundred (100) data were generated. This dataset was used to build an Artificial Neural Network (ANN) model. The ANN model was validated using the data extracted from DeltaEC. A good agreement between DeltaEC simulation results and ANN predictions was observed. This study shows that the ANN approach is capable of analysing intricate nonlinear thermo-acoustic issues.

1 Introduction

Waste heat recovery has been identified as one of the promising techniques to further address the issue of energy shortage and environmental pollution [1]. Low-grade heat is produced through a process of fuel combustion or chemical reaction and then dumped into the environment even though it could still play a crucial role in the economy. In this era of industrialization, more attention has been given to the reuse of waste heat generated in

* miniyenkosin@uj.ac.za
** 217008079@student.uj.ac.za
*** ltartibu@uj.ac.za
**** fcbannwart@fem.unicamp.br

different industries and power plants [1]. A bulk of heat energy is dissipated to the atmosphere from many industries together with the hot flue gas emitted from heated products and types of machinery [2]. To improve the energy efficiency of industries, and minimize the production costs, waste heat must be reused [2]. The reduction of such energy losses could contribute to making the whole process more environmentally friendly as heat is regarded as the main factor for the greenhouse effect [2]. The heat energy lost in flue gases cannot be recovered fully and be utilized. However, much of it could be upgraded and recycled by adopting the necessary strategies as described in this research study. The thermo-acoustic approach has been selected to meet the targets of industries due to its simplicity in design, and potential of converting heat power into electric power at a low cost [3].

Thermo-acoustic systems can convert thermal energy into acoustic waves (thermo-acoustic engine) - or mechanical energy - by utilizing what is known as the thermo-acoustic effect. These acoustic waves can be used to generate electric power (thermo-acoustic generator) or to induce cooling (thermo-acoustic refrigerator) [3]. This conversion is due to the thermo-viscous interaction between the acoustic particle oscillating along with a gas medium and the pore internal walls of a porous material under sufficiently high-temperature gradient inserted into a resonant waveguide network. The latter is referred to as regenerator for travelling-wave systems, and as stack for the case of standing-wave ones [3]. Travelling sound-wave consists of the propagation of oscillatory displacement of subsequent acoustic particles accompanied by oscillating temperature and pressure. When the acoustic pressure propagates along a narrow channel, oscillating heat transfer between the surface of channel walls and the acoustic particles is enhanced, in both directions, as long as the power intensities involved are high enough. Hence, the aggregation of the oscillations may produce refrigeration due to the thermo-acoustic effect, or, reciprocally, may be generated by the thermo-acoustic effect promoted by a sufficiently high thermal gradient imposed in the direction of propagation [3]. A two-stage travelling-wave thermo-acoustic generator (or engine) is depicted in Fig.1. It has the following main components: porous medium (regenerator), heat exchangers (cold and hot heat exchangers), and resonance tube.

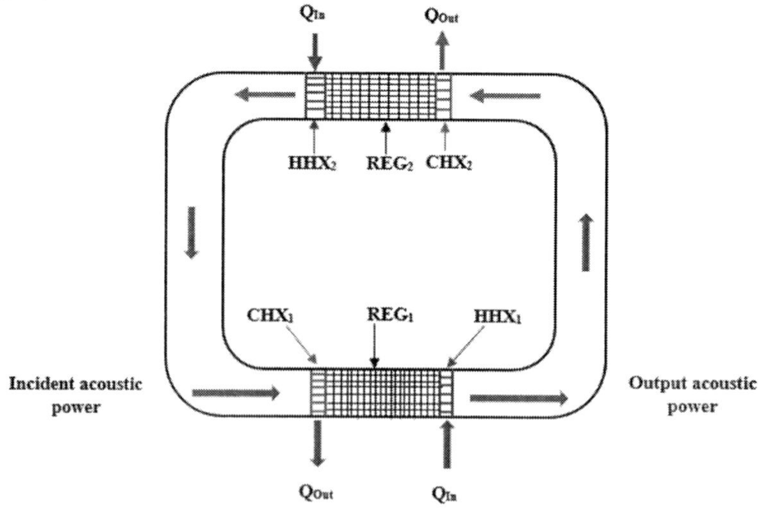

Figure 1: Schematic of a two-stage travelling-wave thermo-acoustic engine.

The porous material (generally sandwiched in-between heat exchangers) are regarded as the main component influencing the efficiency or the performance of the thermo-acoustic systems (refrigerators and engines) [4].

The regenerator is defined as the porous material that permits and restricts the working gas to oscillate while in adjacent thermal contact with the solid internal walls. It has many closely spaced surfaces aligned to the length of the resonance tube [4]. The resonance tube defines the frequency of the system, and allows the sound waves to travel through, and also houses the porous medium and heat exchangers. The main function of the two heat exchangers (hot and cold heat exchangers) is to create a temperature gradient along with the porous medium, which is required to raise the acoustic field that is spontaneously generated. In essence, heat exchangers (HX) play a vital role in both cooling and heating processes. They can have various sizes, shapes and can be made of different materials to transfer thermal energy from one liquid to another. In addition, they must have a high overall heat transfer coefficient with low or no acoustic power dissipation [4]. The effective heat transfer rate occurs between the working fluid and narrow-spaced plates of the HX which are aligned in the direction of the wave propagation [5]. The spaces between the plates of HX and the plate size determines the blockage ratio. A smooth flow of the fluid can be attained when the blockage ratios of the porous medium and HX are similar [5]. The working fluid should have a low Prandtl number, high speed of sound, non-flammable, commercially available, and high thermal conductivity (to increase the thermal penetration depth) [5]. Inert gases meet the above-mentioned requirements and thus provide thermo-acoustic systems with the benefits of being environmentally friendly without ozone depletion or global warming issues.

A significant development in the field of thermo-acoustics has been achieved by Kees de Blok [6], who built a device able to convert low-grade heat energy, in the range of *70 - 200*°C. Multiple regenerator units were proposed [6], to raise the sound power of the thermo-acoustic system at low operating temperature while maintaining higher performance. He also pointed out that the viscous losses can be eliminated by scaling up the regenerator cross-sectional area along the resonance tube. Kees de Blok conducted another experiment on a travelling-wave thermo-acoustic engine with a by-pass configuration, and the air was used as the working gas fluid. A threshold equivalent to a temperature gradient of *65K* was recorded to generate a sound wave within a resonance tube [7]. In a different study, he designed and constructed a four-stage travelling-wave thermo-acoustic generator driven by a temperature gradient of *40K* across each engine stage [7]. Both of his prototypes are suitable for consuming low-grade waste heat.

Artificial Neural Network (ANN) is defined as the information-processing approach that is based on characteristics similar to biological neural networks [8]. ANN have been developed as a generalization of mathematic models of neural biology, based on the following assumptions:

- The processing of information occurs at the elements called neurons [8].
- Signals are carried through neurons over connection links.
- Each connection link consists of associated weight, which is a typical neural network, multiplies the transmitted signals.
- Each neuron carries an activation function (usually nonlinear) to its network input to determine its output signals [8].

A neural network has a large number of simple processing elements called cells, units, neurons, or nodes. Each neuron is joined to other neurons using directed communication links, each carries associated weight [8]. The weights contain details being used by the network to solve a problem. ANN can be considered in a wide variety of problems, such as recalling data or patterns and storing, performing general mappings from input patterns to output patterns, classifying patterns, grouping similar patterns, or providing solutions to constrained optimization issues [8]. Each neuron is consisting of an internal state, (activation level), which is the function of the inputs it has received. Typically, a neuron delivers its activation in a form of signals to various other neurons.

To attain the required correlation between output and input of a network, values must be derived for the activation function and connection weights. The whole process of derivation is named supervised training [8]. ANN are not automated to perform specific work while implemented on the computer. Instead, they are trained to learn patterns used as inputs with respect to the data set. Once the training is completed, new patterns may be displayed to them for classification or prediction [8]. ANN can acquire a knowledge of patterns without direct human control in data from physical models or real systems, computer programs, or other sources. They can deal with many inputs and generate answers that will be suitable for further processing or designers.

ANN have received much attention from researchers worldwide in some engineering fields and thermal science [9]. ANN have been used to solve many intricate issues in air conditioning units, refrigeration, and heat pumps [9]. This application mainly covers the analysis and prediction of the thermal load for air conditioning units and the performance of various refrigeration systems, optimization of operating parameters for refrigeration compressors and thermodynamic performance [9]. Cenk et al [9] applied ANN to heat engineering such as Stirling engines, internal combustion engines, as well as in thermo-acoustic research area (specifically for heat engines). The study describes the development of a gasoline engine using the ANN model to predict emissions and performance of the engine. The results reported show that ANN can be considered as a promising flexible tool for classical modelling techniques. Anas et al. [9] applied ANN to thermo-acoustic refrigerators to predict the cooling load in a standing-wave thermo-acoustic refrigerator under given mean pressure and oscillating frequency. The study describes the prediction of the oscillatory heat transfer coefficient in the thermo-acoustic heat exchangers and reports a good agreement between the experimental and ANN results. Yasar et al. [9] also used ANN to predict the power and torque of the Stirling engines as the responses to operating parameters such as charge pressure, engine speed and hot source temperature. The study reports an excellent agreement between experimental and predicted outputs, which confirms that ANN can be used as an alternative intelligent approach for data prediction [9]. Fatih et al [10] conducted a research study on thermo-acoustic instability prediction by joining traditional numerical methods with soft computing tools. The simulation results of the dynamic neural network and black-box approaches with dynamic fuzzy identification show excellent agreement with the results of Comsol (for the burker-Schumann diffusion flame) and Galerkin solver (for coupled thermo-acoustic engine) [10]. The above-mentioned researchers have demonstrated that ANN models are capable of relating the correlation between the output and input parameters, and predict the performance of energy systems [10].

2 Motivation

ANN is regarded as a data-driven model, and it acquires knowledge from examples where outputs and inputs are imparted to the network repeatedly and sequentially [11]. ANN can acquire knowledge from examples and that makes it efficient problem-solving with significant benefits in some particular aspects. These aspects are learning and recognizing the underlying relations between outputs and inputs without explicit physical consideration [11]. There has been a significant contribution made by researchers in the development of efficient thermo-acoustic generators [11]. The efficiency of the systems and the nonlinearity associated with the working of the systems remain the major potential area of research. Many existing studies have demonstrated the promising performance of ANN as an alternative intelligent tool that can be used to predict future performance in different engineering systems [12].

The driver behind this study is to examine the capability of the ANN model to establish a link between input and output parameters. When sufficient output and input data are provided, the ANN approach can be able to formulate principal correlations between parameters [12]. In this study, a two-stage travelling-wave thermo-acoustic engine has been modelled using DeltaEC. Three (3) parameters affecting the performance of the system have been considered to analyze the system. These parameters include the pressure amplitude, the resonant frequency, the volumetric flow rate, and the input heat. The pressure amplitude and the acoustic power are considered as the main indicators of the performance of the system. The identification of the correlation among the parameters affecting the performance of a thermo-acoustic device and desired performance is the essential foundation of thermo-acoustic components design [12].

3 Detailed description of the proposed approach

This section discusses the method considered for data collection and the technique used to predict future performance in a thermo-acoustic generator.

3.1 DeltaEC Modelling

DeltaEC (Design Environment for Low-amplitude Thermo-acoustic Energy Conversion) has been used by many researchers to build theoretical models and analyze the effect of the geometry configuration of thermo-acoustic systems [13]. This software was developed at Los Alamos National Laboratory by researchers to predict the performance of thermo-acoustic linear systems [13]. It numerically integrates the conservation equations of energy, continuity and momentum. The working principles of the two-stage travelling-wave thermo-acoustic generator are investigated using DeltaEC software, and a code for thermo-acoustic computation is based on the theory of linear thermo-acoustic. It provides a function prototype for characterizing parts in the system such as porous medium, heat exchanger, and resonance tube. Also, the shooting method is adopted to integrate the set boundary conditions and computes the parameters of each segment.
The current study presents the modelling of a two-stage travelling-wave thermo-acoustic system using DeltaEC. Several segments have been selected to develop a travelling-wave looped-tube test rig, as schematically shown in Fig.2.

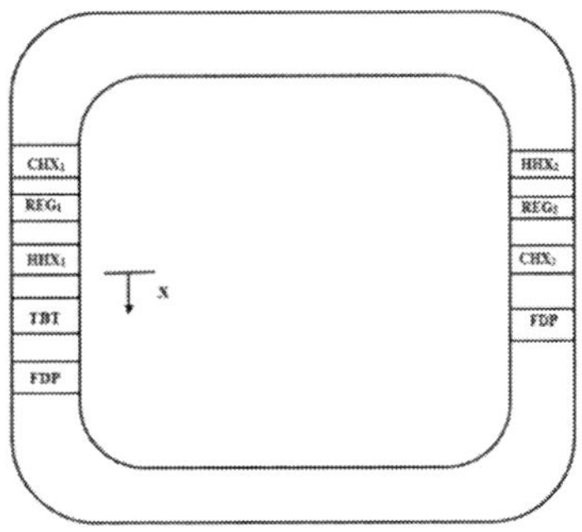

Figure 2: Segments of DeltaEC modelling.

The segments considered include two hot heat exchangers (HHX), two cold heat exchangers (CHX), thermal buffer tube (TBT), feedback pipe (FDP), and two regenerator units. The combined elements in the simulated system have a total length of *3.5m*, and each thermo-acoustic engine stage was positioned *1.595m* apart. An operational sound-wave frequency (f) range for the above resonance length is of *98.9Hz*. This value of sound-wave frequency was estimated by using equation 1 and initialized for the thermo-acoustic system [14]. After performing a simulation on DeltaEC, the response frequency (Guess) was adjusted to *97.052Hz*, and this value is very closer to the estimated resonance frequency.

$$L = \lambda = \frac{c}{f} \quad , \tag{1}$$

with c being the speed of the sound wave (around *346 m/s*), L is the total length of the test rig, and λ is the wavelength. The regenerator and both heat exchangers (cold and hot) were modelled as parallel-plate. The model always starts with a "BEGIN" segment, where the sound-wave frequency, mean pressure, temperature, volumetric velocity characteristics, and acoustic pressure characteristic are defined. The initial values at the "BEGIN" segment are unknowns and the target or guess shooting approach was applied. The reverse polish notation (RPN) segments were utilized as targets and guesses; as a result, the test ring ends with several RPN targets that meet final thermo-acoustic properties to put in the appropriate condition properties in the "BEGIN" segment. The second duct of *1.595m* and the thermo-acoustic core of *0.155m* were added at the end of the first engine stage. The surface temperature for both the cold heat exchanger and hot heat exchangers were set as *1300* and *300K*, respectively. The numerical model has a constant cross-sectional area of $0.008213m^2$. Equally, numbered guesses and targets were used for modelling or predicting the performance of the thermo-acoustic system, as required for a deterministic solution. After each execution round, DeltaEC collates the values of guesses to the targets. If the targets do not correspond, the guesses are adjusted and the integration is recurrent. Several simulations were performed for various input heat to model or predict the performance of a thermo-acoustic engine. The acoustic power, which is regarded as the main indicator of the performance of the device, was recorded after each successful execution of DeltaEC. The simulation commences at the beginning of the thermal buffer tube (TBT), as indicated by variable x in Fig.2. It computes each segment along the resonance looped tube counter-clockwise. The continuous boundary conditions were implemented at $x=0$. The working acoustic pressure amplitude in the thermo-acoustic device was set at *602.28Pa*, the gas inlet temperature at *320K*, the volumetric flow rate amplitude is $1.8676e^{-03}m^3/s$ with a phase angle (*p1* and *U1*) of *87.272°*; the mean pressure range is *1000KPa* with a phase angle of *0°*. The heat input for the first stage was set within the range of *359.48* to *455.75W*, while in the second stage was within the range of *1307.99* to *1656.35W*, and hundred (100) datasets were generated. The frequency, the working pressure amplitude, the volumetric flow rate, the phase angle (p_1 and U_1), the mean temperature, and the heat supplied to the hot heat exchanger were set as guess parameters. The temperatures in the heat exchangers, the magnitude of the pressure at the final segment of each thermo-acoustic engine stage, the volumetric flow rate at the final segment of each thermo-acoustic engine stage, and the total energy flow at the final segment of both engines were set as the target parameters. The details of the DeltaEC model are shown in Table 1.

Table 1: Geometrical parameters of the two-stage travelling-wave thermo-acoustic generator.

Segments	Diameter (mm)	Length (mm)
Ambient heat exchanger	102.26	40
Hot heat exchanger	102.26	20
Regenerator	102.26	95
Duct	102.26	3190
Resonance tube	102.26	3 500

3.2 Determination of samples, training algorithm, and structure for ANN model

Adequate training sample size and depiction of the training sample for the environment of interest are treated as the first and crucial step in developing of ANN model. An ANN has commonly three layers, namely an output layer, an input layer and hidden layers. Theoretically, there is no restriction on the number of hidden layers. Each of the layers is consisting of a certain number of neurons joined together by the links, which are correlated with weights. The number of output and input neurons depends on the complexity of the modelling problem. In the current study, the input layer has three neurons characterizing the three parameters affecting the performance of the device, and one neuron in the output layer characterizing successively acoustic pressure amplitude or acoustic power.
Several ANN models were built within MATLAB. These models have the same number of input and output layers but different hidden layers. The effect of the number of hidden layers on the model performance was analysed by running several simulations to identify the best ANN architecture. Two architecture are described in this study. Hundred (100) data were extracted from DeltaEC software. These data were divided into three, namely test data set, validation data set, and training data set. The training data set was used to fit the weights of the network or for learning purposes whilst the validation data set was used to reduce over-fitting issues that may arise during the training process. The test data was used to evaluate the performance of the trained neural network. The first 70 samples of data (70%) were used to train the network, 15 samples of data (15%) were used for validation, and the other 15 samples of data (15%) was used for testing. The neural fitting app was used to construct and train the network, select data and test its performance using regression analysis and mean square error. A two-layer feed-forward network with linear target neurons and sigmoid hidden neurons was used to fit the data set generated from DeltaEC. The Levenberg-Marquardt backpropagation algorithm was utilized to train both of the neural networks. This algorithm was selected simply because it requires less time to process the data. The structures of the neural network depicted in Fig.3, and Fig.4 represents the ANN models adopted in the experiment. The prediction error (Pe), and average prediction error (APe) were estimated using the following equations [15]:

$$\text{Pe}\,\% = \left|\frac{\text{Predicted results} - \text{Experimental results}}{\text{Experimental results}}\right| , \tag{2}$$

$$\text{APe}\,\% = \frac{\sum_{i-1}^{n} - \text{prediction error}\,\%}{\text{n}} . \tag{3}$$

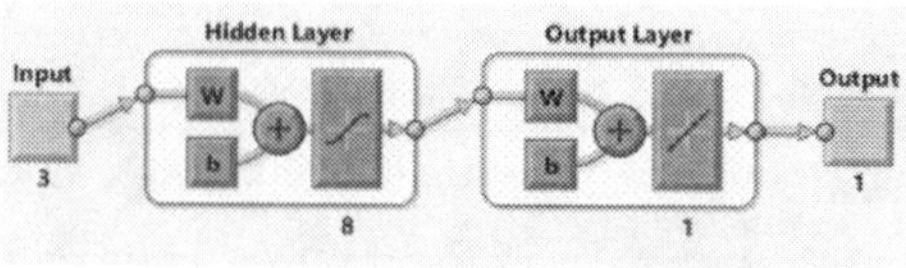

Figure 3: ANN structure (Model 1).

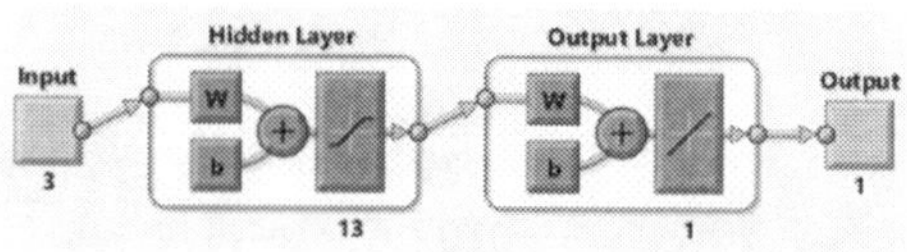

Figure 4: ANN structure (Model 2).

4 Analysis and results

4.1 DeltaEC simulation

The dataset extracted from DeltaEC software is graphically represented in Fig.5. The results presented in Fig.5 reveals that the heat input to the HHX is directly proportional to the output parameters (pressure amplitude, and acoustic power). Therefore, the rise in the heat supplied to the HHX also causes an increase in the total power as well as in the volumetric flow rate. The acoustic power and pressure amplitude were recorded after each successful execution round, and they both have a similar trend as shown in Fig.5.

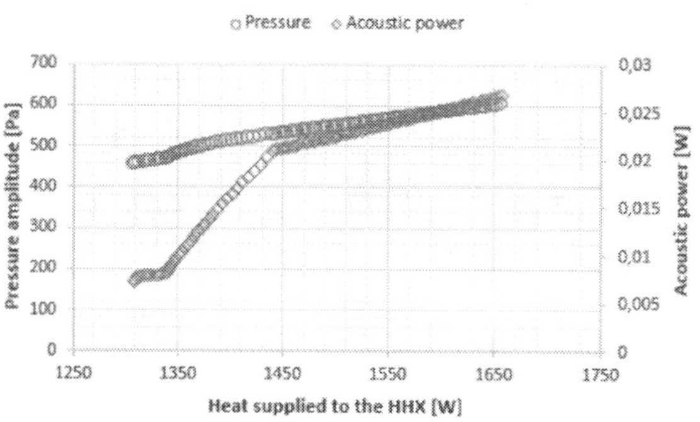

Figure 5: Generated pressure amplitude and acoustic power as a function of input heat.

The simulation was performed by considering various input heat for both of the thermo-acoustic engine stages. The input heat for the first engine stage was set within a range of *359.48* to *455.75W*, while in the second stage was within the range of *1307.99* to *1656.35W*. It is also observed that the output parameters (pressure amplitude and acoustic power) were relatively lower at the end of the first stage as compared with the one at the end of the second stage. The addition of the second thermo-acoustic engine stage increases the pressure amplitude as well as the acoustic power output.

4.2 Analysis of ANN model prediction

The ANN model predicted acoustic pressure amplitude and acoustic power. Both of the models were trained considering different hidden layers. Three criteria were followed to analyse the performance of the network and hence to reach the optimum one among all proposed networks.

The first criterion is to reduce network size and complexity if possible. The second criterion is to select the least mean square error (MSE) during the testing phase. The last criterion is to analyse the regression between the corresponding targets (actual output) and network response from the correlation efficient (R^2) and hence identify the optimum number of hidden neurons. In general, R^2 valves varies between positive one and zero, when R^2 value closer to positive one indicates an excellent positive linear relationship between the targets and network outputs, while the values near to zero indicate a weak relationship [15]. The above process was performed to identify the architecture of the ANN that yield the lowest discrepancy between the actual and predicted values of the acoustic power and acoustic pressure amplitude successively. The average prediction errors for both of the models between predicted and DeltaEC simulation outputs were computed and plotted again the number of neurons in the hidden layers, as depicted in Fig.6 and Fig.7. It was noted from Fig.6 that the average prediction error was minimized from eight (8) to fifteen (15) neurons, while the average error was minimized for thirteen (13) neurons (Fig.7). Therefore, the most optimum networks obtained considering the training performance of the pressure amplitude and the acoustic power have respectively eight (8) and thirteen (13) hidden neurons. The pressure amplitude and the acoustic power network must have three input nodes and one output neuron, which can be represented as 3-8-1 and 3-13-1 networks respectively.

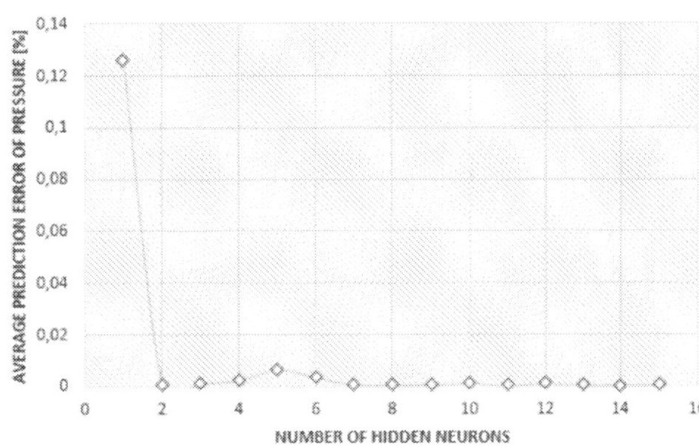

Figure 6: Average prediction error of pressure amplitude [%] versus the number of hidden neurons.

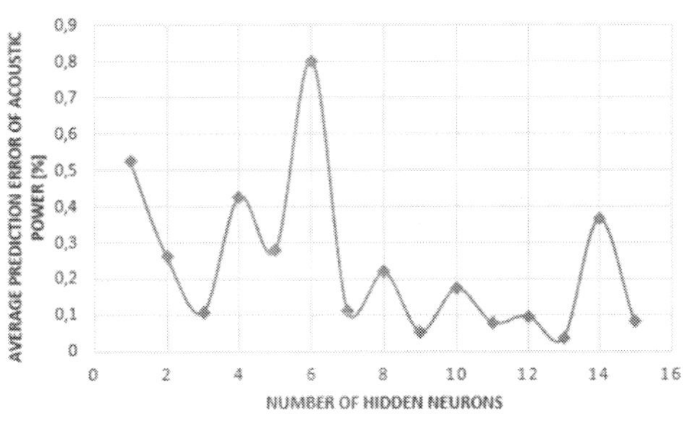

Figure 7: Average prediction error of acoustic power [%] versus the number of hidden neurons.

It was also observed that the performance of the proposed technique (ANN) in predicting the acoustic power and working pressure amplitude in a two-stage travelling-wave thermo-acoustic generator was adequate due to the regression values obtained between target and predicted dataset during validation, training, and testing phases as depicted in Fig.8 and Fig.9.

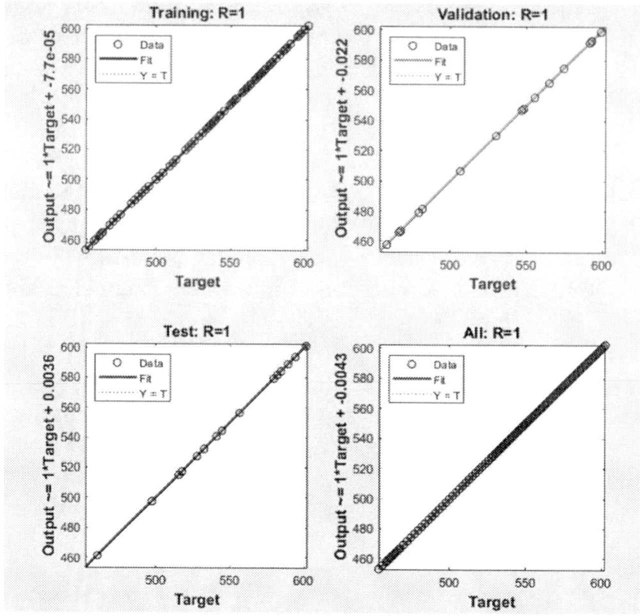

Figure 8: Regression plot for validation, testing, and training phases (pressure amplitude).

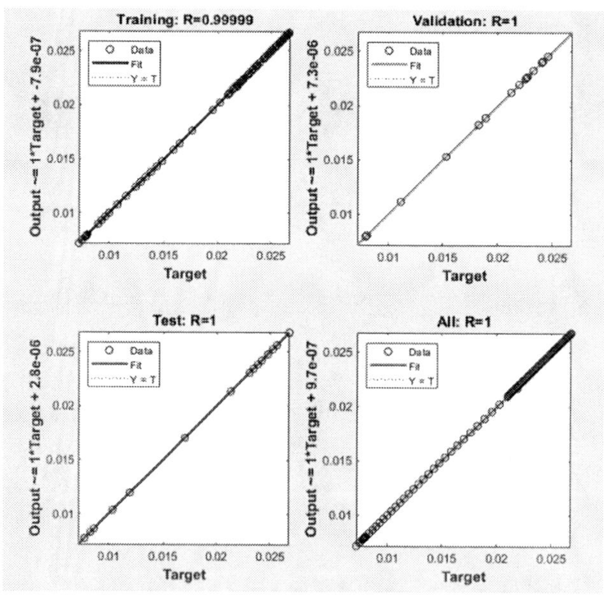

Figure 9: Regression plot for validation, testing, and training phases (acoustic power).

The optimization of the ANN performance was subjected to various constraints such as the construction of the neural network, the amount of both testing and training dataset, and learning parameters. The regression graphs depicted in both Fig.8 and Fig.9 indicate the best performance of the neural network in predicting responses of any new dataset input within the range of our research study. To evaluate the prediction ability of our structural neural network, 15% of the data extracted from DeltaEC simulation were used for the aim of verification. The comparison between the output data extracted from DeltaEC simulation and the results predicted from the ANN model was performed and graphically represented in Fig.10, and Fig.11.

Fig.10 shows that "all" the predicted pressure amplitude values, considering the ANN model match closely the targets (DeltaEC output results). This is evidenced by the highest deviation of 2,5%. A similar observation was made, considering the acoustic power. In the latter case, the highest deviation was 0,016% as shown in Fig.11.

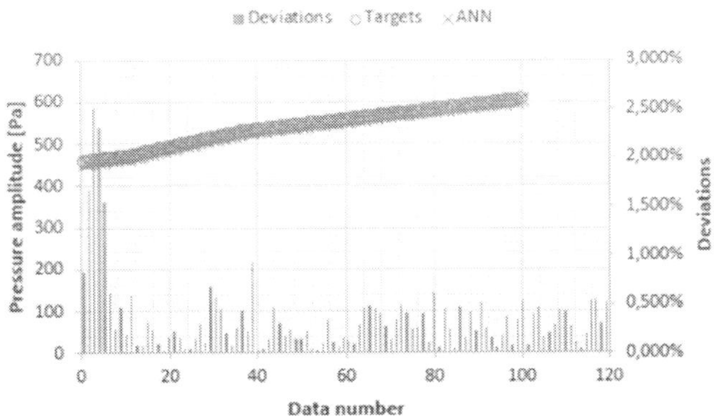

Figure 10: ANN prediction versus output parameters/ pressure amplitude.

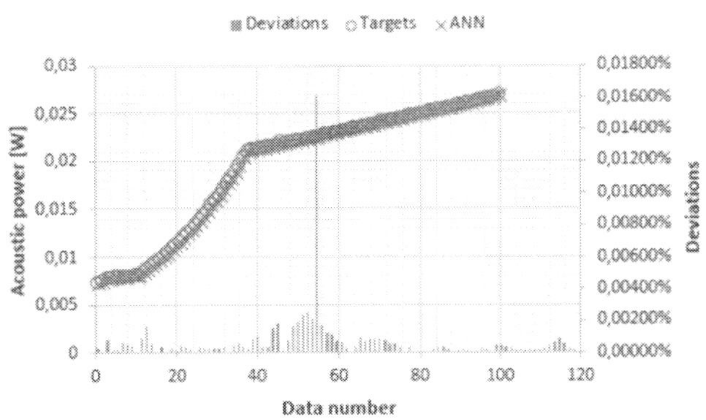

Figure 11: ANN prediction versus output parameters/ acoustic power.

Fig.10 and Fig.11 show the actual dataset of acoustic power and acoustic pressure amplitude extracted from DeltaEC compared to the ANN predictions. It can be seen that the DeltaEC estimation output results for both of the models have a similar trend to the output results predicted by the ANN model. The agreement between these two results shows the potential for the ANN technique for the prediction of the performance of thermo-acoustic generators. The finding implies that the performance of the proposed ANN models is satisfactory in predicting acoustic power and acoustic pressure amplitude. Also, the ANN approach can be used to predict configurations that were not simulated to improve the analysis of the system.

5 Conclusions

This study shows the capabilities of ANN as a flexible modelling technique for estimating the thermo-acoustic output parameters (pressure amplitude and acoustic power) of a two-stage travelling-wave thermo-acoustic engine. The process leading to the identification of the best ANN architecture (3-8-1 and 3-13-1) has been described in order to map the correlation between input parameters and responses. In addition, the influence of input parameters on the responses had been analyzed, and it was noted that the heat supplied to the hot heat exchangers has a significant effect on the output parameters. The rise in the heat supplied to the HHX also causes an increase in the acoustic power as well as in the acoustic pressure amplitude.

The present work has shown the significant benefits of ANN to analyse the relationship between parameters describing thermo-acoustic devices and predicts configurations that were not simulated. The approach proposed in this study could potentially improve the analysis of thermo-acoustic systems and provides a better insight into the issue related to their efficiency. This would result in a potential minimization of time-consuming experiments and provide an alternative modelling approach to the thermo-acoustic community of researchers. Future research work must be extended to investigate the ability of ANN to predict experimental data, which are inherently nonlinear.

References

1. M. Karlsson, M. Abom, M. Lalit, and R. Glav, A note on the applicability of thermo-acoustic engines for automotive waste heat recovery. SAE International Journal of materials and manufacturing, vol. 9, issue. 2, pp. 286 – 293 (2016).
2. M.D. Fahel Bin Noor, Ahsan Habib, and Bijoy Mallick, Heat Storage system: A modern way to reuse and recycle Energy to reduce thermal pollution. International conference on mechanical, industrial, and energy engineering (November 2018).
3. A.A.H. Al-Kayiem, Investigation of Travelling-wave Thermoacoustic Engines with Different Configurations, PhD thesis, University of Glasgow (2017).
4. M. Ngcukayitobi, L.K. Tartibu, S.L. Gqibani, Design and construction of a four-stage travelling-wave thermo-acoustic system for power generation. Proceedings of the ASME 2020 International Mechanical Engineering Congress and Exposition, 6, 16-19 (2020).
5. N.M. Hariharan, P. Sivashanmugam, and S. Kasthurirengan, Influence of stack geometry and resonance length on the performance of thermo-acoustic engine. Applied acoustics, vol.73, issue.10, pp. 1052 – 1058, October (2012)
6. Kees de Blok, "Low operating temperature integral thermo acoustic devices for solar cooling and waste heat recovery". Journal of Acoustic Society of America, vol. 123, issue. 5, pp. 3541. DOI: 10.1121/1.2934526 (2008).
7. Kees de Blok, "Novel 4-stage travelling-wave thermo-acoustic power generator". Conference: ASME 2010 3rd Joint US-European Fluids Engineering Summer Meeting collocated with 8th International Conference on Nanochannels, Microchannels, and Minichannels, vol. 7. American Society of Mechanical Engineers; 2010. pp. 73-79
8. Fausett, Fundamentals of neural networks. Architectures, algorithms, and application. Pearson; 1st edition ISBN-10: 0133341860, pp.9-20 (December 19, 1993).
9. A. Rahman, X. Zhang, Prediction of acoustic-wave parameters of thermo-acoustic prime mover through Artificial Neural Network technique: practical approach for thermo-acoustics. Journal of thermal science and engineering progress, 8, 257-268 (2018).
10. F. Selimefendigil, H.F. Oztop, Soft Computing methods for thermoacoustic simulation, Numerical Heat Transfer Applications, 66(3):271-288 (2014).
11. S. J. Russell, P. Norvig, Artificial intelligence: a modern approach, 3rd edition. Upper Saddle River, NJ: Prentice-Hall (2010).
12. I.A. Basheer, M. Hajmeer M. Artificial neural networks: fundamentals, computing, design, and application. J Microbiol Meth, vol. 43, issue. 1, pp. 3-31 (2000).

13. A. Abduljalil, A. Jaworski, Investigation of thermo-acoustic processes in a travelling-wave looped-tube thermo-acoustic engine, PhD thesis, University of Manchester (2012).
14. P. Novotny, S.S. Hsu, A.B. Wang, T. Vit. Investigation of a travelling-wave thermo-acoustic engine in a looped-tube. EPJ Web of conferences, pp. 67, 02086 (2014).
15. A. A Rhaman, X. Zhang. Prediction of cooling load for a standing-wave thermo-acoustic refrigerator through Artificial Neural Network technique. 9th International conference on applied energy, ICAE2017. 142, 3780 – 3786 (2017).

Optimization of the Pumping Capacity of Centrifugal Pumps Based on System Analysis

Motsi Ephrey Matlakala[1*] and *Daramy Vandi Von* Kallon[1]

[1]Department of Mechanical and Industrial Engineering Technology, University of Johannesburg, South Africa

[*]Corresponding author email: motsiephrey@gmail.com

Abstract. The pumping capacity is the maximum flow rate through a pump at its design capacity. In the process of pumping water and other fluids, pumping capacity is required to accurately size pumping systems, determine friction head losses, construct a system curve and select a pump and motor. Failure to choose the right pump size for pumping system, improper installation and pump operation results into higher consumption of energy. The insufficient pumping capacity affects the plant's operations such as maintenance cost, downtime, loss of production and increase in operating cost. In this study variation of the impeller diameter is used to calculate the new pump curve to improve the pumping capacity. The pumping system is analysed to determine the pumping capacity of the pump. Computational fluid dynamic (CFD) simulations are carried out to determine the performance of the pump and analyses the pumping system to achieve the pumping capacity. Results show that enhanced pumping capacity is achieved at a given impeller design with a specific shift in the pump curve. It is recommended that the pumping capacity can be optimized through trimming of impeller. Trimming of the impeller improves pump efficiency and increases the performance of the pump. In addition, the pumping capacity can also be optimized through the system analysis by adjusting the diameter of the pipes and throttling of the valves. Optimization of the pumping capacity helps with running the pumping system efficiently.

Keywords: Pumping Capacity, System Analysis, Computational Fluid Dynamics, Cost, Impeller, Pump Efficiency.

1. Introduction

Sizing of pumping systems are difficult since the pumps must fit the pumping systems, as a result, it affects the installation, maintenance, and operations of centrifugal pumps and pumping systems. Centrifugal pumps contain two main parts: an impeller (rotor) that rotates at speeds of the motor and imparts centrifugal forces to the production fluid and diffuser (stator) the fixed part that guides the flow to the discharge [1, 2]. Optimization of pumping capacity is carried out through system analysis which involves process analysis, engineering design, maintenance, cost analysis, risk and viability analysis. The impeller of a centrifugal pump regulates performance of pump and determines pumping capacity of the centrifugal pump [3]. Figure 1 shows layout of centrifugal pump with two major components, impeller and volute.

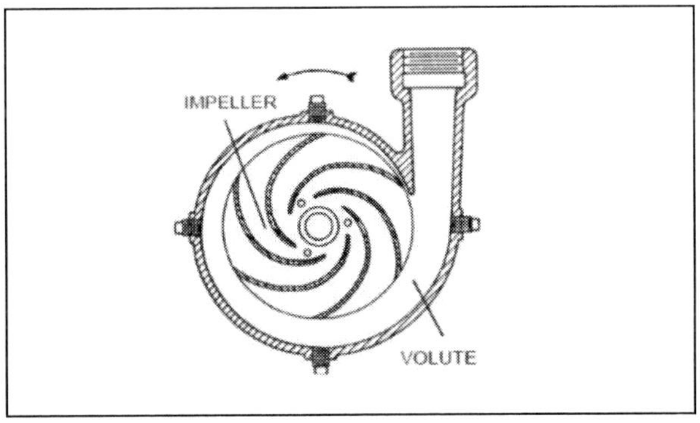

Figure 1. Centrifugal Pump [2]

The design of the impeller diameter has a major effect on the pump efficiency, an entirely new pump can be designed by just modifying the impeller. The pumping capacity is the maximum flow rate through a pump at its design capacity [4, 5]. The speed and diameter of the impeller determine the head or pressure head that the pump can generate [3]. A typical failure mode of centrifugal pumps is caused by the difficulty of installation, operation and maintenance of the pump. The challenge of implementing proper maintenance plans for the pumping system affects the efficiency of the pump. The centrifugal pump efficiency can drop below 5% of required operation duty point, consuming more energy and cost a company more than usual [6]. In the present study we have identified that companies find it challenging to design, install, operate and maintain centrifugal pump of which it affects the efficiency of centrifugal pump. The aim of the paper is to carry out an investigation with the aim to improve the efficiency and reliability of centrifugal pumps through better design of impeller, installation, operation and maintenance practices. This paper present affinity laws to help with predicting the performance of a centrifugal pump and to determine whether a change in the impeller design is a limiting factor. The parameters of the pumping system are considered to establish the operating points of centrifugal pumps.

2. Methodology

2.1. Conceptualization

A multiple method was used to identify potential causes of centrifugal pump failure and dropping of efficiency of the pump. A review of grey and white literature was undertaken to identify the challenges of installation, operation and maintenance of centrifugal pump. System analysis of centrifugal pumps is done through engineering design where specification for the pump is provided based on the pumping system. Affinity laws are presented to help with predicting the performance of the centrifugal pump. A representative variation of pump impeller diameters was used to prove affinity laws on the performance of centrifugal pump. Cost analysis is also taken into consideration to analyze the risk and viability if centrifugal pumps are not restored to their original efficiency [6, 7]. Simulation is a technique used to manage and improve performance of centrifugal pump. It also helps with time saving and cost for companies. Simulation process can help industries to effectively manage and maintain centrifugal pumps.

2.2. Design and Simulation of Centrifugal Pump

In the process of pumping water and other fluids, pumping capacity is required to accurately size pumping systems, determine friction head loss, construct a system curve and select a pump and motor to achieve the pumping capacity [4, 5]. The most critical area that contribute to performance of centrifugal pump is wet area of the pump which include impeller. An efficiency of centrifugal pump can be improved by trimming of impeller diameter even although reducing of impeller diameter from the original design can results in low pressure, flow and power consumption. Therefore, the trimming of the impeller should not be more than 75% of a pump original diameter [1, 3]. Different performance can be achieved and reducing impeller size allows the pump to reach specific. The affinity law can be shown in two ways, by keeping the rotational speed and diameter of impeller constant. Impeller trimming adjusts the centrifugal pump head and flows to the actual needs. Simulation is conducted varying impeller diameters to improve efficiency of centrifugal pump. Design of centrifugal pump with simulation save time and cost of companies because designs are tested on software to identify errors and failures before actual design is made.

3. Results and Discussions

3.1. Pumping System Design

To meet the requirement of pumping systems, Total Dynamic Head (TDH) is used. [8, 9]. The Total Dynamic Head (TDH) is the combined total head of the elements, pipe friction losses (H_f), static head (H_s), and velocity head (H_v). The sum of these three elements of the total head is represented in equation (1):

$$Total\ Head = H_s + H_f + H_v \qquad (1)$$

A pumping system was designed based on system specifications and assumptions that the flow rate varies from 0.05 to 0.4 m3 m^3/s. The material of the pipe is commercial steel pipe; therefore, pipe roughness is, ε = 0.045mm, see Table 1 [8, 10]. The final results are summarised in Table 2.

Table 1. Absolute Roughness of Steel Pipe Material

Pipe Material	**Absolute Roughness, ε**	
	Feet	**Microns**
Drawn brass or copper	0.000005	1.5
Commercial steel	0.000150	45
Wrought iron	0.000150	45
Cast Iron	0.000850	260

The parameters used to calculate the pumping system are as follow:
Pipe Diameter, D = 0.3 m,
Length of the pipe of the system = 45,
The static head (Hs) = 30m.
Equations (2) to (8) are used to calculate the pumping system:

$$\mathrm{H_f} = \frac{\mathrm{f\,L\,v^2}}{\mathrm{2gD}} \tag{2}$$

$$Area\ of\ the\ Pipe = \frac{\pi D^2}{4} \tag{3}$$

$$Velocity = \frac{Flow\ Rate}{Area} \tag{4}$$

$$Reynold's\ Numbe = \frac{Velocity \times Diameter}{Kinematic\ Viscosity\ of\ Water} \tag{5}$$

$$f = \frac{0.25}{\left[log\left(\frac{\varepsilon}{3.7d} + \frac{5.74}{R_e{}^{0.9}}\right)\right]^2} \tag{6}$$

$$\mathrm{H_f} = \frac{\mathrm{f\,L\,v^2}}{\mathrm{2gD}} \tag{7}$$

$$\mathrm{H_v} = \frac{\mathrm{v^2}}{\mathrm{2g}} \tag{8}$$

The calculations of total head were carried out at different flowrates varying from 0.05 m^3/s to 0.4 m^3/s with the pipe diameter of 0.3 m and length of 45m. The final results are summarised in Table 2. It can be seen that an increase in the flowrate increases the total head of pumping system 1.

Table 2. Pumping System Calculation Results of System 1 With A Pipe Length of 45m and Diameter Of 0.3m

Flow	Velocity	Reynold Number	Friction Factor	Head Sys
m3/s	m/s			m
0.3	4.24	712408.77	0.0145	32.92
0.25	3.54	593673.98	0.0148	32.05
0.2	2.83	474939.18	0.0151	31.33
0.15	2.12	356204.39	0.0156	30.77
0.1	1.41	237469.59	0.0164	30.35
0.05	0.71	118734.8	0.0182	30.1

The same procedure was followed to calculate pumping system 2 at 0.35 diameter of the pipe and length of 45m. The results for pumping system 1 and system 2 were plotted on the same graph in Figure 2 to illustrate how the system curve will change when increasing the pipe diameter from 0.3 to 0.35 m. Figure 2 shows that the pumping capacity can be achieved by adjusting the diameter of the pumping system at a constant length of the system. Increasing the diameter of the pipe in a pumping system, the head reduces thus increasing the flowrate. Due to this finding, the centrifugal pump is tested at varying impeller diameters.

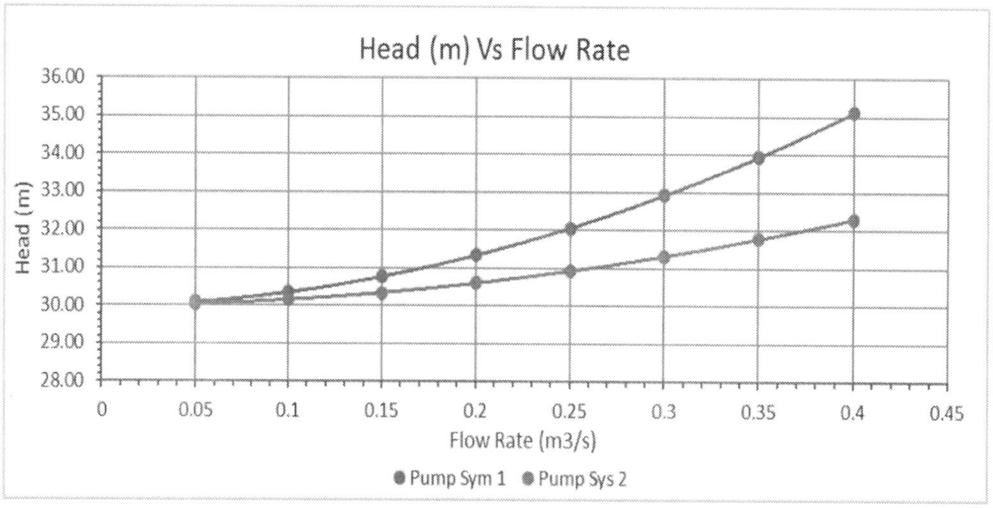

Figure 2. Pump system curve 1 and 2

3.2. Centrifugal Pump Design at Varying Impeller Diameter

The optimization of the pumping capacity of the centrifugal pump is done by trimming the impeller to adjust the head (m) and flowrate (m^3/s) [6]. Assumptions were made to determine the effect of the impeller diameter on the pumping capacity of the centrifugal pump. Impeller diameters of 350mm, 344mm and 338mm were used to do the calculations. The pump specification to suit the existing pumping system design with the impeller diameter of 350mm at the flowrate varying 0.05 m^3/s to 0.4 m^3/s are shown in Table 3.

3.2.1 Affinity Laws for centrifugal pumps

Affinity laws only apply to radial and axial pumps [3, 11]. The formulas applied for affinity laws in constant impeller diameter and constant rotational speed are represented in equations (9) and (10):

$$\frac{Q_1}{Q_2} = \frac{D_1}{D_2} \qquad \frac{H_1}{H_2} = \left(\frac{D_1}{D_2}\right)^2 \qquad \frac{P_1}{P_2} = \left(\frac{D_1}{D_2}\right)^3 \tag{9}$$

$$\frac{Q_1}{Q_2} = \frac{n_1}{n_2} \qquad \frac{H_1}{H_2} = \left(\frac{n_1}{n_2}\right)^2 \qquad \frac{P_1}{P_2} = \left(\frac{n_1}{n_2}\right)^3 \tag{10}$$

Table 3. Pump Specifications

Flow	Total Heat	Water Power
m^3/s	M	kW
0.4	31.36	123.07
0.35	31.8	109.19
0.3	32.18	94.71
0.25	32.5	79.71
0.2	32.76	64.28
0.15	32.97	48.51
0.1	33.11	32.49
0.05	33.2	16.29

The impeller diameters are varied from D_0 as the original and two other values D_1 and D_2 as variations from D_0. The specifications of the pump (Table 1) is at the original impeller diameter of 350 mm. The head and flowrate of the centrifugal pump was calculated varying discharge flowrate, see the calculation summary of the result in Table 4.

Table 4. Affinity Law Calculation Results

Total Pump Head (m)			Flow (Ml/d)		
350 mm	344 mm	338 mm	350 mm	344 mm	388 mm
32.18	31.09	30.01	0.3	0.295	0.289
32.5	31.39	30.06	0.25	0.246	0.241
32.76	31.65	30.56	0.2	0.197	0.193
32.97	31.85	30.77	0.15	0.147	0.145
33.11	31.989	30.882	0.1	0.098	0.096
33.2	32.073	30.965	0.05	0.049	0.048

Affinity laws were applied to determine the pumping capacity of centrifugal pumps. The results were summarised in a table format in Table 4. The head and flowrates were compared at the variation of the impeller diameters of 350 mm, 344 mm and 338 mm. The results obtained from the calculations in Table 4 and Figure 3 shows that the trimming of the impeller reduces the flowrate and the head of the pump. The plotted graph, Figure 3, shows that the trimming of the impeller diameter reduces the head and the flow rate of the centrifugal pump. It was observed that it is important to trim the impeller diameter with the limit of 75% of the original diameter to keep the centrifugal pump operation at the duty point. The duty point of the pump was calculated at two different pumping systems to determine the pumping capacity of the centrifugal pump. The duty point is being shown with a red dot for system 1 and yellow dot for system 2. The two pumping systems in Figure 3 give the operation duty point of the centrifugal pump as shown in Table 5.

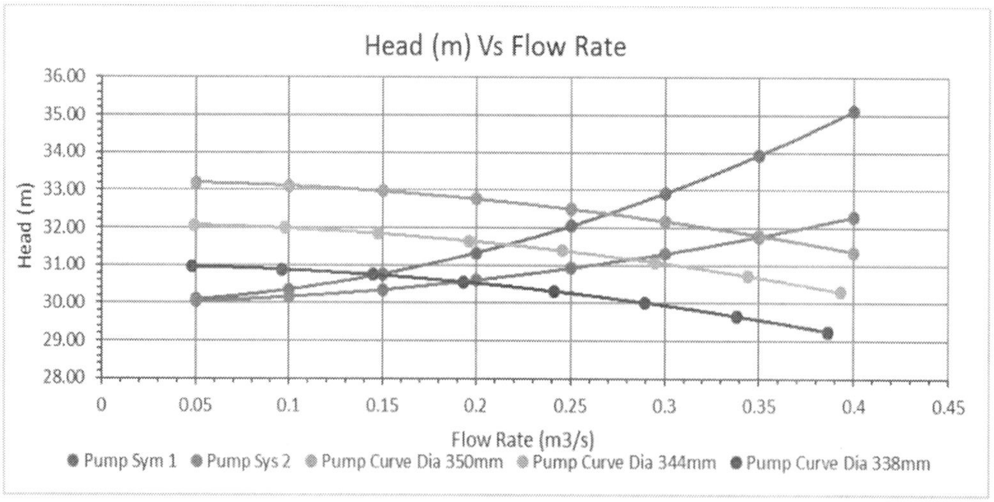

Fig. 3. Head Vs flow rate pump and pumping system curve.

Table 5: Duty Points of the Centrifugal Pump Operating at Pumping System 1 and 2

Impeller Diameter	System 1 Length 45mm		System 2 Length 40 mm	
	Flow Rate	Head	Flow Rate	Head
350	0.27	32.4	0.35	31.8
344	0.215	31.6	0.295	31.2
388	0.15	30.8	0.195	30.6

It can be seen in Table 5 that at the system 1 the flowrate is lower compared to flow rate at system 2 and the head at system 1 is higher as compared to head at system 2. It means to increase the flowrate of the centrifugal pump, reduce the length of the pipe of the pumping system and to increase the head of the pump, increase the length of the pipe of the system. The pumping capacity of the centrifugal pump can be achieved by either increasing the length of the pumping system or trimming of the impeller diameter.

3.3.Simulation of the Impeller Diameter

The diameter of the impeller has an impact on the performance of the pump [12]. The impeller trimming adjusts the centrifugal pump head and flows to the actual needs. The trimming of the impeller should not be more than 75% of a pump original diameter. The simulation of the centrifugal pump was performed with the impeller diameter varying at 350mm, 344mm and 338mm. The study parameters of the centrifugal pump are assumed to be: rotational Speed = 1800 rpm, Pressure outlet = 15000 Pa, and Velocity = 0.6 m/s. The results were obtained after the analysis has been carried out. The results were taken from counterflow through the pump in the midplane view, see Figures 4 – 6.

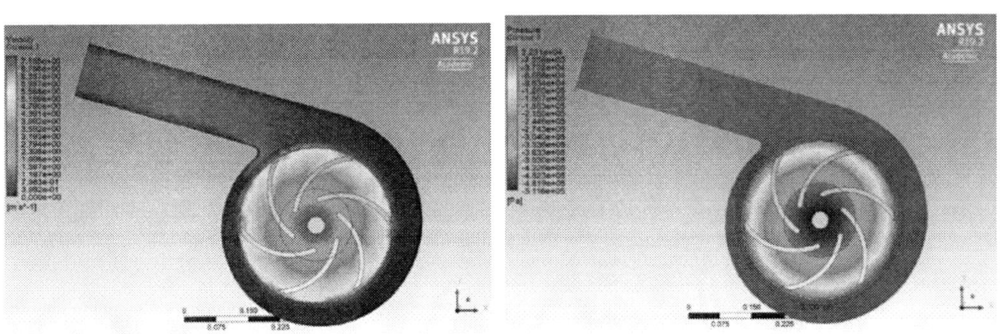

Fig. 4. Velocity and Pressure counter flow through the pump in the mid plane view for 350 mm Impeller Diameter

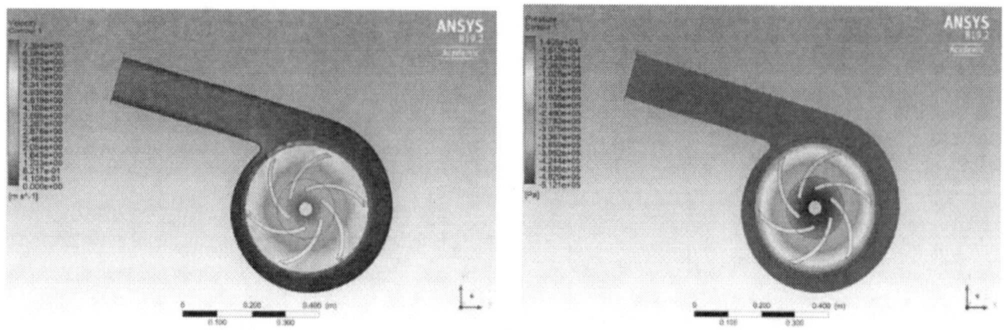

Fig. 5. Velocity and Pressure counter flow through the pump in the mid plane view for 344 mm Impeller Diameter

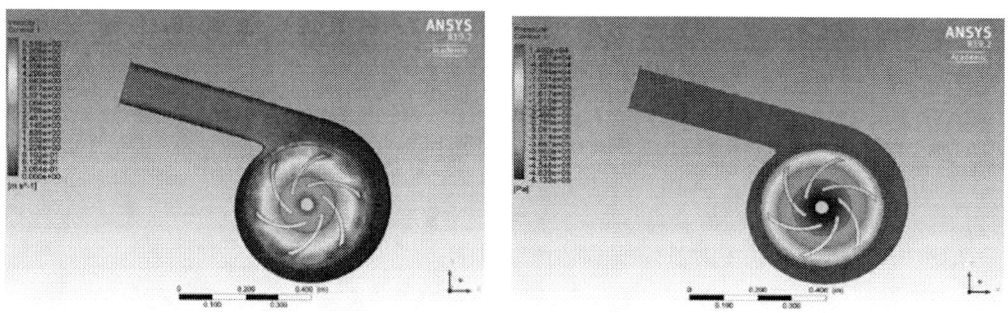

Fig. 6. Velocity and Pressure counter flow through the pump in the mid plane view for 338 mm Impeller Diameter

Table 6: CFD Results of Variation of Impeller Diameter

Impeller Dia (mm)	Maximum Pressure (kPa)	Maximum Velocity (m/s)	Flow Rate (m3/s)	Head (m)	Water Power (W)	Eff (%)
350	20.59	8.79	0.20	2.09	4156.8	91.59
344	14.08	7.39	0.17	1.44	2389.6	52.66
338	14.02	5.52	0.13	1.43	1777.3	39.17

The variation of impeller diameter influences the performance of the centrifugal pump as is indicated in Table 6. The pressure and velocity of the centrifugal pump were found to be increasing with increase of the impeller diameter, see Figure 7. It can be observed that the pressure rises more than the velocity. More pressure is achieved at the bigger impeller diameter of 350mm because of decrease in space inside the pump casing. It appears that the assumption made for the impeller diameter of the model causes the pump to operate efficiently.

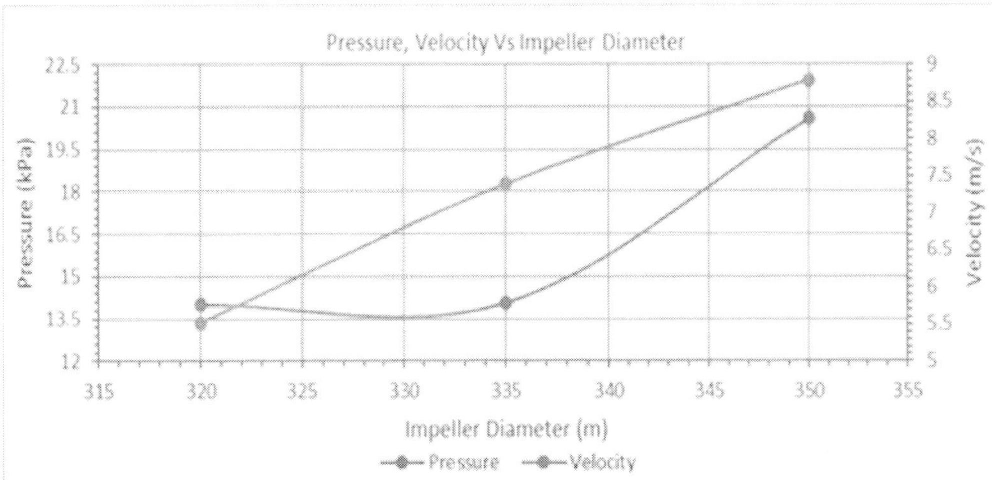

Figure 7. Pressure, Velocity Vs Variation of Impeller Diameter Simulation.

The efficiency of the centrifugal pump in Figure 8, increases with increase of impeller diameter. According to the selection made for the parameters of the model, the pump is found to be more efficient with reducing of the impeller diameter. The Best Efficiency Point (BEP) of the pump is at the maximum flowrate and efficiency of 0.203m3/s and 91.59% respectively. Affinity laws of pumps states that the pump can only be trimmed up to 75% of its original diameter, which means according to these results there is no further trimming that can be done because it will affect the performance of the pump.

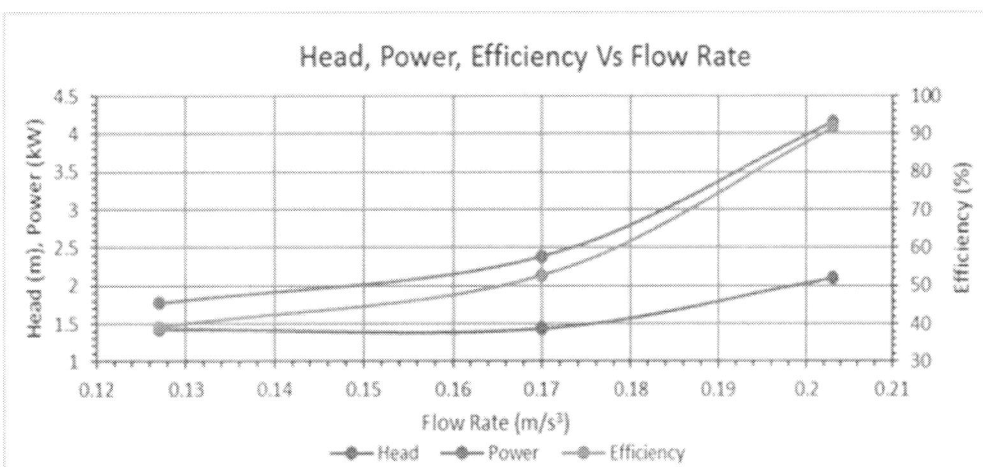

Figure 8. Head, Power, Efficiency Vs Flow Rate for Variation of Impeller Simulation

The simulation results show that the efficiency of the centrifugal pump in Figure 8 increases with increase of impeller diameter. According to the selection made for the parameters of the model, the pump is found to be more efficient with reduction in impeller diameter. The Best Efficiency Point (BEP) of the pump is at the maximum flowrate and efficiency of 0.203m3/s and 91.59% respectively, see Figure 8.

4. Conclusion

The paper presents two ways to improve efficiency of centrifugal pumps, which are systems analysis, trimming of impeller design using affinity laws and simulation. As previously discussed, reduction of the impeller diameter gives a new operation point of the centrifugal pump, therefore, it is imperative to consult with the manufacturers before trimming the impeller because the reduction of impeller affects the flowrate and head. From literature review it was found that the impeller can be reduced up to 75% not above so that the centrifugal pump can continuously operate efficiently. Also reducing the impeller changes the length and overlap of vanes, the width at the impeller exit and often the discharge angle as well. The total pumping system head increases when the flowrate increases and the total head of the pump decreases with the increase of the pump flowrate. To increase the pumping capacity, ensure that the flowrate is kept lower as possible to achieve a higher pressure and head at the centrifugal pump discharge. According to the calculations results illustrated in Figure 3, increasing the length of the pumping system at the constant diameter of the impeller, will reduce the flowrate while increasing pressure and head of the pump. Trimming of impeller diameter improves performance of centrifugal pump. The energy charge of the centrifugal pump is estimated to be R/kWh = R0.82. Monthly saving if the centrifugal pump is restored to original efficiency is R16523.83 therefore the saving cost if the pump is restored is R1,080,820.99.

5. References

[1] M.E. Matlakala, D.V.V. Kallon, S.P. Simelane, P.M. Mashinini. Impact of Design Parameters on the Performance of Centrifugal Pumps. Procedia Manufacturing. Volume 35, 2019. Pp 197 – 206.

[2] M. E. Matlakala, D. V. V. Kallon, "Influence of impeller Blade Count on the Performance of Centrifugal Pump," SAIIE NeXXt (2019)

[3] M. E. Matlakala, Kallon D.V.V, Mogapi K.F, Mabelane I.M, Makgopa D.M. Influence of Impeller Diameter on the Performance of Centrifugal Pump. IOP Conference Series: Materials Science and Engineering. Volume 655 (1). Pp.009-012 (2019)

[4] Hydraulic Institude and Europump, "Pump Life Cycle Costs," *A guide to LCC Analysis for Pumping System,* p. 1, January (2001)

[5] M. E. Matlakala, Kallon D.V.V, Nkoana K.F, Mafu B.D, Mkhwanazi S.B. Effect of Suction Diameter Variations on Performance of Centrifugal Pump. Open Innovation (IO). Pp 170-173 (2019)

[6] M. E. Matlakala Kallon D.V.V. Effect of Discharge Diameter on Centrifugal Pump Performance. SAIIE NeXXXt (2019)

[7] M. E. Matlakala. A Computational Model for the Efficiency of Centrifugal Pumps. Dissertation Submitted to the University of Johannesburg, (2020)

[8] L. Bachus and A. Custodio, Know and Understand Centrifugal Pumps, First Edition ed., AE Amsterdam: Elservier, (2003)

[9] W. Randall and P. Whitesides, "Basic Pump Parameters and the affinity Laws," *PDHonline Course M125 (3 PDH),* p. 4, (2008)

[10] A. Yunus and M. Cengel, Fluid Mechanics Fundamentals and Applications, 738 ed., New York: McGraw Hill, (2006)

[11] M. H. Savar, "Improving the centrifugal pump by trimming impeller," vol. 249, pp. 654-659, (2009)

[12] D. Ajith, "Design and Analysis of Centrifugal Pump Impeller Using Ansys Fluent," *Engineering and Technology Research, (*2017)

An extension of the multiple marker algorithm for study of phase separation problems at the mesoscale

Quinn G. Reynolds[1,2,*], *Oliver F.* Oxtoby[3], *Markus W.* Erwee[1], and *Pieter J.A.* Bezuidenhout[1]

[1]Mintek, 200 Malibongwe Drive, Randburg, South Africa
[2]University of Stellenbosch, Matieland, Stellenbosch, South Africa
[3]ENGYS, John Archer Way, London, United Kingdom

Abstract. Multiphase fluid flow is an active field of research in numerous branches of science and technology. An interesting subset of multiphase flow problems involves the dispersion of one phase into another in the form of many small bubbles or droplets, and their subsequent separation back into bulk phases after this has occurred. Phase dispersion may be a desirable effect, for example in the production of emulsions of otherwise immiscible liquids or to increase interfacial surface area for chemical reactions, or an undesirable one, for example in the intermixing of waste and product phases during processing or the generation of foams preventing gas-liquid decoupling. The present paper describes a computational fluid dynamics method based on the multiple marker front-capturing algorithm – itself an extension of the volume-of-fluids method for multiphase flow – which is capable of scaling to mesoscale systems involving thousands of droplets or bubbles. The method includes sub-grid models for solution of the Reynolds equation to account for thin film dynamics and rupture. The method is demonstrated with an implementation in the OpenFOAM® computational mechanics framework. Comparisons against empirical data are presented, together with a performance benchmarking study and example applications.

1 Introduction

The problem of separating two immiscible fluids after they have become mixed arises in many branches of process engineering. In particular, phenomena related to contact between two or more immiscible fluids are critical for industries such as petrochemistry, extractive metallurgy, chemical engineering, and many others.

In the oil and gas industry, mixing of immiscible liquid phases is generally undesirable and results in intermixing or contamination of product streams. The contaminant is usually water dispersed in a hydrocarbon phase, and specialised settling equipment is required to separate the two phases to an acceptable degree; this must occur by recombination of droplets of the contaminant phase with each other and the bulk phase by settling and coalescence processes. Improving the performance of such separation equipment may have significant process cost and efficiency benefits, and perhaps unsurprisingly a great deal of both fundamental and experimental research has been conducted in this area (for example [1] and

*e-mail: quinnr@mintek.co.za

contemporary papers). In the field of pyrometallurgical engineering, phase mixing and separation are similarly associated with undesirable effects - for example slag foaming, in which gas bubbles are generated by or injected into the process and build up at the upper surface of the viscous slag phase, or slag-metal entrainment, in which small droplets of one phase become separated from the phase interface due to mechanical shear or other disturbances [2–4]. The latter problem is of particular concern in the operation of smelting furnaces, as entrained product particles discarded in a waste phase may represent a significant economic loss or environmental hazard. Experimental studies have demonstrated that the recombination of molten metal droplets dispersed in a molten slag phase may take a considerable amount of time, even if the droplets are of the order of several millimetres in size [5].

Existing research into the computational fluid dynamics (CFD) modelling of phase separation may be broadly grouped into one of two approaches. *Unresolved* methods seek to represent droplets of one phase entrained in another using a continuum approximation and model only the bulk motion and large-scale flow structures of the fluids involved. Models of this type are exemplified by the Euler-Euler method for multiphase flow, in which a separate set of Navier-Stokes equations is solved numerically for each phase present and interactions between the phases (such as intermixing, drag, and droplet break-up and coalescensce) must be accounted for using empirical closures such as population balance models. Such methods have great utility for engineering calculations as they are robust and relatively low in computational complexity, but a drawback is that the closures used are generally problem-specific and must be determined for each case of interest. Examples of work using unresolved methods for the study of unit operations in extractive metallurgy can be found in [6] and [7].

By contrast, *resolved* methods attempt to model the individual entrained droplets and some or all of their interactions with the surrounding phase directly. Within resolved methods, Euler-Lagrange methods treat droplets as discrete particles with trajectories computed from force balances, and pure Eulerian methods model the fluid inside entrained droplets using the same governing equations as the surrounding phase while completely resolving all phase interfaces. The former includes discrete phase models and CFD-coupled discrete element methods and has been widely applied to engineering problems such as the study of bubble plumes in reaction vessels [8], while the latter includes level set and multiple-marker volume of fluids (VOF) methods and is used primarily in the detailed study of individual droplet collisions [9–14]. Euler-Lagrange methods require empirical closures between the particle and fluid models similar to those used in unresolved methods, and as a result are subject to many of the same limitations. Pure Eulerian methods are arguably the most accurate, but are prohibitively computationally expensive due to their need to resolve detail across a large range of length scales (typically from nanometres in the thin films between coalescing droplets up to centimetres or metres for the vessel containing the phases).

Although they are not suitable for industrial applications, pure Eulerian methods may be used to perform numerical experiments and determine parameters required for the empirical closures in more scalable methods. Since experimental data on materials for phase separation problems is occasionally unreliable or unavailable (for example in the molten state at high temperatures for pyrometallurgical applications), such bootstrapping of empirically-based simulations using results obtained from more fundamental numerical models may be the only option available to researchers; further development of pure Eulerian methods for phase separation problems is therefore of practical as well as academic value.

In order to address some of the shortcomings of existing pure Eulerian methods, we present an extension to the multiple-marker method which improves computational performance in problems with large numbers of dispersed phase particles present.

2 Algorithm description

The modelling of droplets or bubbles during close interactions is a challenge for standard VOF methods. Once two particles of a dispersed phase approach each other at distances on the order of the local mesh resolution, the particles will immediately merge together. This is physically unrealistic; in reality a thin film of fluid of the surrounding continuum fluid will form between the particles, and the evolution of this thin film delays the coalescence until it is able to rupture. In order to overcome this the multiple marker method [10, 12] was developed, in which each droplet or bubble of the dispersed phase is assigned its own unique phase in the VOF algorithm. This method prevents premature coalescence of two colliding particles while maintaining an accurate description of the fluid flow physics throughout the model domain. The multiple marker method is typically combined with semi-analytical subgrid models governing the film behaviour [14], and information from the subgrid models is then used to determine the point of film rupture and particle coalescence.

The main limitation of the multiple marker algorithm is its lack of scalability. It is effective for the study of interactions and collisions between pairs or very small groups of dispersed phase particles, but once the number of particles (and hence marker phases) in the simulation increases, the underlying VOF algorithm on which it is based becomes prohibitively expensive.

In the new approach presented here, which we refer to as the *dynamic multiple marker* (DMM) method, this is addressed by dynamically reassigning particles of the dispersed phase to a small number of marker phases such that no adjacent particles use the same phase. This is achieved at each time step in a transient VOF calculation by identifying collections of mesh cells belonging to each discrete particle of the dispersed phase, tagging pairs of particles which are in close proximity to one other, and using this information to generate a graph representing the pairwise contacts for the set of all particles in the simulation. The nodes of this graph may then be associated with a set of marker phases (much smaller than the number of particles) using an appropriate graph colouring algorithm. Finally, all mesh cells associated with the dispersed phase are reassigned to their new marker phase before resuming the VOF calculation. This is combined with a refined implementation of a thin film subgrid model describing film drainage which enables prediction of film rupture between any pair of particles in the simulation. When rupture is detected, particle coalescence is effected by immediately assigning the pair of particles involved to the same marker phase and allowing the VOF algorithm to merge them together naturally.

An overview of the main steps in the DMM algorithm and film sub-model are shown in Figure 1. In general these steps are executed after the pressure, velocity, and phase field calculations for each time step in a transient VOF simulation are completed. The major algorithm components are described in more detail in the following sections.

2.1 Contact detection and marker phase identification

The contact detection and marker phase assignment part of the DMM algorithm (blue block in Figure 1) begins with a threshold calculation on the marker phase fraction fields. All cells in the mesh are looped over, with each being flagged as either belonging (true) or not belonging (false) to the dispersed phase according to a threshold limit specified by the user; this is typically 0.5.

A second loop is then performed in which all *unique, contiguous* sets of cells in the dispersed phase are found - in the present implementation this is performed by iterative sweeping of thresholded cells which are directly adjacent to each other, but any appropriate

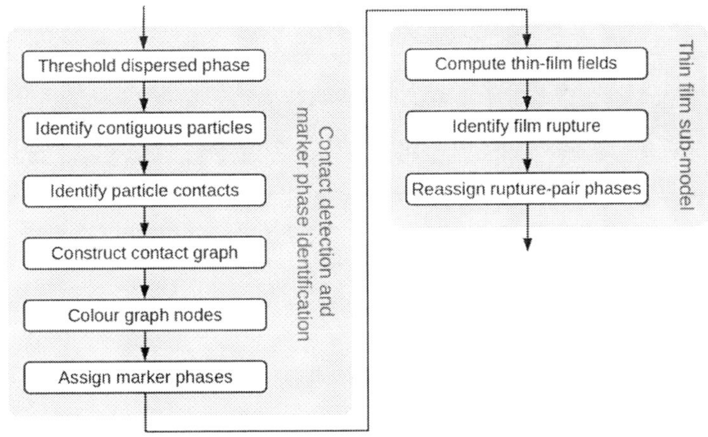

Figure 1: Flowsheet for the DMM algorithm

connectivity-detection algorithm could be used. During this procedure each discrete particle of the dispersed phase is assigned a unique integer identifier, stored at mesh cell level.

Once the particles are identified, contact (or near contact) between pairs of particles is determined by expanding each particle's set of cells in a "halo" around the actual particle using a sweep algorithm. The expanded particles' locations are stored at mesh cell level. A loop over all cells in the mesh is then conducted, and if a particular cell has been flagged as belonging to more than one expanded particle (i.e. the expanded particles overlap in space), contact between the two particles is recorded as the integer pair of their identifiers. The size of the contact detection halo is specified by the user, but is typically on the order of five to ten mesh cells.

Using the collection of particle contact pairs, a contact graph is then constructed. In this graph nodes represent particles of the dispersed phase, and edges represent contact between them. The nodes of the graph are then assigned integer tags in such a way that no pair of connected nodes has the same tag, while attempting to minimise the total number of tags in use – this is the classical graph colouring problem, for which a number of effective heuristic algorithms exist. In the present implementation DSatur [15] is used and provides a balance of simplicity, performance, and near-optimality.

Once the contact graph is coloured, the tag of each node in the graph is used to assign the associated particle of the dispersed phase to a parent marker phase. The marker phases are constructed at mesh cell level using the particles' (or expanded particles') cell sets. Care must be taken in this step to preserve mass continuity and phase fraction summation rules, since certain particles and regions may have been reassigned from one marker phase field to another.

2.2 Thin film sub-model

When particles of the dispersed phase collide, a thin film of the continuum phase is squeezed between them and begins to drain against viscous resistance. At a critical thickness, intermolecular interactions across the film come into play and it becomes unstable and ruptures [16], leading to rapid coalescence of the particles. This critical thickness is typically on the order of 10^{-8} m, and therefore cannot be practically captured within the VOF model.

The Reynolds equation is employed as a sub-grid model governing the film thickness in the present work, and is familiar from lubrication theory [17]. It is a non-linear partial differential equation derived on the assumption that the separation of the particles is much smaller than their radii. A coordinate transform and a volume-based reformulation permit solution on the same mesh as the VOF model, via an iterative coupled approach in which pressure is fed in to the Reynolds equation from the VOF solution. The Reynolds equation is then used to march the film thickness forward in time and additionally to determine a tangential lubrication force which is fed back to the VOF simulation. Interested readers are referred to Musehane et al [14] which contains a full description of this approach.

It should be noted that in many-particle simulations, the volume-based approach for solution of the Reynolds equation has the added benefit of allowing all film equations to be solved simulataneously on the same mesh, and incurs no greater performance penalty than the solution of other CFD variables.

2.3 Implementation details

A technology demonstrator for the DMM algorithm was developed as a numerical solver in the OpenFOAM® [18] open source computational mechanics framework.

A standard multiphase VOF solver available in the OpenFOAM® framework, `multiphaseInterFoam`, was used as a basis for the implementation. `multiphaseInterFoam` uses a mixed PISO-SIMPLE segregated algorithm for the solution of pressure and velocity fields in the fluids combined with stabilised interface compression schemes for the phase fraction fields. The interfacial tension force calculation was extended by introducing smoothing to assist with control of parasitic currents [19], and modified to correctly model a dispersed phase in a continuum.

Droplet identification, contact detection, and marker phase assignment were implemented as separate functions in a new `multiMarkerMixture` code module, extended from the original multiphase solver module. Temporary fields for particle identifiers and phase assignment tags were implemented as boolean or integer lists rather than mesh field objects to reduce storage requirements and improve performance. The particle contact graph was generated in a compressed node-based format from the list of detected contact pairs. The graph colouring algorithm was implemented as a separate solver module, `graphColouring`, to facilitate later modularisation and automatic switching. The new functions in `multiMarkerMixture` were explicitly parallelised using OpenFOAM®'s `Pstream` class objects.

To process the sub-grid thin film model, the Reynolds equation is discretised using standard OpenFOAM finite-volume operators and first order backward-difference time integration. Ancillary extrapolation equations for the film thickness and velocity profile are solved for each pair of indicator phases, and the results assembled onto a single domain on which the Reynolds equation is solved in different areas according to an activation function.

Some care is required to deal with numerical noise in the pressure gradient coming from the VOF solution, caused by both the density gradient between the continuum and dispersed phases, and the pressure jump induced by interfacial tension. While both of these contributions to the tangential pressure gradient used in the Reynolds equation are formally zero, numerically their contribution is not perfectly cancelled. To ameliorate this, the surface tension force term in the momentum equation is subtracted from the pressure gradient, and the result is smoothed in a similar manner to the interfacial tension. This also helps to reduce the noisy effects of parasitic velocities at the surface of the particles.

3 Results and discussion

3.1 Model testing

In this section, the results of some preliminary testing of the DMM method's performance and accuracy are presented.

3.1.1 Performance

In order to evaluate the performance of the DMM method against traditional multiphase approaches for resolved dispersed-phase problems, simulations were performed using a planar two-dimensional metallurgical system in which droplets of dense molten metal settle through a lighter molten slag continuum phase. The parameters of the test case are given in Table 1. A range of initial droplet counts between 1 and 30 were considered – in each case a simulation was executed with both the standard multiple-marker method, in which a unique marker phase must be assigned to each droplet *a priori*, and the new dynamic method. The standard multiple-marker cases were run using OpenFOAM®'s reference `multiphaseInterFoam` solver. In the DMM solver, the thin film sub-model was active but coalescence was disabled to maintain a constant droplet count. Tests were performed on an Ubuntu Linux 18.04 LTS platform equipped with an AMD Ryzen 7 1700X eight-core processor and 16GB RAM. All cases were parallelised on four CPU cores.

Table 1: Model conditions for performance testing case

Parameter	Value	Parameter	Value
Domain width	0.02 m	Domain height	0.025 m
Droplet diameter	0.0025 m	Mesh element size	1.5×10^{-4} m
Simulation time	2 s	Interfacial tension	0.5 N/m
Dispersed phase density	7000 kg/m^3	Dispersed phase viscosity	0.005 Pa.s
Continuum phase density	3500 kg/m^3	Continuum phase viscosity	0.1 Pa.s

The performance of the two methods in terms of execution time required per unit of simulated time (and the resulting performance improvement factor) is shown in Figure 2. It can be seen that the DMM method confers significant advantages when more than three or four particles of the dispersed phase are present. Moreover the scaling of the DMM method at larger droplet counts is essentially flat, implying that much larger numbers of droplets may be simulated while retaining practical run times.

3.1.2 Droplet collisions

To test the thin film sub-model and dynamic coalescence between particles of the marker phase, collisions between droplets of tetradecane in a nitrogen atmosphere were simulated using the DMM solver. This problem has been used as a validation test case in a number of previous publications. Here, we use the setup and initial conditions of Musehane et al [14] in which the droplets are modelled on an axisymmetric two-dimensional mesh. Parameters for the model and cases are given in Table 2. Domain mesh resolution in all cases was 360 x 80 cells, resulting in a mesh element size of approximately 3×10^{-6} m.

In experimental tests [9] coalescence after impact between droplets was observed in cases 1 and 4, but not in cases 2 and 3. Following the approach of Musehane et al, a critical film thickness of 6.8×10^{-7} m was estimated in order to reproduce this behaviour in the DMM

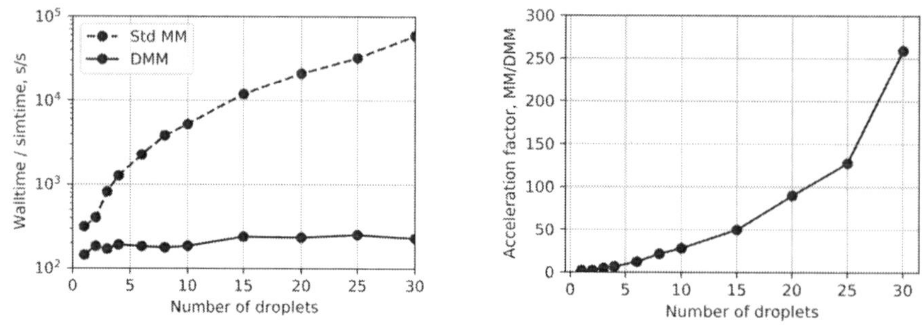

(a) Execution time

(b) Relative improvement

Figure 2: Performance of the DMM algorithm vs standard multiple marker method

Table 2: Model conditions for droplet collision case

Parameter	Value	Parameter	Value
Dispersed phase density	762 kg/m^3	Dispersed phase viscosity	0.00213 Pa.s
Continuum phase density	1.225 kg/m^3	Continuum phase viscosity	1.76×10^{-4} Pa.s
Interfacial tension	0.0265 N/m		
Case 1 closing velocity	0.302 m/s	Case 1 droplet diameter	1.072×10^{-4} m
Case 2 closing velocity	0.240 m/s	Case 2 droplet diameter	1.706×10^{-4} m
Case 3 closing velocity	0.496 m/s	Case 3 droplet diameter	1.676×10^{-4} m
Case 4 closing velocity	0.596 m/s	Case 4 droplet diameter	1.697×10^{-4} m

solver. The evolution of the film thickness between the droplets in each simulation as captured by the film sub-model is shown in Figure 3.

Selected visualisations of the droplet impacts from case 3 and 4 are shown in Figures 4 and 5 respectively. Film thickness and droplet shape behaviour are in reasonable qualitative agreement with the data and simulation results reported by [9] and [14], however more rigorous validation is advised.

3.1.3 Hindered settling

In the final test case the veracity of the DMM method for predicting particle motion in groups of droplets is examined. A planar two-dimensional region containing silicone oil is simulated, and initialised with a number of static water droplets of a fixed diameter. As the simulation proceeds, the droplets settle under gravity toward the bottom of the region. Interactions between nearby droplets and counterflow of the continuum phase progressively reduces the terminal settling velocity as the number of droplets in the region increases – this is referred to as *hindered settling*. The Richardson-Zaki equation (1) is commonly used to characterise this problem, where u_t is the terminal velocity of a single free droplet in the continuum fluid, u_h is the hindered settling velocity, ϕ is the volume fraction of droplets present, and n is an empirical exponent which is related to the Reynolds number of the flow. n typically takes on values between 2.4 and 4.65 [20].

$$\frac{u_h}{u_t} = (1 - \phi)^n \tag{1}$$

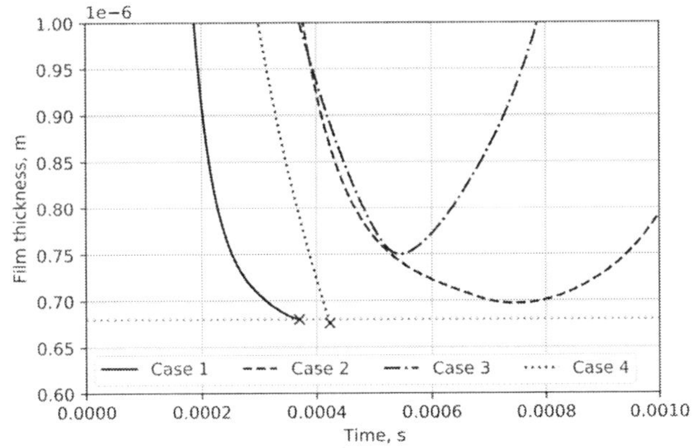

Figure 3: Film thickness as a function of time for droplet collision test cases (x marks point of film rupture and droplet coalescence)

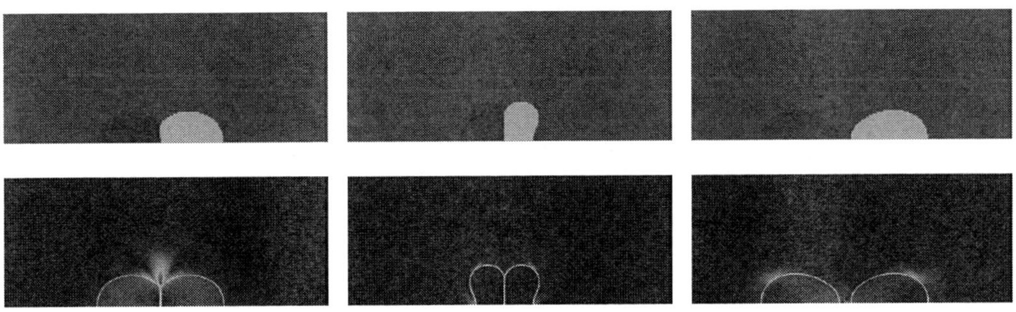

(a) 200 μs (b) 550 μs (c) 1150 μs

Figure 4: Tetradecane droplet collision and rebound case 3, phase fields at top, velocity fields (scale blue 0 m/s to red 2 m/s) at bottom

The parameters for the hindered settling test case are given in Table 3. Droplet volume fractions between 0.075 and 0.35 were simulated using the DMM solver. Periodic boundary conditions were applied on the left and right walls of the region. The thin film sub-model was deactivated since droplet motion, not coalescence, is of interest here. Runs were performed on the South African Centre for High Performance Computing's Lengau HPC cluster. All cases were parallelised on three nodes (72 CPU cores) and took between four and six hours to complete.

An example of the results obtained from the hindered settling simulations is shown in Figure 6, in which the evolution of the velocity field and droplet positions in time can be seen. The structure of the settling band becomes complex over time, with evidence of droplet clumping and bed channelling.

Upon completion each case was post-processed to obtain an average settling velocity of the droplets at each time step. The results were filtered by time (>0.5 s, to allow initial conditions to decay) and vertical position of the droplets (>0.1 m, to prevent settled droplets from skewing the average). The overall average settling velocity, as well as the minimum and

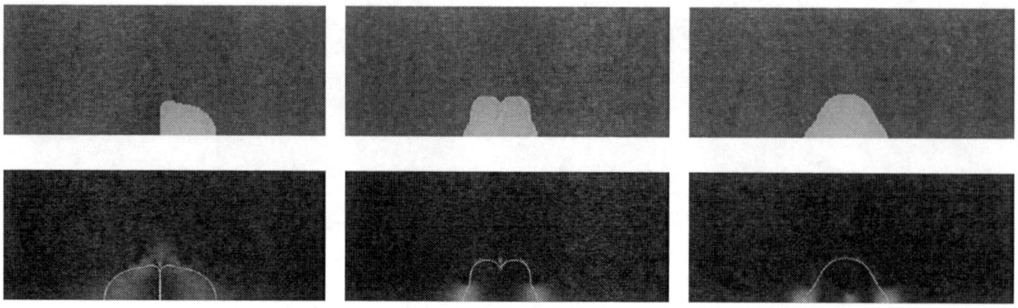

(a) 250 μs (b) 430 μs (c) 610 μs

Figure 5: Tetradecane droplet collision and coalescence case 4, phase fields at top, velocity fields (scale blue 0 m/s to red 2 m/s) at bottom

Table 3: Model conditions for hindered settling case

Parameter	Value	Parameter	Value
Domain width	0.05 m	Domain height	0.2 m
Droplet diameter	0.0025 m	Mesh element size	1×10^{-4} m
Simulation time	8 s	Interfacial tension	0.035 N/m
Dispersed phase density	1000 kg/m^3	Dispersed phase viscosity	0.001 Pa.s
Continuum phase density	800 kg/m^3	Continuum phase viscosity	0.1 Pa.s

maximum values recorded over all time steps, was then calculated. The results are given in Figure 7 and show reasonable agreement with (1).

3.2 Application example

As a demonstration of the full capabilities of the DMM method, an example application in metallurgical phase separation was selected in which a number of small droplets of molten metal must settle through a viscous molten slag layer under gravity, and coalesce into a bulk metal phase. This is a prototypical problem in pyrometallurgical engineering, where undesirable emulsions between liquid phases can easily form during handling and processing of furnace products in the molten state. Of particular interest is the effect of thermophysical properties of the materials on the phase separation behaviour; since these are strongly dependent on chemical composition of the phases as well as their temperature, they are able to vary across wide ranges in pyrometallurgical processes. A further complication is that the properties of the phases affect both the settling rate of the droplets [3] as well as film rupture and droplet coalescence [21]. Meso-scale methods which are able to account for both effects in a single framework are therefore particularly useful in this case.

The parameters chosen for the application study are shown in Table 4, and represent a generic slag-metal separation problem. The film sub-model was active, with a critical film thickness of 1×10^{-5} m.

Periodic boundary conditions were imposed on the left and right sides of the domain, with no-slip walls applied at the upper and lower surfaces. The model region was initialised with 1000 randomly-sized and positioned droplets, and a small layer of bulk metal phase was

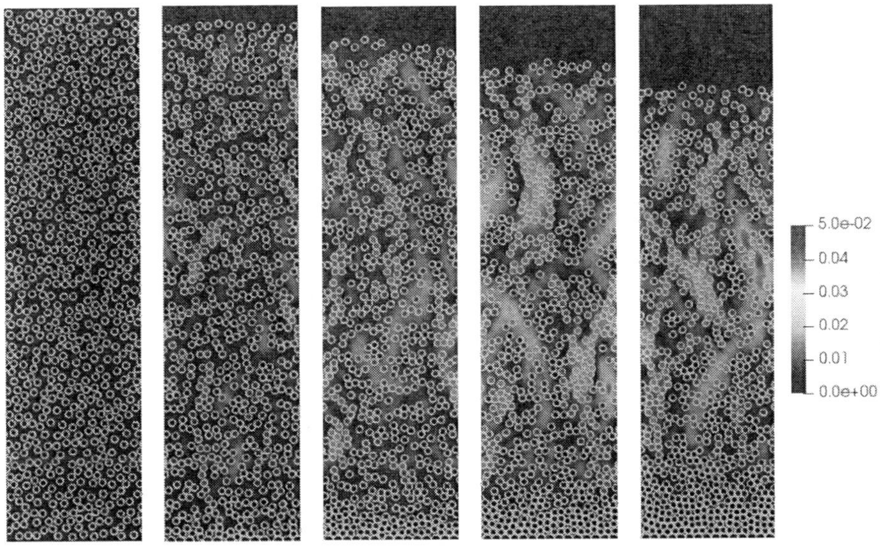

(a) 0 s (b) 2 s (c) 4 s (d) 6 s (e) 8 s

Figure 6: Velocity field for settling of water droplets (outlines) through silicone oil at ϕ = 0.35

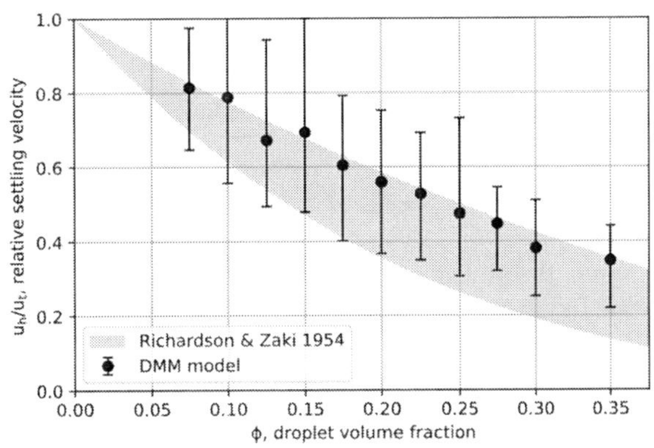

Figure 7: Comparison of empirical and DMM predictions of hindered settling

included at the bottom of the domain. As before calculation runs were performed on the South African Centre for High Performance Computing's Lengau HPC cluster, using three nodes (72 CPU cores) for each case and taking between 18 and 24 hours to complete.

Selected visualisations of the velocity field and phase boundaries from the base case simulation are shown in Figure 8. As time progresses it can be seen that droplets initially separate into a dispersed settling band and a densely-packed layer which accumulates above the metal pool. Within the dense layer adjacent droplets begin to coalesce as they move downward, increasing the average droplet size. The largest droplets at the bottom of the dense layer come into contact with the pool and coalesce with it to join the bulk metal phase. The bulk phase

Table 4: Base case model conditions for phase separation study

Parameter	Value	Parameter	Value
Domain width	0.075 m	Domain height	0.15 m
Droplet diameters	0.002 - 0.004 m	Mesh element size	1×10^{-4} m
Simulation time	4 s	Interfacial tension	0.5 N/m
Dispersed phase density	7000 kg/m^3	Dispersed phase viscosity	0.005 Pa.s
Continuum phase density	3500 kg/m^3	Continuum phase viscosity	0.1 Pa.s

grows steadily at the expense of the settling and dense layers, leading ultimately to separation of the two liquid phases.

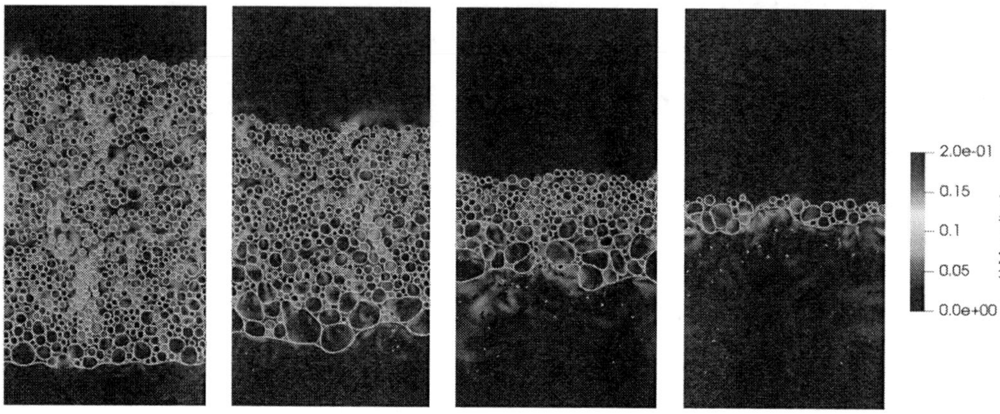

(a) 1 s (b) 2 s (c) 3 s (d) 4 s

Figure 8: Velocity field for settling and coalescence of metal droplets (outlines) through slag

The effect of changing the physical properties of the system was then examined by adjusting their values relative to those given in Table 4 and re-running the model. The total number of droplets remaining in the simulation was recorded at every time step as an indication of how the phase separation is impacted. The results are shown in Figure 9 for slag viscosity and interfacial tension between the phases. It can be seen that in this case the slag viscosity has a moderate effect and influence both settling and film drainage, with higher values causing slower separation. The interfacial tension (which affects only the film drainage) has a marked effect, with lower values significantly increasing the phase separation time; this is to be expected in metallurgical systems which typically have very high interfacial tensions.

4 Conclusions

The dynamic multiple marker method for resolved modelling of multiphase fluid flow problems with dispersed phases was successfully developed and demonstrated. The method extends existing multiple marker algorithms to dynamic situations involving dispersed-phase particle coalescence and breakup, and improves performance to the point where it is practical to consider systems up to thousands of particles. Sub-grid modelling of the behaviour of thin films between dispersed phase particles was implemented via numerical solution of the

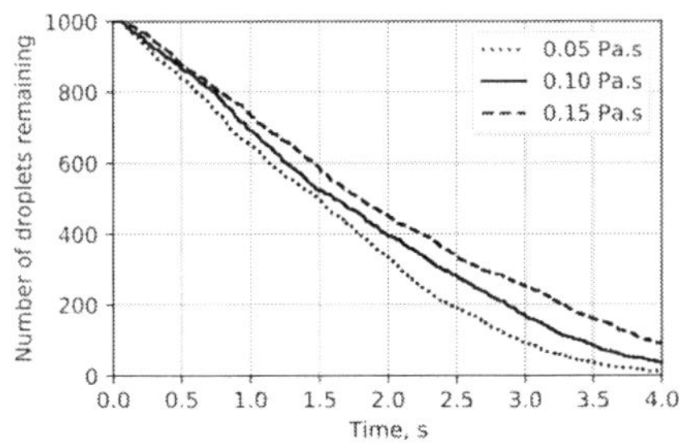

(a) Slag viscosity

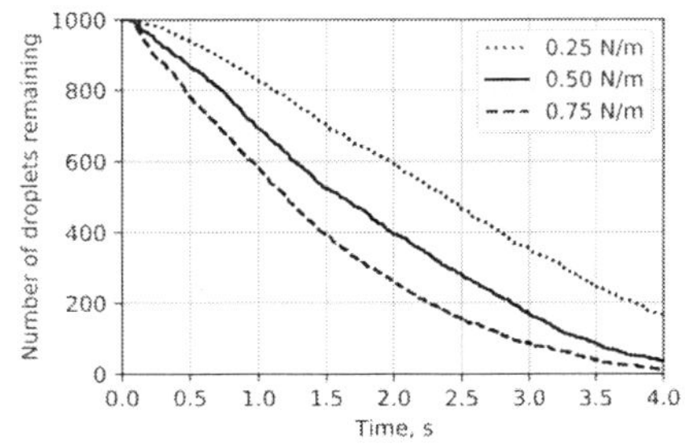

(b) Interfacial tension

Figure 9: Effect of material properties on phase separation dynamics

Reynolds equation, enabling physically realistic control of particle coalescence. Test cases in hindered settling and droplet collisions were seen to be in good agreement with previous work.

It is anticipated that such methods will find some utility in exploring the behaviour of liquid-liquid emulsion and gas-liquid mixing problems which arise in numerous branches of process engineering. They are especially applicable to problems for which experimental data is difficult or impossible to obtain, and may provide a supplemental or alternative route to developing empirical closures for unresolved models. The exploration of a pyrometallurgical slag-metal system demonstrated the new method's capabilities in studying scenarios involving complex phase separation and decoupling.

Naturally much work still remains to be done. The current implementation of the DMM method is limited to two-phase systems with one dispersed phase and one continuum phase, and although the method can in principle be extended to an arbitrary number of each, this will add logical complexity and impact performance. A careful analysis and optimisation of the contact detection and graph colouring algorithms used for marker phase selection is advisable; for heavily parallelised systems with many thousands of dispersed-phase particles, these components start to become significant performance barriers. The accuracy of the thin film sub-model as well as VOF methods in general when simulating materials with extreme physical properties (very high viscosities and surface tensions) also remains an open question, and it is likely that further algorithm development and refinement will be required here together with reliable validation data for problems and regimes of interest. Finally, extension of the thin film sub-models to include additional physics such as local electrostatic effects, non-Newtonian behaviour, and other phenomena is possible in principle and is currently being explored.

Acknowledgements

This paper is published by permission of Mintek. The authors acknowledge the Centre for High Performance Computing (CHPC), South Africa, for providing computational resources to this research project. The authors are also grateful to members of SINTEF's Flow Technology research group in Trondheim, Norway for informal discussions regarding phase separation applications in petrochemical and other industries.

References

[1] E. Barnea, J. Misrahi, Transactions of the Institution of Chemical Engineers **53**, 61 (1975)

[2] R. Jiang, R.J. Fruehan, Metallurgical Transactions B **22**, 481 (1991)

[3] P.K. Iwamasa, R.J. Fruehan, ISIJ International **36**, 1319 (1996)

[4] P. Dayal, K. Beskow, J. Björkvall, D. Sichen, Ironmaking & Steelmaking **33**, 454 (2006)

[5] P.A. Eidem, I. Solheim, E. Ringdalen, K. Tang, B. Ravary, *Laboratory study of slag metal separation for HC FeMn*, in *Proceeding of the Fourteenth International Ferroalloys Congress*, INFACON (Public Organization INFACON XIV, 2015), pp. 190–201

[6] N. Huda, J. Naser, G. Brooks, M. Reuter, R. Matusewicz, Metallurgical and Materials Transactions B **41**, 35 (2010)

[7] M. Sattar, J. Naser, G. Brooks, Procedia Engineering **56**, 421 (2013)

[8] S. Cloete, J. Eksteen, S. Bradshaw, Progress in Computational Fluid Dynamics **9**, 345 (2009)

[9] Y. Pan, K. Suga, Physics of Fluids **17**, 82 (2005)

[10] S. Tanguy, A. Berlemont, International Journal of Multiphase Flow **31**, 1015 (2005)

[11] C. Gotaas, P. Havelka, H.A. Jakobsen, H.F. Svendsen, M. Hase, N. Roth, B. Weigand, Physics of Fluids **19**, 102 (2007)

[12] E. Coyajee, B.J. Boersma, Journal of Computational Physics **228**, 4444 (2009)

[13] M. Kwakkel, W.P. Breugem, B.J. Boersma, Journal of Computational Physics **253**, 166 (2013)

[14] N.M. Musehane, O.F. Oxtoby, B.D. Reddy, Journal of Computational Physics **373**, 924 (2018)

[15] D. Brélaz, Communications of the ACM **22**, 251–256 (1979)

[16] G.D. Mackay, S.G. Mason, Canadian Journal of Chemical Engineering **41**, 203 (1963)

[17] J.D. Chen, Journal of Colloid and Interface Science **107**, 209 (1985)

[18] OpenFOAM, *OpenFOAM 9* (2021), `https://openfoam.org/release/9/`

[19] J.A. Heyns, O.F. Oxtoby, *Modelling surface tension dominated multiphase flows using the VOF approach*, in *Proceedings of the jointly organized 11th World Congress on Computational Mechanics WCCM XI, 5th European Congress on Computational Mechanics ECCM V, and 6th European Conference on Computational Fluid Dynamics ECFD VI* (CIMNE, Barcelona, Spain, 2014), pp. 7082–7090

[20] X. Yin, D.L. Koch, Physics of Fluids **19**, 093302 (2007)

[21] M. He, M. Chen, N. Wang, C. Li, ISIJ International **59**, 973 (2019)

An Experimental Approach to the Application of Digital Image Correlation to Small Punch Creep Testing

Melody van Rooyen[1*], *Thorsten* Becker[1**], and *Mahmoud* Mostafavi[2***]

[1]Department of Mechanical and Mechatronic Engineering, University of Stellenbosch, Private Bag X1, Matieland 7602, South Africa
[2]Department of Mechanical Engineering, University of Bristol, Bristol, BS8 1TR, United Kingdom

Abstract. Deterioration assessment of materials is essential to the continued effective operation of critical components operating, at high temperatures and stresses, in thermal power plants. Within progressive inspection strategies, small punch creep testing (SPCT) presents an effective method for establishing the degree of creep exhaustion of in-service power engineering alloys on account of the small volume of material required (Ø8 mm × 0.5 mm discs). These tests make the application of complex strain fields, analogous to the strain field experienced by the service components, possible. In addition, this work explores the application of a noncontact digital image correlation (DIC) method to the full-field measurement of strains which is not possible through traditional extensometry. The accelerated SPCT tests were performed on 12 % Cr ex-service steels subjected to 600 °C. Experimental methodologies and challenges associated with *ex situ* DIC-SPCT are discussed including equipment and parameter selection, speckle pattern application, preforming requirements and loading method. It is shown that the developed DIC technique using a stereo microscope, airbrushed speckle patterns, sombrero-preformed discs and interrupted tests can be used to compare surface strain development in various aged forms of the ex-service steel. Such an approach is valuable to understanding strain development in small punch tests and to assist in the comparative damage assessment of ex-service power plant materials.

1 Introduction

Small punch testing (SPT) has gained increased attention in the maintenance programmes of aging power plants. Typical constant-force small punch creep testing (SPCT) involves penetrating a disc of material using a hemispherical punch or ball at constant load, high temperatures and within an inert atmosphere whilst punch displacement and/or disc deflection is measured. Such an approach boasts an advantage for life assessment of in-service plant materials due to the small specimen sizes (Ø8 mm, 0.5 mm thickness) that makes sample extraction possible without impairing the operability of the host component

*e-mail: melzvanrooyen@sun.ac.za
**e-mail: tbecker@sun.ac.za
***e-mail: m.mostafavi@bristol.ac.uk

[1]. Site-specific material property assessment of, for instance, welds and heat-affected zone (HAZ) regions, is also possible through this technique.

The popularity of the technique is highlighted in the extensive literature presence of applications to in-service power plant materials. Applications of SPT include tensile property measurement from load-displacement curves of aged rotor steels [2] as well as across different weld regions [3]. This technique was also expanded to characterise the embrittlement of aged materials through the measurement of temperature-dependent fracture toughness and energy [2]. The SPCT has been applied to creep property measurement of numerous new [4], aged base [5] and weldment [6] rotor, steam piping and boiler steels. Several researchers have also employed SPCT within the framework of the European collaborative COPERNICUS project with the aim of assessing the remnant life of X20CrMoV12-1 (X20) and other steels using creep stress exponents and activation energies from Dorn relations [7].

Despite the wide-scale applications of SPCT, the technique involves several challenges that are not normally associated with traditional testing including sensitivities to: geometrical uncertainties in the experimental setup such as the indenter size and eccentricity from the disc centre [8]; upper die clamping conditions [9]; differences in disc specimen thickness with a direct effect on creep rupture times [10]; atmospheric conditions [11]; oxidation [12] as well as friction between punch components and the disc [13].

Although these issues have been largely mitigated through the introduction of a code of practice [14] with strict specifications placed on disc and punch geometry and test procedures, the interpretation of the SPCT data (displacement and load) and its equivalence to standard creep test data (specifically strain and stress) remains the focal point of many research works due to the complex stress and strain states [15]. For establishing the equivalent stress, conversion factors (calculating equivalent stress from load) based on Chakrabarty membrane stretching theory [16] as well as stress distributions calculated from finite element (FE) modelling [13] have been employed. Maximum equivalent strains can be estimated from simple analytical relationships with the central deflection [17], thickness changes at the thinnest point in the disc [18] as well as from FE models [19]. The former two methods only provide single point strain estimates whereas the strain field output from FE analysis is sensitive to numerous unknown parameters, such as friction coefficients and the selected creep model parameters [20]. The current literature presents with few attempts to measure full-field, surface strain maps of SPCT specimens.

To this end, digital image correlation (DIC) is considered an ideal candidate for full-field strain measurement of SPCT disc samples due to its noncontact nature as well as its lack of restriction on measurement length scales [21]. Aydin *et al.* [22] applied two-camera DIC to the *in situ* measurement of the fracture heights of steel sheets during an Erichsen cupping test which shares some similarities with a small punch test. However, no strain distributions were presented from the DIC data. Strain measurement using a stereo microscope DIC technique during a hydraulic bulge test on specimens with dimensions similar to that of SPCT was achieved by Ren *et al.* [23]. Recently, full-field deformations were mapped during room temperature small punch tensile tests using *in situ* DIC [24]. There have been few attempts to extend DIC to strain mapping during SPCT conducted at higher temperatures.

This work investigates the feasibility of using an *ex situ* stereo microscope DIC technique for the measurement of strain fields across a small punch creep tested disc of new and ex-service X20 steel and weldment. This includes: the design of a small punch creep setup together with the design of a stereo microscope DIC setup to capture images with a durable, high resolution speckle pattern for strain measurement; development of a strain-processing algorithm; selection of disc preforming method and specimen loading method (including repeated loading on the same disc specimen or interrupted loading on different disc specimens).

2 Experimental design

Due to the physical difficulties associated with obtaining images on a fine scale in a confined space (< Ø4 mm viewing field) at high temperatures and under argon atmosphere, the present work is confined to DIC images that are captured *ex situ*. The following sections describes the experimental setup required to achieve this, including selected materials and disc geometry, SPCT rig, speckle pattern application and stereo microscope DIC setup and processing methods.

2.1 Materials and specimen geometry

The chemical composition of the X20 material (supplied in pipe sections from a local power utility company) used in this study is given in Table 1. Three material types are outlined: "new" X20 is as-received material with no service exposure; "low" service exposure X20 is an ex-service material exposed to temperatures of 545 °C and pressures of 17 MPa for about 130 000 h of operation; and "high" service exposure X20 is an ex-service material exposed to conditions of 543 °C and 18.1 MPa for approximately 156 000 h of operation. "Low" and "high" service exposure X20 were categorised based on average cavity density counts from scanning electron micrographs with values of 119 ± 48 cavities/mm^2 and 489 ± 451 cavities/mm^2, respectively [25]. Weldment material was extracted from the fusion region between high and low service exposure X20.

Rods of 8 mm diameter longitudinally aligned with the pipe axis were cut from the centre of the pipe thickness as shown in Figure 1. Discs of ~1.0 mm were wire cut from the rods and then polished down using 220 grit followed by 1000 grit silica carbide paper on both sides until an average thickness of approximately 0.5 ± 0.01 mm was measured.

Table 1. Chemical composition of new and ex-service X20 steels in wt% (balance iron)

Material	Elements								
	C	Si	Mn	Cr	Mo	Ni	V	P	S
New	0.21	0.19	0.58	11.6	0.88	0.76	0.25	0.015	0.004
Low	0.19	0.36	0.63	11.7	0.91	0.67	0.32	0.013	<0.005
High	0.17	0.20	0.59	11.6	0.94	0.71	0.34	0.015	<0.005

2.2 Small punch creep testing

2.2.1 Small punch design

The geometric detail of the punch body and a photograph of the punch rig is shown in Figure 2a and b, respectively. The punch was designed in accordance with the CEN CWA 15627 guidelines [14] and consists of an alumina oxide ceramic ball of 2.5 mm diameter and a Nimonic 80A punch body. Load is applied at 2 N/s using an Instron 1341 servohydraulic tester through the punch head onto the ceramic ball, which then causes the disc to bulge into a 4 mm diameter recess. Primary creep initiates when a maximum load is reached and maintained. Temperature is measured by a K-type thermocouple positioned against the disc and is kept within 2 °C of the setpoint as the punch is heated within a single-zone tube furnace. The periphery of the disc is clamped by the upper die as in a classic "small punch bulge test" as outlined in CWA 15627. Argon gas is supplied to the bottom disc surface by a capillary tube at a rate of 0.2 L/min. Deflection of the punch was measured using an LVDT with a resolution of 1 μm.

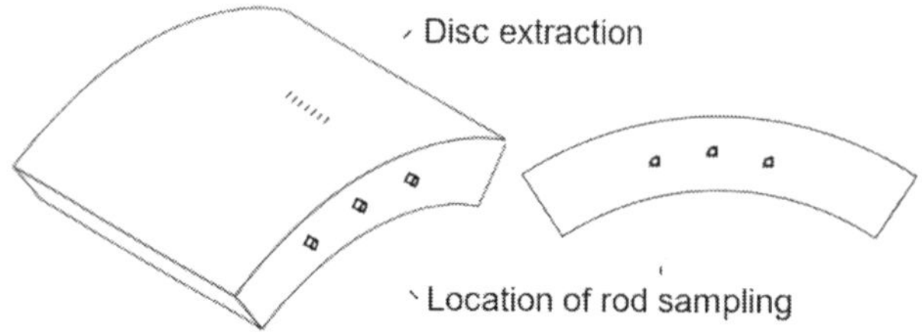

Fig 1. Sampling schematic of rods from which discs are wire cut

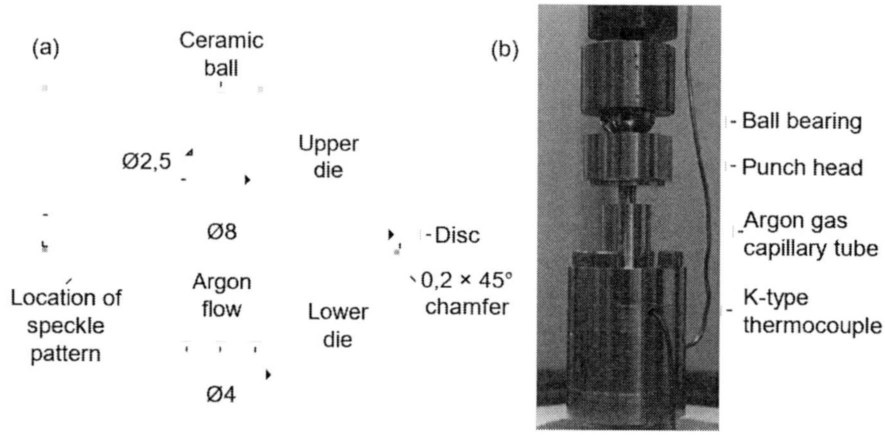

Fig 2. (a) Dimensions of the punch body geometry (supplied in mm) and (b) photograph of small punch creep setup

2.2.2 Preforming

To prevent speckle pattern flaking due to excessive strains, the discs are prebulged into "sombrero" shapes at the same temperature and to the same load used in the subsequent creep test. Such an approach was shown by Ule *et al.* [10] to have no effect on the creep curve development in terms of rupture time and minimum creep rates. Following the prebulging, the load is removed, and the preformed disc is removed for speckle patterning according to the procedure outlined in Section 2.3.

2.2.3 Ex situ specimen loading method

To remove tested discs from the small punch rig for the *ex situ* acquisition of DIC images, continuous loading of specimens is not possible and disc reloading methods are required. Two methods were explored, namely, interrupted tests and retests. Interrupted tests involve the removal of a tested disc after a specified period of constant load application, capturing DIC images as described in Section 2.4 and subsequently inserting another untested, preformed disc to be subjected under the same maximum load and temperature conditions as before for progressively longer times. This method allows semi-continuous monitoring of strain but is specimen intensive. Retesting, on the contrary, involves reloading the same (previously tested) disc after DIC imaging for subsequent testing under the same conditions as prior testing. Although fewer specimens are required through disc reuse, softening or hardening effects on creep strain development are of concern [26]. For both methods, the test is stopped when the load is removed, followed by gradual cooling and removal of the crept disc for DIC imaging.

2.3 Speckle pattern application

The preformed surface of the disc was ultrasonically cleaned with acetone and coated with white VHT Flameproof matte, heat-resistant paint. Because speckle sizes produced from normal spray can application are too large (30 – 80 μm) for the current work, an airbrush was employed. A fine mist of undiluted VHT matte black speckles was applied to the wet paint surface using a generic model SP180KTG airbrush with a 0.2 mm needle, 30 cm stand-off distance, a medium airflow (based on three full turns of the air regulating screw) and 25 kPa working pressure. These settings were selected based on the guidelines offered by Lionello and Cristofolini [27] for applying optimal speckle distributions.

Resulting speckle sizes ranged from 10 – 13 μm as measured using image thresholding in ImageJ [28] of micrographs taken by a Zeiss Imager M2 Axio light microscope at 200× magnification. With the magnification factor of 0.24 pixels/μm encountered in the calibrated DIC images, the average speckle size was therefore covered by approximately 3 pixels as recommended by Schreier *et al.* [29] for optimal correlation. The speckle pattern, as shown in Figure 3b, was air cured for 24 h before testing to harden the pattern for handling.

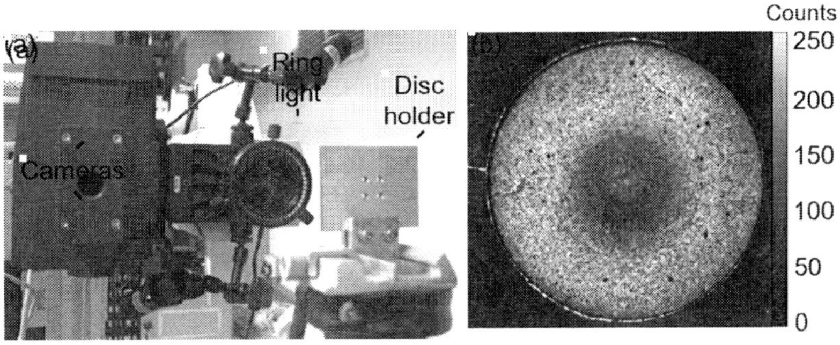

Fig 3. Photograph of the (a) stereo microscope DIC setup and (b) an example view of a speckled preformed disc

2.4 Stereo-microscope digital image correlation setup

The patterned sombrero-shaped disc (shown in Figure 3b) is recorded using the DIC hardware summarised in Table 2. Image acquisition was performed using a Zeiss SteREO Discovery.V12 stereo microscope retrofitted by LaVision to include two cameras and 3D micropositioning translation stages within an integrated system. Diffuse illumination was provided by a 7 W, pure white colour temperature ring light mounted over the microscope lens. Disc samples were mounted on a sample holder fixed to a rigid tripod at about 240 mm from the lens front. A photograph of the setup is provided in Figure 3a. Prior to creep testing, a reference image is taken of the painted, preformed disc with markings made on the specimen holder to assist in aligning the disc to its original (prior to testing) position. Following testing, the creep tested disc is aligned and again imaged, and the process is repeated for subsequent tests.

Table 2. DIC hardware details for small punch creep test

Cameras	Imager M-lite 5M on a Zeiss SteREO Discovery. V12 stereo-microscope
Image resolution	2464 × 2056 pixels²
Lighting	48 LED VisiLight R-LED-66 ring light (7 W)
Lens	Achromat S lens, ×0.3 magnification
Aperture	Open position
Field of view	10.5 × 8.7 mm²
Image magnification	240 pixels/mm
Stereo-angle	5.8°
Free working distance	236 mm
Patterning method	White, heat-resistant base coat, fine black airbrushed speckles
Approximate pattern feature size	3 pixels (13 μm)

2.5 Processing

Due to the inherent complexities of the stress and strain state of small punch testing [17], DIC measured displacements were fitted with a piecewise second-order polynomial over a square window of 11 points using MATLAB ver. R2020a. Prior to fitting, outlier removal based on values greater than three local standard deviations from the local mean was performed to eliminate spurious measurements caused by image artefacts (for instance, dust particles on the microscope lenses). Strains are obtained through differentiation of the fitted displacement fields. A summary of the software parameters is provided in Table 3. For improving the disc alignment relative to the reference disc position, subpixel rigid body shift and rotation correction is performed by using correlation windows of 121 pixels size on larger speckle features on the disc periphery.

2.6 Rigid body displacement tests

In order to characterise displacement variance and bias error in the current setup, baseline noise and rigid body translation tests were conducted, respectively, according to Jones and Iadicola [30] and the processing parameters outlined in Table 3. The maximum applied translations were selected based on the anticipated deformations resulting from the SPCT [31]. Comparisons between applied and DIC in-plane and out-of-plane displacements are shown in Figure 4a and b, respectively. Displacement measurements correspond well in both directions with average root-mean-square values of 8 and 9 μm, respectively.

Insignificant differences were noted in the error analysis for other subset sizes (21 – 81 pixels). Consequently, a subset size of 41 pixels was selected based on the speckle size analysis discussed in Section 2.3. This results on average in 7 × 7 speckles per subset (assuming 50 % speckle coverage) and a displacement measurement spatial resolution of about 250 μm. Self-correlation tests performed on a set of five stationary images using the selected subset size resulted in baseline displacement and strain errors as shown in Table 3.

Table 3. DIC software details for small punch creep test

DIC software	LaVision DaVis, 8.4.0
Subset size	41 pixels (0.17 mm)
Step size	10 pixels (0.04 mm)
Displacement noise floor	In-plane: 0.03 pixels (0.11 μm), Out-of-plane: 0.2 pixels (0.8 μm)
Strain noise floor	32 με (averaged between vertical and horizontal directions)
Subset shape function	First-order
Matching criterion	Zero-normalised sum of square differences (ZNSSD)
Grey level interpolant	Sixth-order spline
Prefiltering of displacements	Spatial piecewise second-order fit with moving mean outlier removal
Strain window	11 data points
Virtual strain gauge (VSG) size	131 pixels (0.55 mm)

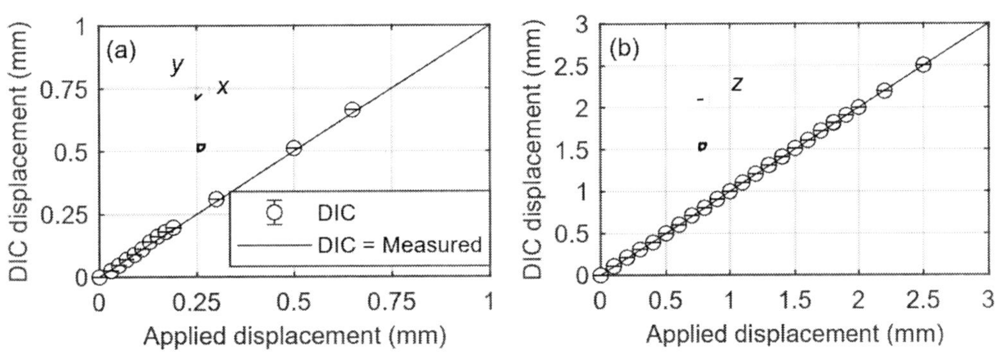

Fig 4. Comparison between DIC measured displacements and applied translations for (a) in-plane and (b) out-of-plane rigid body tests for a 41 × 41 pixels^2 subset and 10-pixel step size.

3 Results and discussions

3.1.1 Interrupted tests versus retests

A demonstration of the *ex situ* small punch strain mapping (in terms of maximum normal strain ε_{norm}) using DIC is shown in Figure 5b-f along with the corresponding time-punch displacement curves (Figure 5a) for interrupted tests of new, low service exposure and weldment X20. A load of 520 N and a temperature of 630 °C was selected in order to accelerate creep curve development within 3 h. The strain maps are measured when the tests are interrupted at the indicated times on the time-displacement curves. Positions (in terms of radial distance, r_{max}, from disc centre) of the local maximum normal strain (ε_{max}) are also indicated in Figure 5b-f.

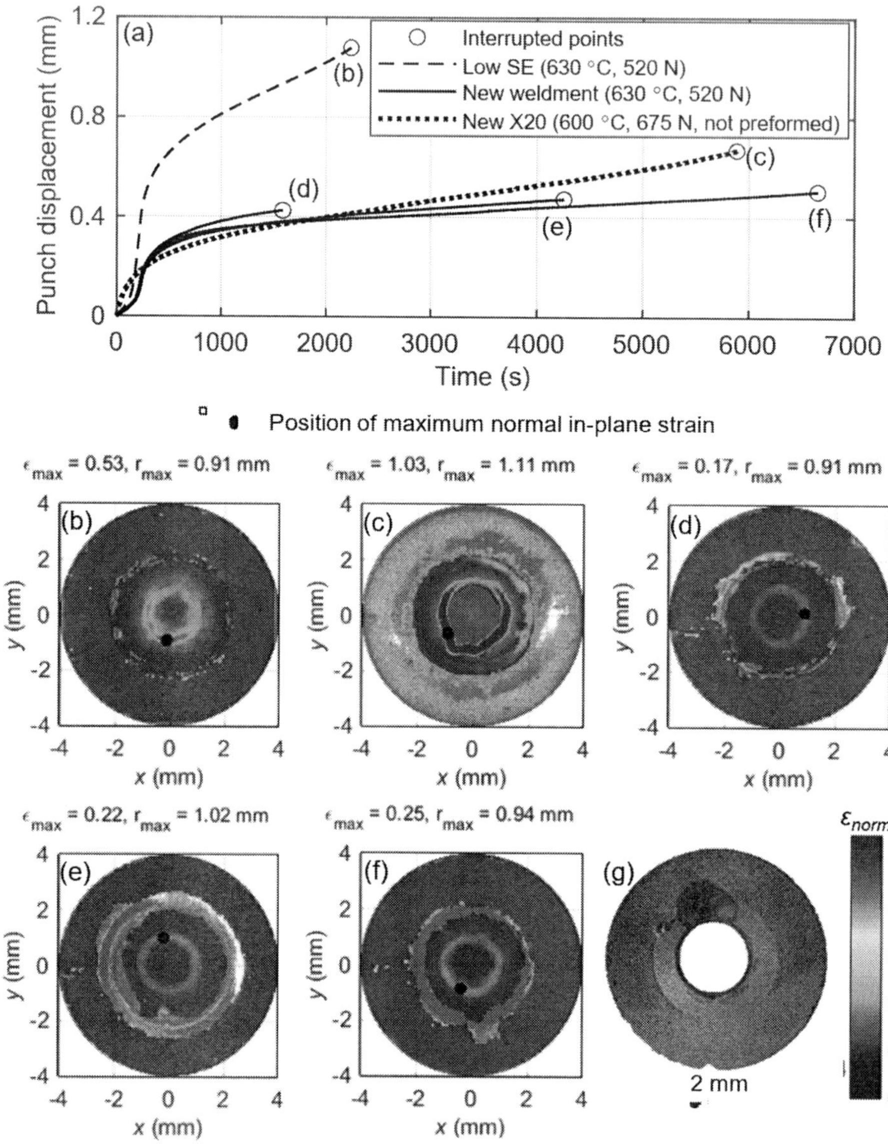

Fig 5. (a) Time-punch displacement curves of interrupted small punch creep tests of new, low service exposure and weldment X20 with corresponding (b-g) maximum normal strain maps. SE = service exposure.

Several observations can be made from the resulting distributions of maximum normal strains on the disc surfaces. Firstly, strains on the disc periphery are near zero due to the clamping mechanism of the upper die. Secondly, a concentrated annulus of high strains forms around the centre of the disc and corresponds to the contact area between the ball and the disc [32]. This ring proceeds to widen and increase in strain magnitude with increasing punch displacement as a larger area of the ball contacts the deforming disc, resulting in a switch from plastic bending behaviour to plastic membrane stretching as the disc deflection exceeds the initial thickness of the disc [17]. Moreover, the low service exposure X20 demonstrates less creep resistance than the weldment for identical testing conditions. As shown in Figure 5b, this results in higher strain values and a thicker annulus than for the weldment strain maps. This observation also confirms the FE predictions of Cortellino *et al.* [33] where peak equivalent plastic strains and simulated creep damage develop at the base of the disc at off-centre contact regions with the punch. This strain concentration also correlates with the locations of through thickness cracks that results in disc failure for high service exposure parent X20 (Figure 5g).

Maximum normal strain maps for retested discs of new X20 weldment are shown in Figure 6. In comparison to the interrupted tests of the new weldment (Figure 5a), the retests result in accelerated creep with shorter times to rupture. For example, the retested sample accelerates into the tertiary regime within 3250 s (0.9 h) whereas the interrupted sample (curve (f) in Figure 5, for instance) remained in the primary creep regime for > 6000 s (1.67 h) of testing. This is possibly due to softening effects brought about by primary creep regeneration which causes increased creep rates following an unloading-loading cycle (point (b) in Figure 6a, for instance) as a result of a reduction in dislocation densities [26].

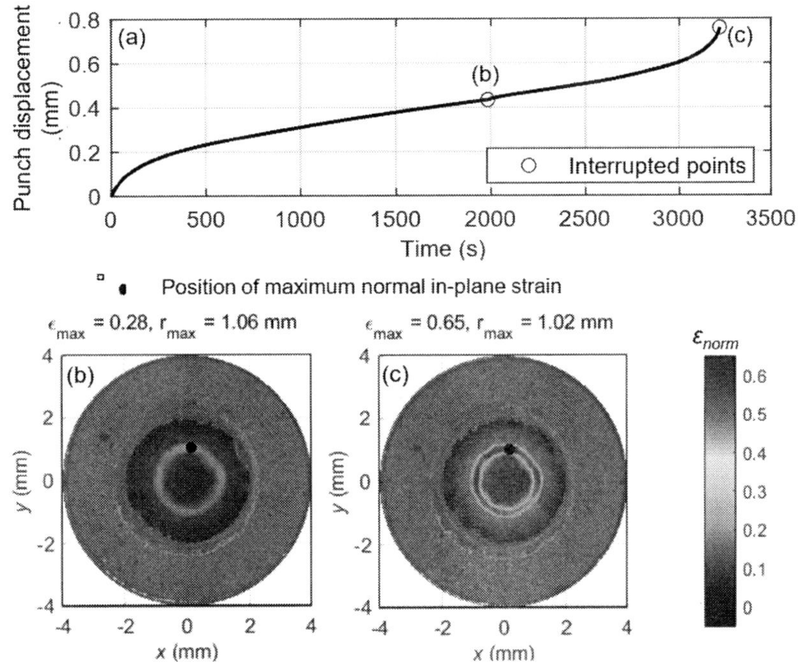

Fig 6. (a) Time-punch displacement curves of a small punch creep retest of new weldment X20 specimen with corresponding (b-c) maximum normal strain maps.

Speckle pattern quality completely degrades for highly strained areas as in the case of Figure 5c where a flat disc of new X20 was subjected to creep loading prior to any preforming. In this case, the central bulge could not be correlated by the DIC software, thereby affirming the need for a preforming process before speckle pattern application. Similar paint degradation was apparent in the retested discs for several load-unloading cycles. Careful handling of retested discs is also necessary due to the embrittlement of the speckle pattern painted layer following repeated testing.

3.1.2 Comparison between new and service-exposed X20

Figure 7 shows the comparison of the maximum normal strain maps of new and high service exposure X20 discs obtained using interrupted tests at 600 °C and 500 N.

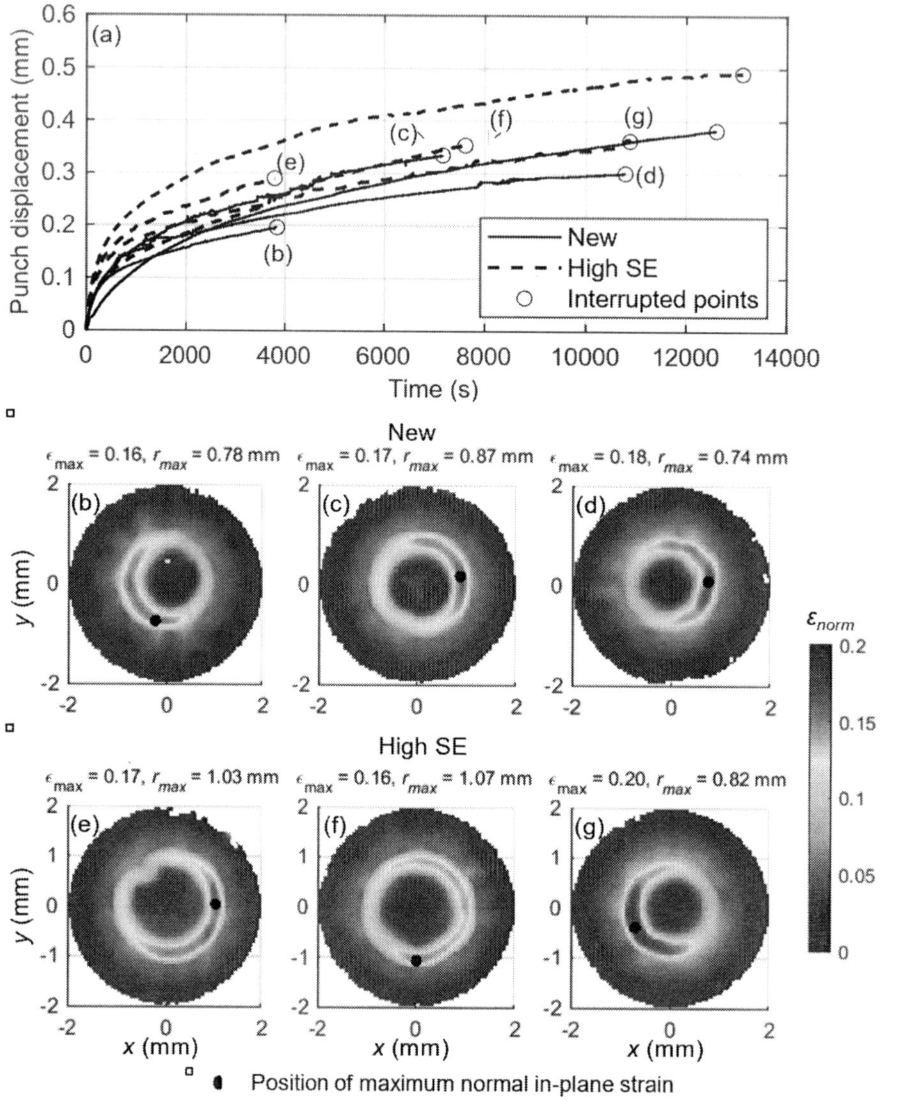

Fig 7. (a) Time-punch displacement curves of interrupted small punch creep tests of new and high service exposure X20 at 600 °C and 500 N with corresponding (b-g) maximum normal strain maps. SE = service exposure.

Firstly, scatter in time-displacement curves is evident, particularly for the high service exposure X20. This could be attributed to nonuniformity of cavity density distributions in the high service exposure material [25] as well as difficulty in aligning the punch ball within the preformed bulge. DIC measured strain maps can also be used to assess the geometric concentricity and uniformity of clamping conditions by considering the deviations from axisymmetric distributions as shown by the unsymmetrical strain concentration rings in Figure 7b-g. Despite these limitations, higher strains within wider annulus strain bands are evident for the service-exposed X20, indicating larger deformations for the same loading conditions. Strain maps obtained for the unlabelled curves in Figure 7a contained outlier values and decorrelated regions due to the degraded speckle patterns at these higher strains and are not presented here.

4 Conclusions and prospects

An experimental technique for the *ex situ* acquisition of strain maps of small punch creep tested power plant steels using DIC is presented. Challenges of speckle pattern resolution, speckle pattern flaking and method of specimen reloading for *ex situ* image acquisition is addressed through airbrushed, heat resistant speckles, preforming into sombrero-shaped discs and interrupted tests of different discs or retests of the same disc, respectively. Interrupted tests are sensitive to differences in local cavity densities of service-exposed material as different discs are employed (extracted from different locations of the same material type). Retests are more material efficient with the use of a single disc and are not sensitive to property variations between various discs. However, softening effects prevent the assessment of the "true" creep response. Larger strains and thicker strain bands are evident for small punch creep tested high service exposure X20 when compared to new X20. Strain maps are also able to indicate deviations from coaxial alignment of the punch head and disc preformed bulge.

The conversion of load-displacement data output from small punch testing into stress-strain relations required by creep analysis has long been a challenge encountered by researchers [17]. This work has highlighted the application of full-field strain mapping to SPCT. Future aims for this technique include combining the DIC strain fields with data from other streams (such as FE stress fields) to aid damage assessment using continuum damage mechanics models. Furthermore, *in situ* combinations of DIC with small punch testing precludes limitations regarding speckle pattern durability, punch ball misalignment, cyclic softening effects and the need for laborious interruptions of creep tests. This forms part of ongoing work aimed at material-efficient damage extraction methods for service-retrieved power plant steels.

The financial support by the National Reserach Foundation, South Africa is thankfully acknowledged.

References

1. J. P. Rouse, W. Sun, and T. H. Hyde, J. Strain Anal. Eng. Des. **48**, 494 (2013)
2. M. Madia, S. Foletti, G. Torsello, and A. Cammi, Eng. Fail. Anal. **34**, 189 (2013)
3. S. Komazaki, T. Kato, Y. Kohno, and H. Tanigawa, Mater. Sci. Eng. A **510–511**, 229 (2009)
4. M. Selent, M. Abendroth, and M. Kuna, Trans. Indian Inst. Met. **69**, 629 (2016)
5. S. Yang, X. Ling, and Y. Zheng, Mater. Sci. Eng. A **685**, 1 (2017)
6. D. T. Blagoeva and R. C. Hurst, Mater. Sci. Eng. A **510–511**, 219 (2009)
7. B. Ule, T. Sustar, T. Rodic, F. Dobes, K. Milicka, V. Bicego, S. Tettamanti, K. Maile, C. Schwarzkopf, M. Whelan, J. Klaput, and R. H. Kozlowski, Technol. Law Insur. **4**, 283 (1999)
8. S. Komazaki, T. Nakata, Y. Kohno, and H. Tanigawa, Exp. Mech. **57**, 487 (2017)
9. D. Andrés and P. Dymáček, Theor. Appl. Fract. Mech. **86**, 117 (2016)
10. B. Ule, R. Strum, and V. Leskovsek, Mater. Sci. Technol. **19**, 1771 (2003)

11. T. Nakata, S.-I. Komazaki, Y. Kohno, and H. Tanigawa, Mater. Sci. Eng. A **666**, 54 (2016)

12. K. Kobayashi and S. Takei, *Influence of Testing Environment on SP Creep Rupture Lives*, in Proc. ASME 2013 Press. Vessel. Pip. Conf., PVP, 14-18 July 2013, Paris, France (2013)

13. T. Nakata, S.-I. Komazaki, Y. Kohno, and H. Tanigawa, Mater. Sci. Eng. A **666**, 80 (2016)

14. CEN, CWA 15627 Workshop Agreement: Small Punch Test Method for Metallic Materials (European Committee for Standardization, Brussels, 2007)

15. T. H. Hyde, M. Stoyanov, W. Sun, and C. J. Hyde, J. Strain Anal. Eng. Des. **45**, 141 (2010)

16. S. Holmström, Y. Li, P. Dymacek, E. Vacchieri, S. P. Je, R. J. Lancaster, D. Omacht, Z. Kubon, E. Anelli, J. Rantala, A. Tonti, S. Komazaki, M. Bruchhausen, R. C. Hurst, P. Hähner, M. Richardson, and D. Andres, Mater. Sci. Eng. A **731**, 161 (2018)

17. Z. Yang and Z. wen Wang, Int. J. Press. Vessel. Pip. **80**, 397 (2003)

18. T. Lee, F. A. Ibupoto, J. H. Lee, B. J. Kim, and M. K. Kim, Exp. Mech. **56**, 395 (2016)

19. P. Dymáček and K. Milička, Mater. Sci. Eng. A **510–511**, 444 (2009)

20. Z. Zhou, Y. Zheng, X. Ling, R. Hu, and J. Zhou, Mater. Sci. Eng. A **527**, 2784 (2010)

21. T. A. Berfield, J. K. Patel, R. G. Shimmin, P. V. Braun, J. Lambros, and N. R. Sottos, Exp. Mech. **47**, 51 (2007)

22. M. Aydin, X. Wu, K. Cetinkaya, M. Yasar, and I. Kadi, Eng. Sci. Technol. an Int. J. **21**, 760 (2018)

23. M. Ren, J. Liang, L. Li, B. Wei, L. Wang, and Z. Tang, Rev. Sci. Instrum. **86**, (2015)

24. V. D. Vijayanand, M. Mokhtarishirazabad, J. Peng, Y. Wang, M. Gorley, D. M. Knowles, and M. Mostafavi, J. Nucl. Mater. **538**, 152260 (2020)

25. M. van Rooyen, T. H. Becker, J. Westraadt, and G. Marx, MDPI Materials. **12**, 3106 (2019)

26. X. Li, S. R. Holdsworth, S. Kalácska, L. Balogh, J. S. Park, A. S. Sologubenko, X. Maeder, S. Kabra, E. Mazza, and E. Hosseini, Mater. Des. **199**, (2021)

27. G. Lionello and L. Cristofolini, Meas. Sci. Technol. **25**, (2014)

28. S. Johannes, A.-C. Ignacio, F. Erwin, K. Verena, L. Mark, P. Tobias, P. Stephan, R. Curtis, S. Stephan, S. Benjamin, T. Jean-Yves, J. W. Daniel, H. Volker, E. Kevin, T. Pavel, and C. Albert, Nat. Methods **9**, 676 (2012)

29. H. Schreier, J. J. Orteu, and M. A. Sutton, *Image Correlation for Shape, Motion and Deformation Measurements: Basic Concepts, Theory and Applications* (Springer, New York:, 2009)

30. E. M. C. Jones and M. A. Iadicola, Eds., *A Good Practices Guide for Digital Image Correlation* (International Digital Image Correlation Society (iDICs), 2018)

31. M. Stoyanov, Investigation of Small Punch Creep Testing, PhD thesis, University of Nottingham, 2013

32. X. Chen, X. Xie, J. Sun, and L. Yang, SAE Int. J. Mater. Manuf. **5**, 345 (2012)

33. F. Cortellino, J. P. Rouse, B. Cacciapuoti, W. Sun, and T. H. Hyde, Exp. Mech. **57**, 1193 (2017)

Identifiability of Tyre Force Contact Prediction from Deformation Measurements

R.G. Gast[1,*], *P.S.* Els[1,**], *S.* Kok[1,***], *D.N.* Wilke[1,****], and *T.R.* Botha[1,†]

[1]Department of Mechanical and Aeronautical Engineering, University of Pretoria
Private Bag X20, Hatfield 0002, South Africa

Abstract. The possibility of accurately inferring the external forces applied to a vehicle can directly contribute to better safety systems and thus lowers the chance of injury or loss of life. These external forces are applied to a vehicle through the tyres and are challenging to measure directly. Still, it is possible to measure acceleration, deformation, or strain on the inner surface of a tyre. These measurements are theorized to be strongly linked to the forces produced by the tyre. However, it is still unknown whether or not one can always identify external forces from internal measurements in this way. Research has mainly focused on obtaining estimates of tyre forces rather than establishing to what extent these tyre forces are identifiable. This paper investigates this by conducting a virtual experiment that simulates known external forces applied to the tyre and computes the strains and displacements inside the tyre. A virtual inverse simulation then recovers the external forces from either the deformation or strain computed on the inside of the tyre. The identifiability of the forces recovered by the virtual inverse simulation is investigated by adding artificial measurement noise and initial guess perturbations to quantify the variance in the identified forces.

1 Introduction

Vehicle safety is improved with the addition of Advanced Driver Assistance Systems which take over from human inputs to prevent or reduce the effect of accidents. These systems rely on estimates of the vehicle state, and thus better control is correlated with better estimates. Tyre forces are the primary external forces applied to a vehicle, especially at lower speeds, but are notoriously difficult to determine. One possibility for determining tyre force estimates requires "intelligent" or "smart" tyres, which have embedded sensors that are used to infer the tyre force - from acceleration, strain or optical sensors [1]. Of all the sensing options, an optical approach is the most attractive as the full-field deformation and strain field of the inner surface of the tyre can be measured. Both these fields are theorized to be strongly correlated to the external forces applied to the tyre. Previous work at the University of Pretoria on an

*e-mail: robin.gast@tuks.co.za
**e-mail: schalk.els@up.ac.za
***e-mail: schalk.kok@up.ac.za
****e-mail: nico.wilke@up.ac.za
†e-mail: theunis.botha@up.ac.za

intelligent tyre concept has lead to the development of the T2-Cam (Tyre-Terrain Camera) system, which uses stabilized stereo-vision cameras to record the inner surface of a rolling tyre [2]. Digital Image Correlation is used to determine the 3D deformation field from a pair of images captured by T2-Cam. The surface strains can be calculated from a pair of reference images of the inside of the undeformed, inflated tyre [2, 3]. Most recently, Pegram et al. showed agreement between point measurements of the strains on the inner surface of the tyre from strain gauges and those measured by T2-Cam [4]. There is limited research regarding whether or not these systems can be used to identify tyre force in general rather than in specific cases. It is unknown whether a measured displacement or strain field corresponds to a unique force distribution at the tyre-road interface. This presents a challenge for developing a model to predict tyre forces from measurements. In this study, a virtual experiment simulates known external forces applied to the tyre and computes the strains and displacements inside the tyre. A virtual inverse simulation is then conducted to investigate the identifiability of the external forces given either the computed deformation or strain fields on the inside of the tyre. The study includes aleatoric uncertainty by adding artificial measurement noise and initial guess perturbations to quantify the variance in the identified forces.

There is little research regarding whether or not these systems can be used to identify tyre force in general rather than in specific cases, and it is not currently known whether a measured displacement or strain field corresponds to a unique force distribution at the tyre-road interface. This presents a challenge for developing a model to predict tyre forces from measurements. In this work a virtual inverse simulation is solved to recover forces on the outside of a tyre, which correspond to either a deformation or strain field virtually measured on the inner surface of the tyre.

In Section 2 the process of investigating the possibility of determining external tyre forces from internal measurements is described using a standard finite element code. A suitable candidate tyre is chosen to base the study on for which an appropriate axis system is defined. The process of acquiring geometry is described and a "test section" is defined to simplify the study. Section 2.4 describes the development of a coarse finite element model of the tyre which is used to generate virtual experiments for the inverse analysis that follows. An optimisation-based approach to identify tyre forces is described in Section 2.5 using the same previously-described model, and is then used to quantify the sensitivity of the problem to measurement noise and an initial guess in Section 3. The results are discussed in Section 4 and show that the identifiability of total forces and moments is high even in the presence of measurement noise and a poor initial guess, but the distribution of forces can not be identified if the initial guess is poor. Therefore in Section 5 it is recommended for future work to focus on regularization methods.

2 Method

2.1 The tyre investigated

To simplify the problem, a tyre is investigated which does not have steel bands running from bead-to-bead and, therefore, will have macroscopic material properties closer to rubber. Additionally, the effects of tread block deformations are removed by choosing a slick tyre with no tread blocks. The tyre being studied is a Trelleborg TM700 280/70R16 agricultural tyre (Figure 1), the same tyre used by Pegram et al. [4] except with the tread lugs shaven off. This tyre does not have steel cords running from bead-to-bead and is, therefore, suitable for the analysis. The tyre geometry has already been captured with a combination of profile gauges and laser scanning and converted into a Computer-Aided Design (CAD) model for previous studies

Figure 1: A Trelleborg TM700 with the tread lugs removed.

2.2 Axis System

The description of forces and moments transferred to or from a tyre requires defining an appropriate axis system. For the investigation of tyre behaviour, an axis system that is centred around the tyre is preferred. For this study, the TYDEX (TYre Data EXchange Format) CAxis system is adopted as shown in Figure 2. In the TYDEX C-axis system, the x-axis is defined as the line in the wheel plane parallel to the ground plane. The y-axis is perpendicular to the wheel plane and therefore is the spin axis, and the z-axis follows from the right-hand rule. The slip angle is the included angle between the tyre heading and the direction of travel. The tyre inclination, or camber angle, is also shown. This is the angle between the ground plane normal and the tyre spin axis normal.

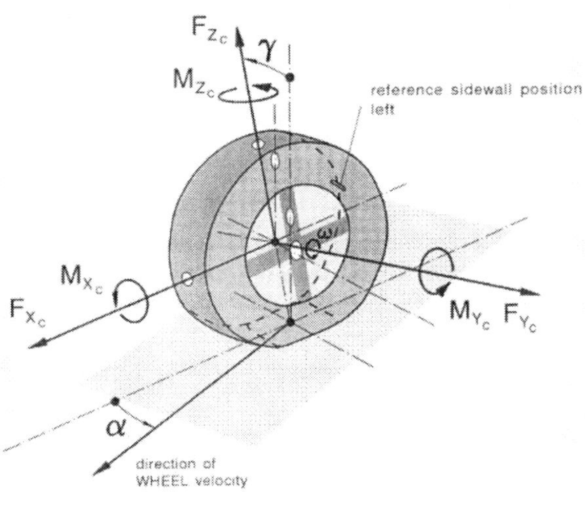

Figure 2: TYDEX C tyre axis system - adapted from Van Oosten et. al. [5]

2.3 Geometry

The tyre inflation pressure places it in a state of tension and allows the tyre carcass to bear significant loads as it acts as a ring under tension. The portion of the sidewall which is opposite an applied radial load is in further tension, and contributes significantly to the vertical stiffness of an inflated tyre. The portion of the sidewall adjacent to radial loads can deform significantly as it naturally buckles, but it does not contribute substantially to the stiffness. This region near to the road will be studied without the inflation effects present, as it is theorized that this represents a scenario in which the applied forces are more readily identified than one in which the aforementioned effects of inflation and the sidewall stretching are present. This is done by modelling the problem as a 90° arc of the tyre which is in contact with the road surface and subject to no internal pressure.

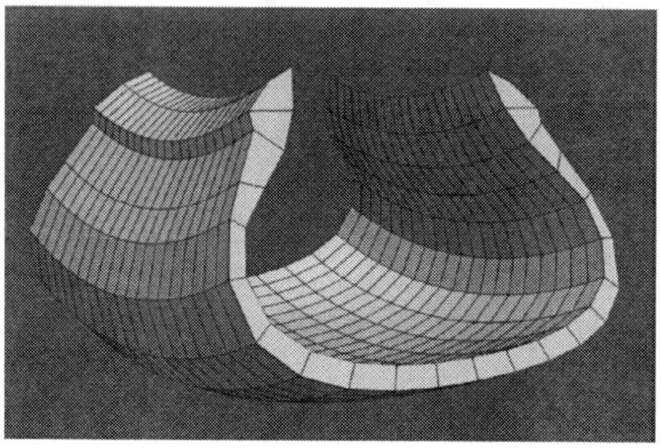

Figure 3: Geometry of the test section.

2.4 Virtual Experiments

One can use a virtual experiment to generate plausible inputs for the virtual inverse simulation. This virtual experiment must reflect the reality of measuring deformations and strains on the inner surface of a tyre, and therefore one must understand the experimental setup used to do this. Laboratory tests using T2-Cam allow for the measurement of strain (ϵ_{xx}, ϵ_{yy}, ϵ_{xy}) and deformation (U_x, U_y, U_z) on the inside of the tyre using stereo-vision cameras The total forces and moments acting on the tyre for a particular loading condition can be measured with a wheel-force transducer. The distribution of the contact forces between the tyre and the contact surface are measured in the vertical direction in the laboratory with a Tekscan Pressure Mapping Sensor 8001 [6], but sensors for measuring the longitudinal and lateral pressures are not widely available. Estimating the distribution of these forces may significantly impact other research areas, such as terramechanics, where the ability to predict both the pressure and sinkage of a tyre substantially improves the possibility of predicting tractive effort. Since it is not practical to measure the distribution of the forces on a tyre, this study is instead conducted using "virtual experiments". A Finite Element model of the test section is used to generate plausible displacements and strains on the inside of the tyre given some loading condition. This is limited to the quasi-static loading of the un-inflated test section against a smooth, flat surface with hard contact. This loading condition generates loads in the z-direction only. The open-source FEM code CalculiX [7] was chosen to perform the virtual experiments. These virtual experiments will be used as a proof of concept to determine to what degree the forces produced by a tyre can be identified by the displacement or strains measured inside the tyre. The test case allows for one to investigate this with some simplifications.

2.4.1 Pre-Processing

The mesh of the test section is generated with the open-source tool Gmsh [8] with 440 incompatible mode eight-note brick elements (C3D8I) to facilitate using a non-linear hyperelastic material model. The material is modelled as Neo-Hookean with an equivalent Young's Modulus of 10 MPa and a Poisson's ratio of $\nu = 0.499$. The material was chosen to be on the low end of the scale of typical rubber stiffness's found in tyres, as it is expected that it will better expose the sensitivity of the problem. The boundary conditions at the bead are modelled as a fixed displacement, which has been shown to have a negligible effect on the tyre's vertical stiffness compared to a full contact simulation of the rim-bead interaction [9]. The boundary condition between the tyre and the road is iteratively applied by first displacing the node on the tread which most violates the road for a given total displacement, running this simulation to determine the new penetration with the road, and then repeating this process until no tread nodes penetrate the road. If a node presents an opposing reaction force in the z-direction at any stage, it is released. This simulation allows one to determine the strains and displacements inside the tyre, given some loading, representing what one can measure in the laboratory on an actual tyre.

2.4.2 Typical Virtual Experiment Results

Figure 4 shows various results of a virtual experiment in which the test section was displaced 30mm, which corresponds to typical displacements that the tyre might sustain under heavy loads. Figures 4a and 4c indicate that in the absence of inflation pressure, the test section bulges in the middle, causing a loss of contact in the centre of the test section. The reaction forces at the tyre tread resulting from the displacement are shown in Figure 4b, where the total vertical reaction force was 359N spread across two distinct areas of contact. The average nodal force over these nodes in contact is 25.68N with a standard deviation of 7.27N. Figures 4d, 4e and 4f show the three components of percentage surface strain calculated from the full-field displacement measurements. These indicate how rapidly the strain dissipates from the inside of the tyre to the outside of the tyre.

2.5 Virtual Inverse Simulations

Given some measured displacement $\boldsymbol{U}$ or strain ϵ inside the tyre, the corresponding external forces, $\boldsymbol{x}$, that caused them must be recovered. The challenge is that the forces must be recovered at a different location to where the displacements and strains are measured. One can investigate the identifiability of these forces with a similar Finite Element model but instead prescribing forces at the tread instead of a displacement and attempt to reproduce the displacements and strains that were "'measured" in the virtual experiment.

An optimization-based approach is suitable for iterating through applied forces on a FEM model until the measured displacements and strains match and is an appropriate tool for determining the sensitivity of the problem. An initial guess, $\boldsymbol{x}_0$, of the external forces applied to the outside of the tyre is required to solve the virtual inverse problem. The resulting displacements or strains are computed and compared to those computed in the virtual experiment. When good agreement is found the procedure terminates, else the optimizer updates the forces and repeats the process until convergence is reached. The locations at which forces

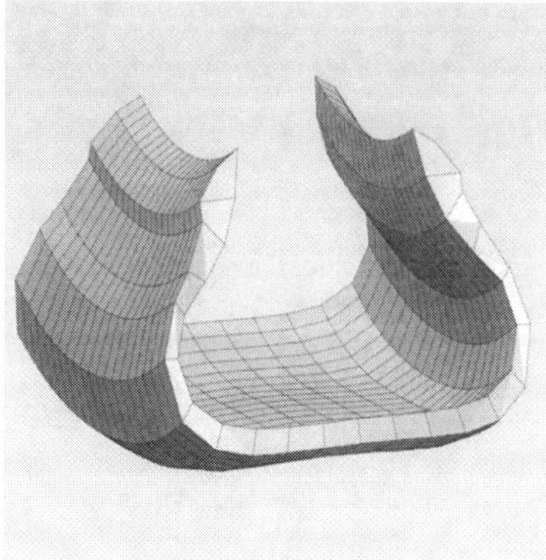

(a) Deformed geometry.

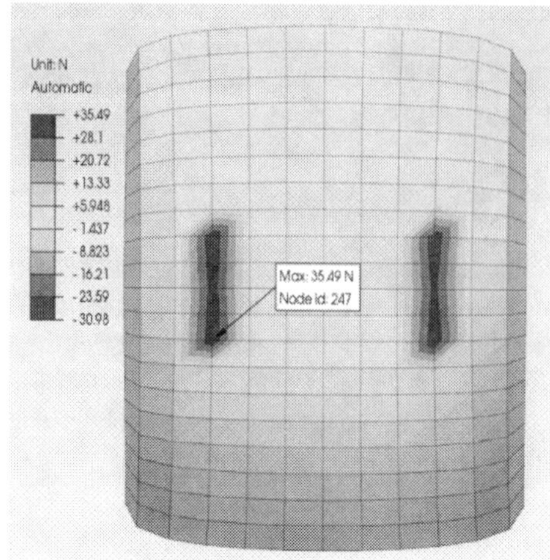

(b) Vertical reaction forces.

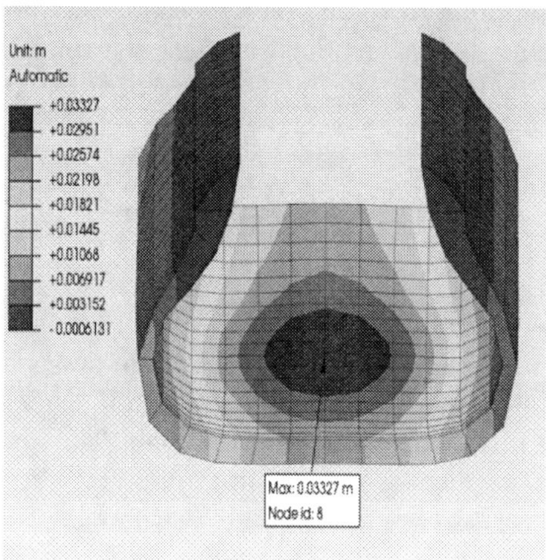

(c) Vertical deformation.

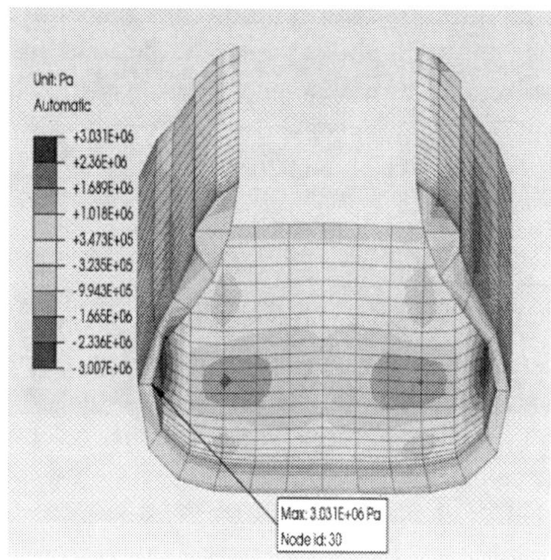

(d) ϵ_{xx}.

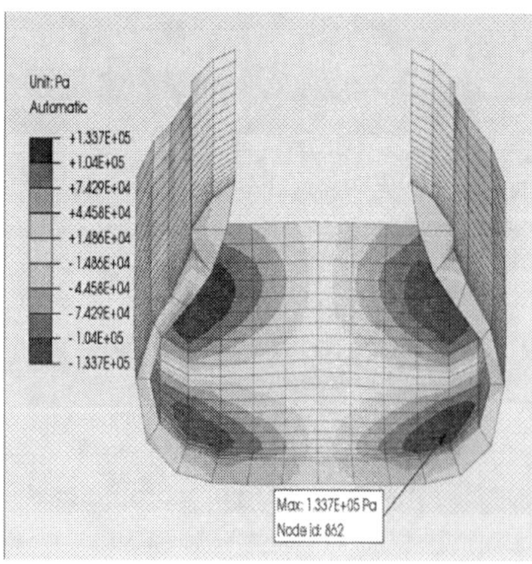

(e) ϵ_{xy}.

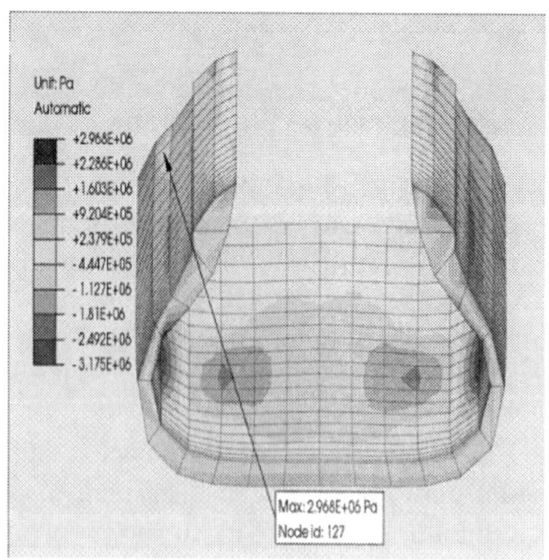

(f) ϵ_{yy}.

Figure 4: Results from the virtual experiment for a 30mm displacement.

can be applied are restricted to those where contact does occur to reduce the dimensionality of the problem from 210 unknowns to 14. The unconstrained Nelder-Mead optimization algorithm is used to solve the optimization problems

$$\underset{x}{\text{minimize}}\, f(\boldsymbol{x}) = \sqrt{\frac{\sum (U_x - U_x^*(\boldsymbol{x}))^2 + \sum (U_y - U_y^*(\boldsymbol{x}))^2 + \sum (U_z - U_z^*(\boldsymbol{x}))^2}{N}}, \text{ and} \quad (1)$$

$$\underset{x}{\text{minimize}}\, f(\boldsymbol{x}) = \sqrt{\frac{\sum (\widetilde{\epsilon_{xx}} - \widetilde{\epsilon_{xx}^*}(\boldsymbol{x}))^2 + \sum (\widetilde{\epsilon_{yy}} - \widetilde{\epsilon_{yy}^*}(\boldsymbol{x}))^2 + \sum (\widetilde{\epsilon_{xy}} - \widetilde{\epsilon_{xy}}^*(\boldsymbol{x}))^2}{N}} \quad (2)$$

where $\boldsymbol{U}$ and $\boldsymbol{\epsilon}$ are the N nodal displacements (U_x, U_y, U_z) and N nodal strains $(\widetilde{\epsilon_{xx}}, \widetilde{\epsilon_{yy}}, \widetilde{\epsilon_{xy}})$ on the inner surface of the test section from the virtual experiment, while $\boldsymbol{U}^*(\boldsymbol{x})$ and $\boldsymbol{\epsilon}^*(\boldsymbol{x})$ are the N nodal displacements and N nodal strains $(\widetilde{\epsilon_{xx}}(\boldsymbol{x}), \widetilde{\epsilon_{yy}}(\boldsymbol{x}), \widetilde{\epsilon_{xy}}(\boldsymbol{x}))$ on the inner surface of the current iteration of the virtual inverse problem. The nodal strains $((\widetilde{\epsilon_{xx}}, \widetilde{\epsilon_{yy}}, \widetilde{\epsilon_{xy}}), (\widetilde{\epsilon_{xx}}(\boldsymbol{x}), \widetilde{\epsilon_{yy}}(\boldsymbol{x}), \widetilde{\epsilon_{xy}}(\boldsymbol{x})))$ are extrapolated from the strains at the integration points of each element to the nodes on the inner surface of the tyre because T2-Cam reports surface strains. The design vector $\boldsymbol{x}$ is the forces applied to the outside of the test section, and $f(\boldsymbol{x})$ is the objective function used in the optimization. Equations 1 and 2 are displacement-based and strain-based virtual inverse problems treated as different methods in this study but can be combined into one objective function for a third formulation.

The sum of design vector $\boldsymbol{x}$ which is solved for in the virtual inverse problem is the total force acting on the tyre, F_z, and each force in the design vector produces can produce up two moments M_x, M_y. If the total vertical force recovered in the virtual inverse simulation correlates with the appropriate virtual experiment, then it indicates that the forces that were recovered are statically equivalent to the true answer. The sum of each of these moments M_x, M_y is close to zero in the virtual experiment therefore if the result of a virtual inverse simulation corresponds to this it indicates that on average the forces are symmetrical but may still vary. The design vector $\boldsymbol{x}$ is a distribution of force and has a particular shape, which is desirable to recover in addition to the total forces and moments. This distribution of forces recovered from the virtual inverse simulation can be compared to the virtual experiment as a distribution of errors. A distribution of errors which is zero-mean with a low standard deviation relative to the largest nodal force predicted by the virtual experiment indicates that the design vector $\boldsymbol{x}$ is a similar distribution to the true answer.

3 Results

3.1 Ideal Virtual Inverse Simulations

The best-case scenario to recover forces from measured displacements or strains would occur if there is no appreciable measurement noise and the initial guess is good. Such a scenario is a useful baseline for comparisons to inverse solutions that follow. Two virtual inverse simulations like this were run, one using the displacement-based method and the other with the strain-based method. In both cases the initial guess was chosen to be 90% of the known answer from the virtual experiment and therefore has the correct shape but lower magnitude so as to avoid numerical instability. Figure 5 shows the results of solving for the contact forces using both virtual inverse methods. The net force in the displacement-based method was 359N and 357N for the strain-based method. The standard deviation of the distribution of forces from the virtual experiment was 7.27N (25.68N mean), while that of the displacement-based and strain-based methods was 7.39N (25.66N mean) and 7.37N (25.52N mean) respectively. In both cases, the objective function converged to a zero error (within the limits of machine precision), and although the distribution of the force is not perfect, it is a reasonable reflection of the expected forces. In both cases a small moment about x and y was present due to the errors in the force distribution, however, they are all under 0.9Nm.

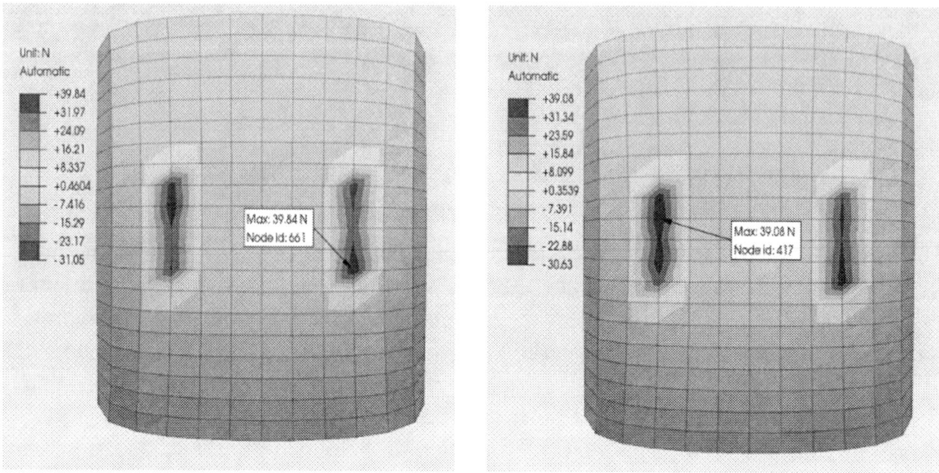

(a) Displacement-based. (b) Strain-based.

Figure 5: Identified Vertical Reaction Forces.

3.2 Effects of Measurement Noise

As this study is focused on the identifiability of the forces resulting from a virtual inverse simulation, comparisons between the two methods will be made for comparable levels of uncertainty. The effect of these uncertainty levels was investigated by sampling from zero-mean 1mm or 0.04% standard deviation for the displacement-based and strain-based virtual inverse simulations, respectively. This noise was added to the appropriate measurement from the virtual experiment. The current strain or displacement is compared to a noisy version of the virtual experiment at each iteration. The initial guess remained as 90% of the known answer from the virtual experiment, as the intent of this portion of the study is aimed at understanding the effect of noise on the identifiability of external forces.

Table 1 shows the total vertical force and both moments (M_x, M_y) from the virtual experiment and the predictions from both virtual inverse simulations with and without measurement noise. The forces and moments for the noisy simulations were averaged over 10 runs. This level of noise does not significantly impact the forces' identifiability, since the total vertical force is predicted well, and there are negligible moments predicted where one expects zero moments. Table 2 shows the RMS error in the displacement and strain inside the test section with and without additive noise. In cases where the displacement-based formulation was used, the RMS error on the strains is not significantly worse than if a strain-based method is used and vice-versa.

The force error at each node from the virtual inverse simulation to the reference force from the virtual experiment is calculated for both the strain and deformation formulations.

A histogram of these errors is presented in Figure 6. Both distributions are zero-mean, but the displacement-based method's standard deviation was 2N whilst that of the strain-based method was 3.2N.

Table 1: Forces and moments predicted by virtual inverse simulations with and without noise.

	F_z N	M_x Nm	M_y Nm
Reference	359.08	0	0
Displacement-based	359.23	-0.840	-0.019
Strain-based	356.71	-0.15	-0.25
Displacement-based + $\mathcal{N}(0, 1)$	358.28 ±0.4	-0.84 ±0.1	-0.020 ±0.05
Strain-based + $\mathcal{N}(0, 0.04)$	357.40 ±0.9	-0.2 ±0.02	-0.27 ±0.01

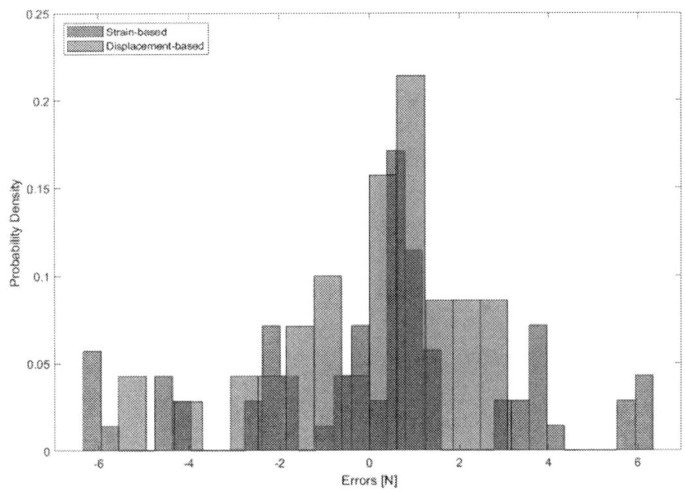

Figure 6: Spread of the normalized errors on the forces recovered by the virtual inverse simulations.

Table 2: Average root mean square error for the virtual inverse simulations with and without measurement noise.

	U_x mm	U_y mm	U_z mm	ϵ_{xx} μ	ϵ_{yy} μ	ϵ_{xy} μ
Displacement-based	0.0075	0.012	0.049	116.60	172.07	138.53
Strain-based	0.030	0.19	0.24	485.70	560.16	252.52
Displacement-based + $\mathcal{N}(0, 1)$	0.010	0.040	0.070	131.08	171.22	147.54
Strain-based + $\mathcal{N}(0, 0.04)$	0.27	1.93	2.25	4538.7	5089.14	2530.81

3.3 Effects of Initial Guess

Suppose the design domain contains many local minima. In that case, an initial guess which is close to the answer will not correctly expose the sensitivity of the forces recovered by the virtual inverse problem. To investigate these, ten runs are performed for each method where the initial guess is randomized but remains on average 90% of the answer. For each virtual inverse simulation, samples drawn from a Gaussian random variable with zero-mean and 4N standard deviation are added to the initial guess, after which the usual procedure is followed. The standard deviation of the uncertainty added to the initial guess is more than 10% of the maximum nodal force in the virtual experiments of 35.5N. In addition, similar virtual inverse simulations are run with a "uniform" initial guess, where the initial guess is the same at every node and corresponds to the test section's vertical stiffness multiplied with the total displacement and divided by the number of nodes in contact. The uniform case is statically equivalent to the true answer but can severely over- or under-estimate local forces at a particular node. The same level of artificial measurement noise used in Section 3.2 was retained for all simulations.

Table 3 shows the average vertical force and moments predicted by the virtual inverse simulations over ten runs, each with two different initial guesses. Although both guesses for both methods show larger moments than when an ideal initial guess is used, the total vertical force is still predicted within 6N (1.6%) in all cases. Table 4 shows that both initial guesses for both methods converged to solutions that have very low RMS error on the displacement and strains on the inside of the test section, regardless of whether a displacement-based or a strain-based method was used. However, when considering the force distributions shown in the histograms of the normalized errors, the variance in the forces predicted locally has become very large. Figure 7a shows that an initial guess of 90% of the true forces with a large uncertainty still produces a zero-mean distribution of normalized errors, but the variance is around four times as high. The standard deviation for the displacement-based and strain-based methods with this initial guess is 16.4N and 11.6N, respectively. Figure 7b shows that the variance is even worse when a uniform initial guess is used, and a bi-modal distribution with high variance appears. For the latter, he standard deviations of the normalized errors for the displacement-based and strain-based methods are 16.8N and 13.3N, respectively.

Table 3: Forces and moments predicted by virtual inverse simulations with different initial guesses.

Initial Guess	Method	F_z [N]	M_x [Nm]	M_y [Nm]
Reference		359.08	0	0
90% + $\mathcal{N}(0,4)$	Displacement	358.6 ±4	-0.933 ±1.3	0.03853 ±0.4
90% + $\mathcal{N}(0,4)$	Strain	356.6 ±5	-1.239 ±2	-0.1884 ±1
Uniform + $\mathcal{N}(0,4)$	Displacement	364.0 ±2	-1.694 ±2	-0.6371 ±0.2
Uniform + $\mathcal{N}(0,4)$	Strain	357.1 ±2	-1.659 ±0.5	0.07057 ±0.4

4 Discussion

In Section 3.2, it is clear that measurement noise does not significantly impact the identifiability of the total vertical contact forces in this test case. For a noise level representative of the

Table 4: Average root mean square error for the virtual inverse simulations with different initial guesses.

Initial Guess	Method	U_x mm	U_y mm	U_z mm	ϵ_{xx} μ	ϵ_{yy} μ	ϵ_{xy} μ
90% + $\mathcal{N}(0,4)$	Displacement	0.03227	0.08697	0.1997	284.4	351.9	226.2
90% + $\mathcal{N}(0,4)$	Strain	0.05158	0.4087	0.4987	911	1029	487.2
Uniform + $\mathcal{N}(0,4)$	Displacement	0.1041	0.6276	0.8744	1569	1895	1082
Uniform + $\mathcal{N}(0,4)$	Strain	0.0446	0.4657	0.4825	930	1027	401.8

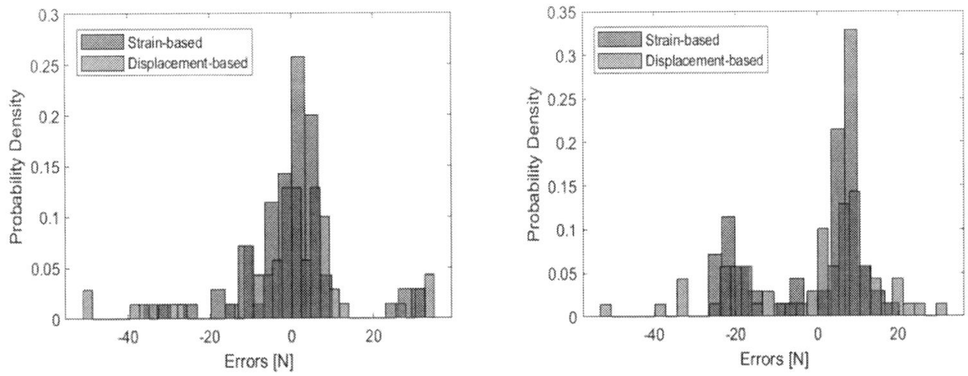

(a) 90% + $\mathcal{N}(0,4)$.

(b) Uniform guess + $\mathcal{N}(0,4)$.

Figure 7: Spread of the local normalized errors on the forces recovered by the virtual inverse simulations.

uncertainty in the deformation measurements, one can expect that the standard deviation of the uncertainty in the force distribution is less than 6% of the total force predicted for the test case. Comparable levels of uncertainty in the strain measurements produce slightly higher but similar levels of uncertainty in the force distribution at 9%. In all cases, the total force predicted by the virtual inverse problem matches the virtual experiment. The standard deviation of the normalized local errors on the forces identified is within 10% of the maximum local force. For this case, it is possible to recover the force distribution over the contact patch.

Section 3.3 shows, however, that the identifiability of the individual contact forces diminishes significantly in the presence of a poor or non-physical initial guess. Despite this, the total force is still acceptably predicted by the virtual inverse problem, even when the distribution of those forces is far from the known solution from the virtual experiment. It is difficult to identify the true solution under non-ideal circumstances, especially if one does not have a good initial guess. Saint Venant's principle dictates that in this case, it becomes increasingly difficult to distinguish between different statically equivalent loads the further away the strains and displacements are observed [10].

The identifiability of the contact forces from deformation measurements is dominated more by the initial guess rather than the measurement noise. In Section 3.2, the initial guess is chosen with the correct shape but a lower magnitude than the true solution, while in Section 3.3, the initial guess can have a non-physical shape. However, the magnitude is still on average lower than the true solution. There is clearly some evidence that if one intends to recover the distribution of contact forces, then the initial guess must be regularized in some way to avoid converging to one of the many local minima that result in non-physical solutions in the solution space.

5 Conclusions

This work aimed to determine the degree to which external forces on a tyre can be identified from the deformations or strains on the inside of the tyre. In the ideal case, where no measurement noise is present, and the initial guess represents the true forces, both the distribution and the total forces on the tyre can be readily identified. If measurement noise is added, which corresponds to what can be expected in laboratory tests, then one can still expect to recover the total force on the tyre, but the distribution of these forces becomes harder to identify. Although the distribution of the forces is more difficult to identify when measurement noise is present, the spread of the errors on the forces is still low relative to the total force predicted. The identifiability of the distribution of the external forces becomes very low once the initial guess used in the virtual inverse problem is not representative of the true answer. However, it is still possible to accurately estimate the total force produced by the tyre in this case. These results indicate that there may be many local minima in the solution space of external forces, many of which have a similar total force but are non-physical for a tyre interacting with the road. It is recommended that future work focuses on methods for regularizing the initial guess used in such a way as to avoid converging to solutions where the distribution of forces is non-physical.

References

[1] The European Commission (2005). Intelligent tyre for accident-free traffic. Finland. https://cordis.europa.eu/project/id/IST-2001-34372.

[2] Gutherie, AG (2016) 3D Computer Vision Contact Patch Measurements Inside Off-Road Vehicle Tyres, Master's thesis, University of Pretoria.

[3] Feldesi, F (2018) Three-dimensional Contact Patch Strain Measurement Inside Rolling Off-Road Tyres, Master's thesis, University of Pretoria.

[4] Pegram, MS, Botha, TR and Els, PS (2021). Full-field and point strain measurement via the inner surface of a rolling large lug tyre. *Journal of Terramechanics*, 96, 11-22.

[5] van Oosten, JJM, Unrau, H-J, Riedel, A and Bakker, E (1997). TYDEX workshop: standardisation of data exchange in tyre testing and tyre modelling *Vehicle System Dynamics*, 1997, 272–288.

[6] Tekscan, Inc (2021). Accessed 1 July 2021, https://www.tekscan.com/products-solutions/pressure-mapping-sensors/8001.

[7] Dhondt, G and Wittig, K (1998) CalculiX: A Three-Dimensional Structural Finite Element Program.

[8] Geuzaine, C and Remacle, J-F (2009) Gmsh: A 3-D finite element mesh generator with built-in pre-and post-processing facilities *International journal for numerical methods in engineering*, 2009, 1309 – 1331.

[9] Stallmann, MJ (2013) Cost Effective FTire Parameterisation Methods for Ride Simulations with Large Off-The-Road Tyres, PhD Thesis, University of Pretoria.

[10] Gere, JM and Goodno, BJ (2013). Mechanics of Materials, 8th Eds. Cengage Learning.

Statistical Design of Experiments: An introductory case study for polymer composites manufacturing applications

Natasha Botha[1,2,*], *Helen M.* Inglis[2,**], *Roelof* Coetzer[3,***], and *F. Johan W.J.* Labuschagne[4,****]

[1]Operational Intelligence, Next Generation Enterprises and Institutions, Council for Scientific and Industrial Research, South Africa
[2]Department of Mechanical and Aeronautical Engineering, University of Pretoria, South Africa
[3]Department of Statistics, North-West University, South Africa
[4]Department of Chemical Engineering, University of Pretoria, South Africa

Abstract. Statistical design of experiments (DoE) aims to develop a near efficient design while minimising the number of experiments required. This is an optimal approach especially when there is a need to investigate multiple variables. DoE is a powerful methodology for a wide range of applications, from the efficient design of manufacturing processes to the accurate evaluation of global optima in numerical studies. The contribution of this paper is to provide a general introduction to statistical design of experiments for a non-expert audience, with the aim of broadening exposure in the applied mechanics community. We focus on response surface methodology (RSM) designs — Taguchi Design, Central Composite Design, Box-Behnken Design and D-optimal Design. These different RSM designs are compared in the context of a case study from the field of polymer composites. The results demonstrate that an exact D-optimal design is generally considered to be a good design when compared to the global D-optimal design. That is, it requires fewer experiments while retaining acceptable efficiency measures for all three response surface models considered. This paper illustrates the benefits of DoE, demonstrates the importance of evaluating different designs, and provides an approach to choose the design best suited for the problem of interest.

1 Introduction

Experimental and numerical work are key parts of any research or industrial project. The engineers and scientists conducting these investigations frequently use some statistical analysis to analyse and report their results. However, not all these investigators have been trained in statistical methods, resulting in a lack of knowledge of how to properly design an experiment or apply the correct statistical tools when there are multiple interacting variables [1]. As a result, investigators often use the one-factor-at-a-time approach which is commonly taught during their training or practised in industry [1].

*e-mail: nbotha1@csir.co.za
**e-mail: helen.inglis@up.ac.za
***e-mail: roelof.coetzer@nwu.ac.za
****e-mail: johan.labuschagne@up.ac.za

In an experimental design we need to identify the factors, levels and response variables. A **factor** is the input variable of interest. If we consider the example of melt compounding of polymer composites, as in the study by Ujianto *et al.* [2], the factors might include temperature and screw speed. The **levels** are the number of different variations we consider for each factor, for example, with three levels we would consider a low, medium and high temperature. A **response variable** is the output variable of interest, such as material strength.

In a **one-factor-at-a-time approach**, the experimenter varies one factor while keeping the other factors of interest fixed. This is illustrated in figure 1(a) for a design space with two factors. Such an approach requires a large number of experimental runs, but does not give insight into the interactions between the variables or identify the optimum conditions [3].

To mitigate these challenges we consider an approach from the field of statistics, statistical design of experiments (DoE). DoE is the collective name for statistical optimisation techniques which are commonly employed to plan an experimental program with a sufficient number of experiments in order to make sound scientific conclusions from the data [3, 4]. It achieves this by optimising the design process, screening for the most influential factors and determining their interactions while establishing and maintaining quality control [3]. The goal of a DoE is therefore to provide a near optimal and efficient design which can be performed with the minimum number of experimental runs.

Bessa *et al.* [5] demonstrated the power of DoE for machine learning in non-linear material simulation. They developed a numerical database of constitutive response curves for microstructural representative volume elements, using microstructure descriptors, material properties and boundary conditions as input factors for the design. Using a DoE approach ensured that they were efficiently sampling the full non-linear design space to generate training data for their machine learning algorithm, avoiding the incomplete information problem of a one-factor-at-a-time method.

DoE's can be grouped into two design classes, factorial designs and response surface method (RSM) designs.

Factorial designs consider all the possible combinations of factors and levels that are being investigated. As a result these tend to rapidly increase the number of experimental runs required as the number of factors increase. A full factorial design will have n^k experiments where n is the number of levels and k is the number of factors. For certain applications, it is possible to do a fractional factorial design which will reduce the number of experimental runs. Factorial designs are generally used when screening the factors of interest. An example

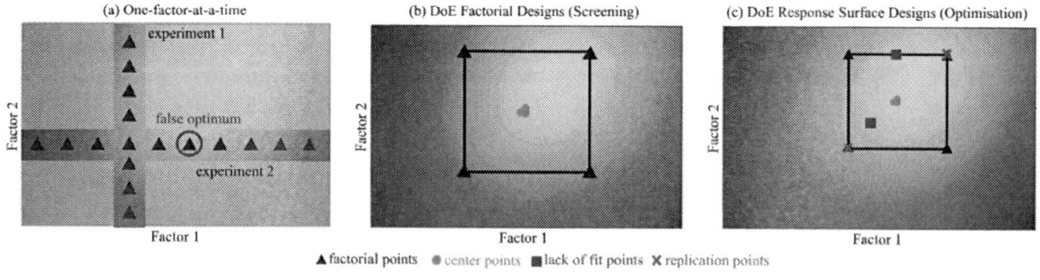

Figure 1. Graphical illustration of the (a) one-factor-at-a-time approach, (b) factorial designs for efficient screening and (c) response surface designs for optimisation. The contour colour gradient represents the response variable (yellow = high value and blue = low value). [Redrawn from [6] under the Creative Commons Attribution 4.0 International License (http://creativecommons.org/licenses/by/4.0/)]

of factor screening considering a full factorial design is shown in figure 1(b) for two factors and two levels.

RSM designs rely on making an initial assumption about the shape of the response surface. The response surface is the response variable plotted as a function of the factors in the design space. An experiment can be efficiently designed to accurately capture the features of the assumed response surface. RSM is employed after factor screening to optimise the product or system. An example of an RSM design is illustrated in figure 1(c) showing a more focused design region compared to the factorial design. Notice that this is the first method which captures an estimate of the true optimum.

The aim of this study is to showcase and discuss several RSM designs. We will explain the design methods and define the criteria which will be used to compare them. We will do this comparison in the context of a case study from the field of polymer composites. The contribution of this paper is in demonstrating how this statistical technique can improve the quality and reliability of experimental or numerical investigations.

2 Theory and Background of Response Surface Methods

In this section we first define the three simplest response surface models. Four classic examples of RSM designs are then introduced and explained: Taguchi Design, Central Composite Design (CCD), Box-Behnken Design (BBD) and D-optimal Design. The criteria for evaluating the efficiency of these methods are defined.

2.1 Response Surface Models

RSM rely on the assumption of a mathematical relationship between the response variable and the factors. In this section we present the three simplest assumptions: a linear model, a two-order linear interaction model and a quadratic model. These are illustrated in figure 2 for a model with two factors. In each of these models y represents the response variable, x_i the factors and β_i and β_{ij} are the regression coefficients. The purpose of an experiment is to characterise the response surface by solving for the number of p regression coefficients.

A **linear model** (also called a linear regression model) for k factors has $p = k + 1$ model parameters and will take the following form:

$$y = \beta_0 + \sum_{i=1}^{k} \beta_i x_i. \tag{1}$$

The **two-order interaction model** (also called an interaction model) for k factors has $p = 1 + (k^2 + k)/2$ model parameters:

$$y = \beta_0 + \sum_{i=1}^{k} \beta_i x_i + \sum_{i=1}^{k} \sum_{j=i+1}^{k} \beta_{ij} x_i x_j. \tag{2}$$

A **quadratic model** for k factors has $p = 1 + (k^2 + 3k)/2$ model parameters:

$$y = \beta_0 + \sum_{i=1}^{k} \beta_i x_i + \sum_{i=1}^{k} \sum_{j=i+1}^{k} \beta_{ij} x_i x_j + \sum_{i=1}^{k} \beta_{ii} x_i^2. \tag{3}$$

2.2 Explanation of Different RSM Design Methods

An overview of four commonly used RSM design methods is given in this section. They are illustrated in figure 3, for three factors, in which case the design space is represented by a cube.

2.2.1 Taguchi Design

The Taguchi design originated as a class of factorial design known as robust parameter design [3, 7]. A Taguchi design tests all combinations of factors and levels, similar to a factorial design, but uses pairs of combinations to reduce the number of experiments required [4]. RSM is applied to the Taguchi design to improve the efficiency of the design by mathematically including both the controllable and noise (or uncontrollable) factors in a single response function to account for uncertainty [3]. Thus the effect of two or more levels of the factors on the response variables can be studied [4, 8]. The two levels of factors can be seen in figure 3(a) with the inner factorial points denoting the controllable factors and the outer factorial points the noise factors. A disadvantage of Taguchi designs is that they do not account for interaction effects between controllable and noise factors [7].

2.2.2 Central Composite Design and Box-Behnken Design

A CCD is a five level factorial design consisting of factorial points at the edges of the design space, axial points at the midpoint of each level (centers of the faces) and a center point in the middle of the design space. A CCD is illustrated in figure 3(b). The CCD is based on a two level factorial design, 2^k, where k is the number of factors. By adding axial and center points to the design, the CCD requires fewer experimental runs compared to a full factorial design while still producing similar results [4]. Axial points are determined by the distance from the design center, α, which provides a measure of rotatability. The rotatability of a CCD provides an indication of the model's ability to ensure the variance of the predicted response is constant and stable when the design is rotated around the center [3]. This rotatibility measure is dependent on the assumed shape of the response surface. For this study, we consider $\alpha = 1$ which assumes a cuboidal shape resulting in a face-centered CCD. The axial points are then in the center of the cube faces as shown in figure 3(b). Center points are added to the design to estimate experimental error. They also provide an indication of curvature, that is detecting quadratic effects [7]. A CCD is generally considered to be efficient when fitting a second order model [3, 4, 8].

An alternative to the CCD is the BBD which is considered to be more economical as it requires fewer experimental runs for the same number of factors [3]. A BBD is based on a 2^k three level factorial design with an incomplete block design [4, 7, 8]. An incomplete block design considers all combinations for a certain number of factors while keeping the remaining factors at the central levels [3, 4, 8]. The BBD is illustrated in figure 3(c). In comparison with

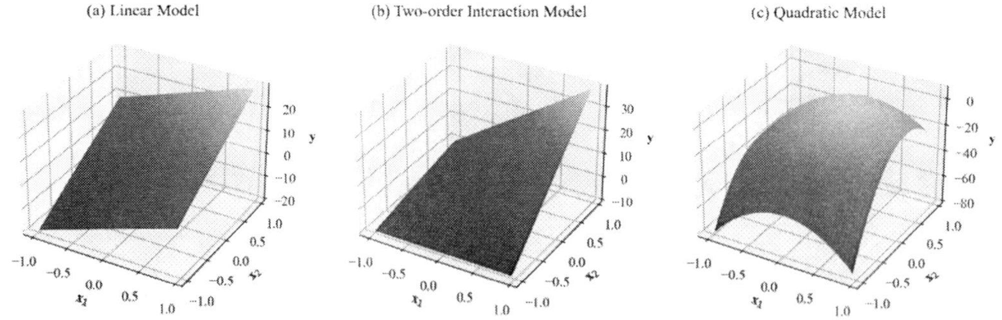

Figure 2. Example response surface plots for the (a) linear (regression) model, (b) two-order interaction model and (c) quadratic model.

the CCD (figure 3(b)) there are no factorial points at the vertices of the design space. Instead, the factorial points lie along a sphere with radius $\sqrt{k}$. This is similar to where the axial points of a spherical CCD would lie if the design were rotated with radius $\alpha = \sqrt{k}$. As with the CCD, center points can also be added to the BBD to estimate experimental error.

2.2.3 D-optimal Design as an example of Optimal Designs

Optimal designs are very flexible and are able to handle different user constraints and more complex experimental design problems, such as: constrained design regions; many factors or levels and limits on the total number of experimental runs [3, 7]. Optimal designs are numerically generated designs which aim to minimise a specific optimality criterion to select a near optimal design [3, 7]. Optimality criteria minimise some mathematical measure which focus on either the estimation of regression coefficients or the prediction of the response variable in the design region [7]. The computational algorithm first constructs a full factorial design to create a set of candidate points. An initial design is then selected at random from the candidate points and design weights ($w_i = 1/p$) are assigned to each design point. The design is then evaluated based on the response surface model and specified optimality criterion. The

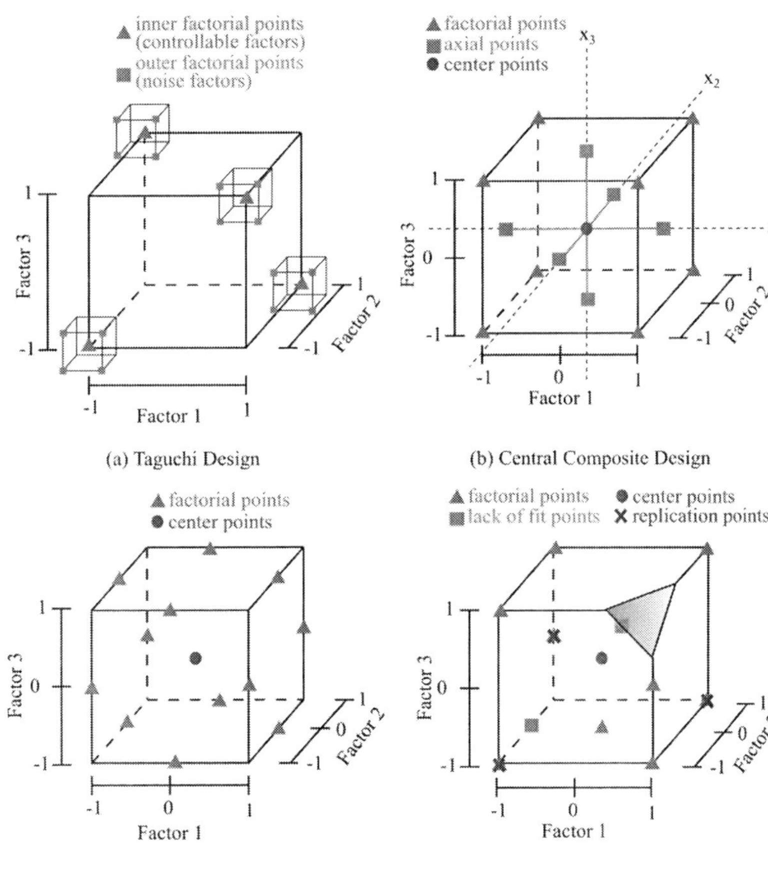

Figure 3. Graphical representation of the three-factor design space for the different RSM designs: (a) Taguchi, (b) Central Composite Design, (c) Box-Behnken and (d) D-Optimal.

optimality criterion is optimised by using a suitable nonlinear optimisation algorithm [3, 7]. The choice of optimality criterion determines the type of optimal design. For this study, we will consider a D-optimal design as it is the most widely used optimality criterion [3, 7]. The interested reader is referred to Montgomery Chapter 11 [3] or Myers *et al.* Chapter 9 [7] for the definitions of the other optimality criteria.

A D-optimal design aims to find a good estimation of the regression coefficients subject to the constraints on the design region and experimental runs [7]. The D-optimal design estimates the regression coefficients by maximising the determinant of $|(\mathbf{X}'\mathbf{X})|$, where $\mathbf{X}$ is the matrix as known as the design matrix. $\mathbf{X}$ is populated with all the combinations of the factor levels (i.e., -1, 0, 1 for a three level design) for each factor and has columns expanded for all the terms in the response surface model. $\mathbf{X}'$ is the transpose of the design matrix. A D-optimal design is illustrated in figure 3(d). The constrained design region is represented by a gray triangle in the top right corner, indicating that no design points can be selected above the triangle. As with the CCD and BBD designs center points can be included for estimation of experimental error. For an optimal design, additional design choices can also be made which include specifying the number of replicates and lack of fit points. Replications (purple 'x' in figure 3(d)) of certain design points can be added if there is a strong interest in a specific combination of factors and levels [3]. On the other hand, if there is a need to test the prediction ability of the chosen response surface model, lack of fit points (green squares in figure 3(d)) are included in the design space [3].

Optimal designs can be classified as either continuous (i.e. approximate) or exact (i.e. discrete) designs [9]. The classification refers to the optimisation approach considered when generating the optimal design. For a continuous design, the aim is to find the minimum number of distinct design points with their assigned design weights ($0 \leq w_i \leq 1$) considering the design region [9]. Exact designs specify the number of experimental runs (N) for the optimisation process at the start [9]. The exact design can be approximated by multiplying the design weights $w_i = r_i/N$ (r_i is the number of replication points) with the total number of experiments (N) [9].

Global D-optimal designs are continuous designs, using the full design space, based on the D-optimality criterion. For a given set of factors, levels and response surface, the global D-optimal design is uniquely defined, and captures the full design response. In this study, each of the experimental designs will be compared to the global D-optimal design for the relevant response surface model. We will compare both the required number of experiments and the efficiency (Sect. 2.3) to the global D-optimal design.

2.3 Design Evaluation

We measure the performance of a design by evaluating its efficiency. The efficiency is an indication of whether a design will be able to produce reliable results under a variety of circumstances [7]. There are two widely used measures of design efficiency: D-efficiency and G-efficiency [3, 7, 9].

D-efficiency is a measure of the precision of estimation of the regression coefficients, β_{ij} [7]. We can understand D-efficiency by considering figure 4(a), which illustrates D-efficiency in two dimensions. In this simple example the area of the ellipse represents the confidence region of the regression coefficients. D-efficiency is achieved when the area of this ellipse is minimised [3, 7], which is equivalent to maximising the determinant, $|(\mathbf{X}'\mathbf{X})|$, where $\mathbf{X}$ is the design matrix containing the factors (x_i) for all levels in the response surface model.

When we want to compare two different designs we need to evaluate the **relative D-efficiency**, which compares the D-efficiency of the current design to that of the global D-optimal design (denoted by subscript *gopt*) by comparing their determinants [3, 9]:

$$D_{eff} = \left(\frac{|(\mathbf{X}'\mathbf{W}\mathbf{X})|}{|(\mathbf{X}'_{gopt}\mathbf{W}_{gopt}\mathbf{X}_{gopt})|} \right)^{1/p}, \tag{4}$$

where p is the number of regression coefficients, and the matrix $\mathbf{W}$ is a diagonal matrix containing the design weights attached to each design point. $|(\mathbf{X}'\mathbf{W}\mathbf{X})|$ is the determinant whose inverse provides insight into the variance of the regression coefficients.

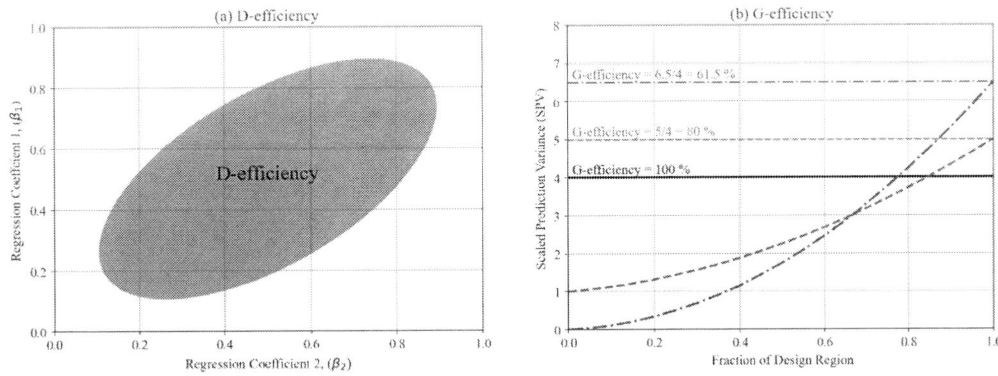

Figure 4. Graphical representation of the (a) D-efficiency where the volume of the ellipse denotes the confidence region of the regression coefficients; and (b) G-efficiency where the dashed and dashed-dot curves are examples, for different designs, of the prediction variance across the design region considering a linear model.

The **G-efficiency** of a design is a measure of the accuracy of the prediction of the response variable in the design region [3, 7]. The G-efficiency is defined as [3, 9]:

$$G_{eff} = \frac{p}{\max(\bar{x}')(\mathbf{X}'\mathbf{W}\mathbf{X})^{-1}\bar{x}} = \frac{p}{\max(SPV)}, \tag{5}$$

where p is the number of regression coefficients and $\bar{x}$ is a vector containing the factor values at which the response variable prediction is made. The denominator is known as the scaled prediction variance (SPV). SPV is the variance of the predicted response at a point in the design region scaled to the number of experimental runs [7].

A simple visualisation of G-efficiency for a linear model with three factors is shown in figure 4(b). Two designs are shown, represented by the red dash-dotted line and the blue dashed line. The y-axis, the SPV, is a measure of how many regression coefficients are required to accurately predict the model response for a given fraction of the design space (FDS), shown on the x-axis. The linear model with three factors has four regression coefficients, i.e. $p = 4$. This is represented by the black line for a G-efficiency of 100 %. The red design has $\max(SPV) = 6.5$ giving a G-efficiency of 61.5 %, while the blue design has $\max(SPV) = 5$ resulting in a G-efficiency of 80 %. This means that the red design would require more replications at the specified design points to achieve similar prediction variance as the blue design. Thus it has a lower G-efficiency. To improve the G-efficiency of the design we want to minimise the variance in the response variable prediction, i.e. change the design such that at an FDS of 1 the SPV is minimised [3, 7].

3 Case Study

In this section we briefly explain the case study. The response surface models are then defined and the global D-optimal and RSM designs generated. Finally, we compare and evaluate the designs in terms of the relative D-efficiency and the G-efficiency, as well as the number of experiments required.

For this study, we used Design Expert [10] to generate the different RSM designs. To develop the global D-optimal designs and to calculate the D-efficiency and G-efficiency we used R [11] with the AlgDesign Package [12].

3.1 Explanation of Case Study

A case study is used to illustrate the differences between the different experimental designs. Ujianto *et al.* [2] aimed to find the optimum compounding conditions for an HDPE/Cloisite 93A/HDPE-g-MA (93/2/5 wt%) polymer composite using an internal mixer. They investigated the internal mixer temperature, rotor speed and mixing time. They chose the lower and upper boundaries for each factor such that the polymer melts without causing material degradation. This paper has been chosen as the case study as it is well suited for RSM designs, allowing us to illustrate the use of DoE in the design of polymer composite manufacturing methods. The paper considers only quantitative (i.e., numerically measurable) factors which significantly simplifies the design approach.

The three factors and two levels considered are summarised in table 1. The authors used a BBD with three center points, requiring 15 experimental runs. For the response variable analysis they applied a quadratic response surface model.

Table 1. Factor and level selection for the case study

Factors	Low	High
Temperature (°C)	150	210
Rotor speed (rpm)	30	130
Mixing time (min)	4	16

3.2 Response Surface Models and Experimental Designs

The Taguchi, CCD and BBD designs are not dependent on the response surface model chosen. In other words, it is not required for the user to specify a response surface model to generate the design. The global D-optimal and D-optimal designs require the user to define the response surface model and will therefore change depending on the selected model. The full set of designs investigated is summarised in table 2. Notice that the Taguchi design is only considered for the linear model and that the CCD and BBD designs show the same number of experiments for each response surface assumption. In the modifications column we indicate if it is the base design (i.e., no additional points added) or whether any points are added to the design. For the Taguchi design additional points cannot be added to the design as it is already a complete design considering all the combinations of factors and levels. Center points (CP) provide a measure to estimate the experimental error and are investigated in all the designs except for Taguchi. For the CCD we can add either axial (Ax) or factorial (Fac) points which provides an indication of consistency and stability in the design (Sect. 2.2.2). For the D-optimal designs either replication (Rep) or lack of fit (LOF) points can be added improving the accuracy of specific factors or the response surface model (Sect. 2.2.3).

Table 2. Summary of the different RSM designs generated using the linear, interaction and quadratic models.

Design Method	Modifications	Number of Experiments for each response surface		
		Linear	Interaction	Quadratic
Taguchi L4	Base design	4	-	-
CCD	Base design	14	14	14
	Add 3 center points (3CP)	17	17	17
	Add 2 factorial points (2Fac)	22	22	22
	Add 2 axial points (2Ax)	20	20	20
	Add 2 factorial and 2 axial points (2Fac, 2Ax)	28	28	28
	Add 3 center, 2 factorial and 2 axial points (3CP, 2Fac, 2Ax)	31	31	31
BBD	Base design	12	12	12
	Add 3 center points (3CP)	15	15	15
D-optimal	Base design	4	7	10
	Add 3 center points (3CP)	7	10	13
	Add 5 replication points (5Rep)	9	12	15
	Add 5 lack of fit points (5LOF)	9	12	15
	Add 5 replication and 5 lack of fit points (5Rep, 5LOF)	14	17	20
	Add 3 center, 5 replication and 5 lack of fit points (3CP, 5Rep, 5LOF)	17	20	23
Global D-optimal		**4**	**8**	**21**

From equation (1) a linear model for three factors will have the following form:

$$y = \beta_0 + \beta_1 x_1 + \beta_2 x_2 + \beta_3 x_3, \quad (6)$$

where x_1, x_2 and x_3 represents the temperature, speed and mixing time, respectively. The relative D-efficiency compares the generated designs to a global D-optimal design as shown in equation (4). The linear model in equation (6) has four unknown regression coefficients to estimate. Therefore, the global D-optimal design is specified as a 2^{3-1} fractional factorial design with equal weights assigned to the four design points ($w_i = 1/4$ for an exact design). A fractional factorial design is very efficient as it covers the whole design region and provides precise estimates for the regression coefficients.

The two-order interaction model for three factors will have the following form based on equation (2):

$$y = \beta_0 + \beta_1 x_1 + \beta_2 x_2 + \beta_3 x_3 + \beta_{12} x_1 x_2 + \beta_{13} x_1 x_3 + \beta_{23} x_2 x_3, \quad (7)$$

where x_{12}, x_{13} and x_{23} represent the interactions between the different factors.This model has seven unknown regression coefficients. The global D-optimal design is therefore a 2^3 full factorial design with equal weights assigned to the eight design points i.e., $w_i = 1/8$.

A quadratic model for three factors will have the following form based on equation (3):

$$y = \beta_0 + \beta_1 x_1 + \beta_2 x_2 + \beta_3 x_3 + \beta_{12} x_1 x_2 + \beta_{13} x_1 x_3 + \beta_{23} x_2 x_3 + \beta_{11} x_1^2 + \beta_{22} x_2^2 + \beta_{33} x_3^2, \quad (8)$$

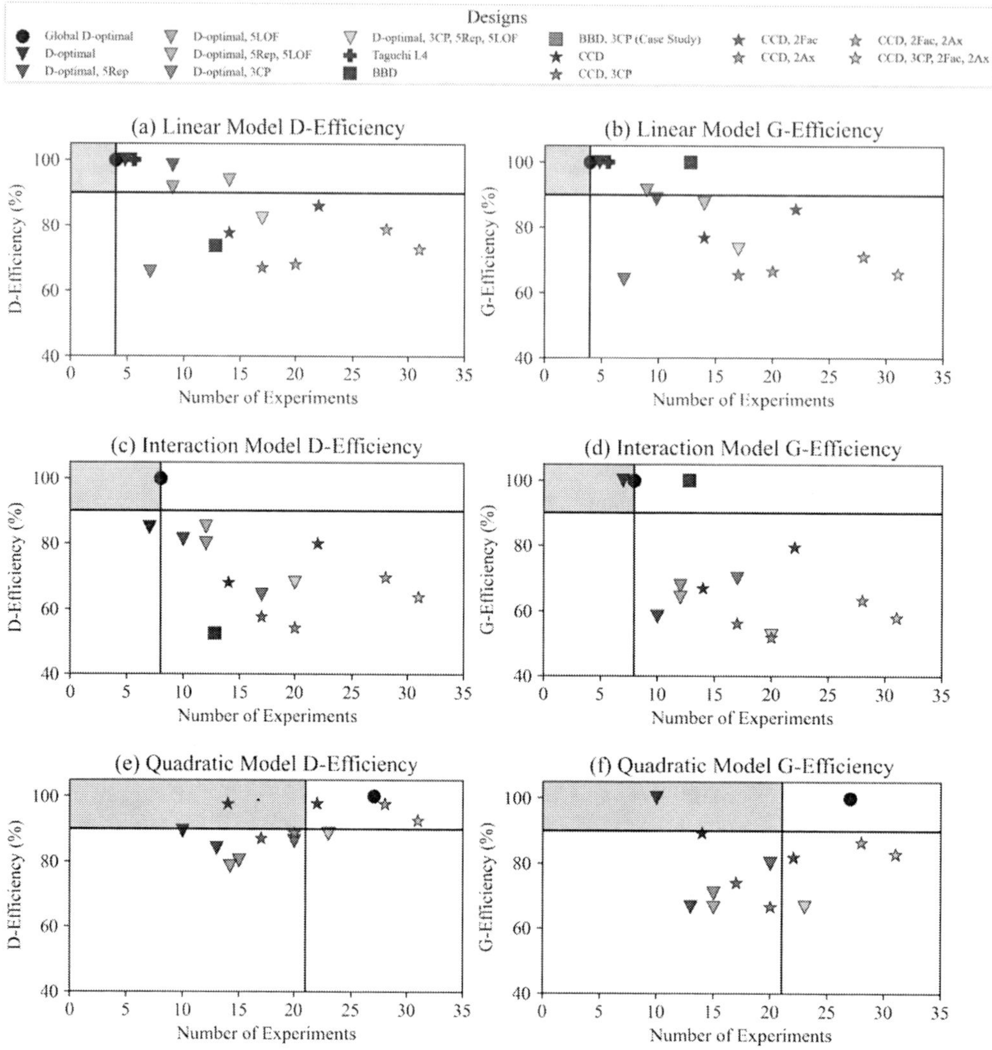

Figure 5. Calculated D- and G-efficiencies for the generated designs considering the (a) and (b) linear model, (c) and (d) two-order interaction model, and (e) and (f) quadratic model. The horizontal line indicates a 90 % efficiency and the vertical line denotes the number of experiments required by the global D-optimal design. Data points are slightly perturbed horizontally if they lie at the same location.

where x_1^2, x_2^2 and x_3^2 represent the quadratic terms. The quadratic model has 10 unknown parameters to estimate. For the global D-optimal design we need to consider three levels to capture the quadratic effects. The global D-optimal design is a fraction of the 3^k full factorial design. However, the design is made up of three parts with different design weights assigned to each. The center point has a weight equal to 0.066, the 2^3 factorial points have a total weight equal to 0.424, and the remainder of the points have total weight equal to 0.510 (see Atkinson *et al.* [9], Chapter 11, table 11.6).

3.3 Results and Discussion

The relative D-efficiency and G-efficiency for the different designs, including the global D-optimal design, are shown in figure 5, plotted against the number of experiments for all three response surface models. In this case study we considered efficiencies ≥ 90 % to provide an indication of good performance. This threshold is shown with a horizontal line. The vertical line shows the number of experimental runs required by the global D-optimal design. The gray area represents the region for a good design where we have high efficiency with fewer experiments than the global D-optimal design. To differentiate between designs at the same location we have incorporated a slight horizontal perturbation for these data points.

For a linear model the global D-optimal design requires 4 experiments. Both the Taguchi and D-optimal designs have a relative D-efficiency and G-efficiency of 100 % and also only require 4 experiments. These designs are therefore most efficient for estimating the linear regression coefficients, and predicting the response variable with minimum variance. They are also equivalent to the global D-optimal design with the exact same number of experiments. BBD has a 100 % G-efficiency, but a less than acceptable relative D-efficiency. It also requires three times the number of experiments compared to the global D-optimal, Taguchi and D-optimal designs. BBD is a less than ideal design to consider, but a viable option if the objective of the study is to predict the response variable. CCD generally provides the worst efficiencies and the highest number of experiments. It is therefore not a viable option when only considering linear effects.

For the two-order interaction model the global D-optimal design requires 8 experiments. It is clear that all the designs are sub-optimal in terms of their efficiencies. However, the D-optimal designs are a viable option as they still have relatively high D-efficiencies (in the 80 % range) requiring fewer experiments. BBD has a 100 % G-efficiency, but a very low relative D-efficiency. As with the linear model, it is an option if response variable prediction is the focus. CCD is not an option to consider for an interaction model as it requires the most experiments and has the worst efficiencies.

To investigate the quadratic effects, the global D-optimal design requires 21 experiments. Some CCD and D-optimal designs require fewer experiments while retaining acceptable efficiency measures.

Adding center points (CP) generally reduces the relative D-efficiency and G-efficiency for all designs and response surface models, while increasing the number of experimental runs by three. However, there is value in adding center points (CP) as it provides stability in the prediction variance and is a measurement of experimental error [7]. When adding replication (Rep) and lack of fit (LOF) points to the D-optimal design the decrease in both relative D-efficiency and G-efficiency is still within the acceptable range with only an additional five experimental runs needed. For CCD we note an improvement in relative D-efficiency and G-efficiency when adding factorial (Fac) points but at the cost of a larger number of experimental runs.

The case study design, a BBD with three center points, generally provided the worst relative D-efficiency for all three response surface models. This indicates that it is not able to estimate the regression coefficients precisely. On the other hand, it generally has an acceptable G-efficiency (in the 80 % range). In other words, it is able to predict the response variable with low variance in the design region. The aim of Ujianto *et al.* [2] was to analyse the effects of the compounding conditions. They would, therefore, have benefited from a design which is able to provide a better estimation of the regression coefficients, such as D-optimal.

4 Conclusion

In this study we have provided an overview of statistical design of experiments (DoE), comparing a range of different response surface method (RSM) designs — Taguchi Design, Central Composite Design (CCD), Box-Behnken Design (BBD) and D-optimal Design — for linear, two-order interaction and quadratic response surface models. We have explained the concepts of relative D-efficiency and G-efficiency, and evaluated the RSM designs using these measures for a case study considering the manufacturing design of polymer composites.

The D-optimal design was found to be more efficient in estimating the regression coefficients, while minimising the response variable prediction variance across the design region. It is considered to be a good design option for all three response surface models. When only considering main or linear effects of the factors, a Taguchi design is also a good choice. For quadratic effects, CCD is a viable option although it does require more experimental runs than a D-optimal design. Additional points (i.e. center, lack of fit and replication) can be considered, although these negatively impact the designs' relative D-efficiency and G-efficiency. It is up to the investigator to decide how many experimental runs are feasible and whether estimation or prediction is more important.

This paper makes a contribution to the computational and applied mechanics community by clearly showing the benefits of considering an experimental design approach, reducing the number of experimental runs while obtaining statistically relevant results. It has also shown the importance of evaluating different designs to determine which one is best suited for the problem of interest.

References

[1] M. Tanco, E. Viles, L. Pozueta, *Advances in Electrical Engineering and Computational Science* (Springer Science and Business Media B.V., 2009), chap. Comparing Different Approaches for Design of Experiments (DoE)

[2] O. Ujianto, M. Jollands, N. Kao, IOP Conference Series: Material Science and Engineering **319**, 012019 (2018)

[3] D. Montgomery, *Design and Analysis of Experiments*, 8th edn. (John Wiley & Sons, Inc., 2013)

[4] T. Gündoğdu, I. Deniz, G. Çalışkan, E. Şahin, N. Azbar, Critical Reviews in Biotechnology pp. 1–21 (2014)

[5] M. Bessa, R. Bostanabad, Z. Liu, A. Hu, D. Apley, C. Brinson, W. Chen, W. Liu, Computer Methods in Applied Mechanics and Engineering **320**, 633–667 (2017)

[6] G. Bowden, B. Pichler, A. Maurer, Scientific Reports **9**, 11370 (2019), https://doi.org/10.1038/s41598-019-47846-6

[7] R. Myers, D. Montgomery, C. Anderson-Cook, *Response Surface Methodology: Process and product optimization using designed experiments*, 4th edn. (John Wiley & Sons, Inc., 2016)

[8] J. Wang, W. Wan, International Journal of Hydrogen Energy **34**, 235 (2009)

[9] A. Atkinson, A. Donev, R. Tobias, *Optimum Experimental Designs, with SAS* (Oxford University Press, 2007)

[10] Stat-Ease, Inc., *Design-Expert® Software (version 8.0.0.)* (2009), https://www.statease.com/software/design-expert/

[11] The R Foundation, *The R Project for Statistical Computing (version 3.6.1)* (2019), https://www.r-project.org/

[12] R. Wheeler, *Package 'AlgDesign' (version 1.2.0)* (2019), https://github.com/jvbraun/AlgDesign

Using Full Field Data to Produce a Single Indentation Test for Fully Characterising the Mooney Rivlin Material Model

John D. Van Tonder[1,*], *Martin P.* Venter[1,**], and *Gerhard* Venter[1,***]

[1]Department of Mechanical and Mechatronic Engineering, Stellenbosch University

Abstract. A theoretical testing method for fully characterising the Mooney-Rivlin hyper-elastic material model is proposed by capturing full-field data, namely displacement field and indentation force data. A finite element model with known parameters will act as the experimental model against which all data will be referenced. This paper proposes a method of inverse finite element analysis operating under the assumption of equally objective function optimal planes or "hyper-planes". The paper concludes that the Mooney-Rivlin material model can theoretically be fully characterised in a single indentation test by applying methods discussed in the paper when using full-field data operating under the assumption of hyper-planes.

1 Introduction

Hyper-elastic materials have found themselves in numerous applications from simple water tight seals to primary components in soft-robotics [11], simulating the behaviour of these materials for improved design requires the use of a hyper-elastic material model. A method of performing inverse finite element analysis (FE analysis) on an indentation test may improve the process of isolating the correct material parameters since this loading method induces complex load cases in the deformed material. Inverse FE analysis also has the advantage in that it may reduce the engineering time for obtaining a suitable material model for FE analysis [10].

This paper will demonstrate that when operating under the assumption that there exists linear planes in the material parameter design space wherein the parameters have a fixed relationship that results in the objective functions for each data type are equally optimal [1],[7]. The Mooney-Rivlin three-parameter model can be fully characterised from full-field data acquired from a single indentation test as when evaluating the material parameters along the intersection line between the two planes for the same indentation test at different depth the parameters are convergent.

2 Methods and Materials

2.1 Analysis Method

Understanding how well the inverse FE analysis method performs under ideal circumstances, as in the case of FE model data, provides valuable insight on which factors are the most

*e-mail: 20380283@sun.ac.za
**e-mail: mpventer@sun.ac.za
***e-mail: gventer@sun.ac.za

relevant before attempting to do the same analysis on physical test sample data. Once the proposed method has demonstrated its effectiveness on accurately isolating the correct material parameters for a given material model based on simulated digital image correlation (DIC) data, as will be demonstrated in this paper. The proposed method can then be applied to physical test data. The evaluated material model in this paper will be the Mooney-Rivlin three-parameter model as seen in equation 3 which applies strain energy density to assess material behaviour [2].

$$W = C_{10}(I_1 - 3) + C_{01}(I_2 - 3) + C_{20}(I_1 - 3)^2 \quad (1)$$

The analysis is broken up into three distinct phases operating under the core assumption that there exists regions or planes in the design space wherein the objective function is equally optimal as mentioned in the introductory section. This means that on these planes the displacement field or indentation force would be identical even if the design variables are not correct, these planes will be referred to as hyperplanes throughout the rest of the paper. Evidence supporting the assumption can be shown from [7] who identified that simulated indentation force had non-unique sets of material parameters since they would result in identical indentation force being witnessed. [4] and [1] both witnessed that optimising the objective function for a single dataset would not always provide a good fit for another dataset or loading case. With [1] showing that the absolute values of the material parameters not being strongly correlated but their product and intern their ratio having a strong correlation.

Phase one would be locating the low objective function regions where the hyperplane can be found. This must be done for both the displacement field and indentation force data under a single variate optimisation scheme using DOT [8] as the optimiser. Once this optimization has converged this point will be assumed to lie on the hyperplane, these points will be identified for both datasets.

Phase two is where the equations of the hyperplanes are identified, this method incorporates a central composite design as recommended by [9] for choosing appropriate sample points for evaluating the response surface, this ensures the second-order model (Equation 2) has a valid dataset to perform the least-squares regression. Evaluating the principle plane using stationary point analysis on the second-order model's parameters yields a linear surface equation for the hyperplane.

$$\begin{aligned} f(x_1, x_2, x_3) = a_0 + a_1x_1 + a_2x_2 + a_3x_3 + a_{11}x^2{}_1 + a_{22}x^2{}_2 + a_{33}x^2{}_3 \\ + a_{12}x_1x_2 + a_{13}x_1x_3 + a_{23}x_2x_3 \end{aligned} \quad (2)$$

Phase three is where the intersection between the hyperplanes from the displacement field data is evaluated resulting in a single line in hyperspace which can be seen as one dimensional since all parameters can be coupled to a single variable α, the hyperplanes can be seen in Figures 3a and 3b. The optimisation algorithm now used is the method of Golden Sections analysing the displacement field for the same experimental model at a different indentation depth, this results in a third hyperplane that rotates about the "ideal" point of the data set making the error profile along the one-dimensional line is strictly uni-model.

2.2 Experimental Model

The geometry under investigation was a 20x20x5 mm cuboid with a cylindrical indenter placed lengthwise along the top surface's centreline with a length of 30 mm with a radius of 3 mm. Because this geometry is symmetric about its centre planes it can be taken advantage of as only half the geometry needs to be simulated, significantly reducing the computational cost. Applying symmetry conditions results in a simulated sample with dimensions 10x10x5 mm and the indenter being placed along the top surface's edge as seen in Figure 2.

As in the physical test the indentation depth will be physically controlled therefore to replicate this in the FE model displacement control for an overall indentation depth of 3 mm over 1 second to induce material deformation revealing the material behaviour. The choice behind this geometry stems from the research of [4], the reasoning behind this geometry was for situations in which small biological tissue samples would need to be evaluated. These samples would not only be in short supply but due to their limited size, it would not be feasible to clamp the samples and perform uni-axial or biaxial tensile tests which opened the question as to whether or not the Mooney-Rivlin material model could be fully characterized with a single sample of simple geometry and relatively simplistic loading application.

The model also simulates the situation in which digital image correlation (DIC) would be used for extracting the displacement field data which is achieved by only evaluating the surface nodes and elements on a single face of the sample as seen in Figure 2.

In-order to generate the theoretical experimental data the same model used in the inverse FE analysis was run once with significantly denser mesh and only surface data on a single face was extracted mimicking the type of data that would result from the use of DIC.

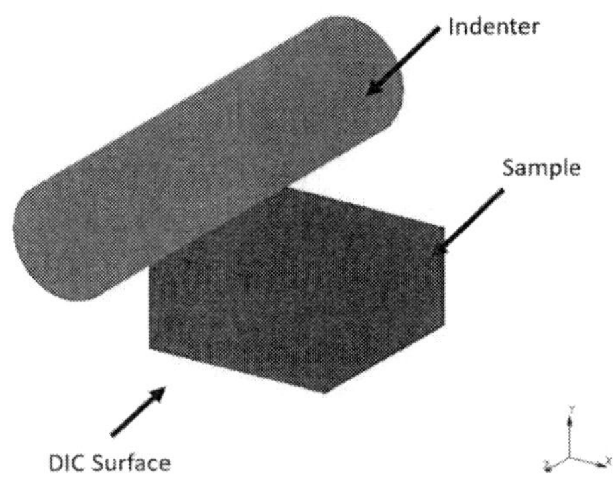

Figure 1: Simulated sample

$$W = C_{10}(I_1 - 3) + C_{01}(I_2 - 3) + C_{20}(I_1 - 3)^2 \tag{3}$$

2.3 Numerical Pipeline

The program used to perform the FE analysis in this investigation was MSC Marc Mentat 2021, this software was favourable since it can be controlled in either batch mode which allows for a procedure script that updates the model to be passed onto the software or direct control through python during operation using the PyMentat modules [5]. Marc Mentat also has a strong python-support library for post-processing of the results files with their built in library. In the numerical pipeline of this investigation, the batch mode method of control was used to updated the material parameters and run the simulation.

The initial optimisation phase was handled by the software package DOT (Design Optimization Tools) of which the sequential quadratic programming (SQP) algorithm was used which is a gradient-based optimisation scheme. This phase was performed separately on both the displacement field data and indentation force data until the "hyperplanes" or regions where the apparent non-unique minimum objective values are for either data set were identified.

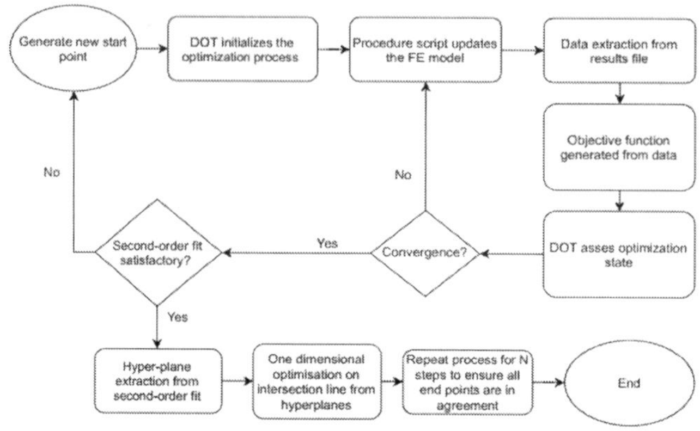

Figure 2: Numerical pipeline flowchart

2.4 Objective Functions

Inverse FE analysis at its core uses some method of optimisation to determine the material parameters of an evaluated sample. This is achieved by minimising an objective function based off error data generated from the experimental and simulation models. Since displacement field and indentation force are very different in terms of their data structure two objective functions were needed.

2.4.1 Displacement Field Objective Function

Root mean square (RMS) was used to determine the error between the simulated and measured displacement fields. This error is represented by a single value for each FE simulation, one for each coordinate direction resulting in three RMS errors as represented by equations 4,5, and 6. Directly summing up these equations will result in a single error value that describes how similar the two displacement fields are, however, this now introduces bias into the objective function favouring the larger displacement errors. The simulation has an indenter that moves in the Y-direction and therefore it is expected that Y-displacements will always be larger than displacements in the X and Z-directions. An attempt to remove this bias was to normalize the displacements with their respective maximum displacements in that direction, this is represented by equation 7. [10] used the same set of equations when characterising a non-linear orthotropic by means of bubble inflation tests.

$$e_x(m) = \sqrt{\frac{\sum_{i=1}^{N}\left(d_{\text{exp\_}x}(m) - d_{\text{sim\_}x}(m)\right)^2}{N}} \tag{4}$$

$$e_y(m) = \sqrt{\frac{\sum_{i=1}^{N}\left(d_{\text{exp\_}y}(m) - d_{\text{sim\_}y}(m)\right)^2}{N}} \tag{5}$$

$$e_z(m) = \sqrt{\frac{\sum_{i=1}^{N}\left(d_{\text{exp\_}z}(m) - d_{\text{sim\_}z}(m)\right)^2}{N}} \tag{6}$$

$$e_{disp} = \frac{1}{M}\sum_{i=1}^{N} \frac{e_x(i)}{\max\left(\left|d_{\exp\_x}(i)\right|\right)} + \frac{e_y(i)}{\max\left(\left|d_{\exp\_y}(i)\right|\right)} + \frac{e_z(i)}{\max\left(\left|d_{\exp\_z}(i)\right|\right)} \tag{7}$$

2.4.2 Indentation Force Objective Function

The objective function for the indentation force was simpler than the displacement field in that there is only one point of measurement, in one direction for each indentation depth and not a "field". It was chosen that any force between the simulation and experimental would increase with the absolute difference between the two measurements. This means that the lower limit for the force objective function would be zero when the measured forces are equal. The force objective function is as follows:

$$e_{\text{force}} = \sum_{i=1}^{m} \left| f_{\text{exp}}(i) - f_{\text{sim}}(i) \right| \tag{8}$$

3 Results

3.1 Phase One Convergence Points Results

After the SQP algorithm in DOT identified many convergence points for each respective dataset it was found, as expected, that none of the starting points converged to the same point in design space across different datasets. This was also true within each dataset itself that it would be extremely unlikely that two starting points in the displacement field or indentation force data would converge to the same point as seen in Figures 3a and 3b. However, these points did display the core behaviour behind this paper that the assumption of a hyperplane, where any point on the plane would result in identical material behaviour for the given data type, is valid since it was possible to fit the first-order surface through the converged points with almost all of these points being on that plane. The correlation coefficients for the linear fits on the two hyperplanes were $R_{force} = 0.999$ and $R_{disp} = 1.000$ which is a near-perfect correlation score providing further evidence supporting the assumption of hyper-planes being present in the data sets. It is also worth noting that none of the starting points did converge to correct material parameters as prescribed in the high density mesh finite element model which was used to generate the synthetic experimental data in this research paper. These findings agree with what was witnessed in the literature.

3.2 Phase Two Central Composite Design Results

The results of phase two were that a second-order model based on equation 2 was successfully fit with the central composite design points for both the displacement field and indentation force datasets. The respective R-squared value for a central composite design centred at an arbitrarily chosen convergence point for each fit was $R_{disp} = 0.981$ and $R_{force} = 0.978$ and adjusted R-squared values of $Adj.R_{disp} = 0.948$ and $Adj.R_{force} = 0.938$ which all indicate a strong correlation with the second-order model. Although these values are dependent on where in design space the central composite design is chosen, the fit values do not differ significantly from the above values, they can be seen as a representative of what can be encountered when performing this analysis.

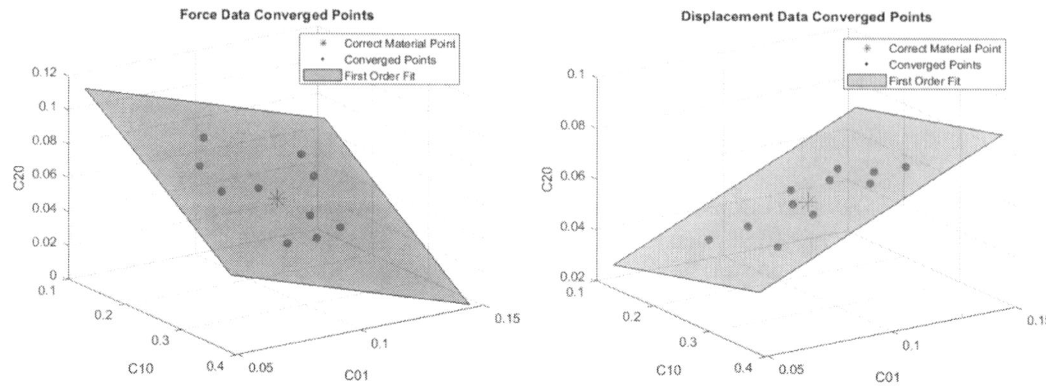

(a) Front view of the force hyperplane showing converged design points

(b) Front view of the displacement hyperplane showing converged design points

Figure 3: Force convergence with linear fit

3.3 Phase Three Golden Section Results

The results for phase three were that the method of golden sections was successfully able to find the optimal point that fit both the displacement field and indentation force data simultaneously when following the one-dimensional line that is generated at the intersection point between the two linear surfaces, also referred to as hyperplanes. This final convergent point from the golden section phase also represents the correct material parameters for the Mooney-Rivlin model with a total magnitude of 2.32% error from the prescribed values and was absolutely convergent at a single point along the intersection line as seen in Figure 4b. This error can be further reduced by repeating phases two to three by implementing a central composite design having a smaller radius centred at the converged material point.

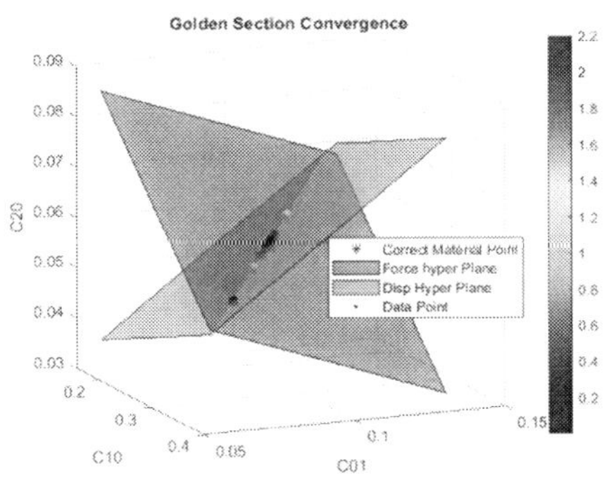

(a)

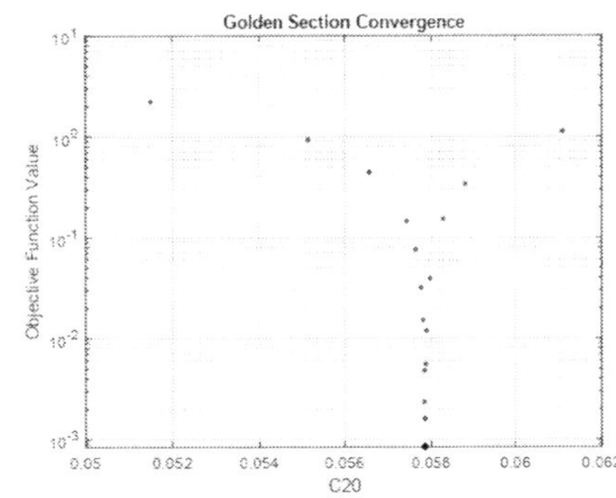

(b) Golden section optimisation along intersection line between material upper and lower bounds of search space

Figure 4: Golden section optimisation on intersection line between the two hyperplanes

4 Conclusion

In this paper, a method of performing indentation tests for fully characterising the three-parameter Mooney-Rivlin model. This method is based on the assumption that within the data sets there are hyper-planes with fixed relationships between the material parameters wherein the objective function is equally optimal. It was demonstrated for both the force and displacement field data that this assumption is applicable since the linear plane was able to be fit through the converged points for random starting positions in the design space with near-perfect correlation coefficients of $R_{force} = 0.999$ and $R_{disp} = 1.000$.

When optimising parameters along the intersection line of the two hyper-planes it was demonstrated in section 3.3 that evaluating the indentation force for a different depth results in the objective function being convergent to a single point. The results, therefore, demonstrate that operating under the assumption of hyperplanes in the data sets the Mooney-Rivlin Material model can be fully characterised with a low level of error of only 2.32%.

References

[1] Kyle Murdock, Caitlin martin, Wei Sun, Journal of the Mechanical Behavior of Biomedical Materials **77**, 148–156, 2017

[2] Mooney,M.(1940).A theory of large elastic deformation, Journal of Applied Physics **11**

[3] Du Toit, Viljoen, *Thesis (MEng)–Stellenbosch University, 2018.*

[4] Bresler, Franciena Helena, *Thesis (MEng)–Stellenbosch University, 2019.*

[5] MSC Software,Marc ® 2020 Python : Tutorial and Reference Manual, 2020.

[6] subprocess,Python 3.9.4 documentation,

[7] Jensen Aw Chee Seng, *Thesis–Liverpool John Moorse University, 2015.*

[8] Vanderplaats Research & Development, Inc.(2001), Design Optimization Tools User Manual

[9] Myers, Raymon H. and Montgomery, Douglas C. and Anderson-Cook, Christine M.,*Response Surface Methodology: process and product optimization using designed experiments, Fourth Edition* (John Wiley & Sons, Inc, New Jersey, 2016), pages 47-55

[10] Jekel, Charles F. and Venter, Gerhard and Venter, Martin P., Structural and Multidisciplinary Optimization **4**,pages 927-935

[11] Coyle, Stephen,Majidi, Carmel,LeDuc, Philip,Hsia, K. Jimmy,Extreme Mechanics Letters **22**,51-59,2018

Development and validation of a dynamic model for a two-stage speed-reducer

Nicholas J. Tutt[1,*], *Martin P.* Venter[1,**], and *Daniel N.J.* Els[1,***]

[1]Stellenbosch University, South Africa

Abstract. This paper presents the development and validation of a six degree of freedom (DOF) dynamic model of a two-stage parallel shaft gearbox, without flaws, which is able to determine the reaction of gearbox components to varying torque inputs and loads. The model utilises flexible shafts and gears rather than using a rigid assumption, to further understand the effect of varying mesh stiffness. The paper replicates the results presented by Diehl and Tang, and improves the number of frequencies that can be analysed. The gear meshing frequencies were expected to dominate the result, however due to the use of a sinusoidal approximation of the varying tooth mesh frequency, the presented model shows the additional gear generated frequencies are present and analysable in the data.

1 Introduction

Large scale two-stage speed reduction gearboxes are a key component in the operation of air cooled condensers common in minimal water loss systems for utility scale power generation. The varying loads experienced by the fan and gearbox, due to the surrounding structures and the local wind patterns prevalent in this application, are of concern towards the operation and life expectancy of such gearboxes. This paper will develop and validate a six degree of freedom (DOF) dynamic model of a two-stage parallel shaft gearbox, without flaws, which is able to determine how the gearbox's components react to varying torque inputs and loads. The model utilises flexible shafts and gears rather than using a rigid assumption, to further understand the effect of varying mesh stiffness.

Numerical models of multi-stage gear systems have been developed to determine the effect of internal defects to the noise and vibration level of an operational gearbox [1]. These flaws amongst many include: tooth wear, fatigue failure and manufacturing errors [2]. This aids in the the study of gear faults and the development of condition monitoring methods [3].

Widely used standards such as AGMA 2101-D04 [4], empirically develop gear design safety factors to account for all parameters and loads that influence gear life. These safety factors consider dynamic factors that cover the rotational speed and shock loads which although useful, under steady operation conditions. It is found that the full effect of dynamic loads are not completely covered by the standards. The amount of unexpected failures of large scale air cooled condenser fans' gearboxes requires research into the causes of this unexpected

*e-mail: nicholas.tutt@gmail.com
**e-mail: mpventer@sun.ac.za
***e-mail: dnjels@sun.ac.za

emergent behaviour. The implementation of the proposed numerical model will provide a tool to do an in-depth analytical analysis into the causes and thus improve the understanding of how torque interacts with a gearbox.

The six DOF model for a two-stage speed reducer, utilising a similar method described in Diehl et al. [5], presented here, is developed. It is then demonstrated that this model can be used as a parametric tool to aid the design of new gearboxes. The research focuses on utilising flexible shafts and gears to give further insight into the dynamic behaviour of the gearbox components.

2 Gear tooth mesh stiffness

Gearbox dynamics are characterised by time varying gear mesh stiffness and non-linear backlash [1, 3, 6–9]. The inclusion of such properties ensure that the respective models fully describe the discrete motion of the gears.

Bai et al. [7] review that backlash always exists within gear transmissions due to lubrication, tooth wear, assembly errors, etc. However, backlash is mainly prominent at zero load and load reversals and thus does not form part of this paper's scope.

Time varying mesh stiffness is a main property included in this paper, especially with regards to how it affects the dynamic behaviour of a gearbox [10]. The interaction described between teeth is shown in figure 1, where the forces between the teeth are affected by the effective stiffness that each tooth contributes in reaction to the applied torque load.

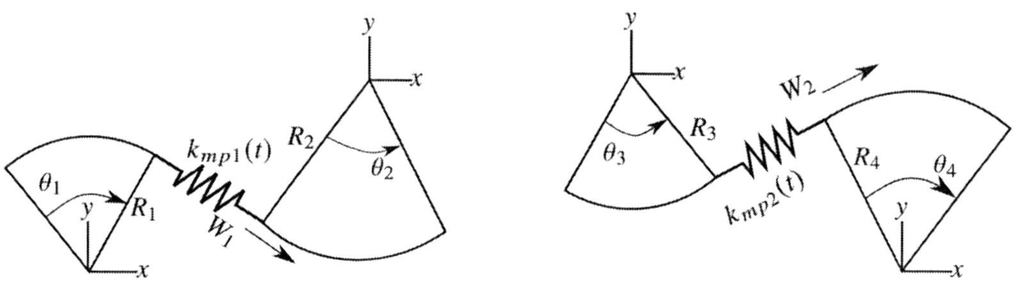

Figure 1. Two-stage meshing freebody diagram showing the forces W_1 and W_2

The resistance force, W_i, generated between the meshing gears is proportionate to the relative displacement of the driving and driven gear in the direction of the meshing line. This is shown in equation (2) with R referring to the gear radius, θ the relative gear position and i the respective gear stage. The resistance force is given in equation (1); where $k_{mp}(t)$ is the varying mesh stiffness function that cycles according to the mesh frequency ω_i, at an amplitude of k_v and a mean stiffness of k_m. The mean mesh stiffness may be determined through experimentation [10], finite element investigation [11], or following standardised procedures such as those outlined in ISO 6336-1:2019 [12].

$$W_i(t) = k_{mpi}(t)\delta_i, \tag{1}$$

$$\delta_i = R_{pi}\theta_{pi} - R_{gi}\theta_{gi}, \tag{2}$$

$$k_{mpi}(t) = -k_v cos(2\pi f_{m_i} t) + k_m \tag{3}$$

$$i = 1, 2$$

The simplified periodic mesh stiffness, k_{mp}, characterised by figure 2 is approximated by equation (3). Equation (3) is calculated with respects to the meshing frequency, f_{m_i}, of the respective gear pairs. The meshing frequencies represent the rate at which tooth pairs mesh and is calculated using equation (4). Where n_{P_i} and n_{G_i} represents the respective speed of the pinion or gear for the different stages, and N_{P_i} and N_{G_i} represent the number of teeth of the respective pinion and gear.

$$f_{m_i} = \begin{cases} N_{P_i} \times n_{P_i}/60 \\ N_{G_i} \times n_{G_i}/60 \end{cases} \quad [\text{Hz}] \tag{4}$$

The ideal mesh stiffness profile is currently being experimentally measured. The measurement utilises a set of large module spur gears loaded under quasi-static conditions and recorded using digital image correlation technology. The results will further advise the best method of simulating the mesh stiffness' periodicity.

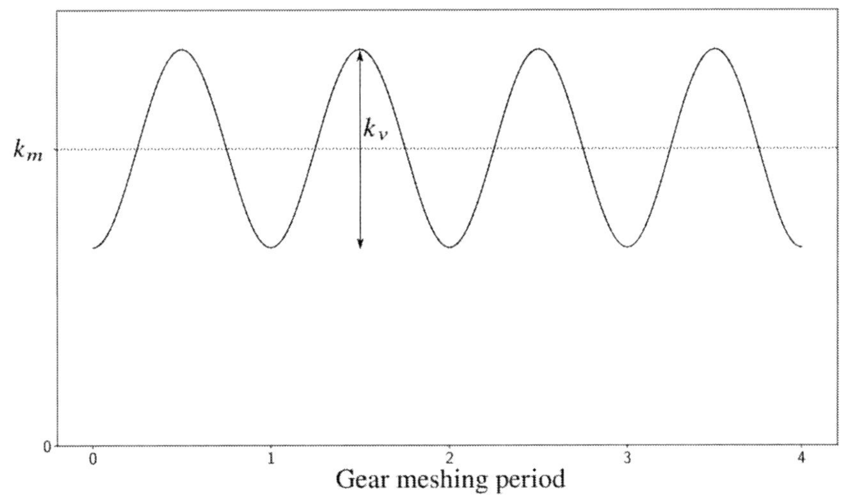

Figure 2. Visualisation of equation (3), which approximates the periodic gear mesh stiffness.

3 Model development

A model of the two-stage gearbox is used for concept analysis and validation. Currently gearboxes are designed using standards [13] that are predominantly empirically derived. Although useful for gearboxes, operating at constant speeds with constant loads, the standards fall short in circumstances where there is continuous torque variation coupled with non-uniform shaft bending.

A model which describes the dynamic response of large scale gearboxes under varying dynamic load conditions is needed. A physical gearbox generates the following frequencies [14]: rotational frequency f_r, meshing frequency f_m, tooth repeat frequency f_{tr}, and assembly phase passage frequency f_{ap}. Where the minimum frequency is the tooth repeat frequency and the maximum is the mesh frequency. The developed model would need to be able to predict these frequencies under normal operating conditions.

The model presented in Diehl and Tang [5], when analysed, only showcase the meshing frequencies as major contributing factors to the vibration response of the gearbox. The proposed model will aim to ensure that the frequencies outlined in Winterton [14] are present when analysing the gearbox's vibrational response.

The proposed model will utilise the same parameters as those outlined in Diehl and Tang [5], shown in table 1. The module of the gears is specified as 1.59 mm, the speed of the driver is set at 820 rpm and the power of the driver is approximately 2.2 kW.

Table 1. Gearbox model parameters

	Number of teeth	Inertia [kgm^2]	Damping [Ns/rad]	Stiffness [Nm/rad]
Driver, I_D		7.51e-3		
Pinion 1, I_1	32	8.06e-5		
Gear 1, I_2	80	2.97e-3		
Pinion 2, I_3	48	4.00e-4		
Gear 2, I_4	64	1.22e-3		
Load, I_L		2.73e-5		
Shaft 1, k_{t1}, c_{t1}			0.0926	8.20e3
Shaft 2, k_{t2}, c_{t2}			1.41	39.9e3
Shaft 3, k_{t3}, c_{t3}			0.517	12.6e3
Mesh stiffness, k_m				4.62e6

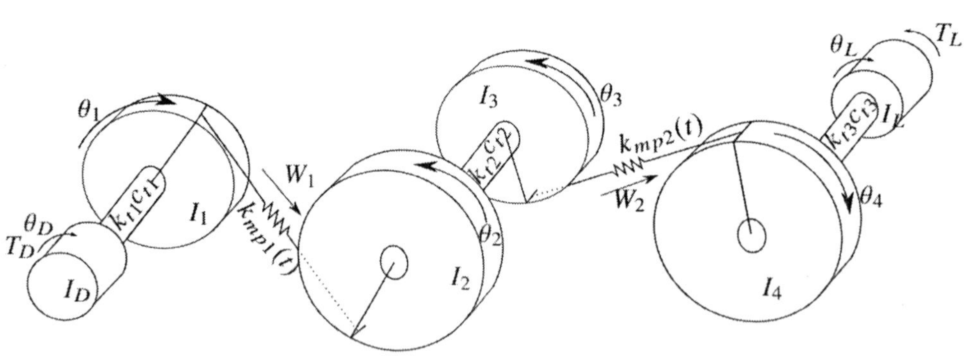

Figure 3. Diagram of the six degree of freedom two stage dynamic gearbox model.

$$I_D\ddot{\theta}_D + c_{t1}(\dot{\theta}_D - \dot{\theta}_1) + k_{t1}(\theta_D - \theta_1) = T_D \tag{5}$$
$$I_1\ddot{\theta}_1 + c_{t1}(\dot{\theta}_1 - \dot{\theta}_D) + k_{t1}(\theta_1 - \theta_D) = -W_1R_1 \tag{6}$$
$$I_2\ddot{\theta}_2 + c_{t2}(\dot{\theta}_2 - \dot{\theta}_3) + k_{t2}(\theta_2 - \theta_3) = W_1R_2 \tag{7}$$
$$I_3\ddot{\theta}_3 + c_{t2}(\dot{\theta}_3 - \dot{\theta}_2) + k_{t2}(\theta_3 - \theta_2) = -W_2R_3 \tag{8}$$
$$I_4\ddot{\theta}_4 + c_{t3}(\dot{\theta}_4 - \dot{\theta}_L) + k_{t3}(\theta_4 - \theta_L) = W_2R_4 \tag{9}$$
$$I_L\ddot{\theta}_L + c_{t3}(\dot{\theta}_L - \dot{\theta}_4) + k_{t3}(\theta_L - \theta_4) = -T_L \tag{10}$$

The model, depicted in figure 3, incorporates the dynamics that occur between meshing gears through the inclusion of the meshing forces equation (1). The gear dynamics are assumed to be dominated by the changing mesh stiffness that occur between engaged tooth pairs as discussed in section 2 and further in Chaari et al. [15], Philippe [16] and Pedrero et al. [17].

4 Model simulation

Equations (5)-(10) are formulated as a set of matrices (11)-(13). The system of equations are written in Python 3.6, utilising methods from the SciPy version 1.5.2 package [18].

There are six different methods (either implicit or explicit methods) of solving an ordinary differential equation (ODE): the explicit Runge-Kutta methods ('RK45','RK23','DOP853') are primarily used for non-stiff problems, whereas the implicit methods ('Radau','BDF','LSODA') are favoured for stiff problems. The stiffness of a problem is not fully defined, however is rather an indication of how fast certain terms within the problem lead to rapid variation of the solution [19].

The specific methods of integration utilised are the explicit Runge-Kutta method of order 5 ('RK45') and the implicit Backwards Differentiation Formula (BDF) which is a multistep variable order (1 to 5) integration method based on a backward differentiation formula for derivative approximation [20].

$$\mathbf{I} = \begin{bmatrix} I_D & \cdots & 0 \\ \vdots & \ddots & \vdots \\ 0 & \cdots & I_L \end{bmatrix} \tag{11}$$

$$\mathbf{C} = \begin{bmatrix} c_{t1} & -c_{t1} & \cdots & \cdots & 0 \\ -c_{t1} & c_{t1} & \cdots & \cdots & 0 \\ \vdots & \vdots & \ddots & & \vdots \\ \vdots & \vdots & & c_{t3} & -c_{t3} \\ 0 & 0 & \cdots & -c_{t3} & c_{t3} \end{bmatrix} \tag{12}$$

$$\mathbf{K} = \begin{bmatrix} k_{t1} & -k_{t1} & \cdots & \cdots & 0 \\ -k_{t1} & k_{t1} & \cdots & \cdots & 0 \\ \vdots & \vdots & \ddots & & \vdots \\ \vdots & \vdots & & k_{t3} & -k_{t3} \\ 0 & 0 & \cdots & -k_{t3} & k_{t3} \end{bmatrix} \tag{13}$$

$$\mathbf{I}\ddot{\theta} + \mathbf{C}\dot{\theta} + \mathbf{K}\theta = \mathbf{T} \tag{14}$$
$$\theta = \begin{bmatrix}\theta_D & \cdots & \theta_L\end{bmatrix}^T \tag{15}$$
$$\mathbf{T} = \begin{bmatrix}T_D & \cdots & T_L\end{bmatrix}^T \tag{16}$$

Due to the potential stiffness when solving equation (14), the problem was first solved with the RK45 method, followed by the BDF method. The BDF method allows for smaller integration time steps to be taken - preventing rapid variation of the calculated solution.

The resultant velocity of the last shaft $\dot{\theta}_L$ is processed by utilising the Fast Fourier Transform (FFT) algorithm that discretises the velocity signal into its different sinusoidal components and calculates their respective powers. When doing so, it is important to take note of the possibility of alias frequencies in the processed signal, thereby calculating erroneous signal components. Therefore, the inverse of the selected integration time step should be greater than double the maximum frequency expected of the system [21].

As mentioned in section 3, the expected frequencies are the: mesh, rotational, tooth repeat and assembly phase passage frequencies. Due to the meshing forces being a major factor in the modelling of the gearbox, it is expected that the meshing frequencies and their multiples will dominate the generated FFT.

5 Model results

The frequency analysis presented in Diehl and Tang [5] show that the meshing frequencies are major contributors to the gearbox model's vibration signal. Their results do not specify whether there are other gear generated spectra displayed in their model, or in their experimental data.

Following the properties outlined in section 3, the model was run for 100 s, with a constant drive and load torque applied. The presented model's FFT results were plotted using the velocity of the last degree of freedom during the model's steady operation. Therefore there were no transient auxiliary frequencies included.

To ensure that minimal alias frequencies were present in the data, the FFT was sampled at 20 kHz, with a frequency resolution of 2.4 HZ/bin. Before calculating the FFT, the signal was multiplied by a corrected amplitude Hanning window function [22] to compensate for the amplitude reduction. The Hanning window forces the sample to zero at the start and end of the selected window, thus reducing signal leakage.

The results from the presented model show that on a large scale, figure 4, the meshing frequencies and their multiples are indeed prominent. And when looking at a smaller scale, figure 5, the smaller gear generated spectra are present without being overshadowed.

6 Conclusion

The objective of this article is to replicate the six DOF model presented in Diehl and Tang [5] and improve the number of frequencies one could analyse from the model. The developed model shows that nearly all the gear spectra can be read from the processed velocity data generated by the model. When analysing the data on the small scale it was shown that the meshing frequencies do not dominate the calculated frequencies as previously expected, and therefore, the objective of this paper has been met.

Following this article the current model will be further developed to include more translational elements such as those of the bearing locations. By doing so, the effect of the movement caused by meshing can be analysed and the effect thereof can be better understood.

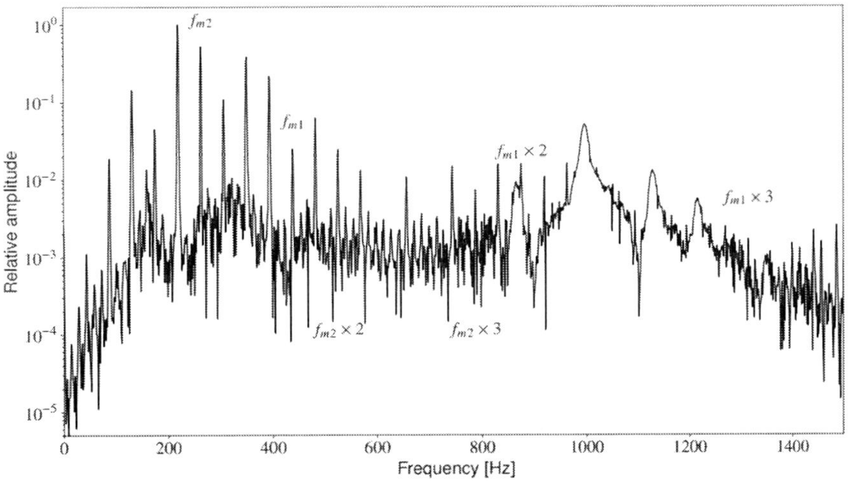

Figure 4. Replicated six degree of freedom model's FFT data (relative to largest amplitude), on a large scale [0 to 1500 Hz] showing the meshing frequencies and their multiples. f_{m1} - first meshing frequency, f_{m2} - second meshing frequency

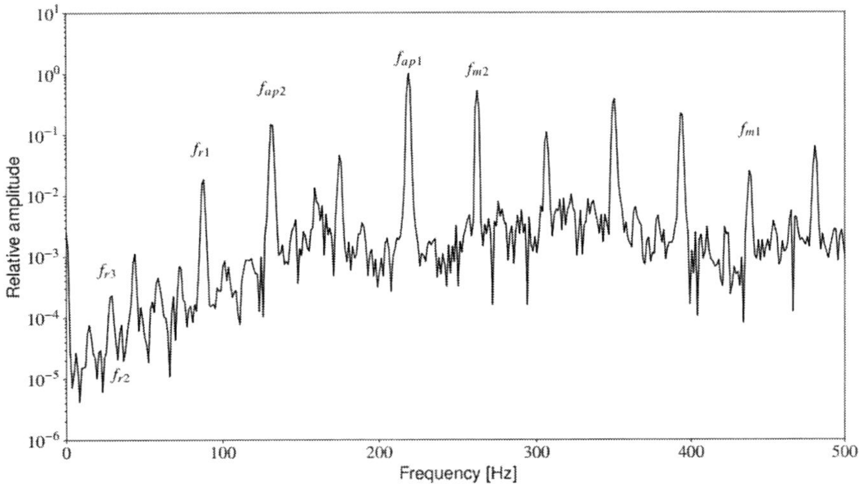

Figure 5. Replicated six degree of freedom model's FFT data (relative to largest amplitude), on a small scale [0 to 500 Hz] showing three of the gear generated spectra as specified in Winterton [14]. f_{m1} - first meshing frequency, f_{m2} - second meshing frequency, f_{r1-3} - rotational frequencies, f_{ap1-2} - assembly passage frequencies

References

[1] Shengxiang Jia, Ian Howard, and Jiande Wang. The Dynamic Modeling of Multiple Pairs of Spur Gears in Mesh, Including Friction and Geometrical Errors. *The International*

Journal of Rotating Machinery, 9(6):437–442, 2003. ISSN 1023-621X. .

[2] Xihui Liang, Ming J. Zuo, and Zhipeng Feng. Dynamic modeling of gearbox faults: A review. *Mechanical Systems and Signal Processing*, 98:852–876, jan 2018. ISSN 10961216. . URL http://dx.doi.org/10.1016/j.ymssp.2017.05.024.

[3] Farag K. Omar, Kamal A.F. Moustafa, and Samir Emam. Mathematical modeling of gearbox including defects with experimental verification. *Journal of Vibration and Control*, 18(9):1310–1321, 2012. ISSN 10775463. .

[4] ANSI/AGMA 2001-D04 Fundamental Rating Factors and Calculation Metrhods for Involute Spur and Helical Gear Teeth. Technical report, American Gear Manufactures Association, Alexandria, Virginia, 2004.

[5] Edward J. Diehl and J. Tang. Predictive modeling of a two-stage gearbox towards fault detection. *Shock and Vibration*, 2016, 2016. ISSN 10709622. .

[6] Grzegorz Litak and Michael I. Friswell. Vibration in gear systems. *Chaos, Solitons and Fractals*, 16(5):795–800, 2003. ISSN 09600779. .

[7] Zheng Feng Bai, Xin Jiang, Fei Li, and Ji Jun Zhao. Research on Vibration Responses of Gear System with Backlash. pages 55–63. 2020. . URL http://link.springer.com/10.1007/978-981-15-0238-5{_}6.

[8] M. Karray, F. Chaari, M. T. Khabou, and M. Haddar. Dynamic analysis of bevel gear in presence of local damage in nonstationary operating conditions. *Lecture Notes in Mechanical Engineering*, (207169):325–330, 2018. ISSN 21954364. .

[9] Shengli Zhang, Jiong Tang, and Yu Ding. Modeling and analysis of time-periodic gearbox vibration. *Proceedings of the ASME Turbo Expo*, 3B:1–8, 2014. .

[10] Naresh K. Raghuwanshi and Anand Parey. Experimental measurement of spur gear mesh stiffness using digital image correlation technique. *Measurement: Journal of the International Measurement Confederation*, 111(November 2016):93–104, 2017. ISSN 02632241. . URL http://dx.doi.org/10.1016/j.measurement.2017.07.034.

[11] Ph. Sainsot and Ph. Velex. On contact deflection and stiffness in spur and helical gears. *Mechanism and Machine Theory*, 154, 2020. ISSN 0094114X. .

[12] BS ISO 6336-3:2019 Calculation of load capacity of spur and helical gears. Calculation of tooth bending strength. Technical report, International Organization for Standardization, GENEVA, 2019.

[13] Timothy J. Lisle, Brian A. Shaw, and Robert C. Frazer. External spur gear root bending stress: A comparison of ISO 6336:2006, AGMA 2101-D04, ANSYS finite element analysis and strain gauge techniques. *Mechanism and Machine Theory*, 111:1–9, 2017. ISSN 0094114. .

[14] John G. Winterton. Component Identification of Gear-Generated Spectra. *ORBIT*, pages 11–14, 1991.

[15] Fakher Chaari, Radoslaw Zimroz, Walter Bartelmus, Tahar Fakhfakh, and Mohamed Haddar. Model Based Investigation on a Two Stages Gearbox Dynamics under Non-stationary Operations. *Condition Monitoring of Machinery in Non-Stationary Operations*, pages 133–142, 2012. .

[16] Velex Philippe. On the Modelling of Spur and Helical Gear Dynamic Behaviour. *Mechanical Engineering*, 2012. ISSN 1448-4846. .

[17] José I. Pedrero, Miguel Pleguezuelos, and Miryam B. Sánchez. Load sharing model for high contact ratio spur gears with long profile modifications. *Forschung im Ingenieurwesen/Engineering Research*, 83(3):401–408, 2019. ISSN 14340860. .

[18] Eric Jones, Travis Oliphant, Pearu Peterson, and Others. SciPy: Open Source Scientific Tools for Python. URL http://www.scipy.org.

[19] C. F. Curtiss and J. O. Hirschfelder. Integration of stiff equations. *Proc. Natl. Acad. Sci. U S A*, 38(1950):235–243, 1952.

[20] Elisabete Alberdi Celaya and Juan José Anza. The integration of stiff systems of ODEs using NDFs and MEBDFs. *Journal of Computational Mathematics*, 31(2):175–189, jul 2013. ISSN 02549409, 19917139. URL http://www.jstor.org/stable/43693999.

[21] Richard S. Figliola and Donald E. Beasley. 7.2 Sampling Concepts. In *Theory and Design for Mechanical Measurements*, chapter 7 Sampling, pages 274–275. John Wiley & Sons, Inc., Hoboken, sixth edition, 2015. ISBN 9781118881279.

[22] T. Irvine. Hanning Window Compensation Factor, 2008. URL http://www.vibrationdata.com/tutorials/Hanning_compensation.pdf.

A review of composite leaf springs for automotive vehicles

Brighton Manaba[1*], and *Vuyo T* Hashe[1]

[1]University of Johannesburg, Doornfontein, South Africa

Abstract. In the quest to conserve natural resources and economic energy. Weight reduction has been the focus of auto-manufactures and the aerospace industry in the present day. Weight reduction can be archived primarily by introducing better design optimization, better material, and improved manufacturing processes. This approach by introducing composite materials into automobile industries, which has high strength to weight ratio and excellent corrosive resistance, can fulfil the requirement. This review paper investigates the use of composite materials in the design of leaf springs for automotive vehicles. The paper underscores the suitability of composite materials for leaf spring in automobiles and develops a new leaf spring design that optimizes weight reduction and strength. The review provides details regarding previous efforts conducted on the design of leaf springs while outlining the knowledge gap areas. This review commences with an introduction to leaf springs and a brief history of their development. Followed by the properties that make an excellent leaf spring, this helps narrow down the information required for further developments providing avenues for possible future research.

1 Introduction

Leaf springs are essential elements in the suspension systems of vehicles including sport utility vehicles, trucks, SUVs, and railway carriages. The purpose of leaf springs is to separate and protect the vehicle chassis from vibrations that it experiences as it travels over rough and bumpy terrain [1]. Additionally, they aid in giving control to the wheels during braking acceleration and general vehicle manoeuvres experienced during road undulations [2]. Present-day, the automobile and aerospace industry face rapidly growing competition to improve existing products [3]. Their quest aims at conserving natural resources and economizing energy through weight reduction. Weight reduction can be achieved primarily by the introduction of improved designs that save on material use, better manufacturing processes and better materials. Driven by the goal to reduce vehicle weight, interest has been drawn into the development of composite leaf springs, which have a higher strength to weight ratio than the conventional materials on the market. This paper studies various composite materials and geometries to find the best material and geometry that performs better than any existing leaf springs.

1 Corresponding author: vthashe@uj.ac.za

2 Leaf Springs

Leaf springs were originally coined laminated or carriage springs. They are one of the oldest spring designs to ever existed, dating back to the 17th century [4] The design of leaf springs is quite distinct in that it does not carry the typical helical/ coil shape but rather a thin rectangular cross-section that spans along an arch-shaped length, see figure 1 below.

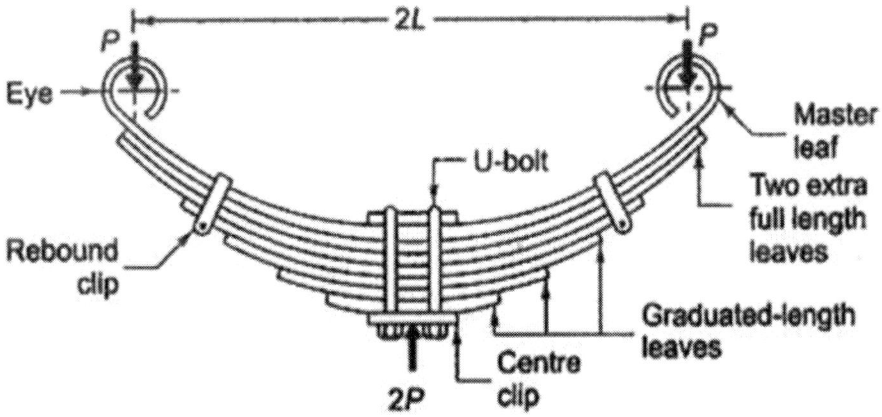

Fig. 1. Leaf springs [5].

To date, various designs of leaf springs have been developed, ranging from the elliptic design, three-quarter elliptic, quarter-elliptic and transverse, to the more modern designs, like the semi-elliptic, the parabolic leaf spring and finally the mono-leaf spring [6]. All these designs had their pros and cons; however, this section will only dissect the modern designs.

2.1 Multi-leaf semi-elliptic

This is the most common design in the history of leaf springs. It utilizes several layers of leaves stacked together, with the longest one on top and progressively shorter leaves following. The longest leaf spring is termed the master leaf, looped at both ends to allow for its attachment to the chassis [7]. The loops are known as the eye; following the master leaf, the second leaf is called the second master leaf. The rest of the other leaves are coined graduated leaves. The leaves are held together by what is not as U- clip; this allows for the distribution of stress among the leaves and prevents other leaves from buckling [8]. This design was advantageous in the sense that it also provided damping action for the vehicle. In addition, t interleaf friction minimized the continued oscillation of the car [9]. However, this design had a drawback; the interleaf friction cannot be controlled and would at times result in unwanted stiction in the motion of the suspension. For this reason, a new design was implemented, the parabolic leaf spring.

2.2 Parabolic leaf spring

This design differs from the multi-leaf spring in that it has fewer leaves [10]. These leaves are also distinctively designed; they have a varying thickness along the parabolic length. This solves for stiction by preventing the leaves from touching except for the ends attached to the chassis and the centre attached to the axle [11]. In addition to the geometry, the design implements spacers that further help avoid any contact between the leaves. The characteristics of this design provided for the reduction in weight as compared to the older versions. It also offered suspension flexibility since interleaf friction is eliminated, translating to improved ride quality. However, this was not enough; further, development was made, which led to the mono-leaf spring design.

2.3 Mono leaf spring

The first mono-leaf spring was design by Chevrolet, a British car manufacturing company [12]. The design was made specifically to solve the problem of weight and stiction caused by inter-leaf friction. To do away with inter-leaf friction, they proposed using only a single leaf instead of several leaves. The single leaf is made so that it can carry both static and dynamic loads without failure and having the capability to store enough shock energy. At the development of this design, a new material type was proposed- the composite material. The introduction of this material type revolutionized the design of leaf springs because it offers superior properties compared to the conventional steel used before. However, in as much as the mono-leaf spring design is ideal, it also provides new challenges; it does not offer the internal damping effects necessary to prevent continued oscillation of the vehicle. The problem, however, can be easily resolved by introducing power dampers.

3 Materials

Leaf spring's material considerations is one of the core areas of focus in this paper. The material used gives the spring its unique properties, which are required for its function. Such properties are maximum tensile strength and minimum modulus of elasticity. These properties provide, for higher specific strain energy (equation 1) one of the crucial properties for a leaf spring [13].

$$S=1/2\ (\sigma^2/\rho E) \qquad (1)$$

This property means that the spring can store more strain energy and protect the chassis from undulations. The review will only focus on composite materials since the development is based on composite materials.

3.1 Composite Materials

Composites materials (composites) are formed by combining two or more materials to form one [14]. The combined materials do not react but instead, maintain a recognizable interface. The development of these materials is motivated by the fact that there is no available homogenous material with the desired characteristics for a given application [15]. For example, fibre materials are used in leaf springs because of the following properties: 1 High strength to weight ratio, 2 Lightweight, 3 Fire Resistance, 4 Chemical and weathering resistance, 5 Design Flexibility, and 6, Manufacturing Economy. Additionally, composite has superior fatigue properties [13], they are even used to repair metallic frames with fatigue damage [14]. In as much as composites are beneficial, they have limitations that include high raw material costs and usually high fabrication and assembly costs. However, the benefits that composites provide are far greater than their shortfalls.

4 Review of past studies based on the analyses of different leaf spring designs

In one of the recent studies made by Mahanthi and Murali [16], they made a comparative study between eight materials, four alloys and four composite materials to check which of the eight materials perform better when it comes to strength to weight ratio. Mahanthi and Murali used Computer Aided Three-Dimensional Interactive Application (CATDIA) to develop similar 3D models for each material and did the analyses using ANSYS's finite element analysis application. The study they made involved applying loads from 1000N to 4000N. Each material was exposed to these four loads, and total deformation and equivalent stresses were noted. Based on the results of static analyses of both steel and composite leaf springs like EN47, KEVLAR, S-Glass Epoxy & E-Glass Epoxy, the conclusion they drew was that Kevlar (composite material) performs best because it had the least weight and performed better in terms of strain energy, total deformation and equivalent stress.

In another study conducted by Prakash [17] in 2017, he limited his analysis to only two materials, steel 65Si7 representing the alloys and E-glass/Epoxy representing the composite materials. Prakash's design imperatives were stresses and deflections, again like Mahanthi and Murali. He did not build the designs but used CATIA (Computer Aided Three-Dimensional Interactive Application) to model designs and ANSYS finite element analysis to perform static analysis on the models. The investigation concluded that steel leaf springs are approximately 70% heavier than composite leaf springs and that they are they are muh more resilient to stressors than steel springs of the same design. Another significant outcome drawn from Prakash's investigation is that composite materials are 5 times more durable than conventional steel springs. In the paper "design and analysis of composite leaf springs" by Pinaknath [18], he investigates the effectiveness of composites leaf springs limiting his materials to only two, E glass /Epoxy (GFRP) for composites and 65Si7 steel for alloys just like Prakash [17]. He made the E-Glass/Epoxy design into a three-layered structure and perfomed both static and dynamic analyses on the models. He then compared his finding against the conventional leaf spring design. The results obtained concluded that composite materials are much stronger as they have less stress and deflections when exposed to loads, the material also possesses useful specific strain energy. Pinaknath also observes that composite materials are 67.88% lighter in weight than steel leaf springs and are an adequate replacement for both stiffness and stress.

Similar to the study carried by Pinaknath, Vivek Rai [19] in the paper he published on the development of composite leaf springs for lightweight commercial vehicles, in place of 65Si7 steel like Pinaknath he uses, EN47 steel. He also used different software to conduct his study, Pro-E wildfire 4.0 for 3D modelling and ANSYS for static analyses. Vivek's objective was to replace the multi-leaf steel spring with a mono composite leaf spring for the same load-carrying capacity and stiffness. To do this, he compared EN 47 steel and composite materials (not specified). The finite element analysis results obtained are shown in Table 2 below.

Table 1. Finite element analysis results [19]

Parameter	EN47	Composite
Load (N)	2446	2446
Max stress (MPa)	738	570
Max Deformation (mm)	25.4	24.6
FoS	1.15	1.57

From the results he obtained in table 2, Vivek concluded that transitioning from conventional steel springs to composite leaf springs was necessary. It is clear from the table that the yield stress and deformation of composite material have no significant difference from that of steel but, the use of composites contribute to a considerable reduction in vehicles weight; the benefits of the composite leaf springs outweigh its shortfalls.

Three years later, after Vivek's study, Sedlacek et al. [20] published a review on the design of composite leaf springs for railway vehicles; their focus was to design and optimize the railway leaf springs done first by numerical simulation then followed by an experimental test. The materials investigated were the glass fibre reinforcement with epoxy resin for composite materials. The material was modelled into the same shape and size as that of a conventional leaf spring.

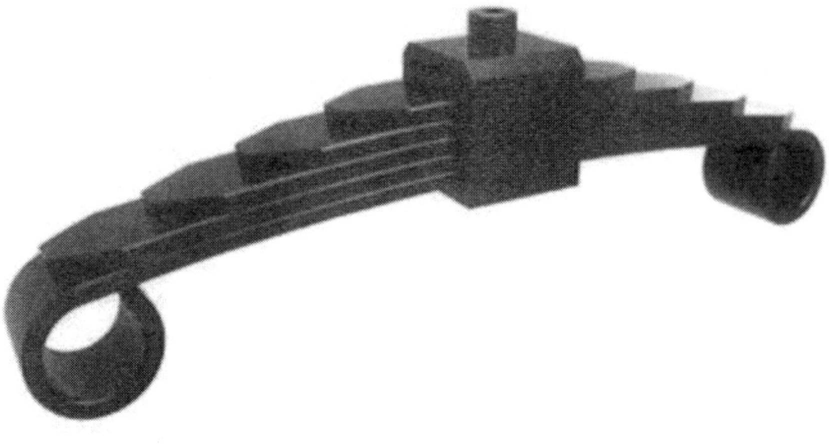

Fig. 2. Conventional Leaf Spring [20]

A maximum stress strength criterion and transverse shear interlaminar criterion were used to evaluate the failure index. From the investigation, the observations were assessed, and the comp

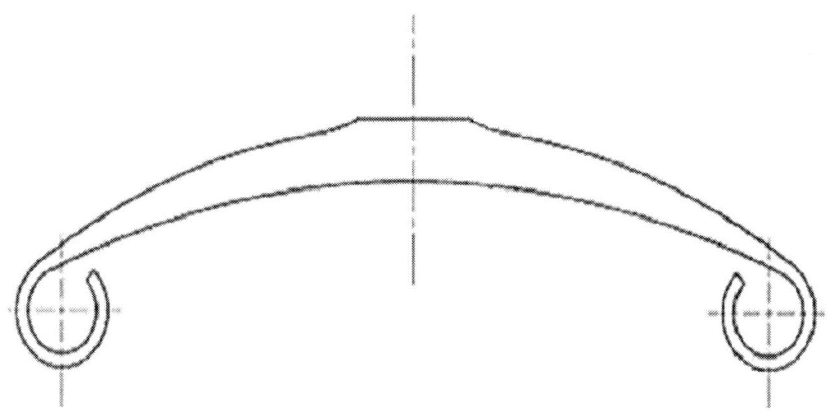

Fig. 3. The geometric parameters of the optimization of the leaf spring that was used [20].

Four years before the work published by Sedlacek et al., Venkatesan and Devaraj [21] conduct a similar study with more specified elements. The study focused on the materials in contrast to Sedlacek et al. study, which focused on geometry instead. The investigation of Venkatesan and Devaraj compared leaf springs made from steel with 0.9 to 1% carbon to those made from a composite material made from Glass/Epoxy reinforcements. Their study further investigated the effect of shape, using uniform cross-section leaf springs and those that are non-uniform (thicken as they approach the middle). The investigation was done using simulative software, 3D modeller for 3D drawing and ANSYS 10.0 for the finite element analyses.

Venkatesan and Devaraj drew two conclusions from their study; 1) the geometry of the leaf springs plays a significant role; it affects the performance of the leaf spring. Composite leaf springs with a thicker centre outperformed the ones with a uniform cross-section for strength. 2) From the comparative study they carried out between composite and steel leaf springs concerning weight and strength, it was observed that composite leaf springs are much lighter and have greater strength to weight ratio than steel leaf springs. Another similar study was carried out by Saini et al. [23] in 2013; their objective was to compare the stresses and weight-saving capabilities of three composite leaf springs made from different materials against steel made leaf springs. Saini et al. used Auto-CAD 2012 for 3D modelling, and the statistical analysis (stress and deformation) was done using ANSYS 9.0 software. The list of the materials and their performances are listed in the table below.

Table 2. Displacements and Von Misses stresses for the same loading conditions [23].

Materials	Displacements (mm)	Stress (MPa)
Steel	10.16	453.92
Graphic epoxy	15.75	653.68
Carbon epoxy	16.21	300.3
E-glass epoxy	15.89	163.22

From the results obtained from their study, only graphite-epoxy has higher Von-Mises stresses than steel, meaning only graphite-epoxy can replace steel from the stress stiffness point of view. A comparative study was also made on the strength to weight ratio. The results they obtained are listed in the table below.

Table 3. % saving of weight by using composites instead of steel [23].

	Materials	% Weight saving
1	Steel	-
2	E-glass epoxy	81.72
3	Carbon epoxy	90.51
4	Graphite epoxy	91.91

Table 4 shows the percentage weight-saving measured against steel. Using either E-glass epoxy or Carbon epoxy of Graphite epoxy, there is 81.72, 90.51, 91.91 cuts in weight, respectively. The conclusion drawn from their results was that composite materials perform better when weight reduction is considered than steel, with graphite-epoxy having the highest performance.

The above review covered the comparison of materials between composite materials and conventional steel materials extensively for leaf springs. The following reviews take a unique route. First, Papacz et al. [24] studied the ability of composite leaf springs to suppress vibrations. Amplitude and reactions of props were measured and recorded. Using these parameters, W. Papacz et al. [3] calculated the coefficients of vibration for composite leaf springs and steel leaf springs. From the analyses they made, they concluded that leaf springs made from composite materials have a more remarkable ability to suppress vibrations than steel springs. They also figured that composite materials are more comfortable to shape into any geometry and can be made into shapes with higher vibration suppression capabilities.

Another unique approach was taken by Sam Varghese et al. They discussed the effects of shapes in the performance of leaf springs. They modelled two composite leaf springs made of glass fibre and with a polymer matrix, which was rectangular shaped (conventional); the other was trapezoidal. 3D models were drawn using solid works software, and the finite element one analysis was done using ANSYS 16.0. Table 5 shows the results of the investigation.

Table 4. [25]

Cases	Deformation(m)	Equivalent Stress (Pa)
Conventional steel leaf spring	0.0004301	1.6757×10^8
E-glass/Epoxy leaf spring	0.0023004	1.3896×10^8
	0.0013804	1.0391×10^8

The conclusion drawn by Sam Varghese et al. was that composite materials reduce the weight by 85% if substituted for steel springs and that trapezoidal leaf springs are much superior to rectangular composite leaf springs.

5 Conclusion

A critical review of the development of composite leaf springs has been presented. The review ultimately concludes that composites are a suitable replacement for conventional steel. This is because of the unique properties that composites possess. The review has also shown ways to assess the performance of leaf springs both experimental and computational ones. The review, however, is subject to limitations in that, most of the work presented focused on computational analysis and less on experimental analysis. Goransson [26] argues that simulations do not precisely simulate real-life events as much as they are essential and accurate. Real-life experiments must be carried out to confirm the results obtained by the simulation. However, there were other unique approaches, one carried out by Frankovský et al.[3], where they investigate the vibration suppression capabilities of composite materials in comparison to other materials. This study calls for the investigation of shapes and the leaf springs' material properties, which is a unique approach to what most researchers are focusing on. Additionally, the review did not discuss much the vibration absorption properties of the springs an effect directly affected by specific strain energy of the material. This calls for further work to be carried out on the shape and other composite material compositions. The generalizability found in this review makes most of the work subject to limitations, for instance, studying one composite material and making conclusions based on the assumption that all composite materials are the same. This study aims to contribute to this growing area of research by exploring these gaps.

1. The performance of other composite materials, since there is a variety.
2. The study of different shapes of leaf springs and looking into their performances (optimization).
3. Study the vibration absorption capabilities of the leaf spring specifically focusing on strain energy for both dynamic loads and statistical loads.

References

1. H. A. Al-Qureshi. (2001). Automobile leaf springs from composite materials. 118, 58–6.
2. J. J. Fuentes, H. J. Aguilar, J. A. Rodríguez, & E. J. Herrera. (2009). Premature fracture in automobile leaf springs. 16 648–655.
3. P. Frankovský, W. Papacz, E. Tertel, & P. Kuryło. (2014). Analysis of the Fatigue Life of Composite Leaf Springs. 611, 346–351. https://doi.org/10.4028/www.scientific.net/AMM.611.346.
4. M. Terrier. (1986). L'invention des ressorts de voiture. 39, 17–30. https://doi.org/10.3406/rhs.1986.4016.
5. Bending stress in leaf spring (2021).
6. M. A. Omar, A. A. Shabana, A. Mikkola, W.-Y. Loh, & R. Basch. (2004). Multibody system modeling of leaf springs. 10, 1601–1638.
7. K. Badshar. (2021). Leaf spring - construction, working, advantages & types. https://www.mechanicalpost.site/2021/02/leaf-spring.html (accessed 2021).
8. M. R. V. Rambade, M. N. S. Dudhale, M. S. K. Valawalkar, M. P. R. Yadav, & M. M. M. Patil. Design. Analysis and Fabrication of Mono Leaf Spring using S-Glass Composite. (n.d.).
9. S. Tripathi & R. K. Ambikesh. (2018). A study of composite materials for leaf spring. (n.d.).
10. P. Solanki & A. K. Kaviti. Design and computational analysis of semi-elliptical and parabolic leaf spring. 5, 19441–19455.
11. A. Atig, R. Ben Sghaier, R. Seddik, & R. Fathallah. (2018). A simple analytical bending stress model of parabolic leaf spring. 232, 1838–1850.
12. S. Bhattacharjee, S. Kanitkar, N. Kalasapur, & V. Patel. Composite leaf spring. 2014 (2013).
13. S. Mouleeswaran. (2012). Design, manufacturing and testing of polymer composite multi-leaf spring for light passenger automobiles - a review (IntechOpen, 2012).
14. https://www.hkdivedi.com/2017/10/bending-stress-in-leaf-spring.html (accessed August 9, 2021).
15. T. W. Clyne & D. Hull, *An introduction to composite materials* (Cambridge university press, 2019).
16. D. L. Mahanthi & C. V. S. Murali. (2017). Design and analysis of composite Leaf Spring for light Weight Vehicle. 4, 147–152. https://doi.org/10.22161/ijaers.4.3.23.
17. P. Shakt. Vehicle leaf spring design, analysis & comparison by using e-glass / epoxy and steel 65si7. 4 (2017) 105–110.
18. P. Dewanji. (2016). Design and analysis of composite leaf spring. 177–183.
19. V. Rai, Development of a Composite Leaf Spring for a Light Commercial Vehicle (Tata Magic). (2013).
20. F. Sedlacek, P. Bernardin, & V. Lasova. Design of a composite leaf spring for railway vehicles. (2016) 493. https://doi.org/10.2507/27th.daaam.proceedings.074.
21. M.Venkatesan & D. H. Devaraj. Design and analysis of composite leaf spring in light vehicle. 2 (2012) 213–218.
22. M. Venkatesan & D. Helmen Devaraj, AH021213218. 2 (n.d.) 213.
23. P. Saini, A. Goel, & D. Kumar. (2013) Design and analysis of composite leaf spring for light vehicles. 2 (2013) 1–10.
24. W. Papacz, E. Tertel, & P. Kuryło. (2014). Performance comparison of conventional and composite leaf spring. (2014).
25. S. Varghese. (2019). Structural design and analysis of leaf spring made of composite material. 7, 3397–3409. https://doi.org/10.22214/ijraset.2019.5558
26. B. Göransson, No title. (2001).

Collision Prediction and Optimal Trajectory Generation for Collision Avoidance Systems in Trackless Mobile Machines

Jesse Declercq[1,*], *Theunis* Botha[1,**], and *Herman A.* Hamersma[1,***]

[1]University of Pretoria
Department of Mechanical & Aeronautical Engineering
Hatfield, Pretoria, 0002, South Africa

Abstract.
The continued high number of fatalities associated with Trackless Mobile Machines (TMMs) in South Africa has led to the introduction of Collision Avoidance System (CAS) regulations in the Mine Health and Safety Act in 2015. These regulations have engendered the profusion of technologically-immature CASs from third-party vendors — all of which are centred on automatic stopping and braking systems. These braking systems often result in trivial or ineffective solutions, proving costly to mining operations. The amalgamation of braking and steering control in CASs may substantially increase the solution space, providing far safer and more efficient manoeuvres. The contribution made by steering control is evaluated through the development and implementation of a recursive non-linear collision prediction estimator and optimal trajectory generation model. Two independent optimal trajectory generation models are proposed to compete against one another in an attempt to synthesize the safest, most predictable, and efficient trajectory. These models are differentiated primarily by their path planning approaches: lattice optimization and Monte Carlo hyper sampling. The collision prediction estimator and trajectory generation models are evaluated in simulation using the Earth Moving Equipment Safety Round Table (EMESRT) mining interaction scenarios. Initial results indicate increased CAS solution spaces when adding steering control in collision-avoiding scenarios, providing safer and more effective solutions in high velocity vehicle interactions.

1 Introduction

The field of autonomous mobility has experienced radical expansion in recent years, with implementations progressing in commercial, industrial and passenger vehicles. This expansion has been incentivised with the potential reduction of mortality rates as human error is eliminated from public roads [1]. The Department of Mineral Resources' Occupational Health and Safety 2017 annual safety report recorded that 30-40% of mining industry fatalities were caused by mobile equipment, while 9% of the fatalities were caused by Trackless Mobile Machines [2], [3]. These concerning statistics lead to the amendment of the South African Mine

*e-mail: Jesse.Declercq@tuks.co.za
**e-mail: Theunis.Botha@up.ac.za
***e-mail: Herman.Hamersma@up.ac.za

Health and Safety Act (MHSA) legislation in 2015, compelling mine employers to introduce automatic emergency braking and proximity detection system at each mine where significant risk of collision exists. These systems must at least include: (a) the automatic detection of a pedestrian or TMM within its vicinity and (b) in the event where no action is taken to prevent a potential collision, further means to retard the TMM until a safe speed is reached [4] [5].

The Earth Moving Equipment Safety Round Table (EMESRT) developed a performance requirement in 2016 in order to augment the interpretation of operation controls in several TMM potential unwanted events [6] However, the MHSA's proposal of proximity detection systems is unclassified and neglected by EMESRT performance requirements, leaving substantial gaps in the collision prediction performance domain. In addition, in the South African mining industry, all current CASs developed by third party vendors make exclusive use of Collision Avoidance By Braking (CABB) systems due to the requirement by the MHSA legislation. These systems are often insufficiently tested or developed to address the significant and unique challenges associated with the practical implementation of such systems in mining environments [5]. Due to the single Degree-Of-Freedom (DOF) constraint, these CABB solutions operate in highly restricted solution spaces, often leading to trivial or energy-inefficient solutions. These solution spaces may be expanded by improving sensing ability, decision-making and road design, however, the solution space is still bound by the system's single DOF manoeuvrability, irrespective of system intelligence. Therefore, by introducing steering control as an additional DOF to the system, the solution space will be expanded significantly, potentially providing far safer and more efficient manoeuvres [7].

2 Collision Prediction

Collision avoidance systems make use of proximity detection systems as instructed by the MHSA in order to detect pedestrians, other TMMs and potential hazards on the road. Once a potential hazard has been detected, this information needs to be relayed to the CAS in a manner in which the appropriate control level can be articulated. This paper presents a probabilistic non-linear collision prediction model which integrates robust TTC calculations with non-linear state estimations and a lucid collision probability method.

2.1 Non-linear State Estimation

All collision prediction techniques rely on a form of vehicle modelling in order to predict future vehicle states. These states are in turn used to predict whether two vehicles or objects will, in future, occupy the same space in time. Some vehicle modelling techniques make use of rudimentary particle kinematic models whilst others make use of more sophisticated kinematic or dynamic vehicle models. The vehicle models are non-linear, capable of accurately tracking the vehicle's motion during cornering. Conversely, the particle model is linear, leading to large prediction errors during cornering manoeuvres. These errors become even more evident during large time steps — which are critical for collision prediction applications. Therefore, the motion modelling is limited to the kinematic or dynamic vehicle models.

These vehicle model, however, assume ideal initial state inputs and do not consider noisy or inaccurate state inputs. These state inputs are acquired from copious sensor and communication network measurements which are subject to vibrational and signal noise as well as accuracy limitations. Henceforth, there is a level of measurement uncertainty when considering the state input to the vehicle model for state estimation. State estimation filters are designed to synthesize better state estimates from noisy or inaccurate sensor measurements by minimizing the error covariance of the estimated state given the noise distributions. As

https://doi.org/10.1051/matecconf/202134700032

aforementioned the motion of a vehicle is inherently non-linear during cornering and therefore, the state estimation filters must be capable of tracking non-linear manoeuvres. Filters which are capable of such tracking are the particle [8], unscented Kalman [9] and extended Kalman filters [10].

Particle filters are excellent when dealing with highly non-linear motion and when the state transition function is easy to compute. However, due to the expensive integration process within the vehicle motion models, the particle filter is infeasible due to the linear coupling of computation time and particle count [8]. Although the Extended Kalman Filter (EKF) and Unscented Kalman Filter's (UKF) accuracy is highly comparable, their computational efficiencies vary greatly. The UKF often takes more than double the computation time of the EKF [9]. Therefore, the choice between the two filters is one made solely on the computational resources available.

2.2 Time-To-Collision

The TTC metric identifies how much time the operator has to evade any potential oncoming collisions, as well as the CAS's required control level according to the EMESRT performance requirements. Although this paper focuses exclusively on level 9 control applications, the TTC metric built into the collision prediction model considers all possible future control level functionality. In the proposed collision prediction model, the TTC metric is calculated using two state estimation stages. The first stage makes use of a fixed step state series, where future states are estimated at a set of fixed time step intervals. The second stage makes use of a dynamic step state series using a calculated integration step size, where future states are estimated at a step size determined by a defined spatial resolution and the relative velocity between the host TMM and potential hazard. The time domain used for the second phase is determined using a standard bisection approach. This time domain is the domain in which the most accurate TTC will be determined. The TTC is then determined using the time in which the points of contact of the two bodies are closest to one another as seen in Figure 1.

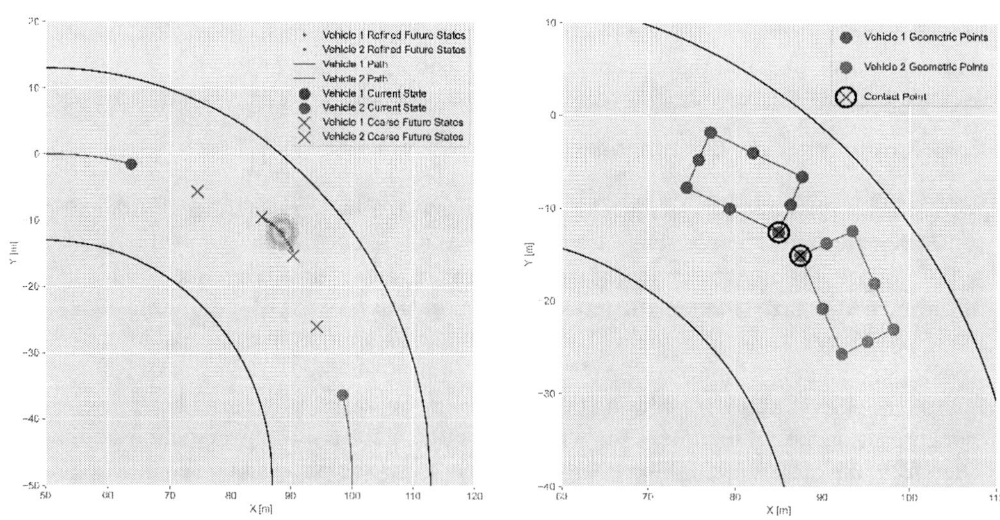

Figure 1: The collision prediction for a head-on collision around a bend. The multivariate Gaussian is shown using the filter error covariance matrix. The second plot indicates the eight contact points considered in collisions and the actual point of contact in the scenario.

Using the mean state estimate and error covariance matrix from the non-linear filter, the probability of collision between the two contact points can be determined.

2.3 Collision Probability

A probabilistic collision prediction model is capable of accounting for measurement uncertainties and noise present in sensors and communication systems. This is accomplished by exploiting filter state distributions. The mean estimated state vector and covariance matrix are used to model multivariate Gaussian's for each state estimate as per the filter. These distributions are used to calculate how likely the two vehicles or bodies are going to occupy the same spatial domain in time.

The probability of collision, $\mathcal{P}$, is calculated by integrating the joint distribution's probability over a safety region, parameterized by ℓ, as shown in Equation 1. The probability of the joint distribution is determined by the product of the likelihood, $\mathcal{L}$, in Equation 2

$$\mathcal{P} = \int_{\bar{y}\in\ell}\int_{\bar{x}\in\ell} P(\Delta\mathbf{x}_k|\mathbf{x}_{k,1},\mathbf{x}_{k,2})dxdy, \tag{1}$$

where the posterior probability is calculated using Equation 2

$$P(\Delta\mathbf{x}_k|\mathbf{x}_{k,1},\mathbf{x}_{k,2}) = \prod \mathcal{L}(\Delta\mathbf{x}_k|\mathbf{x}_{k,1},\mathbf{x}_{k,2})dA, \tag{2}$$

according to the assumed multivariate Gaussian state distribution defined by Equation 3.

$$\mathcal{L}(\mu,\Sigma) = \frac{1}{(2\pi)^{n/2}|\Sigma|^{1/2}}\exp\left(-\frac{1}{2}(x-\mu)^T\Sigma^{-1}(x-\mu)\right). \tag{3}$$

The multivariate Gaussian parameters are that of the non-linear filter's state estimation mean and error covariance matrix. A collision is flagged as true if $\mathcal{P} > 1 - \rho$, where ρ is the allowable probability of intersection. Once the collision is flagged as true, the final TTC is then relayed back to the CAS for possible control level evaluation.

3 Trajectory Generation

Trajectory generation is the development of a joint time-space path, planned according to a set of directives to accomplish an objective. In this paper, trajectory generation is split into two key phases: path planning and velocity profiling. Path planning is responsible for the spatial path development in which the vehicle must follow in order to safely achieve an end goal state. Two path planning approaches are proposed: the Dense Lattice sampling (DLS), Monte Carlo Lattice Hyper Sampling (MCLHS) models. Velocity profiling is the construction of a velocity map propagating longitudinally along the planned path which adheres to vehicle dynamic constraints. Consequently, the velocity profiling phase is contingent upon the path planning phase. The lateral, longitudinal, and combined loading effects are taken into account in order to ensure that neither the friction nor roll-over limits are exceeded [11].

3.1 Path Planning

Path planning aims to synthesize a smooth, collision-avoiding path along which vehicle stability is maintained. The intrinsic non-convexity and high dimensionality of the solution space proves challenging for most optimization algorithms to traverse. The DLS model proved the most effective solution for the defined problem statement from literature; however, often results in either high curvature or collision inducing paths due to the independent curvature and collision avoidance objective functions failing to converge on global optima. Therefore, a novel MCLHS model was proposed to overcome these pitfalls, by propelling particles through the search space using a hyper sampling methodology and a chained Bayesian backpropagation.

3.1.1 Dense Lattice Sampling

The DLS model generates a lattice-bound path according to a set of obstacle avoidance and curvature objective functions [12]. The DLS model achieves an optimal path by independently minimizing both the obstacle avoidance and curvature objective functions by changing the nodal selection and seeded path offset respectively. Formally, the DLS model uses a non-parametric trajectory optimized for road geometry, obstacles and higher order motion directives according to a reference frame as shown in Figure 2.

Obstacle avoidance is prioritized over curvature minimization in the optimization execution order. Obstacle avoidance is achieved by making use of a cost function across all edges $e_{n_k \to n_{k+1}}$ within the lattice as described by Equation 4

$$\underset{\{e_{n_k \to n_{k+1}}\}}{\operatorname{argmin}} \sum_{k=0}^{K-1} w_d d(e_{n_k \to n_{k+1}}) + (1 - w_d)\, o(e_{n_k \to n_{k+1}}) + \delta_{\text{obstacle}}(e_{n_k \to n_{k+1}}), \tag{4}$$

where $d(e)$ is responsible for penalizing long edges, $o(e)$ for penalizing lateral offsets, and the weighting scalar, w_d, determines the trade-off between these two terms. The term $\delta_{\text{obstacle}}(e)$ — set to infinity — is responsible for preventing paths which result in collisions, ensuring edges which lead to obstacle nodes will never be sampled. Once the obstacle-avoiding path of lowest cost is chosen, known as the seeded path, it is then optimized for curvature, using Equation 5

$$\underset{\{o_i\}}{\operatorname{argmin}} \sum_{i=0}^{N-1} w_\kappa \|\kappa_i\| + (1 - w_\kappa) |o_i| . \tag{5}$$

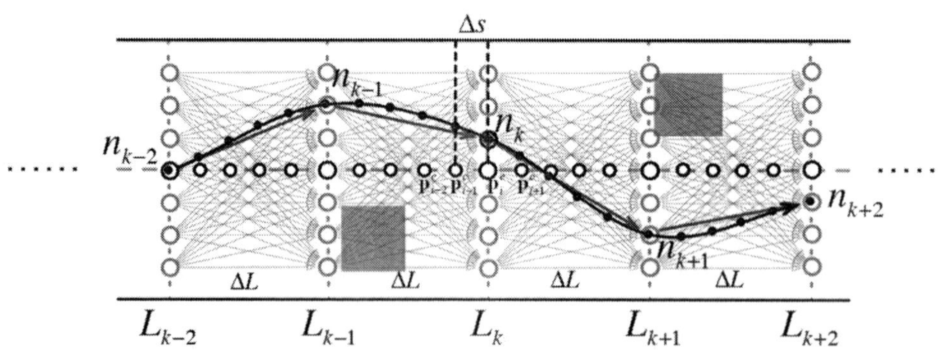

Figure 2: A sample dense lattice indicating the edge ($e_{n_k \to n_{k+1}}$), layer (L_k), node (n_k) and sampling distance (Δs) formation. This diagram also indicates the obstacle detection and avoidance capabilities within the generated seeded path [12].

where the weight w_κ trades off between minimizing cumulative curvature κ and minimizing cumulative lateral offset o. This curvature optimization is performed by the Levenberg-Marquardt (LM) algorithm, however, the Sequential Least-Squares Quadratic Programming (SLSQP) algorithm proved to be far more effective in implementation. Lastly, a clamped cubic B-Spline is then used to smooth the discretized trajectory before being passed to the velocity profiler [13].

3.1.2 Monte Carlo Lattice Hyper Sampling

The novel MCLHS model makes use of a Monte Carlo path exploration technique coupled with Bayesian backpropagation. A series of particles are set to explore the reference path by means of a dense lattice. The connection between nodes from two different layers — "nerves" — are defined by probabilistic weights according to agility directives and lane biases. The goal of the MCHLS model is to overcome the considerable downfalls in most path planners such as DLS, adaptive potential fields, and curvature optimization-based models. Curvature optimization-based models often rely on global solutions attained through gradient method convergence — which is not always possible due to the non-convexity of the solution space. Instead, the MCLHS model propels a collection of particles propagating through the lattice based on a sampling methodology where particles decide on which nerve to choose based on its probabilistic weight, $w_{m:n}$, from the weighting matrix, $\overline{\mathbf{W}}$ as defined in Equation 6.

$$\overline{\mathbf{W}} = \begin{bmatrix} \{w_{0:0} & w_{0:1}\}_{0,0} & \{w_{1:0} & w_{1:1} & w_{1:2}\}_{0,1} & \cdots & \{w_{n:n-1} & w_{n:n}\}_{0,n} \\ \{w_{0:0} & w_{1:1}\}_{1,0} & \{w_{1:0} & w_{1:1} & w_{1:2}\}_{1,1} & \cdots & \{w_{n:n-1} & w_{n:n}\}_{1,n} \\ \vdots & \vdots & \vdots & \vdots & \vdots & \vdots & \ddots & \vdots \\ \{w_{0:0} & w_{0:1}\}_{l,0} & \{w_{1:0} & w_{1:1} & w_{1:2}\}_{l,1} & \cdots & \{w_{n:n-1} & w_{n:n}\}_{l,n} \end{bmatrix} \tag{6}$$

All possible nerves in which the particle can sample — according to the agility directive — sum to one and are known as nerve groups. Nodes which are classified as obstacle nodes have all corresponding nerves connecting them set to zero so that they cannot be sampled. The weight of each nerve chosen in the particle's path is updated according to how likely this path's curvature is the lowest amongst all possible paths. Using a Bayesian approach the posterior probability is defined by Equation 7 [14]

$$p(w_i|\kappa_o) = \frac{p(\kappa_o|w_i)p(w_i)}{p(\kappa_o)} \tag{7}$$

where

$$k_o = \{\kappa_p \in \mathbf{K} : \kappa_p < \kappa_i \forall \kappa_i \in \mathbf{K}\}. \tag{8}$$

The prior probability, $p(\kappa_p)$, is challenging to obtain due to its lattice architecture dependence. Nonetheless, this prior is only necessary for normalization and therefore the normalization constant, α, is substituted across nerve groups instead. Henceforth, the weights are constantly renormalized at each nerve group after backpropagation. By evaluating the structure of the lattice, the likelihood, $p(k_o|w_i)$, can be found by simply applying the product of the chosen weights after weight i until the end of the path which resulted in curvature κ_p. The Bayesian backpropagation is then represented mathematically in Equation 9

$$p(w_i|\kappa_o) = \alpha \cdot p(w_i) \cdot \prod_{j=i+1}^{l} p(w_j). \tag{9}$$

This is done for every particle after it has fully propagated through the lattice and has obtained a resultant curvature. The final nerve, however, is not accounted for in this Bayesian backpropagation. Instead, $p(w_l)$ is determined using Equation 10

$$p(w_l) = \begin{cases} \frac{1}{N_p}, & \text{if } \kappa_p < \kappa_o \\ 1, & \text{if } \kappa_p \geq \kappa_o \end{cases} \tag{10}$$

where N_p is the total number of particles which have already explored unique paths. After a particle has completed its path, the path's curvature is then calculated using Mengers Equation. Finally, the path with the lowest curvature is then selected and a clamped cubic B-Spline is then used to smooth the discretized trajectory before being passed to the velocity profiler [13]. Comparing the DLS and the MCLHS models using five randomized obstacles is shown in Figure 3. The DLS was modelled using an optimized w_d and w_κ parameters. The DLS model in Figure 3 (a) fails to converge upon the optimal solution using both the SLSQP and LM algorithms. The MCLHS converges upon the optimal solution within 100 particles. The particle paths are indicated in Figure 3 (b) highlighting the Bayesian backpropagation of the model.

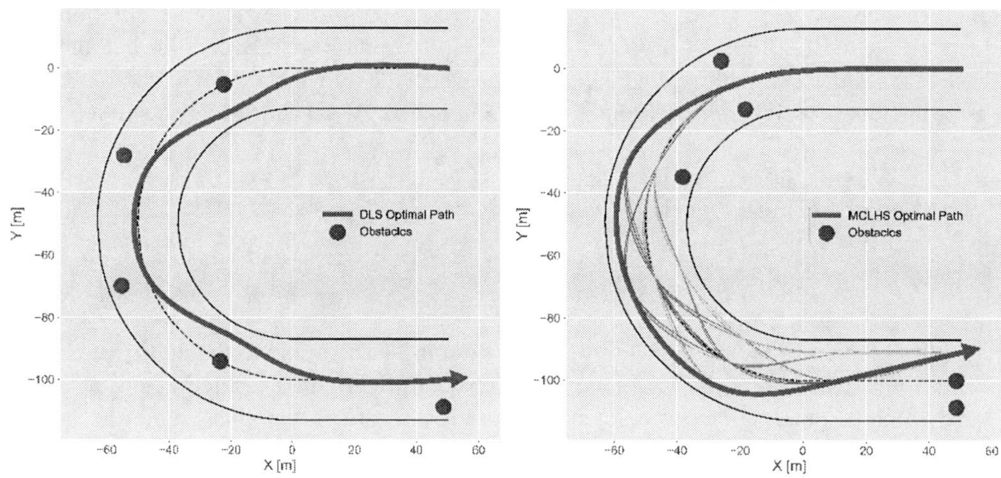

(a) DLS (b) MCLHS

Figure 3: The comparison between the DLS and the MCLHS using five randomized obstacles with the DLS model indicated in (a) and the MCLHS in (b).

3.2 Velocity Profiling

Once the vehicle's path has been planned, the coupled velocity profile is developed in order generate the final trajectory. The velocity profile is developed in the spatial domain using only an initial vehicle state, planned path and its associated curvature. The maximum allowable speed is dictated by the minimum between the lateral acceleration and legal speed limit. The friction circle represents a tyre's maximum grip in both the longitudinal and lateral directions. In order to account for the mining haul trucks susceptibility to roll-over, the friction circle is often modelled as an ellipse rather than a circle, narrower in the lateral direction [11].

A lookahead distance is used to avoid the maximum velocity vector's discontinuous nature. Simply, the lookahead distance x_{ld}, is the maximum expected braking distance $x_{b,local}$ at a given current speed v_c. The lookahead distance introduces a search space from the current vertex to the lookahead distance vertex. At each vertex the local maximum allowable speed v_{local} is defined. The distance required to brake from the current velocity v_c to the propagated local maximum velocity v_{local} is calculated if $v_c > v_{local}$ and termed $x_{b,local}$. If Δx_{local} is larger than the vertices longitudinal difference $\Delta x_{c\rightarrow p}$, then braking is flagged as required.

This search space is utilized to explore longitudinal acceleration solutions at the current longitudinal position. The longitudinal acceleration $\ddot{x}$ for the velocity profile can be calculated using Equation 11, where the maximum allowable deceleration and acceleration is defined by b_{max} and a_{max} respectively.

$$\ddot{x} = \begin{cases} \mathcal{I} b_{max} & \text{if } x_{b,local} + x_\epsilon > \Delta x_{c\rightarrow p} \\ \mathcal{I} a_{max} & \text{elif } v_{max,c} > v_c + v_\epsilon \\ 0 & \text{otherwise} \end{cases} \tag{11}$$

The allowable longitudinal acceleration as a percentage of the maximum allowable longitudinal acceleration is termed the intensity, denoted by $\mathcal{I}$. The longitudinal acceleration intensity can therefore be calculated by considering the friction ellipse as shown in Equation 12

$$\mathcal{I} = \sqrt{1 - \left(\frac{v_l^2 \kappa_l}{\ddot{y}_{lim}}\right)^2}, \tag{12}$$

with x_ϵ and v_ϵ is a safety distance and speed respectively. If no longitudinal acceleration or braking is required, the previous velocity will be continued as the expected velocity. Therefore, the velocity profiler requires the initial velocity of the vehicle as the trajectory is generated. The expected speed vector $\mathbf{v}_e$ is then generated for each longitudinal vertex along the planned path, allowing for a spatial domain for the velocity profile. The velocity profile ensures safe and legal trajectories for the ego vehicle and maintains vehicle stability. In order to execute this trajectory, however, a controller is required.

3.3 Dynamic Controller

The dynamic controller is responsible for executing the velocity profile as well as following the desired planned path. This dynamic controller is a simplified model of the steering controller proposed by T.R. Botha in 2011 [15]. This steering controller makes a combined use of yaw $e_{yaw}(t)$ and lateral offset e_{lat} errors. The yaw component ensures the vehicle's path is parallel to that of the planned path and the lateral offset component ensures the vehicle does not drift from the path due to cross-track errors [15]. The yaw component is given a proportional, integral and derivative term, whereas, the lateral offset component is only given the proportional term. The required steering rate for the dynamic controller is therefore calculated in Equation 13

$$\dot{\delta}(t) = f_{p,yaw}(\dot{x})e_{yaw}(t) + f_i(\dot{x}) \int e_{yaw}(t)dt + f_d(\dot{x})\frac{de_{yaw}(t)}{dt} + f_{p,lat}(\dot{x})e_{lat}(t). \tag{13}$$

The proportional (f_p), integral (f_i) and derivative (f_d) gains are all functions of velocity. This modelling choice was made as steering becomes more sensitive as vehicle speed increases [15]. The predicted lateral acceleration due to the calculated steering angle is compared to the allowable lateral acceleration. Should the predicted lateral acceleration exceed the allowable lateral acceleration, the steering angle δ_t is adjusted using Equation 14

$$\delta_t < \tan^{-1}\left(\sqrt[-2]{\left(\frac{v_t^2(1 - \frac{l_r}{l_r+l_f})}{\ddot{y}_{roll,t} l_f}\right)^2} - \frac{l_r}{l_r + l_f}\right). \tag{14}$$

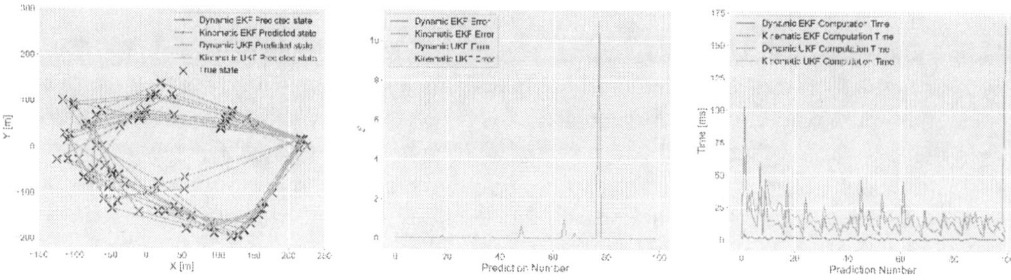

Figure 4: The comparison between the four different non-linear state estimators across positional estimates, relative errors, and computational speeds.

The longitudinal distance between the centre of gravity and the front and rear axles, denoted as l_f and l_r respectively, are simple geometric measurements, playing a crucial role in the maximum allowable steering angle. This Equation is derived using both centripetal acceleration and vehicle kinematic relations. It is capable of determining the maximum allowable steering angle to prevent roll-over. Equation 14 is in fact deterministic in implementation; the maximum allowable steering angle is always desired if triggered in extreme manoeuvres.

4 Preliminary Results

A realistic, physics-based simulation environment was developed in order to validate the collision prediction and trajectory generation model's performance. The geometric parameters of the CAT 789D mining haul truck was used alongside the validated tyre and dynamic vehicle parameters of the Land Rover Defender 2013 110 from the Vehicle Dynamics Group at the University Pretoria. A series of simulations were performed in order to evaluate model capabilities. These simulations provided initial potential value in both the collision prediction and trajectory generation model for steering and braking based CASs.

4.1 Non-linear State Estimation

The UKF and EKF were coupled with the kinematic and dynamic vehicle models to create four different non-linear state estimators. They were evaluated across two metrics: accuracy and computational speed. The accuracy of the filter is measured using the relative error between the estimate and ground truth positional and yaw state components. The computational speed is evaluated as the time taken to complete a single prediction given a static time step size.

Using randomized paths and velocities, a total of 100 prediction steps were used across various time step sizes. The accuracy measurements between the UKF and EKF were superfluous as indicated in Figure 4. At low noise/uncertainty, both filters performed identically. At high noise/uncertainty, the UKF boasted a 0.4 - 0.9% higher accuracy than the EKF, with the dynamic vehicle model always outperforming the kinematic model as expected. However, as seen in Table 1 and Figure 4, the EKF boasted a nearly 6x faster computation time compared to the UKF when using the dynamic model and a near 10x faster computation time when using the kinematic model.

Table 1: Non-linear state estimator computation time comparisons, where the time step Δt is measured in seconds and all statistical measures are evaluated in milliseconds.

	Dynamic Models				Kinematic Models			
	EKF		UKF		EKF		UKF	
Δt	μ	σ	μ	σ	μ	σ	μ	σ
1	6.96	3.22	169.77	46.35	0.48	0.52	7.88	2.21
5	10.65	4.42	25.79	64.64	0.80	0.50	10.08	2.70
10	10.05	5.30	28.04	66.54	0.71	0.49	11.36	2.51
20	14.50	16.50	34.94	80.56	1.00	0.85	17.80	5.17

4.2 Trajectory Generation

The trajectory generation model was evaluated using an L1 head-on collision interaction scenario according to EMESRT requirements. This interaction scenario is used to evaluate the ability of the path planner to develop a collision avoiding smooth path, the velocity profile to synthesize a coupled safe trajectory and the dynamic controller to execute the final trajectory. Furthermore, a comparison between CABB and the proposed CAS is compared in this interaction scenario to advocate the potential value added by the addition of steering control. The performance of the MCLHS model coupled with the velocity profiler and dynamic controller in an L1 interaction scenario is indicated in Figure 5.

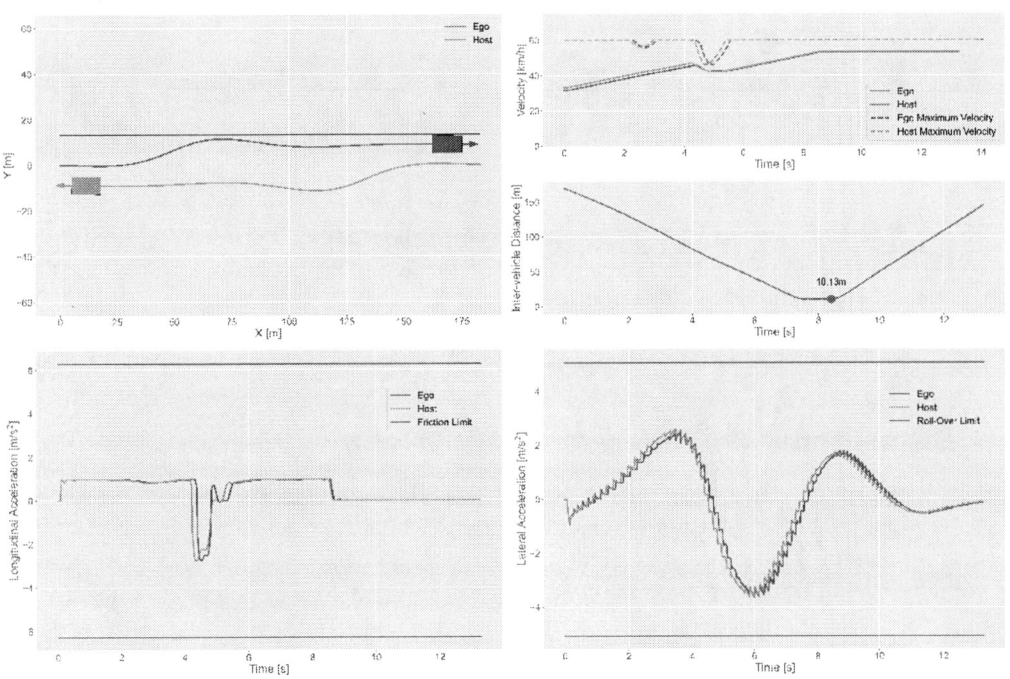

Figure 5: The simulation results in a L1 EMESRT interaction scenario using the MCLHS model. These plots indicate the planned path, velocity profile intervehicle distance and combined loading effects on each vehicle.

The model yielded exceptional results as the collision was avoided and vehicle stability was maintained whilst complying to left-lane standard in South African traffic law. The results are promising; however, additional EMESRT interaction scenarios are required to be evaluated. These interaction scenarios should be simulated many times in order to develop a statistical analysis of the model's performance.

4.3 Steering Contribution

The potential value added by steering control was evaluated over two key performance metrics: time to avoidance and distance to avoidance over a range of velocities. The results in Figure 6 highlight how the avoidance distance increases quadratically during braking manoeuvres and increases linearly during steering manoeuvres. When analysing the avoidance time, braking manoeuvres increased the time required linearly, whereas steering manoeuvres decreased the time required exponentially. The transition speeds between the avoidance time and distance, differ significantly; the transition speed when analysing the avoidance distance is 33km/h and 24km/h when comparing the avoidance time. The shaded regions of the graph indicate which manoeuvre proves to be more effective with respect to its relative metric. Clearly, at higher speeds steering manoeuvres are superior to braking due to the decreased required response time and distance. At lower speeds, braking may prove to be more effective if the collision may not be averted with steering or the response time required is exceptionally low.

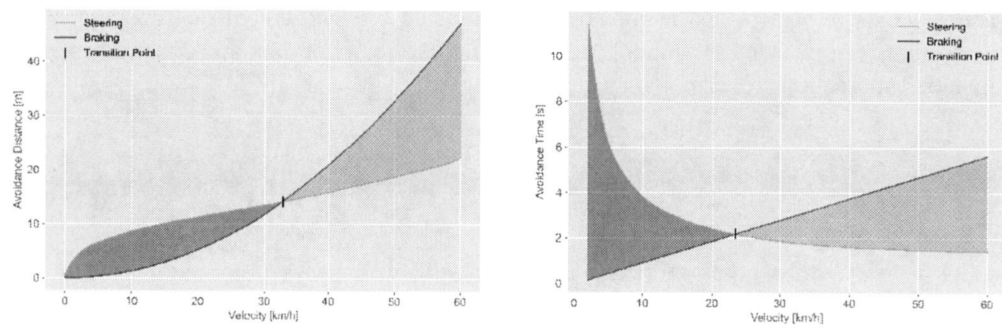

Figure 6: Comparison between the braking distance and steering distances and times to avoid collisions in a head-on collision scenario. The distances are the minimum distance the vehicle travels longitudinally to avoid the collision.

5 Discussion

This paper introduces a probabilistic non-linear collision prediction and optimal trajectory generation model for collision avoidance systems which take advantage of the addition of steering control. The collision prediction model incorporates non-linear state estimation with probabilistic TTC's and geometric point-contacts in order to flag potential collisions. The trajectory generation model is split into path planning and velocity profiling. This model creates a cohesive spatial path and velocity profile for the vehicle to follow, avoiding collisions and maintaining vehicle stability.

Initial results illustrate the collision prediction model's competence during non-linear manoeuvres and high measurement uncertainty. The trajectory generation model's velocity profiler and dynamic controller proved competent in simulation, maintaining vehicle stability and safety in all tests. The MCLHS model overcomes many of the shortcomings synonymous with common path planners — such as the DLS model — and demonstrates predictable and intelligent decisions in collision avoidance scenarios. These models have all shown exceptional value in a small selection of applications; however, a series of simulations and statistical analyses are required before any performance conclusions can be made.

References

[1] M. Green, J. Senders, *Human error in road accidents* (Visual Expert, 2004)

[2] P. Moore, *Mine collision management system testing accelerates with South Africa's 2020 legislation deadline looming* (IM Mining, 2019)

[3] Occupational Health and Safety, Trackless Mobile Machinery **1**, 1 (2017)

[4] Mine Health and Safety Act, Ammendment of Chapter 8 of the regulations relating to machinery and equipment (South Africa Proclamation No. R. 125). Government Gazette 38493:22 **29**, 72 (2015)

[5] C.D. H.A. Hamersma, P.S. Els, *Mining Goes Digital* (CRC Press, 2019)

[6] EMESRT, *Vehicle Interaction Systems*, in *PR – 5A* (EMESRT, 2016), pp. 1–11

[7] J. Jansson, Ph.D. thesis, Linkoping University (2005)

[8] G. Hendeby, R. Karlsson, F. Gustafsson, EURASIP Journal on Advances in Signal processing **2010**, 1 (2010)

[9] M. Wielitzka, A. Busch, M. Dagen, T. Ortmaier, in *Kalman Filters-Theory for Advanced Applications* (InTech, 2018), pp. 56–75

[10] J. Jansson, Ph.D. thesis, Linköping University Electronic Press (2005)

[11] T.D. Gillespie, *Fundamentals of vehicle dynamics*, Vol. 400 (Society of automotive engineers Warrendale, PA, 1992)

[12] T. Gu, J. Snider, J.M. Dolan, J.w. Lee, *Focused trajectory planning for autonomous on-road driving*, in *2013 IEEE Intelligent Vehicles Symposium (IV)* (IEEE, 2013), pp. 547–552

[13] I. Noreen, MDPI Symmetry (2020)

[14] R. Swinburne, Revue Philosophique de la France Et de l **194** (2004)

[15] T.R. Botha, Master's thesis, University of Pretoria (2011)

A Quadratic Elasticity Formulation for Dynamic Interacting Structures in Flow

Ridhwaan Suliman[1,2*] and *Oliver* Oxtoby[3]

[1]CSIR Information and Cyber Security Centre, Defence and Security Cluster, Meiring Naude Road, Pretoria, 0002, South Africa
[2]National Institute for Theoretical and Computational Sciences (NITheCS), South Africa
[3]ENGYS, Studio 20, Royal Victoria Patriotic Building, John Archer Way, London, SW18 3SX, United Kingdom

Abstract. The deformation of slender elastic structures due to the motion of surrounding fluid is a common multiphysics problem encountered in many applications. In this work we detail the development of a numerical model capable of solving such strongly-coupled fluid-structure interaction problems, and analyse the dynamic behaviour of multiple interacting bodies under fluid loading. In most fluid-structure interaction problems the deformation of slender elastic bodies is significant and cannot be described by a purely linear analysis. We present a new formulation to model these larger displacements. By extending the standard modal analysis technique for linear structural analysis, the governing equations and boundary conditions are updated to account for non-linear terms and a new modal formulation with quadratic modes is derived. The quadratic modal approach is tested on standard benchmark problems of increasing complexity and compared with analytical and full non-linear numerical solutions. An analysis of the dynamic interactions between multiple finite plates in various configurations under fluid loading, as well as the effects of the spacing between the structures, is also conducted. Numerical results are compared with theoretical and experimental approaches. The inverted hydrodynamic drafting effect of elastic bodies in an in-line configuration can be confirmed from our numerical simulations.

1 Introduction

Computational mechanics is a growing discipline which uses computational methods to obtain approximate solutions to problems governed by the principles of mechanics. Fluid-structure interaction (FSI) constitutes a branch of computational mechanics in which there exists an intimate coupling between fluid and structural or solid domains; the behaviour of the system is influenced by the interaction of a moving fluid and a flexible solid structure. Examples of such FSI systems include wing flutter on aircraft, aeroelasticity of turbomachinery blades, flows in elastic pipes and blood vessels, heart valve dynamics, swimming of fish and in the processing of paper. In this work, we make use of a blend of mathematical and computational approaches to study strongly-coupled fluid-structure

* Corresponding author: RSuliman@csir.co.za

interaction problems involving long, thin structures.

2 Strongly-coupled fluid-structure interaction approaches

Two computational fluid structure interaction (FSI) approaches are implemented in a partitioned manner to model FSI problems. In the first approach a finite volume method is developed and implemented in *OpenFOAM* by coupling fluid and structural solvers. The finite element method has primarily been used for modelling the mechanics of solids [1]. The finite volume method [2] has traditionally been more dominant in the field of fluid mechanics but has received increased attention for use in solid mechanics over the last two decades. Both schemes can be considered as methods of weighted residuals where they differ in the choice of weighting function [3]. In terms of computational FSI approaches, many recent studies have made use of a single discretisation scheme, either finite element [4-6] or finite volume [7-9], to solve the entire domain. This simplifies the treatment at the interface of the fluid and solid domains. The fluid is assumed to be viscous, incompressible and isothermal with the governing equations given by the continuity and Navier-Stokes equations in an arbitrary-Lagrangian-Eulerian (ALE) reference frame. The solid is assumed to be a homogeneous isotropic elastic solid undergoing large, non-linear deformation with the motion governed by Cauchy's first equation of motion. In this work, we make use of the finite volume method for discretisation of the entire domain. In the second approach a reduced-order modal approach for solving the structural equations is implemented and coupled with an incompressible finite volume fluid solver in *OpenFOAM*. Strong coupling of the fluid and solid domains is achieved by means of a fixed-point solver using a dynamic relaxation parameter based on Aitken's method. It is a simple yet highly robust and efficient approach.

2.1 Governing equations

The equations governing viscous incompressible isothermal fluid flow in an arbitrary-Lagrangian-Eulerian (ALE) reference frame [10] are given by the continuity and Navier-Stokes equations. To close the governing equations, a constitutive relation for stress is required. We assume a Newtonian fluid to relate the stress to strain.

The partial differential equations that describe a homogeneous isotropic elastic solid undergoing large, non-linear deformation are given by Cauchy's first equation of motion [11]. Assuming an isotropic hyperelastic St. Venant–Kirchhoff material model, the stress-strain relationship is given in terms of the second Piola–Kirchhoff stress. The first Piola–Kirchhoff stress is related to the second Piola–Kirchhoff stress by the deformation gradient, which relates quantities in the undeformed configuration to their counterparts in the deformed configuration. The constitutive stress-strain relationship is given in terms of the Green–Lagrange strain, which is related to the displacement field through the gradients of displacement.

The finite volume method of discretisation is used for the spatial discretisation of the equations above. This is done by placing them into weak form or integrating over an arbitrary control volume and applying the divergence theorem of Gauss.

2.2 Modal FSI analysis

Tangential to the full finite volume FSI approach, a reduced-order modal approach to solving the structural equations has been used and coupled with an incompressible finite volume fluid solver in *OpenFOAM*.

For a modal structural analysis, Lagrange's equations of motion in the general case are used, which provides an expression for the eigenvalues or natural frequencies and eigenvectors or mode shapes.

2.3 Fluid-solid coupling

In this work we limit our analyses to incompressible flows and have made use of the incompressible *pimpleDyMFoam* flow solver. In its unmodified form, the coupling between the fluid and solid is done once at the beginning of every time-step when movement of the mesh occurs. The fluid flow provides a traction onto the structure. This results in a deformation of the structure which in turn affects the fluid flow. Coupling conditions for the traction, displacement and velocity are applied to ensure momentum conservation or force equilibrium at the interface, as well as to enforce the kinematic or geometric continuity and no-slip conditions respectively. However, the above conditions require an iterative procedure such that a strongly-coupled solution is obtained.

FSI systems can be solved using a single or monolithic solution method, which is inherently strongly-coupled, or a partitioned solution method, which can be strongly- or weakly-coupled. Monolithic methods ensure stability and convergence of the solution, since all equations are discretised and solved simultaneously. However, this approach may suffer from ill conditioning and convergence is generally slow. Conversely, partitioned approaches allow for the use of two independent solution techniques for solid and fluid equations in isolation. Weakly-coupled partitioned methods may diverge or result in inaccurate solutions when applied to problems where there are strong interactions between solid and fluid domains. The FSI implementation described above is a weakly-coupled one: the modal equations are solved as a boundary condition only once at the beginning of each time-step. If applied to strongly-coupled problems, the solver results in inaccurate solutions or diverges.

A separate coupling algorithm or additional outer iterations between the fluid and solid is therefore required to achieve strong-coupling. The most popular strongly-coupled FSI algorithms either use fixed point iteration or interface Newton–Krylov methods. Both approaches have their own shortcomings: fixed point methods are generally slow to converge as they make use of Gauss–Seidel iterations and methods to speed up convergence are needed, whilst Newton–Raphson methods require the computation of Jacobians that may be difficult to calculate exactly. A strongly-coupled fixed-point solver with dynamic relaxation has been implemented in this work. It is the most basic of the above approaches yet is highly robust and efficient. The fixed point solver uses a relaxation parameter based on Aitken's method. The solver was modified to call the mesh movement algorithm after every iteration, thus allowing a re-calculation of interface forces, displacements and velocities with every iteration hence resulting in a fully-coupled scheme upon convergence.

2.4 Two-dimensional flapping beam

Both FSI approaches have been successfully validated and compared on benchmark cases and the results for a common FSI benchmark problem, proposed by Turek and Hron [12], of an elastic beam in the wake of a cylinder undergoing vortex-induced vibration, are included here. The properties of the fluid and solid are as described in [12]. A uniform constant fluid velocity, Reynolds number Re = 100, was applied at the inlet, while pressure was set to zero at the exit. Flow is assumed to be laminar. Snapshots of the beam deflection and velocity contours at various times are shown in Fig. 1, showing large oscillations of the beam in its second mode of vibration as vortices are shed periodically from either side of it.

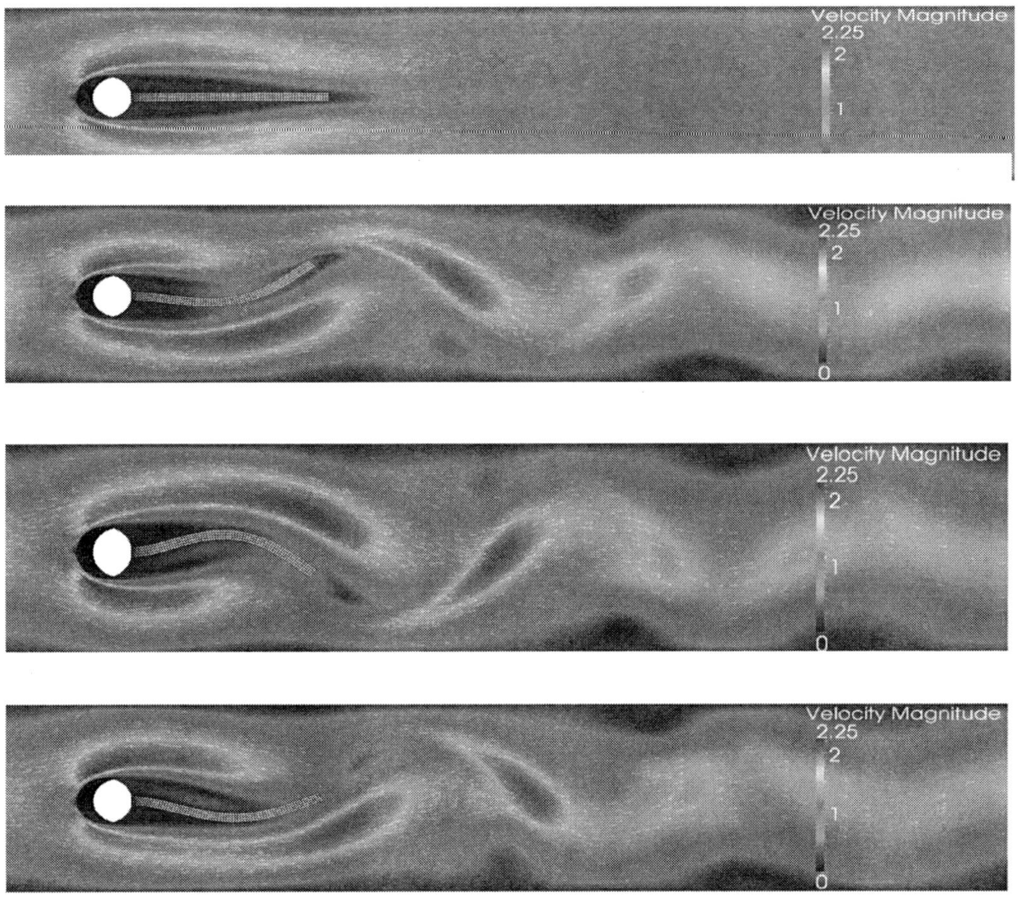

Fig. 1. Velocity contours for a 2D beam oscillating in its second mode of vibration.

The vertical tip displacements of the beam using the modal FSI solver and the finite volume FSI method are compared with the results of Turek and Hron [12] in Fig. 2. There is a good correlation between these results.

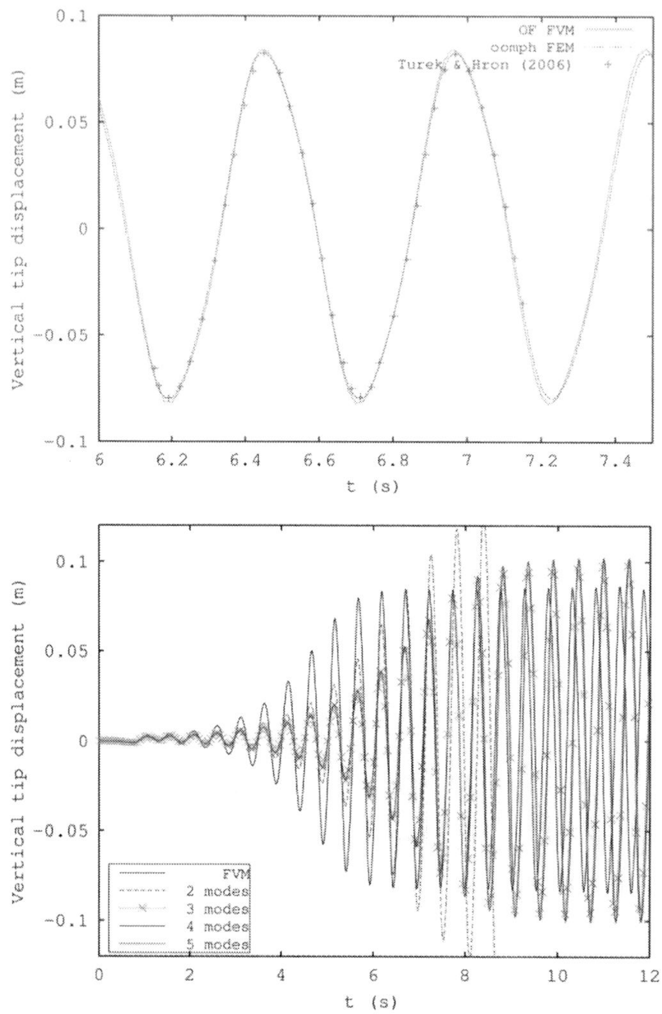

Fig. 2. Vertical tip displacement for the Turek & Hron case using the finite volume method (top) and modal approach (bottom)

The finite volume FSI results, however, required a very fine mesh for the solid domain and consequently a long run-time. Upon further investigation of just a cantilever beam under a tip load it was found that the finite volume method for structural mechanics suffered from a phenomenon of shear locking. When the aspect ratio of the elements (ratio of element width to height) is large, a numerical shear strain component is produced in addition to a bending strain. This absorbs strain energy and results in a decrease in displacement or stiffening, as shown by the results in Fig. 3 (top). In addition, an evaluation of the accuracy of the finite volume scheme shows that the scheme is only first-order accurate, see Fig. 3 (bottom), which also explains the slow convergence in the result.

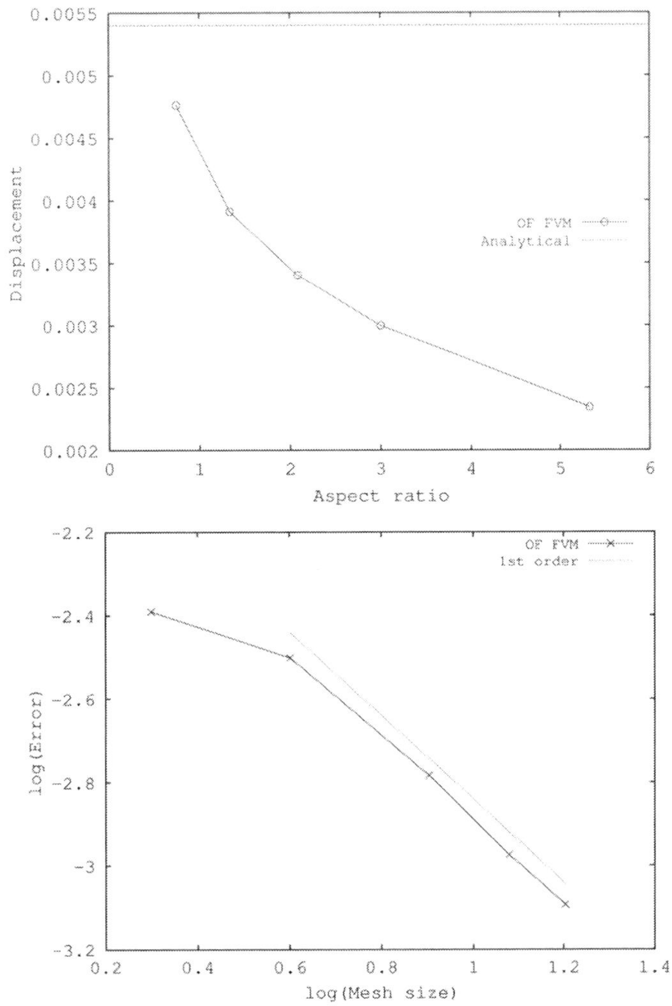

Fig. 3. Analysis of a cantilever beam using the finite volume method: shear locking (top) and order of accuracy (bottom).

3 Stability analysis

The FSI system may become unstable due to the interaction between the fluid-induced pressure and structural rigidity, with flow velocity and structural length being critical parameters that influence the stability of the system. Three distinct theoretical approaches have traditionally been used to study the flapping stability of slender elastic structures immersed in axial flow. These include the traditional hydrodynamic stability theory, which can be viewed as a fluid-centric approach, the structural acoustic or wave-based approach and the aeroelastic theory or structure-centred approach. A number of detailed experimental studies have also been conducted including soap-film, water-tunnel and wind-tunnel experiments. If the flow velocity is below a certain critical value, the structure remains unmoved or in a so-called stretched-straight state. Above the critical flow velocity, the structure undergoes self-sustaining and regular flapping, while at a much greater velocity, irregular and chaotic flapping results. In terms of structural length, it has been shown that below a critical length the structure always remains in the stretched-straight state. Most theoretical studies have been restricted to linear analyses and two dimensions, whilst only a few studies have looked at the effect of multiple slender bodies. The use of computational methods may help to overcome these limitations.

The modal FSI approach was validated by application to a two-dimensional, two degree-of-freedom airfoil. Theoretical methods predict a critical velocity of 51.5 m/s. The results from the transient analysis with an inlet velocity of 51.5m/s show constant amplitude flapping (Fig. 4) which corresponds well with theoretical predictions.

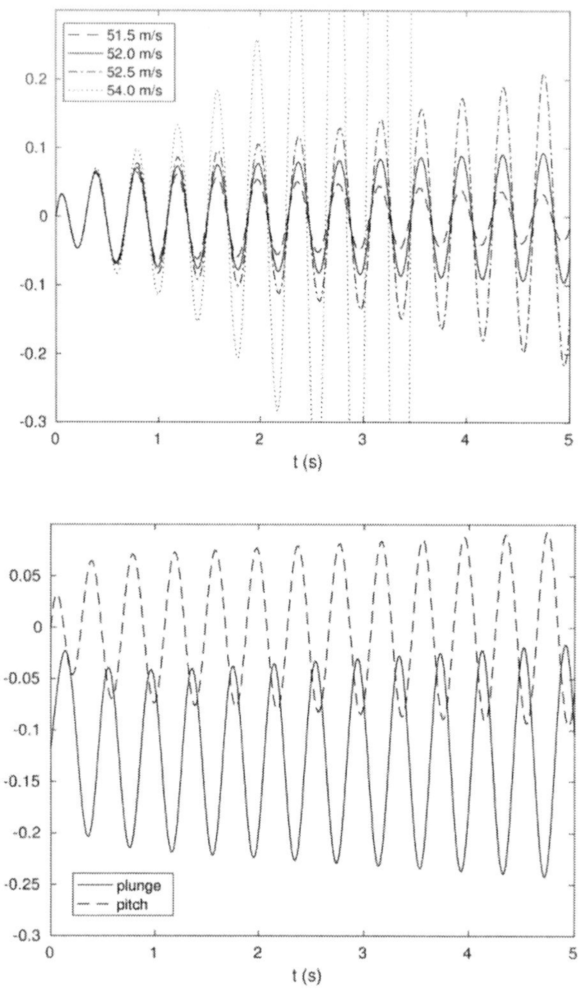

Fig. 4. Flutter analysis of a two-degree-of-freedom airfoil.

A rigorous stability analysis of a two-dimensional finite plate under axial flow is conducted. The plate, which is initially at rest, is clamped at the leading edge and free at the trailing edge. A small initial perturbation is applied and the long-term transient behaviour of the plate is analysed. The bending rigidity of the structure stabilises the system, whilst an increase in aerodynamic pressure due to an increase in flow velocity destabilises the system and results in flutter. Following previous work [13-15], the critical stability curves can be defined in terms of the non-dimensional flag density and non-dimensional velocity. These critical stability curves are shown in Fig. 5. Also shown on this figure are various points that were studied in this analysis. To aid comparison, the flow is assumed to be inviscid in this case.

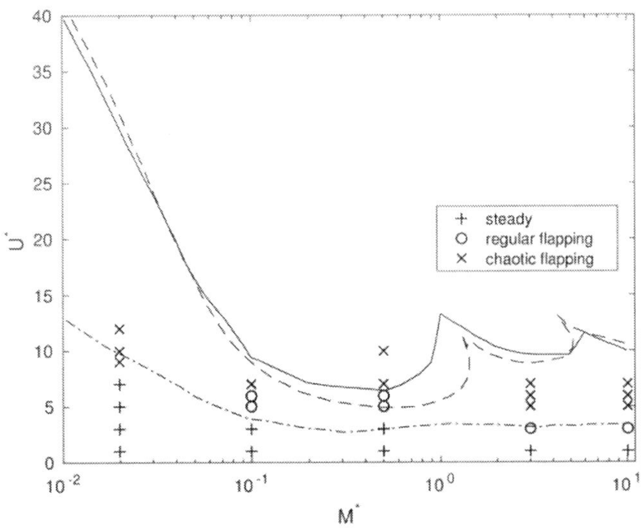

Fig. 5. Critical stability curves for a finite flag under axial flow together with data points in this study.

The critical stability curves shown in Fig. 5 contain branches that correspond to a different mode structure, as also observed experimentally by Eloy et al. [14]. The first branch for M* < 1 corresponds with a flapping of the plate in its second mode of vibration, while the branch for 1 < M* < 8 corresponds with flapping in the third mode and M* > 8 corresponds with the fourth mode. An analysis of the flapping behaviour shows good agreement with the results of Eloy et al. [14] and Michelin et al. [13]. Snapshots of the flapping behaviour of the plate at M* = 0.002 and U* = 9 are shown in Fig. 6. The second mode of vibration dominates this flapping behaviour and can be observed by the single node, or position of zero displacement, along the plate.

 https://doi.org/10.1051/matecconf/202134700033

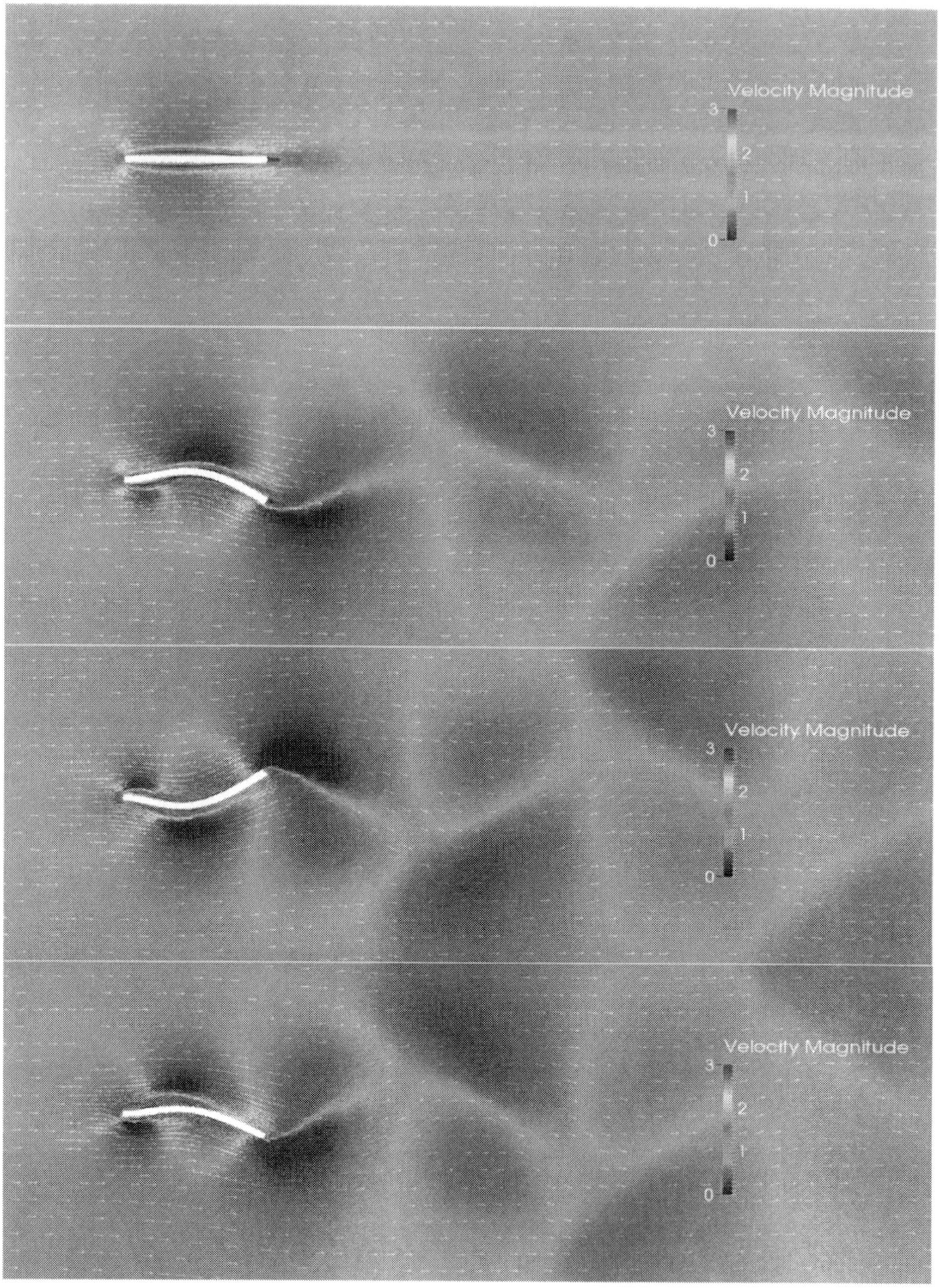

Fig. 6. Snapshots of the flapping behaviour of the plate at M* = 0.002 and U* = 9.

References

1. O.C. Zienkiewicz, R.L. Taylor, *The Finite Element Method: Volume 2 - Solid Mechanics, fifth ed.*, Butterworth-Heinemann, Oxford (2000).

2. S.V. Patankar, *Numerical Heat Transfer and Fluid Flow*, McGraw-Hill, New York (1980).
3. R. Suliman, O. Oxtoby, A. Malan, S. Kok, Appl. Math. Model. **38**, 2265 (2014).
4. B. Hubner, E. Walhorn, D. Dinkler, Comput. Methods Appl. Mech. Eng. **193**, 2087 (2004).
5. W. Dettmer, J. Peric, Comput. Methods Appl. Mech. Eng. **195**, 5754 (2006).
6. P. Texeira, A. Awruch, Comput. Fluids **34**, 249 (2005).
7. A. Slone, K. Pericleous, C. Bailey, M. Cross, Comput. Struct. **80**, 371 (2002).
8. C.G.H. Weller, Int. J. Numer. Methods Eng. **64**, 1575 (2005).
9. R. Suliman, O. Oxtoby, A. Malan, S. Kok, Eur. J. Mech. B Fluids **49A**, 272 (2015).
10. C. Hirt, A. Amsden, J. Cook, J. Comp. Phys. **14**, 227 (1974).
11. G. Holzapfel, *Nonlinear Solid Mechanics: A continuum Approach for Engineering*, John Wiley & Sons Ltd., West Sussex, England (2001).
12. S. Turek, J. Hron, Lect. Notes Comput. Sci. Eng. **53** (2006).
13. S. Michelin, G. Stefan, L. Smith, B.J. Glover, J. Fluid Mech. **617**, 1 (2008).
14. C. Eloy, R. Lagrange, C. Souilliez, L. Schouveiler, J. Fluid Mech. **611**, 97 (2008).
15. S. Alben, M.J. Shelley, Phys. Rev. Lett. **100**, 074301–1 (2008).

Modelling and simulation of ultrasonic inspections in welded rails subjected to practical environmental conditions

Dineo Ramatlo[1*], *Philip* Loveday[2], *Craig* Long[3] and *Daniel* Wilke[1]

[1]Center for Asset Integrity Management, University of Pretoria, Hatfield, 0086, South Africa
[2]School of Mechanical, Industrial and Aeronautical Engineering, University of the Witwatersrand, Johannesburg, Wits 2050, South Africa
[3]CSIR, Manufacturing Cluster, Pretoria, 0001, South Africa

Abstract. A permanently installed Ultrasonic Broken Rail Detection system monitors the Sishen-Saldanha railway line in South Africa [1]. The system detects complete rail breaks at long-range using guided wave ultrasound. For the system to be reliable, its damage detection performance must be evaluated under actual environmental and operational conditions (EOCs). However, obtaining monitoring data containing damage reflections is virtually impossible since detected defects in operational rail track sections are immediately removed and replaced with new rail. Laboratory experiments are also not possible since end reflections from short sections of rail dominate the response. Therefore, damage signals can only practically be obtained from numerical simulations. The simulated damage signals should be realistic and include varying EOCs, especially temperature variations. This paper aims to demonstrate a procedure to model temperature variations in ultrasonic signals. The temperature model and the modelling framework developed in [2] are used to simulate reflections from welds. The framework models the excitation, propagation and scattering of guided waves from discontinuities by employing a hybrid model based on the 3D Finite Element method (FEM) and the 2D Semi-Analytical Finite Element (SAFE) method. The simulated results are validated using experimental measurements collected from an operational rail at different temperatures.

1 Introduction

The Council for Scientific and Industrial Research (CSIR) and Maritime Institute of Technology (MIT) identified the need for a reliable monitoring system for defect detection in rails. The aim is to guarantee a solution to the problem of train derailments caused by broken rails. This has led to a permanently installed Ultrasonic Broken Rail Detection (UBRD) system, which monitors the Sishen-Saldanha railway line in South Africa [1]. The system has been designed to detect complete breaks by transmitting ultrasonic guided waves between permanently installed alternating transmit and receive transducers, spaced

* Corresponding author: dineo.ramatlo@up.ac.za

approximately 1km apart [1,3]. If the receive station does not detect the transmitted signals, an alarm indicates a broken rail, and train operation stops to prevent derailments. Since the installation of the UBRD system on the Sishen-Saldanha Ore line in April 2016, seven rail breaks and several major flaws [1] recorded as false alarms have been reported. Rail breaks can occur under a train and can cause derailment of part of the train. Detecting a crack before the rail breaks would avoid these derailments and also allow condition-based maintenance.

Defects such as cracks in rails can be detected from ultrasonic signals, collected over a specified period. These signals will contain reflections from benign structural features (such as welds) that do not represent damage and potentially small reflections from growing damage. The procedure would then employ appropriate algorithms to distinguish and classify the reflections according to their sources and determine and locate the reflection coming from damage. Liu et al. [4] demonstrated the application of this procedure to pipelines where unsupervised machine learning algorithms were adopted to detect corrosion. Although superficially, the application of this proposed technique to pipe and rail problems may appear similar, there are important differences that present challenges in rail applications. Specifically, in our application, many more modes are excited, and propagation is generally dispersive in our case, making signal processing significantly more challenging.
Furthermore, variations in environmental (e.g. temperature) and operational (e.g. passing trains) conditions may produce significant changes in the ultrasonic signals, thereby masking the damage. The challenge is, therefore, to distinguish between these benign signals and the true damage signals. For both pipelines and rail tracks, obtaining monitoring data for different damage scenarios under varying environmental and operational conditions is virtually impossible since detected defects in sections of an operational waveguide are immediately removed and replaced. The alternative has thus been to carry out a series of laboratory experiments while inducing growing damage on the waveguide test piece. However, laboratory damage experiments are also not possible for rail tracks due to end reflections from short sections of rail dominating the response. Therefore, damage signals for rails can only practically be obtained from numerical simulations.

This paper aims to demonstrate a procedure to model temperature variations in ultrasonic signals. The temperature model and the modelling framework developed in [2] are used to simulate reflections from welds. The framework models the excitation, propagation and scattering of guided waves (GWs) from discontinuities. This is achieved by respectively employing a hybrid model that couples a 3D FEM of a transducer or a 3D FEM of the reflector (weld) with 2D SAFE models of the rail [2,7,9]. The 2D SAFE models capture the semi-infinite nature of the rail. The simulated results are validated using experimental measurements from an operational rail (with a set-up shown in Figure 1a) at different temperatures. It is believed that if the proposed procedure can accurately simulate reflections from welds, then it will also be capable of modelling the reflections from defects. In future, these simulated defect signatures can then be superimposed on measured data to evaluate a monitoring system.

2 Methods

2.1 Modelling of a GW inspection

The modelling framework presented in [2] uses several existing numerical models to effectively create a numerical representation of a complete GW inspection system for a rail depicted in Figure 1a. The system includes a piezoelectric transducer to excite propagating waves in the rail, which acts as a waveguide and aluminothermic welds, which act as reflectors of the propagating waves. The numerical models that form the basis of the modelling framework are the SAFE method for modelling wave propagation; two hybrid methods couple SAFE and traditional 3D FEM to model GW excitation and scattering from reflectors, respectively.

The SAFE method for 1D propagation models the arbitrary but constant cross-section of an infinite elastic waveguide using 2D finite elements, Figure 1a. Wave propagation along the length of the waveguide is achieved by using an analytic treatment for variations in the propagation direction. The SAFE method solves for the wavenumbers and corresponding mode shapes supported by the waveguide. These wave properties are solved at selected frequency points and used to compute other dispersion properties of the waveguide, such as the phase and group velocities. The attenuation in the rail was modelled using damping, and the damping coefficients were determined using the optimization procedure explained in [2]. Details of the SAFE method can be found in references [5,6].

The excitation of the rail using a piezoelectric transducer was modelled using a hybrid model which couples a 3D FEM model of a transducer and a 2D SAFE model of a waveguide, Figure 1a. This model properly accounts for transducer dynamics, which is important when resonant transducers are employed, such as in this case. The model solves the modal amplitudes in the elastic waveguide of the propagating modes, given a voltage applied to the piezoelectric transducer. For this paper, a 17.5 cycle Hanning windowed tone-burst function centred at $35kHz$ was used for excitation. This hybrid model uses an interpolation procedure between the 3D space and the 2D space to calculate the modal amplitudes of the modes excited by the transducer. The method is explained in detail in references [7,8].

The scattering of guided waves from reflectors is modelled using another hybrid model, which couples a 3D FEM of the reflector with two SAFE models to represent the semi-infinite incoming and outgoing rails, Figure 1a. The reflector can be discontinuities such as welds, cracks or other defects that may result in reflections. This hybrid model solves for the modal amplitudes of the reflected and transmitted guided waves by enforcing continuity and equilibrium on the boundaries of the left and right semi-infinite waveguide intersecting the 3D volume of the reflector. The results are presented in reflection and transmission coefficient matrices, R and T (reference [2]), respectively. The hybrid method for scattering from discontinuities is explained in detail in the paper by Benmeddour [9].

The numerical model representation of a GW inspection system for a section of rail depicted in Figure 1a employs the numerical models of wave propagation, excitation and scattering from welds and is depicted in Figure 1b. The response of the waveguide is computed by summing the reflections coming from each weld. The weld reflections are calculated by propagating the waves excited by the transducer to a specific weld, reflecting those waves from that weld, and propagating them back to the transducer. Figure 1c shows a schematic representation of how the reflections from welds B and C can be calculated, respectively. For weld C, the guided waves are transmitted through weld A, reflected from weld C and transmitted through weld A again during backward propagation.

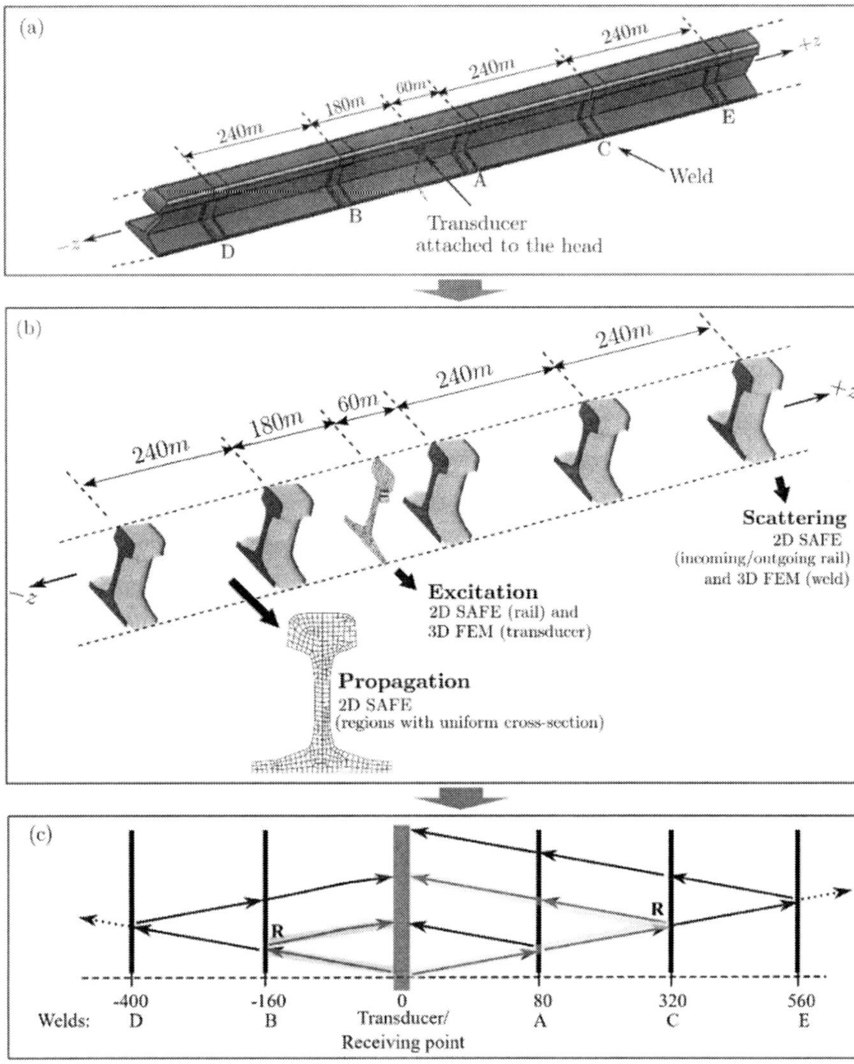

Fig. 1. Modelling of GW excitation, propagation and scattering from welds [2].

2.2 Simulation of a GW inspection for a specific temperature T

When a waveguide in which an ultrasonic signal propagates is subjected to a temperature change, the time of arrivals for the reflections will change. A temperature increase will cause the reflections to arrive late. This is because a temperature change slightly changes the Young's Modulus of the rail leading to a thermal expansion of the rail and changes in the wave velocity. The Young's Modulus of the rail can be related to temperature through:

$$E = E_0 - \frac{\partial E}{\partial T}(T - T_0) \qquad (1)$$

where E_0 and T_0 are respectively the Young's Modulus and temperature of the baseline signal. A change in the Young's Modulus causes a change in the wavenumber (Figure 2) and group velocity of the wave in the rail. If this change is not included in the dispersion compensation procedure, the distance to a reflector will appear to change with temperature. The apparent reflection distance for a specified temperature reading can be evaluated as:

$$z = (1 + 0.00013(T - T_0))z_0 \tag{2}$$

z_0 is the reflection distance in the baseline signal. Equation 2 was deduced after evaluating the influence of temperature on field measurements.

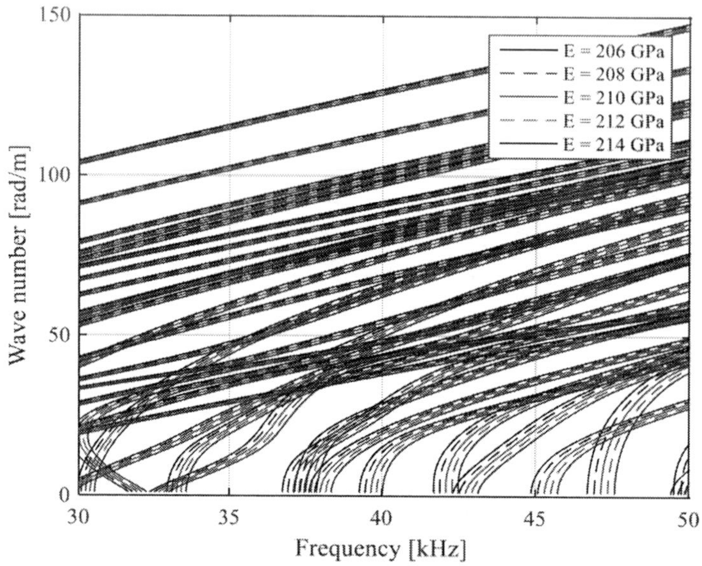

Fig. 2. The wavenumbers of propagating modes for different Young's Modulus.

Figure 3a shows the temperature history for the 50 field measurements in the dataset collected from a field experiment [2,11]. For each experimental measurement, the apparent reflection distance for a weld located at 60m from the transducer (weld A) is plotted in Figure 3b. The apparent reflection distance for welds B, C, D and E are plotted in Figure 4. The results show that the plots for the apparent distances for welds B, D, C and E follow the same trendline as the temperature history instead of the apparent distance for weld A. This is because the reflections for weld A are highly affected by noise in the signals. Figure 4 also shows that the apparent reflection distances simulated using Equation 2 for each measurement compare very well with that extracted from experimental measurements.

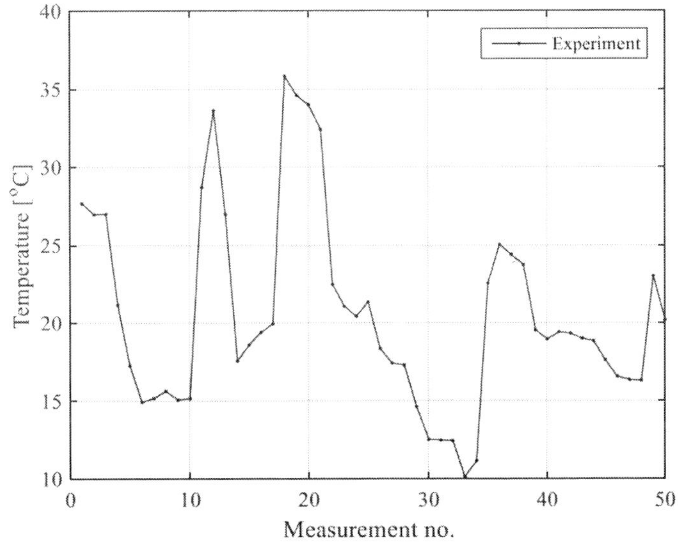

(a)

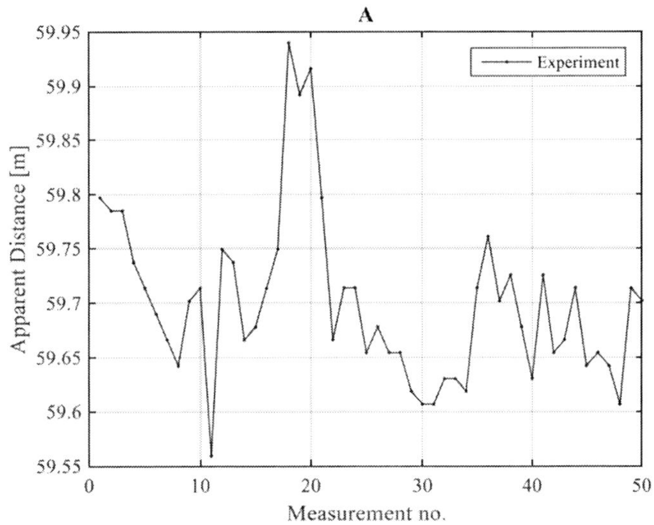

(b)

Fig. 3. (a) The temperature history for a dataset of 50 signals and (b) apparent reflection distances for weld A.

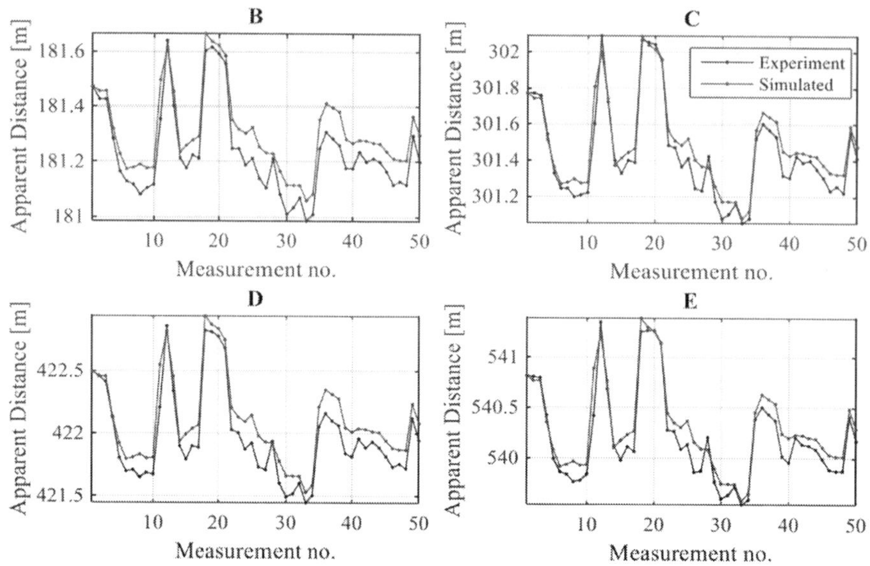

Fig. 4. Apparent reflection distance for welds B, C, D and E.

3 Results

For a welded rail track, the modelling framework in [2] computes the total response due to all propagating modes by summing the reflected guided waves from each weld:

$$U(z,\omega) = \sum_{w=1}^{W} \sum_{i=1}^{n} \alpha_{w,i}\psi_i \tag{3}$$

where $\alpha_{w,i}$ is the amplitude at (location z) of mode i reflected from weld w, and can be computed in vector form as:

$$\{\alpha_w(z=0)\}^T = \left\{\{\alpha_{TX}\} \circ \left\{e^{-j\tilde{\kappa}(z_1-z_0)}\right\}\right\}^T [\boldsymbol{T}] \circ \cdots \circ \left\{e^{-j\tilde{\kappa}(z_{w-1}-z_{w-2})}\right\}[\boldsymbol{T}] \circ \left\{e^{-j\tilde{\kappa}(z_w-z_{w-1})}\right\}[\boldsymbol{R}] \circ \left\{e^{-j\tilde{\kappa}(z_{w-1}-z_w)}\right\}[\boldsymbol{T}] \circ \cdots \circ \left\{e^{-j\tilde{\kappa}(z_0-z_1)}\right\} \tag{4}$$

α_{TX} denotes the modal amplitudes excited by the transducer, and $w = 1,2,3 \ldots W$ indicate the welds A, C, E… and W included in the simulation. The apparent distances z_w for reflections coming from weld w are computed using Equation 2 for a specific temperature condition. For the welds on the left side of the transducer, $w = 1,2,3 \ldots W$ would indicate the welds B, D, F… and W. In Equations 3 and 4, the appropriate modal properties (wavenumbers $\tilde{\kappa}$ and mode shapes ψ) for forward and backward propagation should be adopted as explained in [2].

The frequency-domain response in Equation 3 can be converted to the time domain by taking an inverse Fast Fourier Transform. The distance domain response is achieved by applying a dispersion compensation procedure in reference [10].

The simulated results for three selected temperatures are compared to experimental measurements in Figure 5a, with Figure 5b clearly showing the reflections from welds B, C, D and E. The field experiment was conducted using a pulse-echo piezoelectric transducer to excite the waves in the head of the rail. References [2,11] gives the details of how the measurements were collected for different temperature readings. The reflection between 350m and 400m was not simulated as it is not coming from a weld but a block of mass stuck to the rail. The results show that the field measurements for each temperature condition were well approximated, meaning that the non-robust temperature model is acceptable.

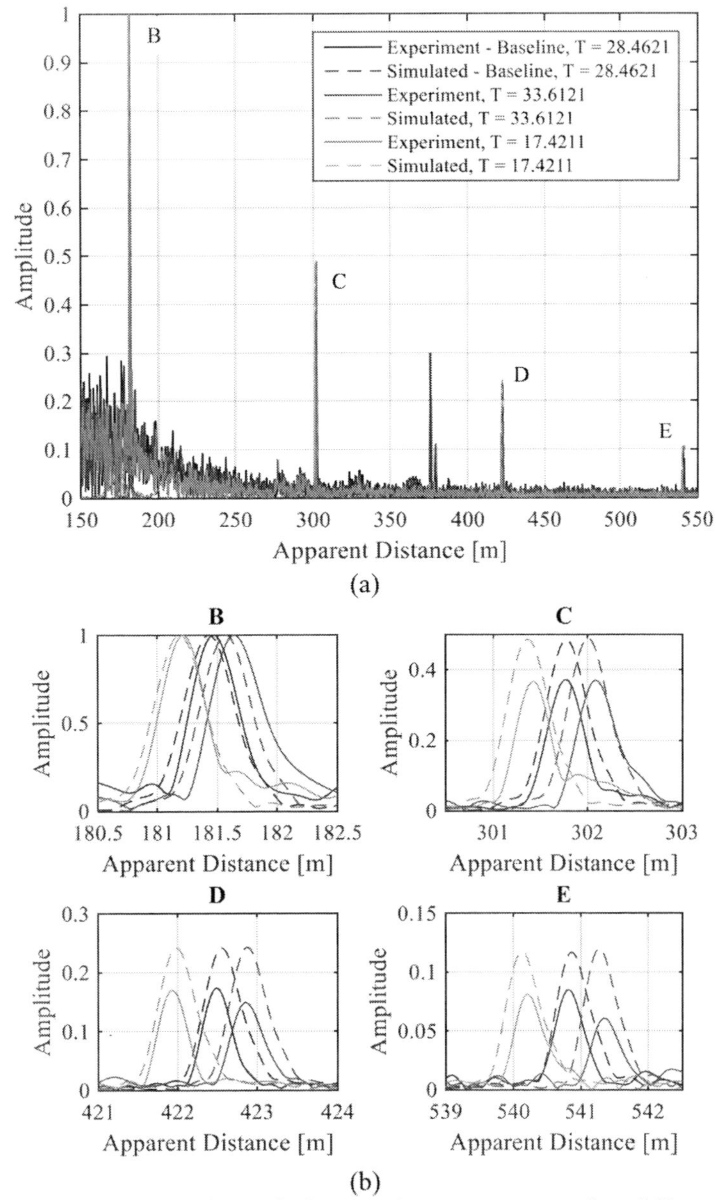

Fig. 5. (a) Comparison of field and simulated measurements for different temperature conditions. (b) Reflections from welds B, C, D and E.

4 Conclusion

This paper aimed to demonstrate a procedure to model temperature variations in ultrasonic guided waves propagating in 1D waveguides. The temperature model was used with the modelling framework developed in [2] to simulate weld reflections in rail tracks. The framework models the excitation, propagation and scattering of GWs from discontinuities by respectively employing a hybrid model that couples a 3D FEM of a transducer with a 2D SAFE model of the rail; a 2D SAFE model of the rail; and another hybrid model which couples a 3D FEM of the weld with two SAFE models to represent the semi-infinite incoming and outgoing rails. The temperature model was used to predict the apparent reflection distances from the welds for different temperature conditions. The simulated results were acceptable though the effect from other EOCs affecting the guided waves was evident. In future, the temperature model will be improved to include other EOCs by employing robust data analysis techniques to the field data. In future, the proposed procedure will be used to model reflections from discontinuities such as damage.

References

1. F. A. Burger, P. W. Loveday, *Ultrasonic broken rail detector and rail condition monitor technology,* In Proceedings of the 11th International Heavy Haul Association Conference, IHHA, Cape Town, South Africa (2017)
2. D. Ramatlo, C. Long, P. Loveday, N. Wilke, A modelling framework for simulation of ultrasonic guided wave-based inspection of welded rail tracks, Ultrasonics, **108** (2020)
3. F. A. Burger, P. W. Loveday, C. S. Long, *Large Scale Implementation of Guided Wave Based Broken Rail Monitoring*, 41st Annual Review of Progress in Quantitative Nondestructive Evaluation, United States of America (2015)
4. C. Liu, J. Dobson, P. Cawley, *Efficient generation of receiver operating characteristics for the evaluation of damage detection in practical structural health monitoring applications*, Proceedings Royal Society, **437** (2017)
5. T. Hayashi, W. J. Song, J. L. Rose, Guided wave dispersion curves for a bar with an arbitrary cross-section, a rod and rail example, Ultrasonics, **41** (2003)
6. L. Gavric, Computation of Propagative Waves in Free Rail Using a Finite Element Technique, J. Sound Vib., **185** (1995)
7. P. W. Loveday, Simulation of Piezoelectric Excitation of Guided Waves Using Waveguide Finite Elements, IEEE Trans Ultrason Ferroelectr Freq Control, **55** (2008)
8. D. A. Ramatlo, C. S. Long, P. W. Loveday, D. N. Wilke, *SAFE-3D analysis of a piezoelectric transducer to excite guided waves in a rail web*, AIP Conference Proceedings, **1706** (2015)
9. F. Benmeddour, F. Treyssède, L. Laguerre, Numerical modeling of guided wave interaction with non-axisymmetric cracks in elastic cylinders, Int J Solids Struct, **48** (2011)
10. P. D. Wilcox, A rapid signal processing technique to remove the effect of dispersion from guided wave signals, IEEE Trans Ultrason Ferroelectr Freq Control, **50** (2003)
11. P. W. Loveday, C. S. Long, D. A. Ramatlo, Ultrasonic guided wave monitoring of an operational rail track, Struct. Health Monit., **19** (2020)

Modelling fluid and particulate flow through a ventriculoperitoneal shunt in a variable temperature environment

Tshiamo Ramokoka[1], and *Muaaz* Bhamjee[1,*]

[1]Department of Mechanical Engineering Science, Faculty of Engineering and the Built Environment, University of Johannesburg, Auckland Park Kingsway Campus, Johannesburg, South Africa

Abstract. One of the most prevalent causes of failure for a ventriculoperitoneal shunt is blockage, the other being infection. This study looks at the blockage of the shunt valve, and whether the occlusion of a shunt valve is accelerated by the presence of an infection. This study assumes that an infection will raise the number of white blood cells contained in the cerebrospinal fluid to fight it and will thus accelerate shunt occlusion. The experiment simulates a shunt system by suspending a shunt valve in a water bath that has a temperature that varies between 37°C and 41°C. A computational fluid dynamics model of the shunt system is used to gain further insight into the flow behaviour under these conditions. The results of the CFD model were validated using the experimental results. There was an average error of 15% between the readings that were obtained in the experiment and the CFD model. The experimental results showed that there was a decrease in the volume flow rate at the outlet of the shunt system, which was not large enough to point towards any blockage. Both the model predictions and the experimental results show that increased temperature and particulate concentration alone do not result in shunt occlusion, particularly at the shunt valve. This result effectively excluded the shunt valve as a region of shunt occlusion due to infection, as an infection occurs due to the growth of bacteria along the surfaces of the shunt system and this bacterial growth is more likely to occur at the proximal and distal ends of the shunt system.

Nomenclature

C_D	-	drag coefficient	μ	-	Dynamic viscosity (kg/(m.s))
C_{vm}	-	virtual mass coefficient	υ	-	kinematic viscosity (m^2/s)
d_p	-	diameter of particulate	p	-	Pressure (Pa)
d_{ij}	-	deformation tensor	ρ	-	Density (kg/m^3)
$D_{T,p}$	-	thermophoretic coefficient	Re	-	Reynolds number

* Corresponding author: muaazb@uj.ac.za

E	-	energy (J)	T	-	Temperature (°C, K)
$\bar{F}$	-	force (N)	$\bar{\bar{\tau}}_{eff}$	-	Shear stress (Pa)
g	-	gravitational acceleration (m/s^2)	$\vec{v}$	-	Fluid velocity (m/s)
h	-	sensible enthalpy of fluid (J/kg)	u_p	-	Particle velocity (m/s)
$\bar{I}$	-	unity tensor	m_p	-	Particle mass (kg)
k_{eff}	-	Effective conductivity	ρ_p	-	Particle density (kg/ m^3)

1 Introduction

A ventriculoperitoneal (VP) shunt is used in the treatment of hydrocephalus, a medical condition caused by the inadequate drainage of cerebrospinal fluid (CSF) from the ventricles [1]. The rate at which the CSF is absorbed is less than that at which the CSF is produced, which results in the accumulation of CSF in the ventricles [1]. The treatment of hydrocephalus by the insertion of a VP shunt is one of the more common treatment methods for hydrocephalus [2]. A VP shunt comprises of three elements: an inflow/ventricular catheter, a valve for unidirectional flow and a peritoneal/outflow catheter [3]. The inflow catheter drains the excess CSF from the ventricles; the valve controls the flow of the CSF through the shunt system and the peritoneal catheter routes the CSF from the valve to the peritoneal cavity, where it is then absorbed [3].

The shunt system can fail in many ways, such as through over-drainage, underdrainage, infection, blockage, leaks, migration, and discontinuity [3], [4]. Unlike the other methods of shunt failure, infection is a non-mechanical problem, as infection is typically caused by the growth of bacteria along parts of the shunt system [5]. Particulates contained in the CSF (i.e., brain tissue, proteins, blood cells) can create a blockage in the shunt system, depending on the concentration of the particulates. An infection will result in an increase of the temperature of the CSF, as well as the amount of white blood cells contained in the CSF. This will increase the likelihood of a blockage occurring in the shunt system.

A CSF analysis is usually done to confirm an infection. A positive CSF analysis will show an increase in the number of white blood cells present ($1000 - 5000\ cells/mcL$) with a high percentage of neutrophils ($> 80\%$), as well as an increased protein concentration [6]. Per Tamburrini, Calderelli and Di Rocco [7], infection is the second most prevalent type of shunt failure in children, second to mechanical failure. Fever is typically a sign of bacterial infection but can be mild or there could be no fever present, which may delay the diagnosis [7].

Studies of CSF shunt failures in children were looked at to determine how prevalent shunt failures due to infection are, as well as to how prevalent reoccurrences of shunt infections are in children. The study by Simon et.al. [8] did a statistical analysis of shunt revisions in children over a four-year period and it was determined that the risk of infection can be attributed to the revision of a CSF shunt. Moreover, it was found that the risk of infection increases with an increase in the number of shunt revisions a patient had, with the risk being 13 times greater in patients who have had 2 or more revisions than those who have had no revisions [8].

A study was conducted in Turkey which was centred around determining the rate of recurring shunt infections after the initial shunt infection has been treated [9]. The study was done over 10 years [9]. For this study, the recovery of 428 patients was tracked [9]. Of these patients, 58 were treated for an infection [9]. The final findings of the study detail a method of effectively reducing the risk of recurring shunt infections. In the case of infection, the total removal of the shunt and insertion of a new external ventricular drainage catheter should be used to treat the infection [9]. In addition, isolating the pathogen responsible for the infection and using antibiotics appropriately will decrease the length of the infection [9].

A study was conducted at the University of Johannesburg in 2017, which looked at the flow of fluid and particulates through a VP shunt [10]. The shunt valve was the point of interest, as it was assumed that the blockage was likely to occur at this region [10]. The only drawback from this study was that the effects of the change in fluid temperature were not studied.

The literature has not completely addressed how a change in the temperature of the CSF (which is a simulation of infection through the symptom that is fever), as well as a change in the concentration of the particulates in the CSF will affect the behaviour of the VP shunt system. This is of interest particularly where such changes in temperature, and subsequent increase in the concentration of particulates, will affect the potential likelihood of the occurrence of a blockage in the shunt system. Thus, the aim of this study is to investigate the flow characteristics of CSF containing particulates through the VP shunt, while the temperature of the CSF is not kept constant. The methodology used for this study include using an experimental test bench to study the flow of the fluid and particulates through a VP shunt, as well as setting up a computational fluid dynamics (CFD) model for studying the flow through the VP shunt. The results of the CFD model will be validated against those of the experiment. This study is only limited to paediatric cases.

2 CFD Model

2.1 Geometry and computational mesh

The geometry of the shunt valve was derived from that of the Medtronic CSF Control Valve, which was used in the experiment. The geometry for the model was modified slightly but has the general shape of the Medtronic shunt valve. Fig. **1** shows the geometry of the CSF Control Valve.

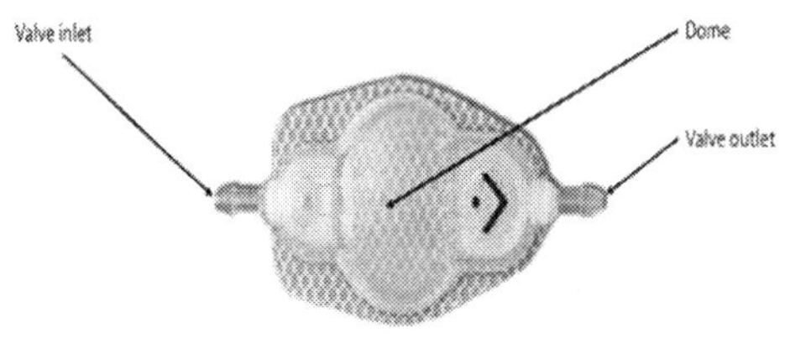

Fig. 1. CSF Control Valve geometry

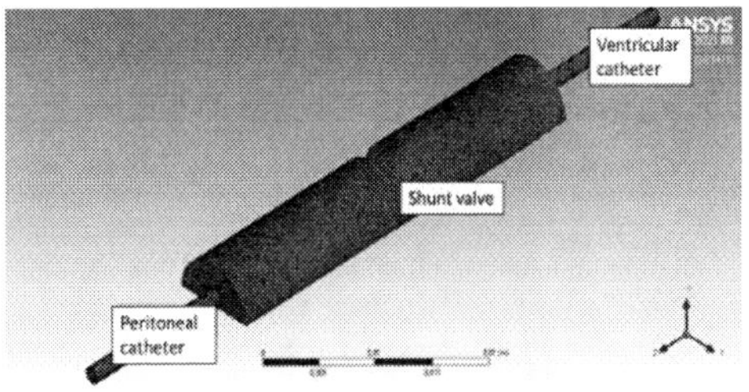

Fig. 2. Mesh

To simplify the model, the shunt valve was drawn as a cylinder cut in half to serve as the dome shaped chamber. A void was introduced into the geometry to model the flow restriction created by the valve which only opens when the differential pressure is enough to drive the fluid through the shunt valve. The dimensions for the model were directly derived from those of the Delta Valve produced by Medtronic [11]. The dimensions were used because the CSF Control Valve and the Delta Valve are very similar in appearance. The diameter of the inlet and outlet catheters was set at 1.5mm, the length of the ventricular and peritoneal catheters was set at 15mm and the length of the shunt valve was set at 14mm. The radius of the hemispherical shape of the valve was set at 4mm.

The mesh comprises of a mixture of tetrahedral and hexahedral cells. The tetrahedral cells were converted to polyhedra to improve the cell quality, as well as reduce cell count and decreases computational power [12]. The element order was selected to be quadratic. Fig. 2 shows the mesh that was created for this model.

A mesh convergence study to ensure that the results were mesh independent [13]. A method presented by Roache was used for the mesh convergence study, which is based on the Richardson Extrapolation [14]. The Richardson Extrapolation, and thus the Grid Convergence Index (GCI), work on the basis that the solution from each grid is within the asymptotic range (GCI $\approx$ 1) [14]. Three meshes of element sizes that differed by a grid refinement ratio of $r = 1.41$ were developed. For the mesh independence study, three meshes were generated, for element sizes $0.5mm$, $0.354mm$ and $0.25mm$. The models were run and the results for each of these meshes used to find the Grid Convergence Index for the fine grid [14]. The GCI was found to be 1.004, which shows that the fine grid (0.25mm) is in the asymptotic range, and thus, the solution is mesh independent. Therefore, the element size used in all subsequent models is $0.25mm$.

2.2 Governing equations

The flow through the shunt system was modelled as steady incompressible flow, as the velocity of the fluid/particulate mixture is quite low, which will result in a Mach number less than 1. Laminar flow was selected for this CFD model as the typical flow of CSF has a Reynolds number that is less than 10 [15]. The flow of fluid through the shunt system is governed by the conservation of mass and conservation of momentum equations [16] as per equations (1) and (2) respectively.

$$\nabla \cdot (\rho \vec{v}) = 0 \tag{1}$$

For a compressible Newtonian fluid, the Navier-Stokes equations are [17]:

$$\rho \vec{v} \cdot \nabla \vec{v} = -\nabla p + \nabla \cdot \left(\mu(\nabla \vec{v} + (\nabla \vec{v})^T) - \frac{2}{3}\mu(\nabla \cdot \vec{v})\bar{I} \right) + \bar{F} \tag{2}$$

The force $\bar{F}$ accounts for source terms such as body forces (due to gravity) as well as momentum exchange from the particulate to the fluid. The energy equation solved by ANSYS Fluent encompasses the transfer of energy due to conduction and viscous dissipation [18]. The equation is written as follows:

$$\vec{v}(\rho E + p) = \nabla \cdot \left[k_{eff} \nabla T + \left(\bar{\bar{\tau}}_{eff} \cdot \vec{v} \right) \right] \tag{3}$$

Due to the temperature gradient between the surroundings and the fluid flowing through the VP shunt system, heat transfer will occur through the walls of the shunt system. The term E in equation (3) can be written as:

$$E = h - \frac{p}{\rho} + \frac{v^2}{2} \tag{4}$$

The *Discrete Particulate Modelling (DPM)* model was selected, to introduce particulates into the VP shunt system. The particulates contained in the CSF are protein, monocytes, and lymphocytes. The lymphocytes and the monocytes have a diameter of $14\mu m$ and $20\mu m$ respectively [19]. The protein cells have a diameter of $3.8nm$ [20].

The physical models that were selected for the DPM were the Drag Force, Virtual Mass Force, Pressure Gradient Force, Saffman Lift Force, and the Thermophoretic Force. The ratio of the density of the fluid to that of the particulate is greater than 0.1, therefore the Pressure Gradient Force and the Virtual Mass Force have a significant effect on the results of the CFD model [18]. The Saffman Lift Force was enabled as it is the force that results from shear and the Thermophoretic Force was enabled to account for the effect of fluid temperature gradients on the particulates [18]. A key assumption that was made was that the particulates have a spherical shape. Thus, the particle force balance is given by [18]:

$$m_p \frac{du_p}{dt} = \frac{18\mu}{\rho_p {d_p}^2} \frac{C_D Re}{24} (v^{\rightarrow} - u_p) + C_{vm} \frac{\rho}{\rho_p} \left(\boldsymbol{u}_p \nabla \vec{v} - \frac{du_p}{dt} \right) + m_p \frac{\rho}{\rho_p} (\vec{v} \nabla \vec{v}) + \tag{5}$$
$$m_p \frac{2(2.594) v^{\frac{1}{2}} \rho d_{ij}}{\rho_p d_p (d_{lk} d_{kl})} (\vec{v} - \boldsymbol{u}_p) + \mathrm{D}_{\mathrm{T,p}} \frac{1}{T} \nabla T + m_p \boldsymbol{g} \frac{(\rho_p - \rho)}{\rho_p}$$

The fluid is assumed to be Newtonian despite the presence of the monocytes, lymphocytes and albumin. The white blood cells (monocytes and lymphocytes) are solids contained in the blood plasma as well as the CSF and, thus, are insoluble. Therefore, they arguably will not affect the viscosity of the CSF, thus, maintaining the Newtonian assumption [21]. However, proteins (albumin) are soluble and will have an impact on the fluid viscosity and violate the Newtonian assumption [21], [22]. In future work a non-Newtonian model will be incorporated to account for impact of the proteins on the CSF viscosity and the viscoelasticity of the blood plasma akin to [23], [24]. To account for the effect of the particles on the fluid a two-way coupling was implemented between the fluid phase and particulates. However, the effect of the particulates on the fluid phase is only via momentum transfer as per equations (2).

This study assumes that an infection will raise the number of white blood cells contained in the cerebrospinal fluid. The white blood cells were modelled as particulates. Thus, in this study the model reduces the biological complexity involved in the process of infection to a.) the change in temperature of the CSF and/or shunt surfaces and b.) an increase in particulates as a consequence of the rise in white blood cells in response to infection. Thus, the biological complexity is reduced to heat transfer and particulate mechanics.

2.3 Material properties

The materials used in this CFD model include water, medical grade silicone and brain matter. The water is representative of the CSF, as they have similar properties [21], [25]. The medical grade silicon is the material used for the components of the shunt system and the brain matter is composed of blood cells and protein. The thermophysical properties for these materials are given in Table 1 [19], [20], [26], [27], [28].

Table 1. Thermophysical properties

Cell type	Density (kg/m^3)	Dynamic viscosity (kg/(m.s))	Enthalpy (J/kg)	Thermal conductivity (W/m.K)	Specific heat capacity (J/kg.°C)
Lymphocyte	1077	-	-	-	-
Monocyte	1077	-	-	-	-
Albumin	0.15	-	-	-	-
Silicone	2300	-	-	0.2	1050
Water	998.2	0.1003	154950	0.598	4200

2.4 Boundary conditions

Fig. 3 shows an image of the boundary conditions for the CFD model. The two primary locations for the boundary conditions are the inlet and outlet of the shunt system. The inlet and outlet of the shunt valve are points of interest because any changes in the flow characteristics can be observed at these locations. A constant temperature condition ($37°C$, $38°C$, $39°C$, $40°C$ or $41°C$) was placed on the walls of the shunt system to model the heat transfer between the shunt system and the surroundings. The boundary conditions for the CFD model are summarised in Table 2.

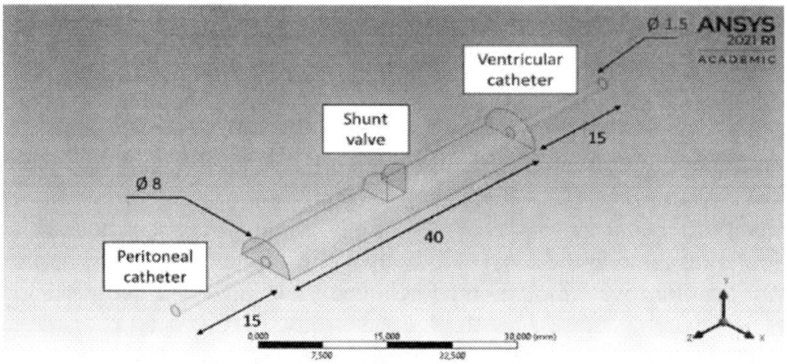

Fig. 3. Boundary conditions

Table 2. Boundary conditions

Boundary	Type	Velocity (m/s)	Pressure (Pa)	Temperature (°C)	DPM type
Inlet	Velocity inlet	0,01572	0	37	Reflect
Outlet	Pressure outlet	-	0	Solution dependent	Escape
Ventricular catheter	Wall	-	-	Constant Temperature	Reflect
Shunt valve	Wall	-	-	varies	Trap
Peritoneal catheter	Wall	-	-	varies	Reflect

When shunt blockage occurs there are also upstream changes that occur, thus, the upstream velocity profile may not be constant. However, a constant velocity profile is used in the models in this study so that we could compare the model predictions directly to the experimental results. In the experimental setup a syringe pump was used to mimic the upstream/inlet flow condition. Whilst not fully representative of the non-constant upstream condition in 'reality', it was representative of the experimental setup against which the CFD predictions were validated.

2.5 Solver settings

The pressure-velocity coupling that was selected is the SIMPLE coupling. The discretization selected for the gradient was "Least Squares Cell-Based". The pressure discretization scheme was "standard". The discretization scheme selected for momentum and energy was "Second Order Upwind". The model was solved as a steady state problem. The tracking scheme used for the DPM is the Runge-Kutta scheme and the source terms were linearised to increase numerical stability for steady-state flows.

3 Experimental Setup and Results

3.1 Experimental setup

The aim of the experiment was to determine whether an increase in temperature and particulate concentration would lead to shunt occlusion. The test bench was set up as per Fig. 4 and the volume to be infused (VTBI) for the syringe pump was set at $100ml/hr$, as the normal flow rate of CSF at $20ml/hr$ was too low to yield meaningful results. The concentration of the particulates/gelatine powder varied between $3g/L$ and $15g/L$, increasing by increments of $3g/L$. The gelatine powder was mixed with distilled water, which was used to mimic CSF with particulate matter.

In the experiment gelatine was used to represent the lymphocytes, monocytes and the albumin for the following reasons: i.) because the gelatine would not go into solution at the temperature range that we operated at, ii.) it was easy to clean between experimental runs and iii.) it was not abrasive, thus, minimising potential damage to the tubing. Water was used represent the CSF because water is close in viscosity [21] and density [22] to CSF.

The volume CSF exiting the shunt system at the end of the time interval was recorded and then converted into a volume flow rate, which was used to determine whether the shunt is blocked. The premise behind this is that the lower the volume flow rate, the greater the chance that the shunt is blocked.

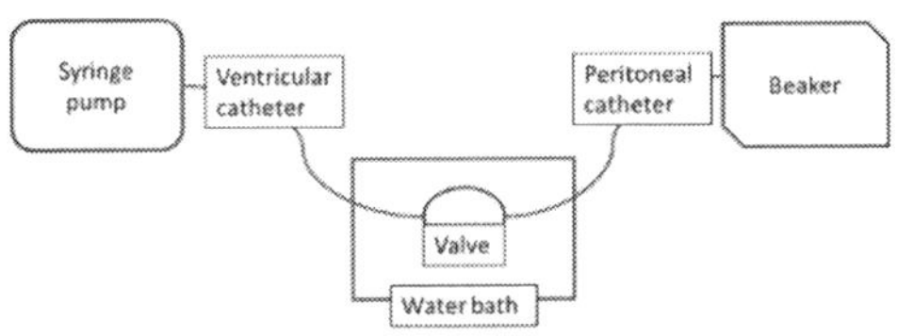

Fig. 4. Test bench schematic

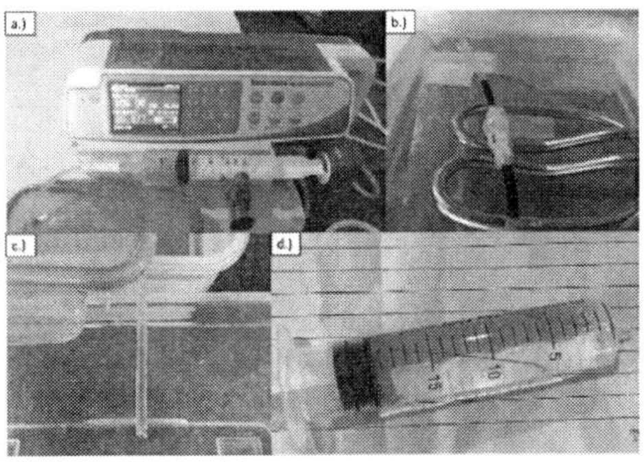

Fig. 5. Experiment

The experiment was carried out in 15-minute intervals per increase in temperature and per increase in concentration. The first readings were taken at $37°C$, then the next readings at $38°C$, $39°C$, $40°C$ and finally $41°C$. The temperature was increased up to $41°C$ as any fever above $41.7°C$ is considered as life-threatening [29]. The shunt valve was immersed in a water bath, with the temperature changed after readings have been taken for each particulate concentration entry. The shunt valve and catheters were flushed out with clean water after every reading to create a fair testing environment. A syringe pump was used to introduce CSF into the mock shunt system. The fluid/particulate mixture at the outlet was collected in a container and was transferred into a syringe with markings for easier reading of the volume. An image of the experimental setup is shown in Fig. 5.

In Fig. 5, image a.) shows the syringe pump with the syringe full of the gelatine mixture. This mixture flowed through the system at the 100ml/hr flow rate. Fig. 5b.) shows the VP shunt valve immersed in the water bath. Fig. 5c.) shows the outlet of the shunt system, where the fluid/particulate mixture is collected in a container. Fig. 5d.) shows the fluid/particulate mixture collected at the outlet, which was transferred into a syringe to read the volume.

3.2 Experimental results

The volume collected at the outlet at each concentration level and temperature level was plotted on a graph to show the relationship between increasing temperature and volume of the fluid/particulate mixture. The error associated with the results stems from the precision of the measuring instruments used in the experiment [30]. As the smallest graduation of the syringe is $1mL$, the error associated with the measurement of the volume at the outlet is $0.5mL$ [30]. The error bars for the experimental data are also plotted on the graph. Fig. **6** shows the volume of the fluid/particulate mixture at the outlet of the shunt system.

The values obtained for the volume of the fluid/particulate mixture at the outlet of the shunt system are quite accurate for the $6g/L$ and $9g/L$ concentration levels, as the readings are mostly grouped around the $24mL$ mark. This shows that there was little variation in the volume collected at the outlet of the shunt system with increasing temperature. The values obtained at the other concentration levels, i.e., $3g/L$, $12g/L$, and $15g/L$, were more dispersed and therefore less accurate. Therefore, the experiment is repeatable for the two midrange concentration levels, and not the lowest or the highest concentration levels.

The experiment was run as per the setup described in the previous section. The volume flow rates recorded at the outlet are mostly similar to the volume flow rate at the inlet, with some values that fall outside of this range. These results mostly deviate from the original hypothesis, which was that the volume flow rate of the fluid/particulate mixture at the outlet would decrease with increasing concentration and temperature, i.e. the lowest volume flow rate would ideally be found at the $15g/L$ concentration at $41°C$. These results show that there is a low chance of shunt occlusion as the decrease in flow rate is not high enough to show that there is significant particulate accumulation in the shunt system.

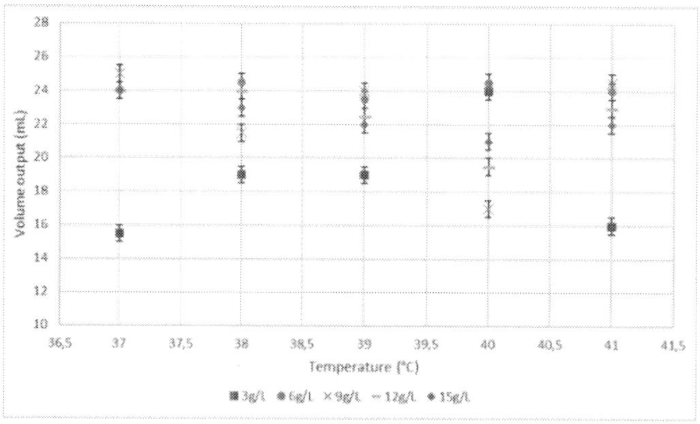

Fig. 6. Volume at the outlet with experimental error

From the data points collected, the conclusion that can be drawn is that the variation in temperature has no effect on the ability of the particles to accumulate along the walls of the shunt system, as each iteration of the results yielded clear fluid at the distal end. This could be attributed to the fact that the gelatine particles sink to the bottom of a container when the fluid is standing. The fluid was regularly agitated throughout the experiment to ensure that the particles do flow through the system, but this made no difference to the concentration of particulates in the fluid at the output. It was found that the particulates would accumulate at the inlet of the shunt valve, but the particulate accretion did not affect the fluid flow significantly.

The accumulation of particles at the inlet of the shunt system was expected, but the effect which it had on the experiment was not. The assumption made was that the particulate mixture would flow through the system even though there were particles at the inlet of the system. This is analogous to a ventricular catheter being slightly blocked but still allowing fluid with particulates through. However, in this experiment, the particles somewhat filtered the fluid out. The densities of the fluid and the particles are different, but the separation of the two mediums was not considered. This showed that the area of the shunt system that is most prone to blockage is the inlet of the ventricular catheter, as per Tamburrini, Calderelli and Di Rocco [7] but also that the effect of the blockage is minor when the test is run for a short period of time. The serious effect of the blockage may take effect over a long period of time and will only then pose a danger to the health of the patient.

4 CFD Model Results and Discussion

4.1 Validation of CFD model

For the CFD model, the first step that was taken was to validate the model. The model was run at every temperature and particulate concentration, and the volume flow rate at the outlet was recorded.

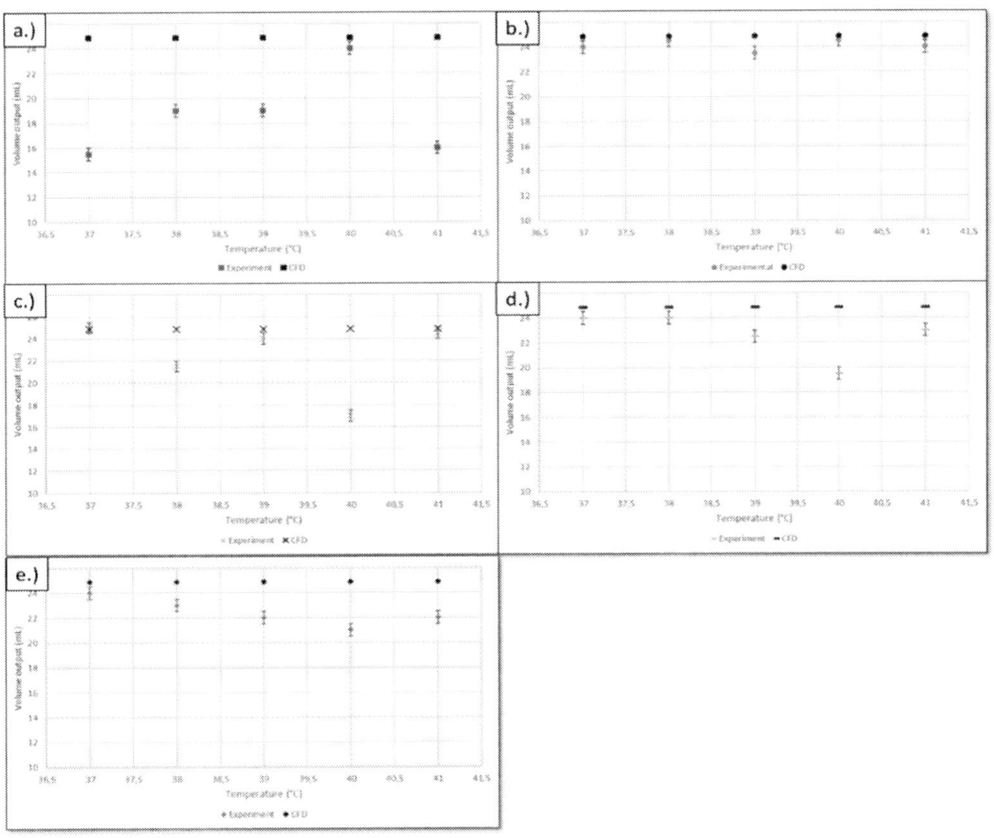

Fig. 7. Graphs of experiment vs CFD for a.) 3g/L, b.) 6 g/L, c.) 9 g/L, d.) 12 g/L and e.) 15 g/L

The volumes obtained at the outlet were compared, as the error due to the measuring instrument for the experiment could be accounted for. As can be seen in Fig. 7, the graphs for the $6g/L$ and $9g/L$ show that the CFD model is a good approximation of the real behaviour of a VP shunt under the test conditions. The differences between the experimental results and the results of the CFD model are minimal. Some of the readings from the CFD model fall within range of the error bars for the readings that were obtained in the experiment. For the other three concentration levels, the CFD model is not a very good approximation of the shunt behaviour as the differences between the results from the experiment and CFD model differ greatly in some instances. There could have been fewer discrepancies with the experiment for the $6g/L$ and $9g/L$ concentrations compared to the other concentration levels. However, the model is still valid.

The relative error associated with this model can be determined using the following formula:

$$\% \, error = \left| \frac{CFD - experimental}{experimental} \right| * 100 \qquad (6)$$

The average relative error associated with the experiment and the CFD model was found to be 15.1%. This average was taken for all the readings based on all the tests that were done. The readings that fall outside of the $23 - 25 \; mL$ band resulted in greater error margins, whereas the readings within the band resulted in lower error margins. Taking the error associated with the experimental results into account, the results obtained from the CFD model are comparable with the results from experiment and the model is valid.

4.2 CFD model results

The most definitive method of determining if there is shunt occlusion is by observing the volume flow rate at the outlet in relation to the inlet. The volume flow rate decreased by $0.45ml/hr$ for all cases obtained from the CFD model, which is not a decrease that points towards a definite blockage in the shunt system as this is only 0.45% of the inlet flow rate. The changes in particulate concentration and temperature did not affect the fluid flow at these conditions.

Three locations in the shunt valve have been studies as they are locations where the flow characteristics may change due to geometry and may be points of interest in terms of potential sites for blockage. The flow patterns within the shunt valve are shown using velocity vectors, which are shown in Fig. **8**. The dominant flow is in the positive z-direction, directing the fluid/particulates towards the outlet. The flow has characteristic recirculation zones, as shown in the circled regions in Fig. **8**, near the inlet (Fig. **8**a), upstream and downstream of the void (Fig. **8**b) and near the outlet (Fig. **8**c),. These recirculation regions are potential sites for entrainment, recirculation and accumulation of particles and subsequently potential sites for blockage.

One key observation is that there is a decrease in particulate loading between the inlet and outlet of the model. As shown in Fig. **9**, the particulate loading is largest near the inlet. Considering the large recirculation zones observed near the inlet, as shown in Figure 8a, the inlet is likely the site with the highest probability for blockage.

This was also observed in the experiments, as there were instances where the fluid collected at the outlet of the shunt system contained almost no particulates and that there was an accumulation of particles at the inlet of the shunt. The models that were run had varying results – in the $3g/L$, $12g/L$ and $15g/L$ concentration bands, the maximum DPM concentration at the outlet of the shunt system tended to decrease with increasing temperature but the opposite was true for the $6g/L$ and $9g/L$ concentration bands.

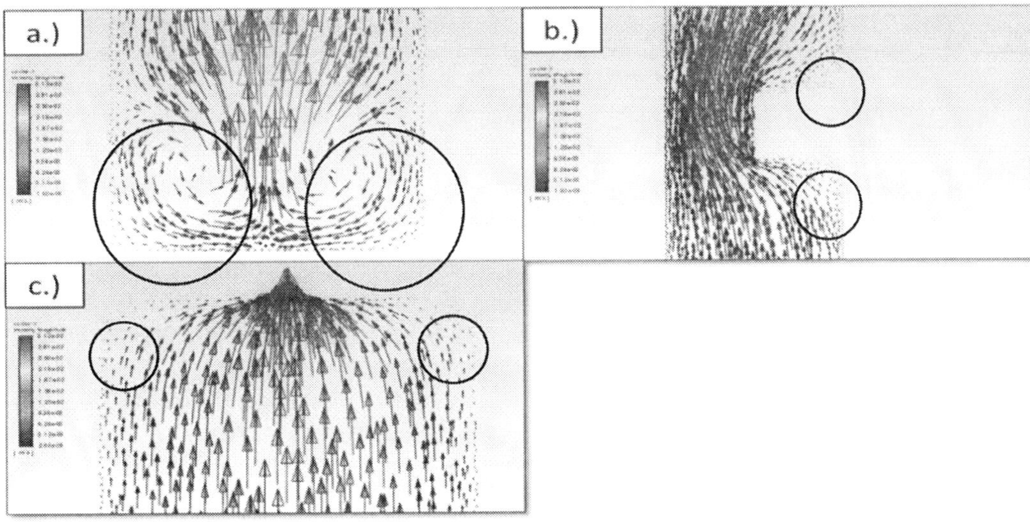

Fig. 8. Velocity vectors for valve inlet (a.), void (b.) and valve outlet (c.)

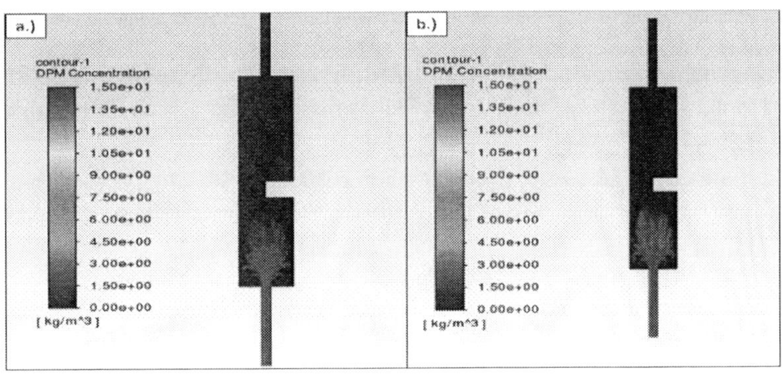

Fig. 9. DPM concentration at valve inlet at 41°C for a.) 6g/L and b.) 9g/L

5 Conclusions

This study investigated whether the occlusion of a shunt valve is accelerated by the presence of an infection using CFD and experiment. The results obtained in the experiment and the CFD model show that there is no real relationship between increasing particulate concentration and temperature, and shunt occlusion at the shunt valve. The general conclusion that could be drawn from the models that were run is that the change in particulate concentration and temperature did not affect the outlet flow rate. This shows that there was no shunt occlusion due to changes in the temperature of the fluid/particulate mixture, as well as the increase in concentration.

The model helped with providing a better understanding of the flow through a VP shunt when the shunt is suspended in an environment with changing temperature, as it effectively showed that increasing temperature and particulate concentration alone do not affect the functionality of the VP shunt system. This result excluded the VP shunt valve as a region of shunt occlusion due to infection, as there would be no bacterial growth in this region. The VP shunt will only malfunction due to infection only when there is a blockage of the proximal or distal catheters [6]. These two regions are more prone to bacterial growth than the shunt valve itself.

These results are valid in as far as the assumption that infection manifests in a.) the change in temperature of the CSF and/or shunt surfaces and b.) an increase in particulates as a consequence of the rise in white blood cells in response to infection. Future work should focus on extending the biological complexity of infection beyond the reductionist assumptions of heat transfer and particulate mechanics. Instead, complexity such as bacterial growth along the shunt surfaces must be modelled to better understand the effect of infection on shunt occlusion. Furthermore, the non-Newtonian behaviour of the fluid in the presence of soluble particles such as albumin should be accounted for in future work.

The Centre for High Performance Computing (CHPC) is acknowledged for the use of the CHPC facilities.

References

1. R. Rizvi, Q. Anjum, *Hydrocephalus in Children*, Journal of Pakistan Medical Association, **55**, 11, (2005)
2. M. Hussain, R. A. Raja, A.-u.-D. Shaikh, M. H. Ali, *Ventriculoperitoneal Shunt Blockage*, Journal of Ayub Medical College, Abbottabad, **24**, (2012)
3. S. Moyle, *Hydrocephalus and Shunts*, Ausmed, 13 April 2017. [Online]. Available: http://www.ausmed.com/articles/hydrocephalus-and-shunts/ [Accessed 7 December 2017]
4. Johns Hopkins University Staff, *Shunt Procedures*, The Johns Hopkins University, The Johns Hopkins Hospital and Johns Hopkins Health System, [Online]. Available: http://hopkinsmedicine.org/neurology_neurosurgery/centers_clinics/cerebral-fluid/procedures/shunts.html [Accessed 7 December 2017]
5. Y. Gutierrez-Murgas, J. N. Snowden, *Ventricular shunt infections: Immunopathogenesis and clinical management*, Journal of Neuroimmunology, **276**, 0, (2014)
6. D. L. Wells, J. M. Allen, Ventriculoperitoneal Shunt Infections in Adult Patients, AACN Advanced Critical Care, **24**, 1, (2013)
7. G. Tamburrini, M. Calderelli,, C. Di Rocco, *Diagnosis and management of shunt complications in the treatment of childhood hydrocephalus*, Reviews in Neurosurgery, **1**, 3, (2002)
8. T. D. Simon, J. Butler, K. B. Whitlock, S. R. Browd, R. Holubkov, J. R. W. Kestle, A. V. Kulkarni, M. Langley, D. D. Limbrick, N. Mayer-Hamblett, M. Tamber, J. C. Wellons, W. E. Whitehead, J. Riva-Cambrin, *Risk Factors for First Cerebrospinal Fluid Shunt Infection: Findings from a Multi-Center Prospective Cohort Study*, The Journal of Pediatrics, **164**, 6, (2014)
9. A. Yilmaz, A. M. Musluman, N. Dalgic, T. Cansever, T. Dalkilic, E. Kundakci, Y. Aydin, *Risk factors for recurrent shunt infections in children*, Journal of Clinical Neuroscience, **19**, (2012)
10. T. Ramokoka, *Modelling fluid and particulate flow through a VP shunt,* University of Johannesburg, Johannesburg, (2017)
11. A. C. Aralar, M. D. Bird, R. D. Graham, B. Koo, M. B. Shenai, P. V. Chitnis, S. Sikdar, *Ultrasound Characterization of Interface Oscillation as a Proxy for Ventriculoperitoneal Shunt Function*, in Proceedings of the 38th Annual International Conference of the IEEE Engineering in Medicine and Biology Society (EMBC), Orlando, (2016)
12. ANSYS Inc., *ANSYS Fluent Meshing User's Guide*, SAS, Inc., Canonsburg, (2016)

13. Autodesk Support, *How to Perform a Mesh Convergence Study*, Autodesk, Inc., 24 May 2015. [Online]. Available: https://knowledge.autodesk.com/support/simulation-mechanical/learn-explore/caas/sfdcarticles/sfdcarticles/How-to-Perform-a-Mesh-Convergence-Study.html [Accessed 23 June 2019]
14. P. J. Roache, *Perspective: A Method for Uniform Reporting of Grid Refinement Studies*, Journal of Fluid Engineering, **116**, (1994)
15. V. Kurtcuoglu, D. Poulikakos, Y. Ventikos, *Computational Modeling of the Mechanical Behaviour of the Cerebrospinal Fluid System*, ASME Journal of Biomedical Engineering, **125**, (2005)
16. Comsol, Inc., *Navier-Stokes Equations*, Comsol, Inc., 22 February 2017. [Online]. Available: https://www.comsol.com/multiphysics/navier-stokes-equations [Accessed 26 September 2019]
17. B. Munson, T. Okiishi, W. Huebsch , A. Rothmayer, *Differential Analysis of Fluid Flow*, in Fundamentals of Fluid Mechanics, Sixth Edition, Singapore, John Wiley & Sons, Inc., pp. 307-308, (2009)
18. ANSYS Inc., *ANSYS Fluent Theory Guide*, SAS IP, Inc., Canonsburg, (2016)
19. Faculty of Biological Sciences, University of Leeds, *White blood cells*, University of Leeds, [Online]. Available: https://www.histology.leeds.ac.uk/blood/blood_wbc.php [Accessed 13 September 2019]
20. A. Tojo, S. Kinugasa, *Mechanisms of Glomerular Albumin Filtration and Tubular Reabsorption*, International Journal of Nephrology, **2012**, (2012)
21. I. Bloomfield, I. Johnston, L. Bilston, *Effects of protein, blood cells and glucose on the viscosity of cerebrospinal fluid*, Pediatric Neurosurgery, **28**, 5, (1998)
22. A. Tajima, H. Nakata, S. Lin, V. Acuff , J. Fenstermacher, *Differences and similarities in albumin and red blood cell flows through cerebral microvessels*, **5**, 2, (1992)
23. C. Kotsalos, J. Latt, B. Chopard, *Bridging the computational gap between mesoscopic and continuum modeling of red blood cells for fully resolved blood flow*, Journal of Computational Physics, **398**, (2019)
24. C. Kotsalos, J. Latt, J. Beny, B. Chopard, *Digital blood in massively parallel CPU/GPU systems for the study of platelet transport*, Interface Focus, **11**, (2021)
25. A. Lui, T. Polis, N. Cicutti, *Densities of cerebrospinal fluid and spinal anaesthetic solutions in surgical patients at body temperature*, Canadian Journal of Anaesthesia, **45**, (1998)
26. D.P. Agamanolis, *Chapter 14: Cerebrospinal Fluid*, NEOMED, August 2016. [Online]. Available: http://neuropathology-web.org/chapter14/chapter14CSF.html [Accessed 24 September 2019]
27. Thermtest Europe, *Thermal Conductivity of Water*, Thermtest Europe, 11 February 2021. [Online]. Available: https://thermtest.se/thermal-conductivity-of-water#:~:text=According%20to% 20 literature1%20the,K%20at%2020%20C2%B0C [Accessed 31 July 2021]
28. AZoM.com, *Silicone Rubber*, AZoNetwork, [Online]. Available: https://www.azom.com/properties.aspx?ArticleID=920 [Accessed 31 July 2021]
29. California Pacific Medical Center, *Caring for Your Child's Fever*, Sutter Health, 2015. [Online]. Available: http://www.cpmc.org/advanced/pediatrics/patients/topics/fever.html [Accessed 27 March 2018].
30. Advanced Instructional Systems, Inc. and the University of North Carolina, *Measurements and Error Analysis*, Advanced Instructional Systems, Inc. and the University of North Carolina, 2011. [Online]. Available: https://www.webassign.net/question_assets/unccolphysmechl1/measurements/manual.html [Accessed 26 February 2020].

Discrete element model calibration for industrial raw material simulations

Johan Bester[1], *Philip* Venter[*1], and *Martin* van Eldik[1]

[1]School of Mechanical Engineering, North-West University, Private Bag X1290, Potchefstroom, 2520
URL: www.engineering.nwu.ac.za

Abstract. The use of computational fluid dynamics in continuous operation industries have become more prominent in recent times. Proposed system improvements through geometric changes or control strategies can be evaluated within a relatively shorter timeframe. Applications for discrete element methods (DEMs) in real life simulations, however, require validated material-calibration-methods. In this paper, the V-model methodology in combination with direct and bulk calibration approaches were followed to determine material model parameters, to simulate real life occurrences. For the bulk calibration approach a test rig with a containment hopper, deflection plate and settling zone was used. Screened material drains from the hopper, interacts with the deflection plate, and then settles at the material angle of repose. A high-speed camera captured material interaction with the rig, where footage was used during simulation validation. The direct measuring approach was used to determine particle size, shape and density, while confirming friction and restitution coefficients determined in the bulk calibration method. The test was repeated and validated for various geometrical changes. Three categories of validation were established, namely particle speed assessment, -trajectory assessment and -plate interaction assessment. In conclusion, the combination of direct and bulk calibration approaches was significant in calibrating the required material model parameters.

1 Introduction

The relevance and application of discrete element methods (DEMs) in industry has increased significantly in recent years. The use of an accurately calibrated model decreases the time required for studies into operational and design changes. In particular, these models aid in identifying root causes for reduced efficiency and/or equipment life without physically interfering with plant production. This reduces the total cost of the investigation when compared to the conventional trial and error approach [1].

A DEM is normally used in studies where the importance of identifying bulk material behaviour is important [2]. Therefore, DEMs are typically applicable to production industries, and more specifically, the steel industry.

[*] Corresponding author: Philip.Venter@nwu.ac.za

Within the steel industry, iron making is the process of converting raw materials into molten iron. This is typically done through a blast furnace, where raw material is loaded from the top whilst hot air and pulverised coal is injected and combusted at the tuyeres, just above the hearth [1].

The raw materials that are typically a calculated mixture of metallurgical coke, sinter, ore and additives, are loaded into the top of the blast furnace via 2 methods; 1) a skip loading system and 2) a conveyor loading system [1]. With the conveyor loading system, hoppers are filled with their designated materials and then drained by loading the material onto a continuously moving steep inclined sidewall conveyor. This conveyor then transports the material to the top of the blast furnace for discharge. Some of the process problems that typically occur at this loading point is constant spillage and abrasion wear on belt edges.

The bulk coke material handling internal to the iron making process was identified as a possible case study. This study will demonstrate that a DEM is capable of simulating the bulk material behaviour when coke is discharged from a hopper onto a steep inclined sidewall conveyor.

There are numerous software packages that can be utilized in a study relating to a DEM. Simcenter STAR-CCM+ software package is used in this study due to the coupling of a DEM into a multiphase environment being readily available to the user [3].

2 Literature Survey

A paper by Coetzee [4] was aimed at critically evaluating different validation and model calibration techniques and determining if there is a definite model parameter or technique to be used for the most accurate results. The parameters evaluated were: particle shape, size, density, stiffness, rolling resistance, inter-particle and boundary friction coefficients (both static and dynamic), coefficient of restitution, cohesive properties and adhesive properties.

Thompson [5] investigated the calibration of different DEM model parameters for dry and wet granular materials, in order to determine if the model is capable of replicating material bulk behaviour. Marigo and Stitt [6] used a rotating drum as a validation mechanism for a DEM model developed for the evaluation of the influence that the detail of the particle shape has on bulk material behaviour. The only parameter that [6] identified as critical in addition to those mentioned by [4] was the Poisson's ratio.

Grima and Wypych [7] investigated the sensitivity of the variation of various DEM parameters on the impact reaction force of polyethylene pallets on a plate. This study attempted certain scaling rules in order to reduce the number of particles in the domain and in doing so minimise the computational time.

Further, Horn [8] conducted a study on the calibration of relatively large ore aggregate particles by using a large-scale shear box. The parameters identified to be critical in addition to those mentioned by [4] [5] and [6] was bulk density, material porosity, Young's modulus and the internal friction angle. This study used small scale tests for validation of the calibrated DEM model and it was concluded that the model replicated the material bulk behaviour within an acceptable margin.

3 Experimental Setup

The design of the test bench and the calibration methodology was based on work by Quist and Evertsson [9]. The referenced study was aimed at developing a calibration framework that would not only assist in validation but also ensure accuracy of results.

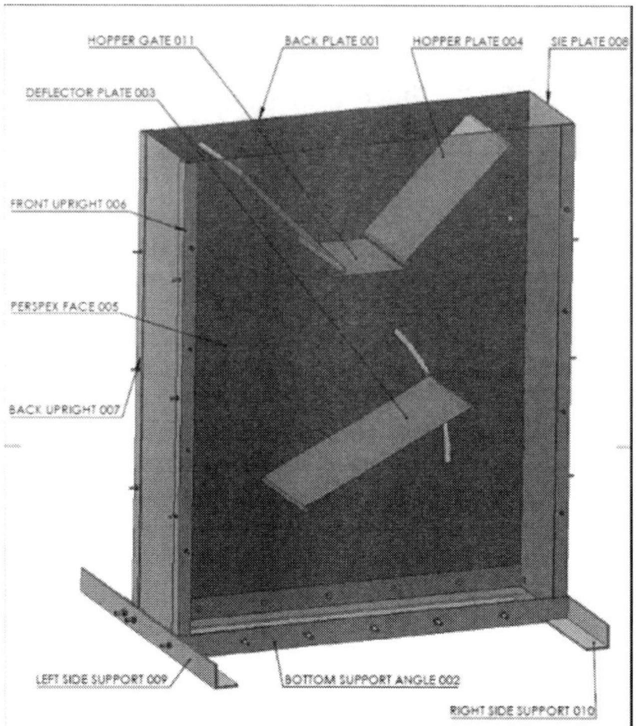

Figure 1: Test rig setup used in this study

The calibration framework developed by [9] and referred to as the "V-model" comprised of three levels, each linking between the simulation and experimental domains. The calibration process starts at the first level, which is the single property lab test. This level correlates with the direct measuring approach of [4]. These properties are then included in the simulation domain and a first level validation is done.

The second level is defined by multiple flow regime experiments. This level determines if the bulk flow behaviour produced by individually calibrated parameters correspond within a degree of accuracy to an experimental test. This level correlates well with the bulk calibration approach of [4]. It is a common occurrence in this level to find that the parameter adjustments are required before the model correlates well with the experimental results. It is advised that during this stage, the experimental setup should be altered to ensure there is still an acceptable degree of accuracy between the experimental and simulation domain [9].

The third level is defined by an industrial scale experiment. For this level the model parameters are validated further with actual material handling processes. At this point, it can be determined if the model parameters are within an acceptable region of accuracy and thus deemed calibrated.

The relevant experimental setups to determine the parameters mentioned in the literature survey can typically be categorized into two approaches; the *Bulk Calibration* and *Direct Measuring* approach [4].

Table 1: Approaches to determine DEM parameters used in this study [4]

Bulk Calibration Approach	**Direct Measuring Approach**
Coefficients of Restitution (Particle-Boundary)	Particle Shape
	Particle Size
Static Friction Coefficient (Particle-Particle)	Particle Density
	Static Friction Coefficient (Particle-Boundary)

4 Parameter Calibration

4.1 Direct Measurements

4.1.1 Particle Size

Throughout literature the size of the particles in the DEM simulation is considered as an important parameter that has to be included in the calibration of the model. In most cases the size of the particle has to be scaled up to reduce the number of particles in the simulation so that the computational time can be reduced.

Roesller and Katterfield [10] stated that the total degrees of freedom are affected when the particles are scaled and that the scaled particles should be calibrated with regards to the material bulk behaviour. In laboratory simulations with a small number of particles it is possible to size the particles closer to the actual value, but when industrial size simulations (typically hopper or silo discharge) are simulated, the user is forced to reduce the number of particles in the simulation.

Shigeto et al. [11] investigated the influence of scaling up fine power particles for a screw conveyor application. In this report spherical particles were used and it was found that particles could be scaled up to 4 times while still producing accurate material bulk behaviour.

Grima and Wypych [7] determined that care should be taken when scaling rules are applied to particles when impact force is evaluated. The authors concluded that the size of the particles had an effect on the resolution of the result and therefore if the particle scaling was too great a variation in the impact area would be obtained [7].

Xie et al. [12] used a similar conveyor design set up as [7] in order to investigate the wear process of a conveyor transfer chute. It was concluded that the particle size should not be scaled by more than a factor of 2 when conveyor transfer chutes are modelled.

Measurement

Coke particles were obtained and screened according to normal processes. Additional screening was required due to size limits on the experiment test rig. The following size distribution was obtained:

Table 2: Particle size distribution obtained for experiment

Particle Size Distribution of Coke						
	>100 mm.	**80-100 mm.**	**60-80 mm.**	**40-60 mm.**	**35-40 mm.**	**30-35 mm.**
% Coke per size	0%			83%	11%	6%
Modelled Size	80mm			60mm	40mm	35mm

4.1.2 Particle Shape

The most common particle that is used when setting up a DEM analyses is a spherical particle shape. This is due to a spherical particle shape providing sufficient contact detection criteria which allows the solving time of the simulation to be reduced [13].

The study of [13] investigated the difference in DEM parameters when spherical clumps of 2, 4 and 8 spheres were used. It was concluded that a manually generated spherical clump particle replicated the material behaviour more accurately than a single sphere particle.

Pasha [14] investigated the difference in modelled bulk material behaviour if spherical particles with an additional friction factor are used vs. optimised particle clumps without a friction factor. It was stated that particles comprising of 5 spheres in a clump will be able to produce adequate results for material bulk behaviour [14].

Markauskas and Kacianauskas [15] developed a particle simulation with the use of a hopper draining test and simulated particles shaped like axisymmetric clumps. It was found that even though the shape closely resembled the actual particles, the model still required some calibration with regards to other parameters in order to accurately replicate the material bulk behaviour [15].

Measurement

For the particle shape calibration, 14 random particles were selected from the screened samples. The particles were placed on a grid representing a 60mm x 60mm block. A photo was taken and imported into a CAD program after which representative shapes could be drawn on scale. The result is shown in the figure below.

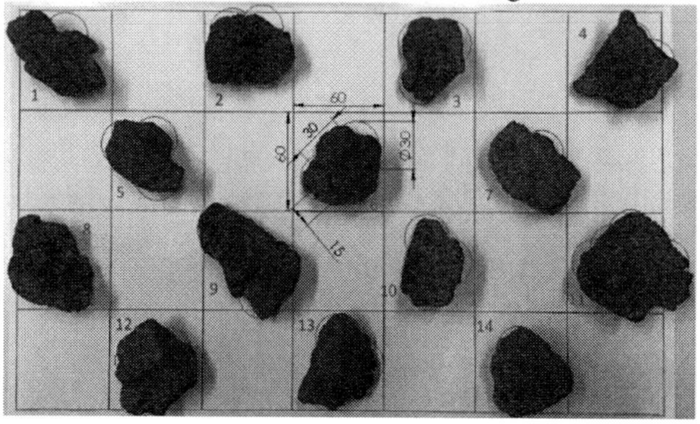

Figure 2: Visual shape analyses

It followed that most particles could be represented by a triangular cluster of spheres, and for simulation purposes it was therefore decided to follow this modelling approach. The size distribution of the simulated particles was obtained by changing the diameter of the spheres within the cluster according to the total particle diameter required.

Figure 3: A modelled particle in STAR-CCM+

4.1.3 Particle Density

In industry, material density is generally defined as the bulk density. The bulk density is defined as the mass of a sample per unit volume.

Horn determined the specific particle density of a medium size ore aggregate (40mm) by measuring the volume of water being displaced when a known mass of material was submerged in the container [8]. The bulk density was then determined by filling the large-scale shear box tester volume with the material, measuring the mass of the container and dividing it by the known volume of the shear box tester container. Additional properties like the void ratio and porosity were then calculated with the use of the particle density and the bulk density. These values were then used for the calibration of the DEM particle density.

Measurement

Since the volume of a particle could not be calculated as easily as the mass of the particle, a test setup was required. A particle sample is submerged in water whilst the displacement of the water is measured. The assumption is made that the particles don't absorb water. From the assumption, the volume of the coke particle is equal to the volume of displaced water.

The results from this test are given in the table below:

Table 3: Particle density calculation

Particle Density Calculation		
Weight of Container	0.390	kg
Weight of Material without Container	1.328	kg
Volume of water without material sample	0.003	m^3
Calculated volume of material	0.001	m^3
Particle Density	1021	kg/m^3

The calculated particle density was determined to be 1021 kg/m^3. This was determined to be within the typical range for coke (800-1200 kg/m^3) processed at this specific plant. Therefore, this value was deemed accurate for modelling purposes.

The solid material bulk density is defined by [8] as the mass of the sample divided by the volume which the material occupies. From this definition it is clear that the bulk density also includes the voids that are present between the particles as illustrated by Figure 4.

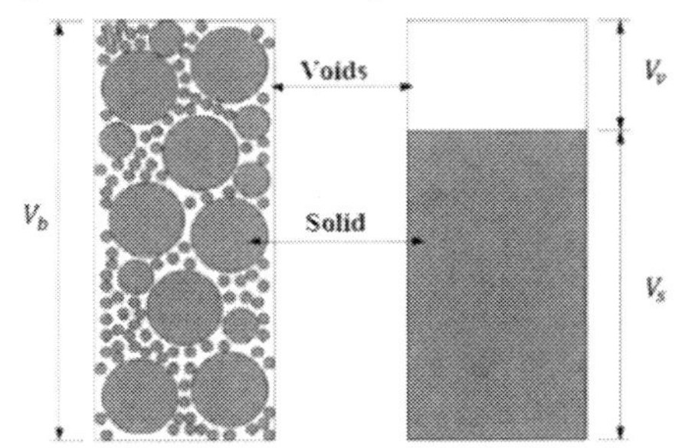

Figure 4: Bulk volume, voids volume, solid volume

In order to calculate the bulk density, the sample was placed within a known volume and weighed. The bulk and particle densities are then used to determine the void fraction e. and porosity n. by using the equations below;

$$e = \frac{\rho_p}{\rho_b} - 1 \tag{1}$$

$$n = 100\left(\frac{e}{e+1}\right) \tag{2}$$

The bulk density was determined to be 428.4 kg/m^3, with the void fraction and porosity being 1.39 and 58.1% respectively. These values were within the 400-600 kg/m^3 design limits.

4.1.4 Particle-Boundary Friction Coefficient

The particle-boundary friction coefficient is a parameter that is used to describe a particle's resistance against motion when it is in contact with a boundary surface. Failure to calibrate this value correctly will lead to errors in drainage time and the velocity of particles exiting the hopper.

Research performed by [8] determined the particle-boundary friction coefficient by placing a particle on the surface of the boundary to be assessed. The boundary surface was then lifted until the particle started to slide. The angle (θ) at which the particle slid was then recorded and a friction coefficient (μ_b) was calculated with the use of this angle and the equation below.

$$\mu_b = \tan(\theta) \tag{3}$$

Measurement

The deflector plate of the test rig is used for this calculation. Due to the nature of the problem to be modelled, only 2 materials are included in the boundaries. These are the rubber (conveyor material) and liner plate (liner in coke hopper).

These materials are individually fastened to the test rig deflector plate after which a particle is placed on the plate. The deflector plate is rotated until the particle starts to slide. The angle at which the particle slides is recorded with a protractor. The test is repeated 20 times and an average particle-boundary coefficient was calculated for the interaction between coke-conveyor and coke-liner.

Table 4: Particle-boundary friction coefficient results

Particle-boundary friction coefficient results		
Material	**Average angle deg.**	**Average friction coefficient**
Rubber	40.23	0.852
Ceramic liner plate	13.5	0.240

4.2 Bulk Calibration

4.2.1 Coefficient of Restitution

The coefficient of restitution is a parameter used in DEM to illustrate the damping mechanism which occurs when a particle makes contact with a boundary surface or another particle.

When no viable test is available to determine the coefficient of restitution between particles, it is recommended to that a default value of 0.3 should be used when simulating hard rock ores [5].

Some researchers followed the drop test to determine the coefficient of restitution but when particles with varying shapes are used a reverse calibration technique was followed [4]. Thus, a simulation model was set up and the coefficient of restitution was changed until it matched the actual drop test result.

A drop test consists of a particle being secured at a known height above the boundary material. The particle is dropped and allowed to bounce on the boundary material while the movement is captured with a high-speed camera. The height the particle obtains after contact is then measured which is then used to calculate the velocity directly after contact. It is assumed that air resistance does not have a significant effect on the particle velocity. This enables the use of equation below for determining the velocity after contact.

$$v_f^2 = v_i^2 + 2g(x_f - x_i) \tag{4}$$

Measurement

An actual material particle does not have an exact spherical shape. To overcome this issue, particles were assessed in order to find the most spherically shaped particles.

The bottom deflector plate was positioned horizontally with a grid fastened to the back plate that consisted of 20x20 mm. blocks printed over the entire area. This grid will enable the measurement of displacement height achieved after impact. This test was only completed for rubber material since the bounce mechanism is not required for the interaction between particles and the chute liner. The particle is released from the trap door position by hand to ensure the motion of the particle is not influenced by the trap door.

Figure 5: Experimental setup for coefficient of restitution calibration

For the simulation model, a similar setup in terms of geometry and drop position was made. The restitution coefficient is changed between tests, whilst a maximum velocity report measures the particle velocity throughout the drop test. The maximum velocity after impact of each coefficient of restitution is recorded.

By noting the maximum velocity after the simulated particle impact, a comparison can be made with the initial velocity calculated from the drop test results. The coefficient of restitution can then be calculated by determining the value at which the simulation and experimental domain deliver similar velocity results.

The drop test was repeated 10 times of which the average results are tabulated below along with the simulation model results.

Table 5: Coefficient of Restitution experimental and simulation results

Coke-Rubber velocity calculation result				
Experimental results			*Simulation Results*	
Particle displacement after impact	0.14	m	CoR = 0.2	0.688 m/s
Gravity	-9.81	m/s^2	**CoR = 0.4**	**1.024 m/s**
End velocity	0	m/s	**CoR = 0.6**	**1.726 m/s**
Initial velocity after impact	**1.657**	m/s	CoR = 0.8	2.454 m/s

As expected, with an increase in the coefficient of restitution, the initial velocity after impact is increased. By analysing the results, it can be determined that a coefficient of restitution of 0.6 will deliver sufficiently accurate results. Note that the value of 0.6 is used for both the normal and tangential components of the interaction since a spherical particle interaction was modelled.

As stated by [5], there is yet to be a test developed that will provide sufficient results for a particle-particle restitution coefficient. According to [5], a typical value for granular material is 0.3 for both the normal and tangential restitution coefficients.

4.2.2 Particle-Particle Friction Coefficient

The particle-particle friction coefficient is a parameter that is used within the DEM analyses which dictates the particle's resistance against sliding when in contact with another. This is a parameter which is not easily measurable and is also reliant on other parameters.

For instance, if a sample of raw material is illustrated via spheres in a DEM analyses these particles will not experience the same interlocking characteristics due to the difference in shape. This parameter is therefore used to simulate a restriction in the movement so that the spherical particles replicate the actual material bulk behaviour.

Measurement

The experimental procedure consisted of a screened batch of coke particles being loaded into a containment hopper which is then drained through the trap door. The particle interaction with the deflector plate is then assessed. The entire draining process is recorded with a high-speed camera and the angle of repose formed at the bottom of the rig is measured.

A geometrically similar simulation model was setup with the calibration parameters determined in section 3.1 whilst the particle-particle friction coefficient was varied until the simulation model matched that of the experimental test.

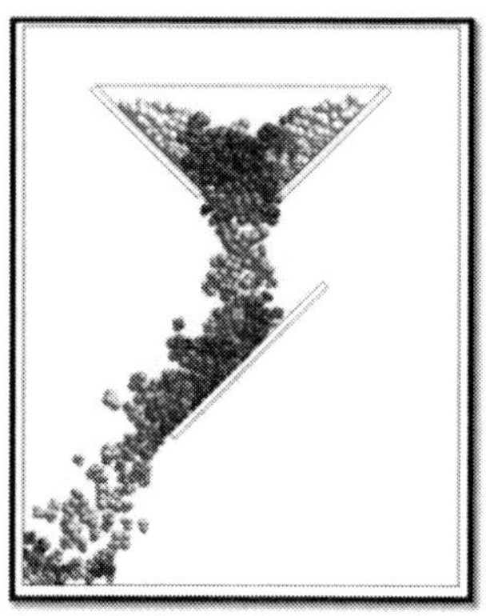

Figure 6: Hopper draining simulation

The following results were obtained from repeating this test 5 times.

Table 6: Hopper draining test results

Hopper draining test		
Test number	**Draining duration**	**Angle of repose**
1	1.872 s	33 degrees
2	1.920 s	40 degrees
3	2.448 s	31 degrees
4	1.968 s	35 degrees
5	1.750	36 degrees

By analysing the hopper draining time in the simulation with the range of friction coefficients assigned, it was concluded that the friction coefficient does not have a significant effect on the hopper draining time.

The second portion of this section was to evaluate the effect the friction coefficient has on the angle of repose. The image of simulation result was captured and imported into a CAD program which allowed the measurement of angles.

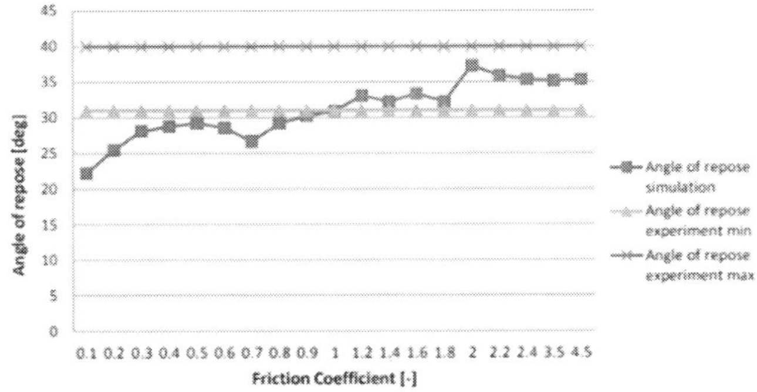

Figure 7: Angle of repose vs friction coefficient

From the figure above, it is clear the angle of repose is influenced by the particle-particle static friction coefficient. Coetzee [4] stated that the angle of repose will start to flatten out, implying that there is a point at which the static friction coefficient would no longer have a significant effect on the angle of repose.

Coetzee [4] suspected that this point was reached, after which 2 additional simulations were done with a 3.5 and 4.5 static friction coefficient for confirmation. As shown in the figure above, this was indeed the case.

By comparison, it is evident that a particle-particle static friction coefficient of 2 matches the experimental results the best.

5 Final Validation

As stated by [9], a final parameter validation has to be done whereby the test rig settings should be adjusted to verify that the simulation still produces a similar result to that of the experimental test.

For this validation, it was decided to adjust the deflection plate to the extreme scenario a horizontal position. It is the furthest point from the initial calibration settings used and would thus be a clear indication of the model parameters are correct.

Figure 8: Final validation simulation model vs experimental model

The experimental test was repeated 3 times and recorded with a high-speed camera. The hopper draining time as well as the angle of repose formed on the deflection plate was measured and average across the 3 tests.

The table below is a summary of the results;

Table 7: Validation test results

Validation test results assessment			
Parameter	**Experimental**	**Simulation**	**Deviation**
Angle of repose (left)	44.57 deg	43.04 deg	-3.55 %
Angle of repose (right)	35.33 deg	32.87 deg	-7.49 %
Hopper drain time	1.8 0 s	1.44 s	0.36 s

The hopper draining time deviation of 0.36 s. corresponds with the findings in 4.2.2 and is thus deemed acceptable. The simulation model also, on average, produced a lower angle of repose for both the left and right side; 3.55% and 7.49% respectively.

From the results it can be concluded that the combination of direct and bulk calibration approaches is significant in the calibration of the required material model parameters.

6 Conclusion

In this paper a combination of the direct calibration and bulk calibration approaches were used to calibrate the material parameters for the Hertz Mindlin contact model. The direct measuring approach was used for the calibration of particle size, shape, density and boundary friction coefficients.

A test rig was built for the bulk calibration approach according to the calibration work done by [9]. The test rig was used for the calibration of the particle-particle static friction coefficient and the boundary coefficient of restitution. The test rig was also used for the final parameter validation and it was determined that the calibration process was applicable in producing a simulation result, which matched the experimental result.

The model accuracy obtained from combining the direct and bulk calibration methods justifies its use in building industrial models related to bulk material handling. The impact of design and operational changes can be quantified without stopping the industrial process. It also enables an iterative design process with the goal of maximising production and the operational life of components.

References

1. R. Fourie, *Discrete element simulation for the loading of a steep inclinde side wall conveyor*, Potchefstroom: North-West University (2018)
2. Siemens Product Lifecycle Management Software Inc., *Simcenter STAR-CCM+ : User Manual.* (2019)
3. Siemens PLM, *Fluid Dynamics Simulation* (2021) Online. Available: https://www.plm.automation.siemens.com/global/en/products/simulation-test/fluid-dynamics-simulation.html. Accessed (29-04-2021).
4. C. Coetzee, *Review: Calibration of the discrete element method,* Powder Technol., 104 (2017)
5. M. Thompson, *From the field of the computer - A case study on the development of input parameters for DEM simulation purposes*, IMHC (2008)
6. M. Marigo, E. Stitt, *Discrete Element Method (DEM) for Industrial Applications: Comments on Calibration and Validation for the Modelling of Cylindrical Pellets* KONA Powder Part. J., 236 (2015)
7. A. Grima, P. Wypych, *Investigation into calibration of discrete element model parameters for scale-up and validation of particle-structure interactions under impact conditions,* Powder Technol., 198 (2011)
8. E. Horn, *The calibration of Material Properties for use in Discrete Element Models*, Stellenbosch: Stellenbosch University (2012)
9. J. Quist, M. Evertsson, *Framework for DEM Model Calibration and Validation* in Proceeding of the 14th European Symposium on Comminution and Classification (2015)
10. T. Roessler, A. Katterfeld, *Scalability of Angle of Repose Tests for the Calibration of DEM Parameters* in ICBMH2016 Conference (2016)
11. Y. Shigeto, M. Sakai, S. Matsusaka, *Development of a large scale discrete element modeling for a fine particle conveying*, Engineers Australia (2011)
12. L. Xie, W. Zhong, H. Zhang, A. Yu, Y. Qian, Y. Situ, *Wear process during granular flow transportation in conveyor transfer,* Powder Technol., **288**, 65 (2016)
13. C. Coetzee, *Calibration of the discrete element method and the effect of particle shape,* Powder Technol., 50 (2016)
14. M. Pasha, C. Hare, M. Ghadiri, A. Gunadi, P. Piccione, *Effect of particle shape on flow in discrete element method simulation of a rotary batch seed coater*, Powder Technol., **296**, 29 (2016)
15. D. Markauskas, R. Kacianauskas, *Investigation of rice grain flow by multi-sphere particle model with rolling resistance,* Granul. Matter, **2,** 143, (2011)
16. S. Timoshenko, *Theory of Elasticity* (1951)
17. M. Paulick, M. Morgenmeyer, A. Kwade, *Review on the influence of elastic particle properties on DEM simulation results,* Powder Technol., **283**, 66 (2015)
18. Q. Li, M. Feng, Z. Zou, *Validation and Calibration Approach for Discrete Element Simulation of Burden Charging in Pre-reduction Shaft Furnace of COREX Process,* ISIJ Int., **53**, 1365 (2013)

MATEC Web of Conferences **347**, 00037 (2021) https://doi.org/10.1051/matecconf/202134700037
SACAM2020

The prediction of the turned machining induced residual stresses in Ti6Al4V: A Critical Surface Integrity Descriptor

Gary Styger[1*], *Rudolph* F Laubscher[1]

[1]University of Johannesburg, Department of Mechanical Engineering Sciences, Corner of Auckland Park and Kingsway, South Africa

Abstract. The surface integrity of a turned machined surface is an essential indicator of a component's fatigue and corrosion performance. A critical descriptor of this property is the residual stress, both on the surface and subsurface of a part. However, experimental determination of vital surface integrity parameters such as surface roughness, hardness, affected microstructure, and residual stresses are costly, time-consuming, and involve the part's destruction. Therefore, predicting these parameters, such as residual stress versus depth, would be of great value and could aid in the correct machining parameters (cutting speed, feed rate, edge tool radius, rake angle, coolant) for the desired part performance being selected. A study was undertaken to determine the influence of a worn tool and multiple cuts on a wide range of cutting speeds on residual stresses induced by machining an outside-turned bar of Ti6Al4V titanium alloy. Thus, a project was begun to build a non-linear dynamic finite element model to predict the residual stresses thus developed due to the machining manufacturing process.

1 Introduction

The aeronautical and medical industries are primarily users of titanium due to their corrosion resistance and strength to weight ratio. Grade 5 titanium is a standard titanium alloy in use in the industry. The manufacturing of titanium components is often challenging. High-speed machining of titanium alloys remains a challenge for the suppliers, customers, machine-tool designers, and cutting insert/tool manufacturers. Metal cutting causes extensive plastic deformations and high temperatures. These coupled mechanical and thermal effects introduce surface changes in the component. This process may affect the workpiece's material properties and residual stress state at the surface and near-surface. The higher cutting speeds found in high-speed machining may result in a generation of tensile residual stresses—the low thermal conductivity of titanium, high-temperature toughness, and chemical reactivity of titanium exacerbate this problem. The machining-induced residual stresses typically are evaluated with non-contact probing techniques such as x-ray diffraction. Numerical predictions of the cutting process may significantly reduce the effort to quantify the residual stress state of a machined surface. Recently, Ozel et al. [1] showed that "useful" results could

* Corresponding author: garystyger@gmail.com

be possible by finite element modelling (FEA) of the residual stress state and comparison to experimental data. They also concluded that the tool micro-geometry influences the residual stress state. This paper reports on the current work, evaluating the geometry change due to the "worn tool" predicted from simulations. Multiple cutting passes were then simulated, considering the updated "worn tool" geometry. The predicted residual surface stresses induced by each "cutting pass" were compared to data from X-ray Diffraction (XRD) results extracted from a 3D oblique outside turned sample.

Outeiro et al. [2] suggest a procedure for extracting the "numerically-predicted" near-surface machining-induced residual stresses with XRD to measure the residual stresses in the sample. The procedure is as follows:

The predicted residual stresses should be extracted from the FE model far enough away from the chip formation zone, suggesting 1.5 mm.

The mean value of the predicted surface residual stress should be evaluated over a finite volume of the work material. The minimal spatial resolution that the X-ray diffraction can achieve is approximately 1 mm in the direction normal to the X-ray beam and a sub-surface 10-20 μm depth in the direction of the X-ray beam.

The predicted residual stresses should be extracted from the FE model after multiple simulated cutting passes.

The tool wear should be monitored during the machining tests, and it must be considered when modelling the residual stresses. This work predicted the wear of the insert with the use of the FEA model.

Besides the above conditions, it must also be noted that FE models predict only the macro (type-1) residual stresses. While a non-destructive technique such as X-ray diffraction measures both macro (type-1) and micro (type-2) residual stresses, the correlation between the predicted and measured residual stress results is challenging and remains a topic of intense research.

2 Experimental work

A 75 mm Grade 5 round bar was turned in a lathe at different cutting speeds (30, 100 and 200 m/min), feed (0.2 mm/rev) and depth of cut (1 mm) under both coolant and dry conditions. A new SANDVIKTM CNMA 12 04 08 H1P (uncoated carbide) insert was used per cutting speed. The geometry of the insert consisted of a negative rake angle of 5 degrees, a relief angle of 5 degrees and a tool nose radius of 0.8 mm. A conventional emulsified mineral oil-based flood cooling system was utilised.

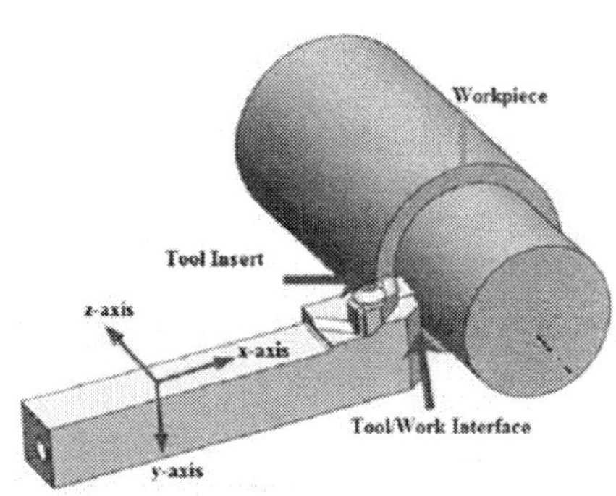

Fig. 1. Turning machining setup [3].

Light optical microscopy with an Olympus BX-51M Optical Microscope with a digital camera was utilised to evaluate the chip morphology. The chip segmentation ratio was measured for each of the cutting speeds. The chip segmentation ratio (G_S) is given by:

$$G_S = (h_{max} - h_{min}) / h_{max} \quad (1)$$

Where h_{max} is the maximum chip thickness, and h_{min} is the minimum chip thickness. G_s = 1 implies a completely segmented or discontinuous chip. Gs = 0 implies a fully continuous chip without any segmentation occurring.

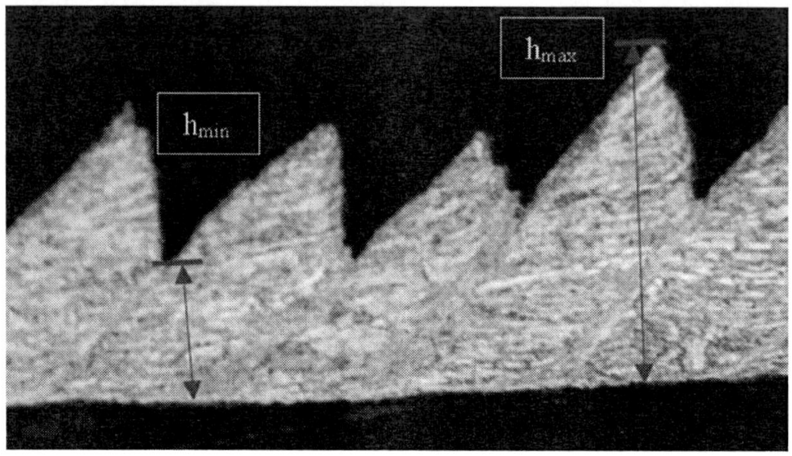

Fig. 2. Distances are required to predict the chip segmentation ratio.

The width of the flank land between the crater and cutting edge (KF) and the distance from the edge to the back crater contour (KB) was measured from the light optical microscopy images, using MIPAR™ Version 3.4.2 [4].

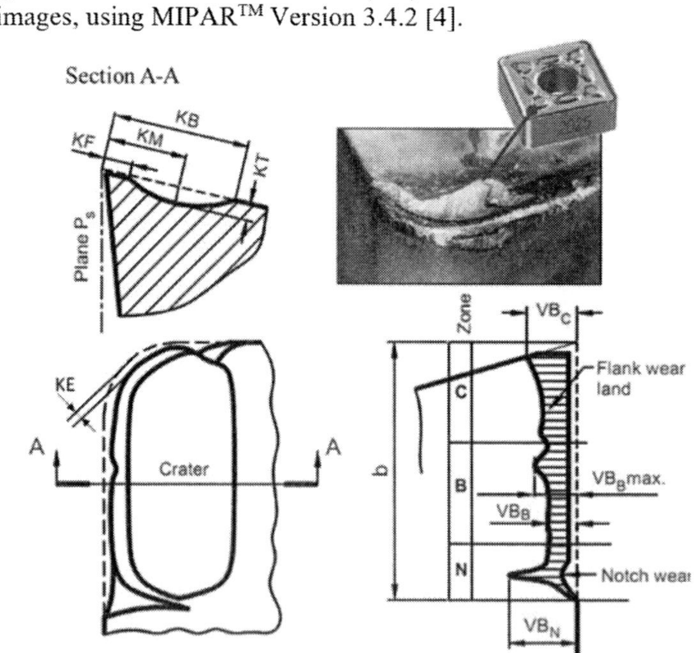

Fig. 3. Typical tool inserts wear patterns [5].

3 Numerical work

Thirdwave Systems AdvantEdge™ Version 7.8 is a finite element analysis software dedicated to certain 2D and 3D machining operations. It helps predict cutting forces, temperatures in the workpiece and the tool, and tool wear and deflection. The solution algorithm is based on a dynamic explicit Lagrangian coupled thermo-mechanical formulation [6]. The model is built by selecting the type of machining operation such as turning, broaching, sawing, or milling. The main interest of the current work is orthogonal turning. The required process parameters are feed rate, depth of cut, length of the cut, cutting speed and initial temperature of the workpiece. Cutting tool geometry, including tool tip radius, rake and relief angle, can all be individually set. The material properties of the tool and the workpiece also need to be defined—AdvantEdge ships with an extensive materials library for both the workpiece and tool materials. Various "thermoviscoplastic" material models are available for use. The included material libraries to be unsuitable in the prediction of the machining-induced residual stresses [7].
In this study, the "Power-Law" phenomenological model material model for Ti6Al4V was used as implemented by Laubscher et al. [7]. Default element sizes from the software were used in this study. The software implements continuous adaptive remeshing which is used to overcome severely distorted elements at the cutting interface during the simulation. The outside turning process was modelled as a 2D plane strain orthogonal model for this work. Further functionality is available via FORTRAN user-subroutines for the material models.

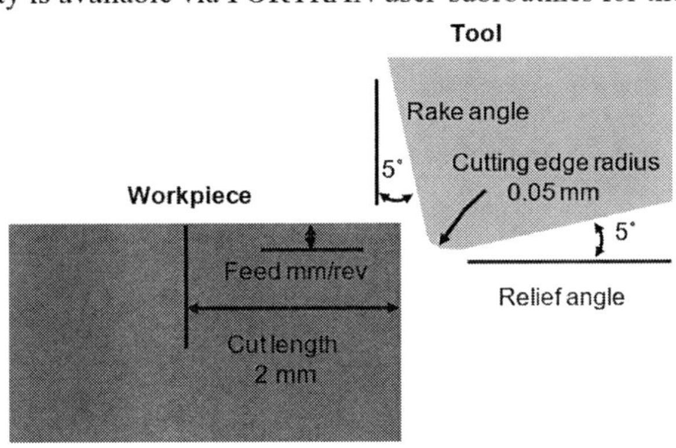

Fig. 4. Setup of the simulation model [7].

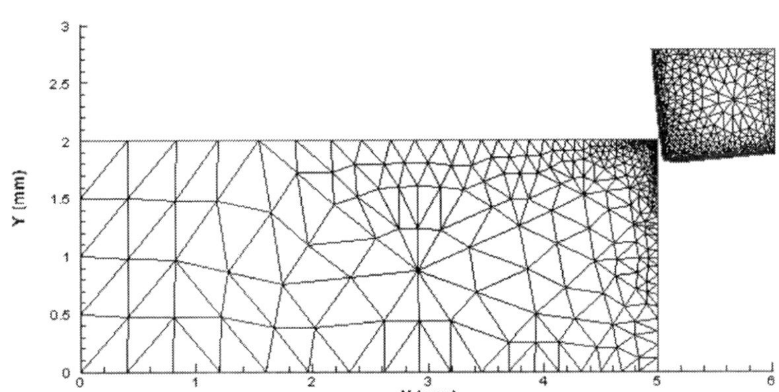

Fig. 5. Initial mesh [8].

A cut depth of 1 mm and a range of different cutting speeds (30, 100 and 200 m/min) were evaluated. The feed rate was 0.2 mm/rev. Conventional flood cooling is modelled with an assumed general convection coefficient of 100 W/m^2.K [7]. Dry cooling is modelled with an assumed general convection coefficient of 20 W/m^2. K. Simulations were run with and without coolant. The Coulomb friction model is implemented in the software [8]. The coefficient of friction of 0.5 was assumed for all simulations.

Prediction of tool wear is available within AdvantEdge. Various models of wear are available. Fortran user-defined wear models can be implemented if standard models do not meet user requirements. Mishra et al. [9] conducted experimental and numerical work to determine the coefficients required for Usai's wear model for dry machining of Ti6Al4V to predict the rake 'crater wear' and 'flank wear'. This wear model and the determined coefficients were implemented in the software's 'wear' simulations for all cutting speeds. The tool's nodal positions are updated based on the "wear rate" from the Usai wear model's calculation.

$$dw/dt = Ke_{(-\alpha/(T+273.15))}\, pV \tag{2}$$

Where dw/dt = wear rate (Volume loss per unit area per unit time), K = 8.13e-8 (1/Pa) – Usai's wear constant, α = 244.63 – Usai's wear constant, V = sliding velocity (length / time).

The user must input the 'cutting time' over which the 'wear' of the tool should be considered. The total cutting length for each cutting speed was 30 mm from the outside turning machining experiments. Therefore, the cutting time was calculated for the numerical model for a cutting length of 30 mm based on each cutting speed. Figure 4 shows an example of the resultant "worn tool" overlaid with the new tool. The "worn" tool geometry was exported from the software and then re-imported into the software for the "1st cut" and "2nd cut" simulations to predict the residual stresses of each pass.

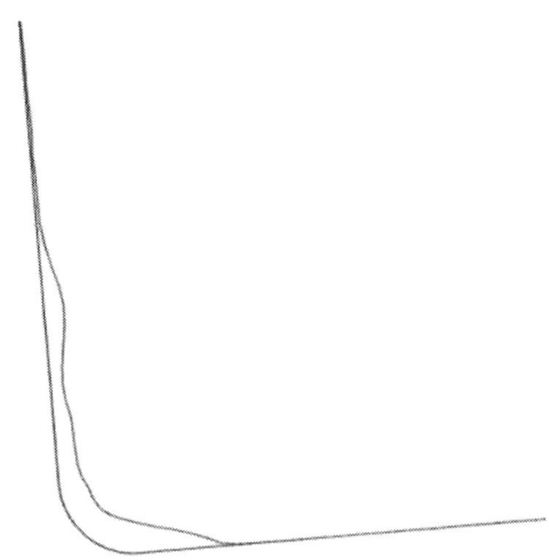

Fig. 6. New tool (blue) and worn tool (red) overlayed

Residual stress can be evaluated after a single or secondary cut by numerical removal of the chip and tool after the prescribed cut length within the software. A procedure allowing for thermo-mechanical relaxation is then conducted. Vibrations are damped out from the dynamic explicit solution, and heat is allowed to dissipate sufficiently for the workpiece to attain the initial prescribed room temperature. Both first and second cut simulations were conducted to evaluate the effect of the machining-induced residual stresses.

4 Results

The results of the chip morphology, tool wear, and XRD residual stress measurements and the numerically predicted residual stresses will be discussed in the following sections.

4.1 Chip morphology

Laubscher et al. [6] describe that titanium alloys develop segmented chips, even at low cutting speeds. The chip segmentation for each of the cutting speeds as well as the coolant and dry conditions for the experimental work, is herein described. The table below indicates the chips are segmented (GS<1) for all cutting speeds. The predicted results indicate a minor deviation of 2% from the Ti6Al4V outside turned bar, running at a cutting speed of 100 m/min without coolant, for the first cut, compared to the measured results. The most significant deviation of 34% was found for the predicted cutting speed of 30 m/min without coolant for the first cut compared to the measured results. The predicted results for the first and second cuts show a reasonable correlation with the measured results, considering that a 3D oblique outside turned sample is compared to a 2D plane strain orthogonal numerical model for predictions.

Table 1. Results of the chip segmentation ratio for the measured and predicted chips.

Cutting Speed (m/min)	Cool-ant/Dry	Measured chip seg-mentation Ratio	Numerical chip segmen-tation Ratio - 1st cut	Numerical chip segmen-tation Ratio - 2nd cut	Devia-tion - 1st cut	Devia-tion - 2nd cut
30	Coolant	0.50	0.61	0.52	22%	4%
30	Dry	0.47	0.63	0.61	34%	29%
100	Coolant	0.55	0.52	0.50	-4%	-8%
100	Dry	0.63	0.64	0.58	2%	-7%
200	Coolant	0.73	0.58	0.52	-21%	-29%
200	Dry	0.55	0.64	0.58	17%	7%

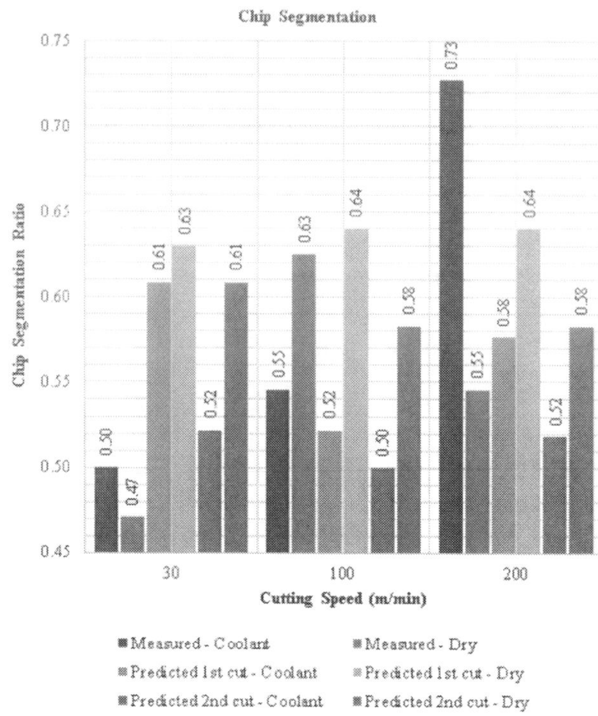

Fig. 7. Results plot of the chip segmentation.

The numerical results predict "segmented chips".

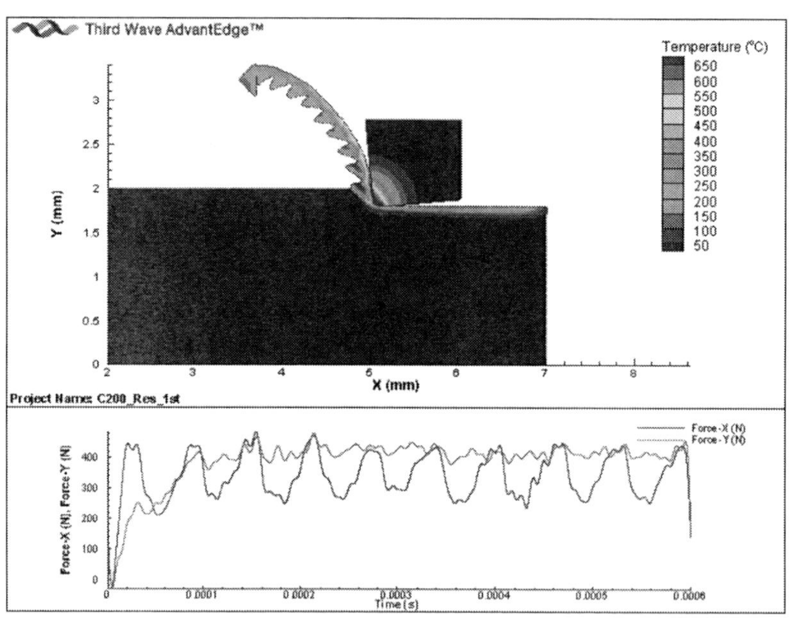

Fig. 8. The numerical model predicts segmented chips as seen in this contour plot of the temperatures and the "cyclic" nature of the cutting forces.

4.2 Worn tool measurements

The worn tool measurements for each cutting speed and the coolant and dry conditions for the experimental work are described herein. The table below indicates the KB and KF distances (described above) for all cutting speeds. Outeiro et al. [2] found that for predicting orthogonal machining of AISI 316L stainless steel, after a second cut and the inclusion of the worn tool's geometry, increased compressive surface residual stresses along the length of the cut. The predicted results indicate the minor deviation of 6% from the Ti6Al4V outside turned bar, running at 30 m/min without coolant, for the first cut compared to the measured results. The most significant deviation of 65% was found for the predicted cutting speed of 30 m/min with coolant, for the first cut, compared to the measured results. The predicted results for the first and second cuts show a reasonable correlation with the measured results, considering that a 3D oblique outside turned sample is compared to a 2D plane strain orthogonal numerical model for predictions.

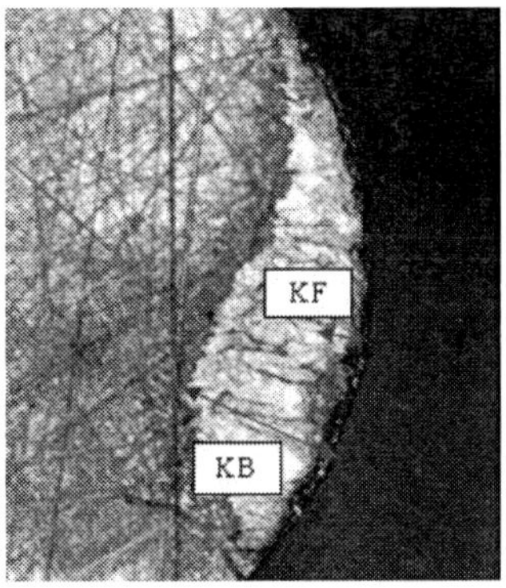

Fig. 9. Image of the tool wear showing the "built-up" edge of titanium in the worn crater.

Table 2. Results of the tool wear distances for KB and KF for the measured and predicted tools.

Cutting Speed (m/min)	Cool-ant/Dry	Measured - KF (mm)	Measured - KB (mm)	Numeri-cal - KF (mm)	Numeri-cal - KB (mm)	Devia-tions - KF	Devia-tions - KB
30	Coolant	0.02	0.14	0.03	0.17	**65%**	21%
30	Dry	0.05	0.20	0.05	0.22	**-6%**	12%
100	Coolant	0.04	0.17	0.06	0.20	38%	15%
100	Dry	0.03	0.14	0.04	0.15	30%	10%
200	Coolant	0.03	0.18	0.03	0.20	10%	8%
200	Dry	0.03	0.21	0.04	0.20	27%	-7%

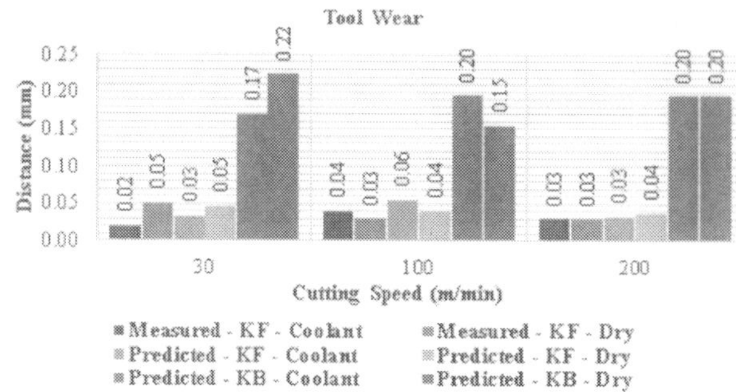

Fig. 10. Results plot of the tool wear.

4.3 Residual stress measurements

Surface residual stresses were evaluated by the Nelson Mandela University (eNtsa) using non-contact X-ray diffraction. The surface residual stress measurements were measured with a ProtoTM i-XRD combo with a copper x-ray source, using a 2mm round aperture. These experimental results were compared with those of the numerical work discussed in this paper.

Table 3. XRD measurement parameters.

Voltage (kV)	Current (mA)	Filters	Bragg angle	Exposure time (sec)	No. Exposures/angle	Psi osc
20	4	Ni	139.69 ° on 213 hkl	3	12	3 °

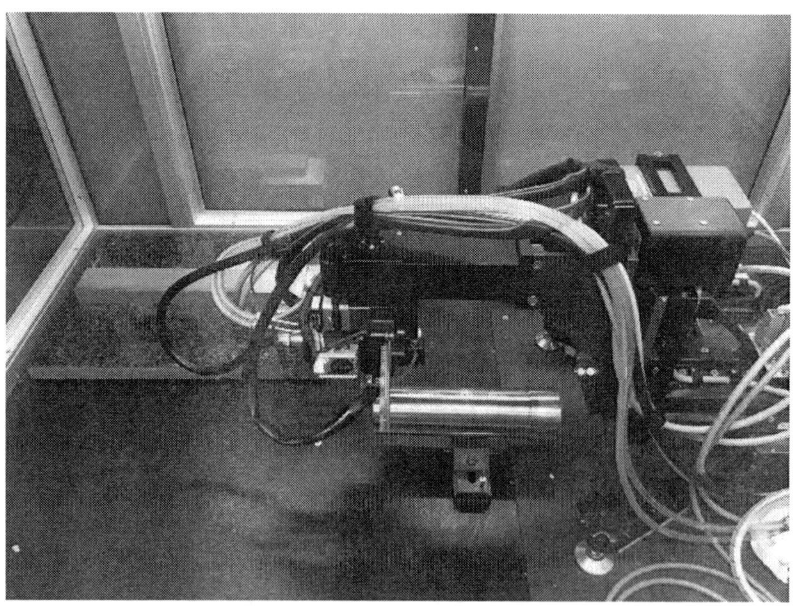

Fig. 11. Experimental setup for the XRD residual stress measurements.

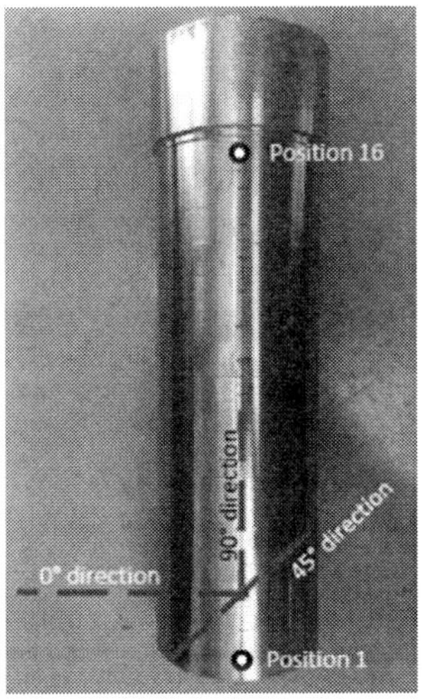

Fig. 12. Direction of the direct stress components.

5 Results

This study's residual stress vs depth was extracted from the centre (1 mm from the chip formation zone) of the total cutting length of 2 mm. This extracted distance is a deviation from the recommended procedure as described by Outeiro [2]. The predicted surface residual stresses were averaged over 18 microns below the surface to compare the XRD measured and predicted residual stresses. These residual stresses were only extracted for a single position along the cutting length. Hence no averaging of the results occurred in the cutting direction. Ozel et al. [1] found the surface residual stresses compressive for Ti6Al4V at different cutting speeds. The predicted results indicate a minor deviation of 11% from the Ti6Al4V outside turned bar, running at 200 m/min with coolant, for the second cut compared to the measured results. The most significant deviation of 56% was found for the predicted cutting speed of 100 m/min without coolant for the second cut compared to the measured results. The predicted results for the first and second cuts show a reasonable correlation with the measured results, considering that a 3D oblique outside turned sample is compared to a 2D plane strain orthogonal numerical model for predictions.

Table 4. Results of the measured and predicted residual stresses.

Cutting speed (m/min)	**Cool-ant/Dry**	**Measured stress (MPa)**	**Stress for the 1st cut (MPa)**	**Stress for the second cut (MPa)**	**Deviation for the 1st cut (%)**	**Deviation for the 2nd cut (%)**
30	Coolant	-400	-357	-456	-11%	14%
30	Dry	-290	-432	-338	49%	17%
100	Coolant	-327	-278	-247	-15%	-25%
100	Dry	-253	-350	-395	38%	**56%**
200	Coolant	-363	-212	-402	-42%	**11%**
200	Dry	-310	-208	-395	-33%	28%

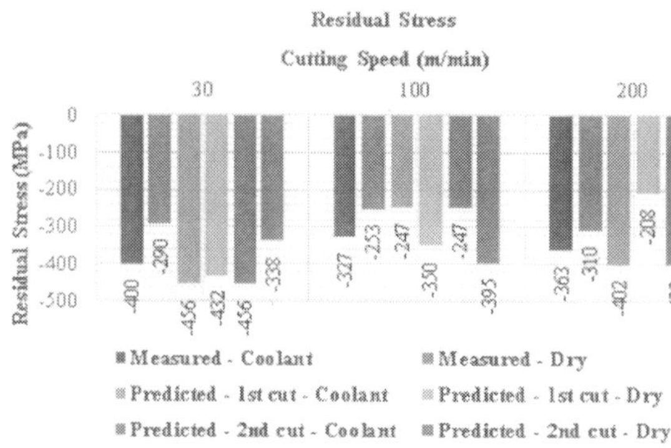

Fig. 13. "Measured" and "Predicted" near-surface residual stresses.

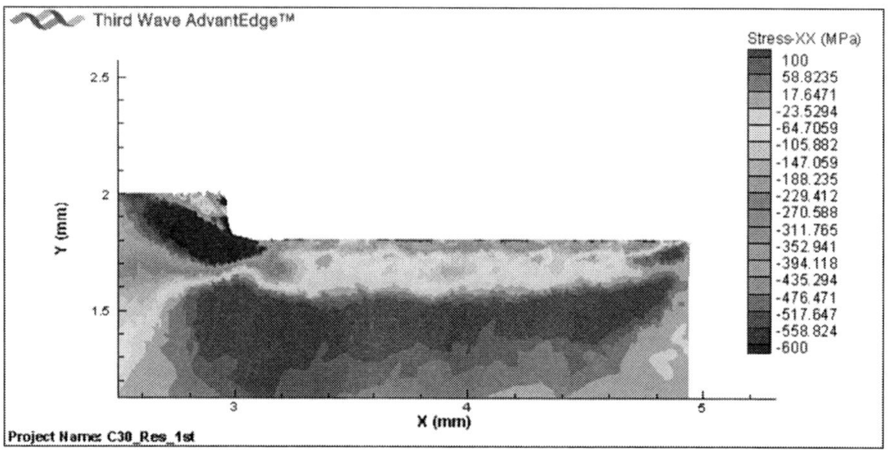

Fig. 14. Contour plot of the residual stresses, in the cutting direction for the cutting speed of 30 m/min with coolant.

6 Conclusions

This paper presents a numerical simulation of outside turning of Grade 5 titanium alloy, considering tool wear and multiple cuts, to predict residual surface stresses. The non-linear and dynamic FEA results are compared and validated against experimental work. A dedicated phenomenological material model had to be used to correctly capture the observed experimental behaviour of the machining-induced residual stresses. This constitutive model is based on experimental work conducted and published data for Grade 5 [7]. Reasonable agreement between experimental work and the simulation results as far as chip morphology and tool wear is achieved. The simulation presents a good approximation of the surface residual stresses at several cutting speeds. It correctly presents a fully compressive residual stress field. The simulation correctly captures that the maximum compressive residual stress is located a finite distance below the surface. It does, however, not accurately capture the cutting speed sensitivity as observed in the literature. The difficulty in predicting the changes in the residual stresses with different cutting speeds are believed to be linked to the thermal modelling component of the simulation that assumes a simple Coulomb friction model with a single value for the coefficient of friction across the insert-to-workpiece interface. A constant overall heat transfer coefficient and an assumed heat transfer over the tool-workpiece interface without considering the effects of temperature changes. A more general-purpose non-linear finite element code where the user controls the thermal aspects (heat generation and dissipation) of the modelling will be required to investigate this further. The modelling procedure, as recommended by Outeiro [2] as followed in this paper, demonstrates the relevance of considering tool wear and multiple cuts before extracting the residual stresses from the numerical model.

References

1. T. Ozel, D. Ulutan, *Predication of machining-induced residual stresses in turning of titanium and nickel based alloys with experiments and finite element simulations*, CIRP Annals, **61**, Issue 1, 547-550 (2012)
2. J. Outeiro, K. Ee, O. Dillon, P. Wanigarathne, I. Jawahir, *Some Observations on Comparing the Modelled and Measured Residual Stresses on the Machined Surface Induced by Orthogonal Cutting of AISI 316L Steel*, Proceedings of the 9th CIRP International Workshop on Modeling of Machining Operations, 11–12 May, Bled, Slovenia, 475-481 (2006)
3. G. Styger, R.F. Laubscher, G. Oosthuizen, *Effect of Constitutive Modeling During Finite Element Analysis of Machining-induced Residual Stresses in Ti6Al4 V*. Procedia CIRP. **13**, 292-431 (2014)
4. J. Sosa, D. Huber, B. Welk, H. Fraser, *Development and application of MIPAR™: a novel software package for two- and three-dimensional microstructural characterisation*, Integrating Materials, **3**, 123-140 (2014)
5. L. Čerče, F. Pušavec, and J. Kopač, *A New Approach to Spatial Tool Wear Analysis and Monitoring*, Journal of Mechanical Engineering **61**, Issue 9, 489-497 (2015)
6. T. D. Marusich and M. Ortiz, *Modelling and Simulation of High-Speed Machining*, International Journal of Numerical Methods in Engineering, **38**, 3675-3694 (1995)
7. R. Laubscher, G. Styger, and G. Oosthuizen, *A numerical analysis of machining-induced residual stresses of grade 5 titanium alloy*, R & D J. South Afr. Inst. Mech. Eng. **30**, 39–46 (2014)
8. AdvantEdge™ *USER MANUAL*, 1-340 (2020)
9. S. K. Mishra, S. Ghosh, S. Aravindan, *Temporal and spatial crater wear prediction of WC/Co tools during dry turning of Ti6Al4V alloy*, Wear, **448–449**, 203-229 (2020)

Numerical Simulation of Blast Loaded CFRP Retrofitted Steel Plates

Mujtaba M. Shuaib[1], *Steeve* Chung Kim Yuen[1*], and *Gerald* N. Nurick[1]

[1] Blast Impact and Survivability Research Unit (BISRU), Department of Mechanical Engineering, University of Cape Town, Private Bag, Rondebosch 7701, South Africa

Abstract. This paper reports on the results of a numerical study to simulate the response of carbon fibre reinforced polymer (CFRP) retrofitted steel plates to applied blast loads using finite element software, LS-DYNA. The results of the simulation were validated against plate response and magnitude of deformation obtained from previous experiments. The uniform blast load was generated in the experiment by detonating a cylindrical charge down the end of a square tube. The finite element code LS-DYNA was used to simulate the structural response of the respective blast structures. For the numerical model, the blast load was simulated using the mapping feature available in LS-DYNA for the multi-material arbitrary Lagrangian-Eulerian (MM-ALE) elements which significantly reduced the size of the air domain in the model. The simulations showed a satisfactory correlation with the experiments for the blast results and post-failure deformations that occurred in CFRP retrofitted steel plates.

1 Introduction

Structural retrofitting with fibre reinforced polymer (FRP) composites has been widely used to repair a range of deteriorated existing structures such as bridges and pipelines in oil and gas industry. FRP retrofitting has several advantages, including the high stiffness- and strength-to-weight ratios, good corrosion resistance, and shape flexibility of FRP laminates [1-3]. The use of FRP for strengthening is common in concrete structures, and it has recently been used to retrofit existing steel structures. Several studies have shown that FRP retrofitting enhances the flexural capabilities, ultimate loading capacity, and lateral impact resistance of steel structural elements [2]. Carbon fibre reinforced polymers (CFRPs) are one of the most commonly used FRPs, and they are preferred for improving the resistance of structural members due to their significantly higher elasticity modulus and tensile strength [4].

FRP retrofitting can provide a viable method for protecting existing structures from extreme loading conditions such as impact and blast. The resistance of FRP retrofitted structures to blast loads has been investigated with a focus on concrete and masonry structural elements such as slabs and walls [5, 6]. For metals, despite significant study on the structural response of fibre-metal laminates (FMLs) to blast loads [7], few studies have been found to investigate the potential of FRP retrofitting to mitigate the damage of blast-loaded FRP retrofitted metallic structures [2, 8]. Bambach et al. [9] conducted experimental and analytical

* Corresponding author: steeve.chungkimyuen@uct.ac.za

studies on the energy-absorbing characteristics of CFRP strengthened aluminium beams under the transverse blast. A remarkable decrease up to 30% was observed in deflections due to CFRP strengthening of the aluminium beams. Previous research [10, 11] has demonstrated through experiment and numerical analysis that reinforcing steel pipes with sufficient CFRP thicknesses considerably increases their resilience to external blast.

Numerical simulation using finite element method (FEM) has been widely used to assess the mechanical response of fibre-based polymer composites to impact and blast loads [7]. For blast loading, research has been carried out using several commercial FEM software packages (e.g., LS-DYNA and ABAQUS) to simulate the response of composite structures to impulsive loadings [12]. Karagiozova et al. [13], Sitnikova et al. [14] and Li et al. [15] investigated numerically the response of fibre-metal laminates to high impulsive blast loading, and found a good agreement between numerical and experimental modes of failure.

This paper evaluates numerically the potential of FRP retrofitting to mitigate deformations in blast-loaded metal plates while maintaining additional strength. LS-DYNA was used to develop a numerical model that was validated with previous experiments. The paper discusses a number of potential finite element methods in LS-DYNA for simulating the response of composite structures subjected to blast loads.

2 Experiment Setup

The configuration of the experiment is shown in Fig. 1. Specimens of CFRP retrofitted steel plate were prepared for experimental testing. In addition to retrofitted specimens, bare steel plates were tested as a baseline to assess the performance of the retrofitted CFRP. The steel plates used were made of high-strength low-alloy (HSLA) steel and cut in 300 x 300 mm quadrangles with 1mm thickness. The steel plates used were made of high-strength low-alloy (HSLA) steel (205.9 GPa modulus of elasticity, 275.4 MPa yield strength, and 353.1 MPa ultimate tensile strength – obtained from uni-axial tensile tests). An area of 200 x 200 mm was exposed to the blast loading where the outer edges were clamped. The uniform blast loads were generated by detonating plastic explosives (PE4) ranging between 7g and 19g in 40 mm cylindrical discs. 1g of explosive was used to attach the detonator to the centre of the disc of explosive. Unidirectional (UD) CFRP composite sheets were used and bonded to one side of the steel plate. Two fibre-configurations of 3 and 4 layers of carbon fibre laid longitudinally were applied in the retrofitted specimens. Epoxy-based resin was used between each layer to maintain adhesion in steel-CFRP and CFRP-CFRP interfaces.

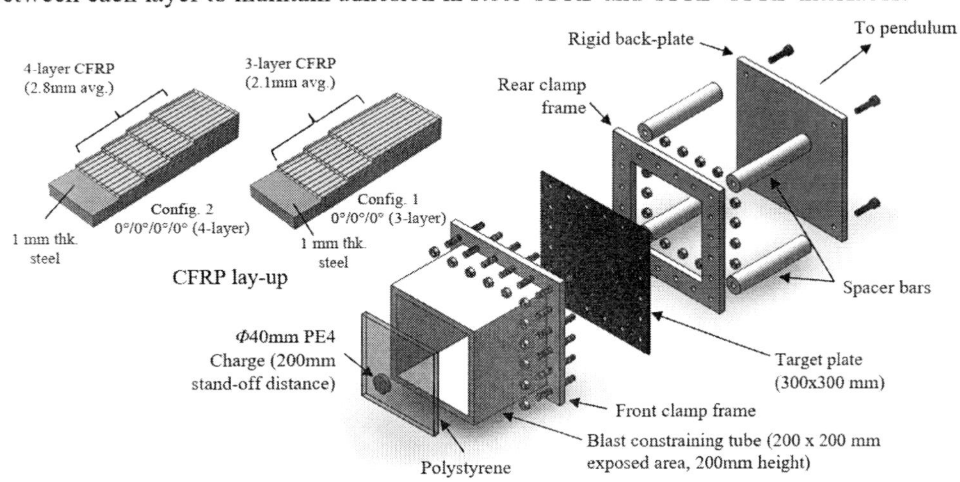

Fig. 1. Schematic representation of the experimental setup

3 Model Description

LS-DYNA finite element software was used to numerically simulate the structural response of CFRP retrofitted steel plates to uniform blast loading. Fig. 2 illustrates the model components, which mainly consisted of a target plate (bare steel or CFRP retrofitted steel plate), an air domain, an explosive charge, a steel clamping frame, and steel bolts for frame fixing. To reduce computational time, one-quarter of the plate was modelled, and one-quarter symmetry boundary conditions were applied to nodes intersecting the symmetry planes. The clamping frame and bolts were considered as rigid parts with fixed boundary conditions.

Blast pressures were applied on a reduced air domain using the mapping feature offered in LS-DYNA for arbitrary Lagrangian-Eulerian (ALE) elements [16, 17]. The ALE mapping approach was applied in a manner similar to the ALE coupling method developed by Salvik [18], where the only the air adjacent to the Lagrangian structures was considered. As illustrated in Fig. 2 (a), the air-blast was first simulated by explicitly modelling the air and explosive with the multi-material arbitrary Lagrangian-Eulerian (MM-ALE). Pressure data from a selected ALE mapping layer in the simulated full air-domain model (Fig. 2(a)) were saved into a mapping output file using the ALE mapping feature. The generated mapping file was then used as an input in the reduced air-domain model, with the saved pressure imposed in the mapping layer (Fig. 2 (b)). Fluid-structure interaction (FSI) was used for communication between the target plate and the ALE domain. To simulate the openings in the blast constraining tube used in the experiment, non-reflecting boundary conditions were assigned at the top and bottom of the air-domain.

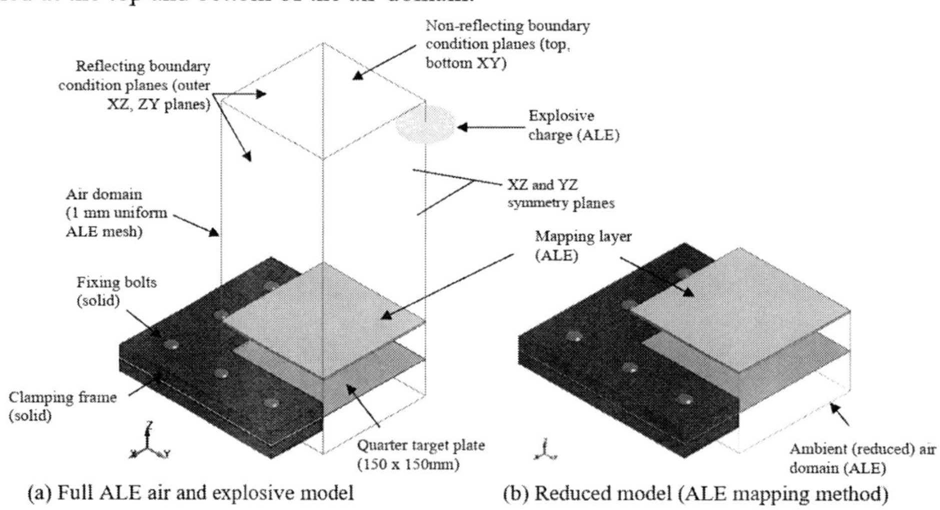

Fig. 2. Model components with quarter symmetry boundary conditions

3.1 Modelling of CFRP retrofitted steel plate

CFRP retrofitted steel plates were modelled using the *PART_STACKED_ELEMENTS (PSE) keyword. The PSE feature enabled for the creation of a stacked model for shell-like structures, which was used to model composite structures such as sandwich plate systems and composite laminates [17, 19]. Fig. 3 illustrates the retrofitted plate components.

The CFRP retrofitted steel plate was modelled as a sequence of shell elements and solid elements using the PSE feature. For plate modelling, the steel and CFRP layers were meshed with 2x2 mm 4-node Belytschko-Tsay shell elements (ELFORM=2) with 1 mm and 0.176 mm thicknesses, respectively. The 2x2 mm element size was selected based on the findings of a previous mesh sensitivity research [20] and numerical simulation [21] of plates subjected to near-field blast loads. To capture layer delamination, the cohesion between the shell layers was modelled by inserting 2x2x0.15 mm 8-node solid cohesive elements between the steel plate and CFRP layers. The steel frame and bolts were modelled with coarse mesh of solid elements with typical side length of 5mm. *CONTACT_AUTOMATIC_NODES_TO_SURFACE / SURFACE_TO_SURFACE cards were used to simulate the contact between the target and the bolts, and the contact between the target and the frame.

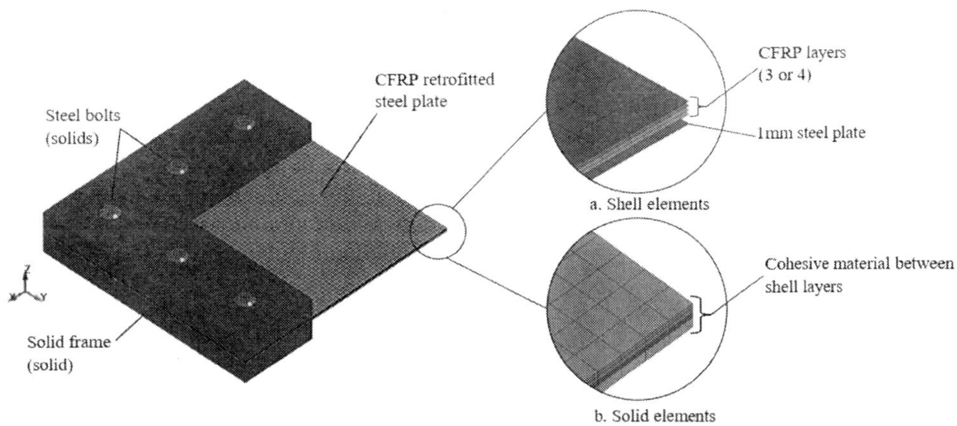

Fig. 3. Components of CFRP retrofitted steel plate modelled using PSE feature

3.1.1 Material properties for steel material

The Johnson-Cook (JC) plasticity material model (MAT_015) was selected to model the steel shell elements. The flow stress was expressed by the JC model as in Eq. 1, where σ_y is the von Mises flow stress, $\bar{\varepsilon}^p$ is the effective plastic strain, $\dot{\varepsilon}^* = \dot{\varepsilon}/\dot{\varepsilon}_0$ is the dimensionless plastic strain rate for $\dot{\varepsilon}_0 = 1.0$ s^{-1}, and T^* is the homologous temperature. The material constants in Eq. 1 are A (the yield stress), B and n (represents the effects of strain hardening), C (the strain rate constant) and m (temperature constant).

The average yield stress of the quasi-static uniaxial tensile tests was used to calculate parameter A. The parameters B and n were calculated by fitting the strain hardening part (which includes A, B and n) of Eq. 1 to the experimental true stress-strain curve, with A set to the average yield stress of steel. The JC input data used for A, B and n 267.8 MPa, 23.4 MPa and 0.65 respectively. Values of the strain rate and temperature constants were taken from Refs. [22, 23] as further data for the steel used were not available. The input data used for C and m were 0.014 and 1.03 respectively.

$$\sigma_y=(A+B\bar{\varepsilon}^{p^n})(1+C\ln\dot{\varepsilon}^*)(1+T^{*m}) \quad (1)$$

3.1.2 CFRP modelling

To simulate the mechanical behaviour of the unidirectional CFRP material, the enhanced composite damage material model (MAT_054) was used. Previous research [24, 25] used the MAT_054 material model to evaluate the response of FRP composite materials to impact and blast loads, demonstrating a good degree of accuracy. In MAT_054, the Chang-Chang failure criterion was implemented, and the corresponding layer in the element was removed after the maximum fibre tensile strain was met. For modelling the cohesive elements, the cohesive mixed-mode material (MAT_138) was selected, which was characterised by a bilinear traction-separation law (TSL) with a quadratic mixed mode delamination criterion. In this study, experiments were not carried out to characterise the mechanical properties of the CFRP and epoxy materials (e.g., tensile strength, maximum tensile strain). The MAT_054 input data for the unidirectional CFRP from Ref. [26] and the MAT_138 input data for epoxy from Ref. [27] were applied in the model.

3.2 Blast modelling – ALE mapping

The ALE mapping method in LS-DYNA was used to simulate the blast load located close to target using the MM-ALE approach while saving significant computational time. Using the mapping approach, blast pressures obtained by explicit MM-ALE modelling of explosive and air were applied to a layer of ALE elements (known as the mapping layer) facing the explosive charge. The pressures of ALE elements obtained in the mapping layer were saved and applied in another model that included the remaining of the air domain and the structure (known as the reduced model). The ALE mapping approach was implemented in two stages: (i) generating pressure data for a selected ALE mapping layer in a full MM-ALE model (explicit air and explosive); (ii) using the pressure data received in the first step from the ALE mapping layer and apply it to a model that includes the remaining MM-ALE air domain (air-only). A different LS-DYNA model was used for each stage.

3.2.1 First stage – Blast pressure generation in ALE mapping

A full ALE air-domain model was modelled to generate the blast pressures resulting from explicitly modelling the explosive with the geometrical properties considered in the experiment. The *INITIAL_VOLUME_FRACTION_GEOMETRY was used to define the air and explosive materials in the same mesh. Fraction background of ALE explosive material was defined as cylinder with 40mm diameter and heights that corresponded to those used in the experiment, as shown in Fig. 2 (a).

A uniform fine mesh of hexahedral elements and side length of 1mm was used to model the air domain (2,430,000 8-node solid element). The element size for the air domain was determined by the findings of a previous mesh sensitivity study [20]. The linear polynomial equation of state (EOS) was used to model air as an ideal gas, while the MAT_008 and John-Wilkins-Lee EOS were used to characterise the PE-C4 explosive. The MAT_008 and JWL EOS input data used were similar to those in Ref. [22].

The mapping layer was situated 18mm away from the target plate and was modelled as a one-element thick layer of ALE elements. *BOUNDARY_ALE_MAPPING (BAM) keyword was used to write a mapping file that stores the blast pressure histories of the mapping layer ALE elements. IVOLTYP=2 was set in BAM keyword to direct the software to write the data of ALE elements, included in a box shape, in the mapping file. A termination time of 0.25ms was used to write the data in the mapping file, which was selected to ensure that the blast pressures are fully generated.

3.2.2 Second stage – Blast pressure application in ALE mapping

In the second stage, the output mapping file generated from the first stage was used as input mapping file in the reduced air-domain model The pressure data saved in the mapping file were imposed in the mapping layer (Fig. 2 (b)) and used as a source for adjoining ALE air elements. In the reduced model, a uniform mesh of hexahedral elements with a side length of 1mm was also used. The ALE air domain surrounded the Lagrangian structure, and fluid structure interaction (FSI) controlled the interaction of the generated blast pressures with the structure. The use of the mapping technique resulted in a significant reduction in model size, with the number of ALE air domain elements reduced to 230,000 (by 95%).

4 Results and Discussions

4.1 Blast pressure results

Fig. 4 illustrates the pressures for the blast wave transmitted by the detonation of a 9g charge at different time instants in the mapped model. It was found from the simulation that the pressure values produced from the reduced air domain and applying the mapped pressure layer are nearly identical to those obtained from the full air domain. The reflected pressures on the plate mid-point recorded from the simulation for 7 g, 9 g and 13 g charges is shown in Fig. 5 (a). In the simulation, pressure curves showed a positive pressure phase with a peak reflected pressure that occurred rapidly (arrival time of 42-45 μs) and for very short durations (200 μs approx.). The predicted peak pressures increased with explosive charge mass, ranging from 87.1 to 147.9 MPa.

The pressures were validated by comparing impulses reflected by the target plate with experimental results. The reflected impulse was evaluated using a rigid target plate in the model. The simulated impulse-time histories for the 7 g, 9 g, and 13 g charges are shown in Fig. 5 (b). The simulated impulse values are shown in Table 1, and they were found to correlate well with the experiment results, with a variation margin of less than 10%.

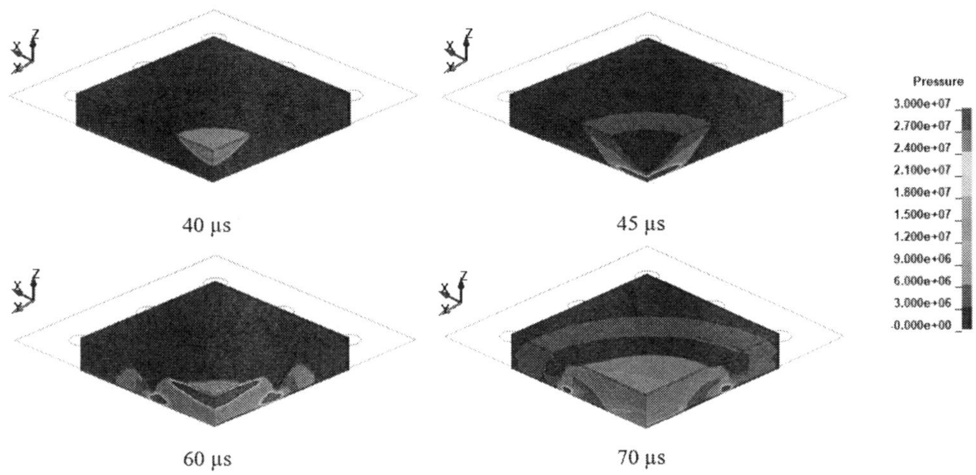

Fig. 4. Blast wave pressures (in Pa) in the mapped model at various time instants for 9g charge mass

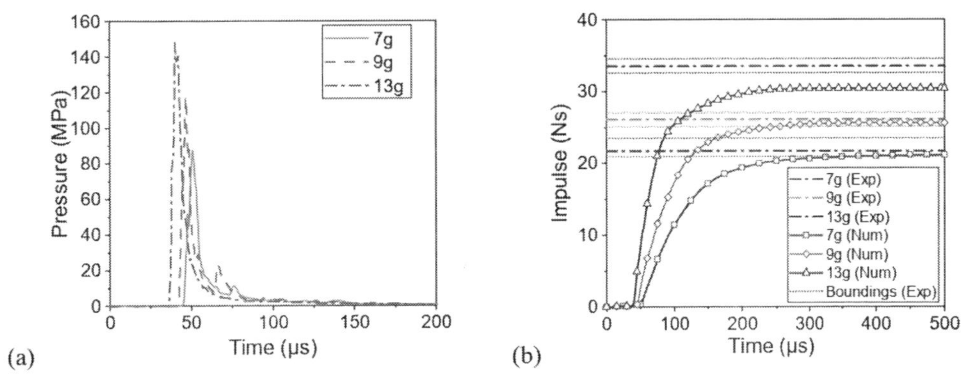

Fig. 5. Simulation results of the blast wave properties: (a) pressure-time histories at the centre of the target plate, (b) Numerical (Num) and Experimental (Exp) impulse-time histories

4.2 Deformations in bare steel plates

The failure mode observed in the experiment and the simulation was the Mode I (large inelastic deformation) in square uniformly blast loaded plates [28]. The simulated deformed shapes of bare steel specimens in the blast testing are plotted in Fig. 6 (a). The plastic strains in the simulation shown in Fig. 6 (b) predicted the yielding regions, including the plate middle region and the at the boundaries, as well as the elongation in bolt holes. Figure 7 (a) depicts the residual deformation of a bare specimen subjected to a 13g explosive mass blast. Fig.7 (b) presents the simulated deflection histories of bare specimens. The simulation showed that steel plate behaved almost linearly with time to a large deformation, reaching a maximum and then exhibiting a minor amount of residual elastic vibration around a mean displacement after around 500 μs. The results of a comparison of the simulated mid-point displacement and the experimental data are shown in Table 2. It can be observed that the results show very good correlation with a margin of variation less than 5%.

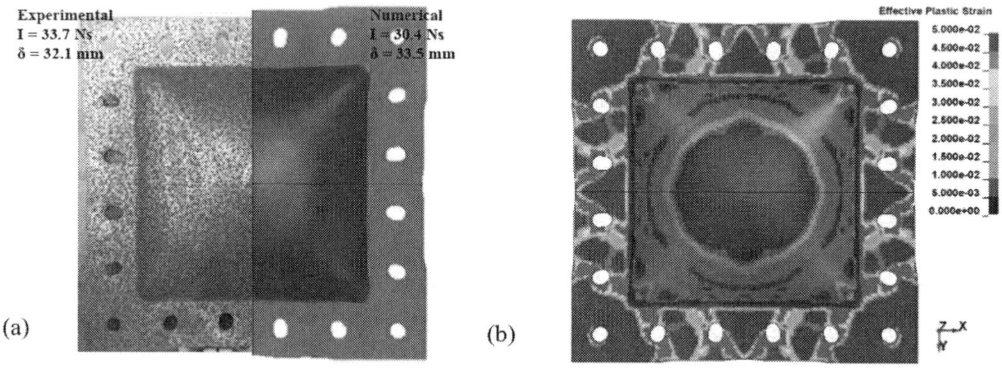

Fig. 6. Comparison of numerical simulation results with experimental results for 13g charge: (a) overlook view; (b) effective plastic strains

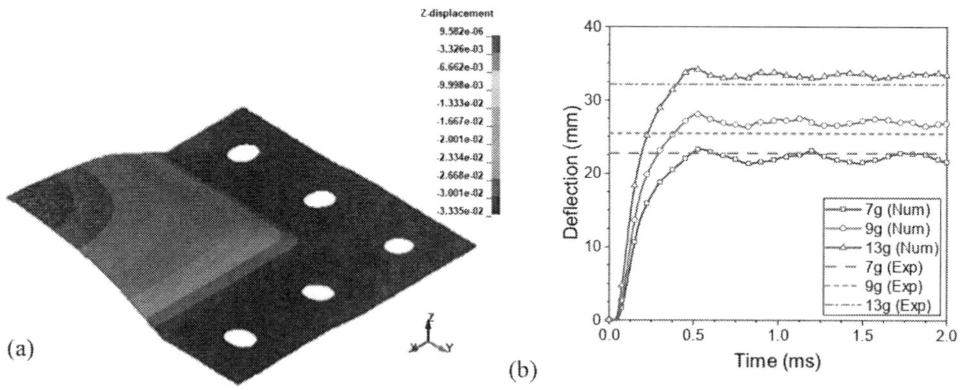

Fig. 7. Numerical results for bare specimens: (a) final deformed shape – quarter plate for 13g (impulse = 30.4 Ns, displacement in metres), (b) mid-point displacement histories

4.3 Deformations in CFRP retrofitted steel plates

4.3.1 Comparison of CFRP retrofitted steel plate deformation

For the CFRP retrofitted steel plates, numerical simulations were performed, and predictions of plate deformation were compared to the experiment for validation. In the experiment and simulation, the steel plate profiles in the CFRP retrofitting specimens all experienced Mode I failure, but with less transverse deflections than bare specimens. Fig. 8 illustrates comparison between the experimental and the simulated deformation of steel in 3-layer CFRP (CB) and steel blast-faced (SB) retrofitted specimens for 13g charge. The experimental results showed that the bonded CFRP effectively reduced the plate deformation. For 3-layer CFRP retrofitted specimens, the plate deflection decreased relative to bare plates by 42% to 52% (9.6-14.1 mm).

Further reduction in deflections was observed in 4-layer CFRP retrofitted plates relative to bare and 3-layer CFRP retrofitted specimens, where 4-layer CFRP retrofitted specimens demonstrated a further reduction in deflection relative to bare plates by 50% to 64% (13.2-19.2mm). The effect of the blast-facing material (i.e., CFRP or steel) was more evident in 4-layer CFRP retrofitting specimens, where the mid-point displacement of SB specimens was 0.1 to 3.3mm lower than that of CB specimens.

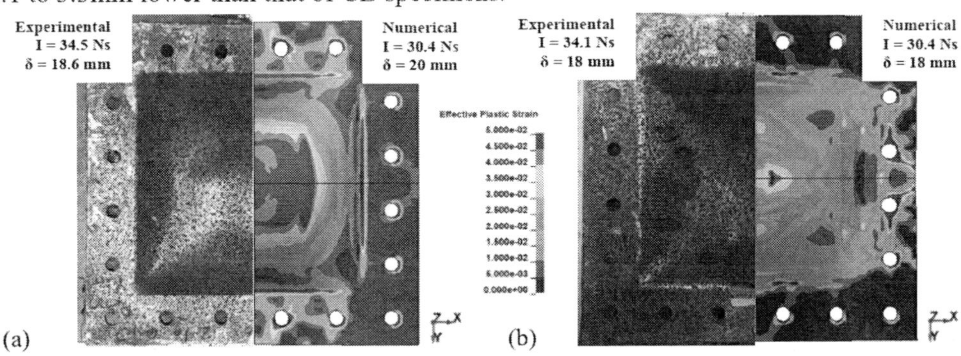

Fig. 8. Comparison of deformation between experimental and numerical results for 3-layer CFRP retrofitted plates for 13g: (a) CFRP blast-faced specimen (bottom); (b) steel blast-faced specimen (top)

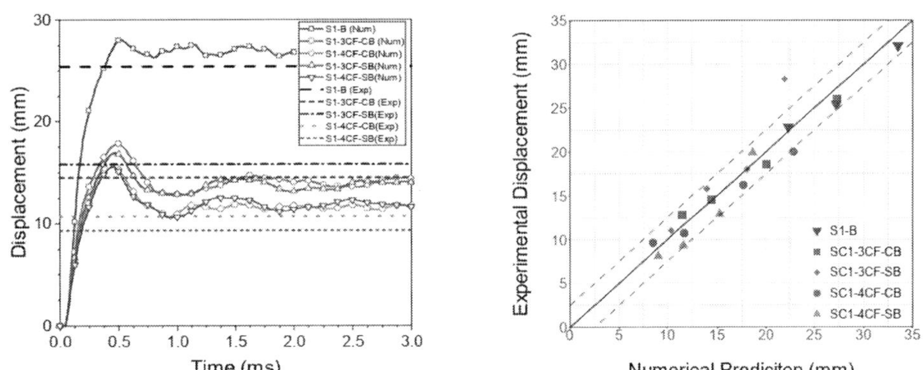

Fig. 9. Numerical results (a) mid-point displacement histories of bare and CFRP retrofitted specimens for 9g charge, (b) comparison of experimental and numerical mid-point displacements

Similar findings were obtained from the simulation in general, as illustrated in Fig. 9 (a) for the comparison of the mid-point displacement histories obtained from the simulations of bare and CFRP retrofitted steel plates for 9g charge. The displacement trends in Fig. 9 (a) indicated that the steel plates in the retrofitted panels obtained lower maximum deformation and residual elastic vibration levels than bare steel specimens due to CFRP retrofitting. The simulation for the mid-point deflection is plotted against the experimental data and is shown in Fig. 9 (b). The numerical model reasonably predicted the mid-point deflection of CFRP retrofitted plates, with 90 percent of the simulated displacements estimating the experimental within a ±2.5mm (or 20%) variation range. The numerical results followed a similar pattern as the experimental data. Table 1 summarises the experimental and numerical results.

Table 1. Summary of experimental and numerical results

Specimen	Mass of PE4 (g)	Blasted side	CFRP No. of layers	Experimental Impulse (Ns)	Experimental mid-point displacment (mm)	Numerical Impulse (Ns)	Numerical mid-point displacment (mm)	Displacement variation (%)
S1-B-07	7	Steel	-	21.7	22.8	21.1	22.3	-2.2
S1-B-09	9	Steel	-	25.1	25.4	25.7	27.2	7.1
S1-B-13	13	Steel	-	33.7	32.1	30.4	33.5	4.4
SC1-3CF-CB-07	7	CFRP	3	20.9	12.8	21.1	11.5	-10.5
SC1-3CF-CB-09	9	CFRP	3	25.5	14.5	25.7	14.4	-1.0
SC1-3CF-CB-13	13	CFRP	3	34.5	18.6	30.4	20.0	7.7
SC1-3CF-CB-16	16	CFRP	3	38.8	26	35.4	27.3	4.8
SC1-3CF-SB-07	7	Steel	3	21.3	11	21.1	10.4	-5.8
SC1-3CF-SB-09A	9	Steel	3	27.6	15.8	25.7	14.0	-11.6
SC1-3CF-SB-09B	9	Steel	3	25.7	14.3	25.7	14.0	-2.3
SC1-3CF-SB-13	13	Steel	3	34.1	18	30.4	18.0	0.2
SC1-3CF-SB-16	16	Steel	3	36.9	28.3	35.4	21.9	-22.4
SC1-4CF-CB-07	7	CFRP	4	23.5	9.6	21.1	8.5	-11.6
SC1-4CF-CB-09	9	CFRP	4	27.1	10.7	25.7	11.7	9.3
SC1-4CF-CB-13	13	CFRP	4	33.9	16.2	30.4	17.7	9.0
SC1-4CF-CB-16	16	CFRP	4	38.6	20	35.4	22.9	14.5
SC1-4CF-SB-07	7	Steel	4	21.1	8.1	21.1	9.0	11.6
SC1-4CF-SB-09	9	Steel	4	25.9	9.3	25.7	11.6	24.9
SC1-4CF-SB-13	13	Steel	4	33.9	12.9	30.4	15.2	18.1
SC1-4CF-SB-16	16	Steel	4	40.1	19.9	35.4	18.6	-6.5

4.3.2 Damage modes in CFRP

For CFRP, the experimental results showed combined FRP failure modes due to uniform blast loads including metal-FRP debonding, delamination/tearing, fibre rupture and matrix failure. Fig 10 (a) shows the typical FRP modes as obtained from the experiment. In CFRP blast-faced specimens, CFRP demonstrated a complex response where metal-fibre debonding, delamination and longitudinal tearing modes were exhibited at lower impulses. For steel blast-faced specimens, the damage shown by CFRP included metal-fibre debonding, longitudinal tearing at the plate boundaries parallel to fibres as shown in Fig. 10 (a). Fibre fracture in the plate boundaries crossing the fibre direction was observed at higher impulses.

The simulated model predicted the major failure modes in CFRP including the matrix failure, FRP fracture and metal-FRP debonding. For example, Fig. 10 (b) depicts the numerical deformation in a CFRP blast-faced retrofitted specimen for 16g, where the model predicted metal-FRP debonding that agreed with experiment observations. Despite capturing the major failure modes, the simulation underestimated FRP fracture at the boundaries observed in steel blast-faced specimens, and metal-FRP debonding was extended across a wider area than the experiment revealed. The discrepancy in CFRP damage observed in the simulation could be attributed to the fact that values used in various parameters of the CFRP and adhesive material models (e.g., failure strain) used in the simulation might differ from the actual mechanical properties of these materials.

However, the simulation was still able to provide a better understand for the failure mechanism. Fig. 11 shows a comparison of the simulated and experimental deformation in steel blast-faced retrofitted specimens for 16g. The CFRP damage occurred in the experiment at the plate boundaries in the form of FRP fracture in the direction perpendicular to fibres and delamination in the fibre direction. The numerical model captured the failure zone, where the damage was indicated by the von Misses stress as shown in Fig. 11 (b), which corresponded with the experimental observation in general.

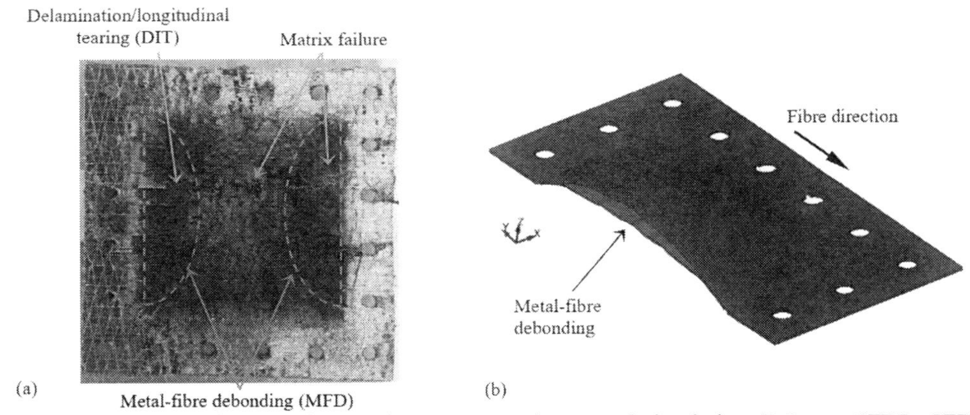

Fig. 10. Comparison of CFRP damage between experiment and simulation (3-Layer CFRP, CFRP blasted-side, 16g): (a) typical damage modes (experimental); (b) damage in cross-section (numerical)

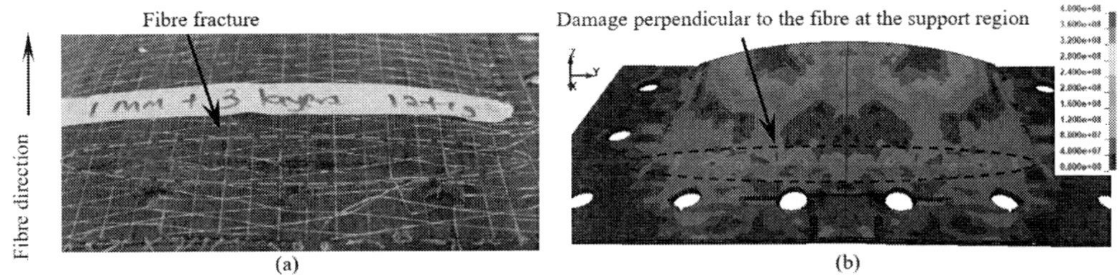

Fig. 11. Comparison of CFRP damage at supports between experiment and simulation (3-Layer CFRP, steel blasted-side, 13g): (a) experiment; (b) simulation

4.4 Comparison of computational time

Fig. 12 compares the CPU (central processing unit) elapsed time required to conduct the analysis using various methods. Using an SMP (Shared Memory Parallel) LS-DYNA solver in a single machine (Intel® Core™ i5), The time required to complete the analysis of 2 msec analysis termination time for the full air-domain model was 730296 CPU seconds (8 days, 10 hours). The use of a reduced air-domain model with mapped layer allowed for a 95% reduction in elapsed time in the full model, as the mapped model required 36139 CPU seconds (10 hours) to complete the analysis of a similar termination time. The effectiveness of the mapping technique was further evaluated by executing an analysis of the whole air-domain model using the LS-DYNA MPP (Massively Parallel Processing) solver on a high-performance computer (HPC). The CPU elapsed time required for 2 msec termination time using MPP solver and 32 cores (4 nodes, 8 cores per node) was 27664 (7 hours, 41 minutes), lowering the analysis time using SMP solver by 96%. Hence, the use of mapping techniques appeared to provide significant computational savings when using an SMP solver.

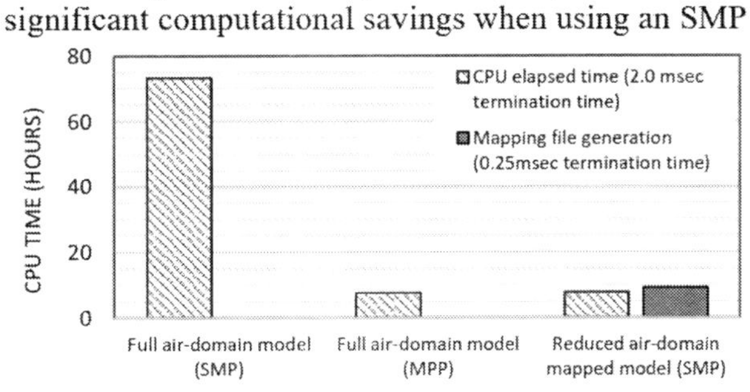

Fig. 12. Comparison of CPU elapsed time for the analysis

5 Conclusions

This paper reports on the results of a numerical study to simulate the response of carbon fibre reinforced polymer (CFRP) retrofitted steel plates due to blast loads using finite element software, LS-DYNA. The results of the simulation were validated against plate response and magnitude of deformation obtained from experiments. When using the SMP solver, the use of a reduced air-domain model with mapped layer allowed for a 95% reduction in CPU elapsed time in the model simulated with fine air mesh. The use of PSE feature provided savings in the time and efforts required to model the composite plates. Displacement histories obtained from the analysis indicated satisfactorily estimates for actual residual deflections with variation range of ±20%. For damage modes in analysed retrofitted specimens, the steel plate exhibited large inelastic deformation which agreed with actual failure modes found experimentally. While for CFRP layers, the simulated model captured the main damage features that include fibre-steel debonding, fibre fracture and delamination in CFRP layers. The use of the LS-DYNA mapping approach for blast application and PSE for composite plate modelling resulted in a satisfactory prediction of CFRP-steel composite plate response under blast while saving significant computational time.

Acknowledgments: Part of the LS-DYNA computations were performed using facilities provided by the University of Cape Town's ICTS High Performance Computing team: hpc.uct.ac.za.

References

[1] L. Hollaway, "A review of the present and future utilisation of FRP composites in the civil infrastructure with reference to their important in-service properties," *Construction and Building Materials,* vol. 24, no. 12, pp. 2419-2445, 2010.
[2] X.-L. Zhao, *FRP-strengthened metallic structures*. CRC Press, 2013.
[3] A. I. Danilov, "Some Aspects of CFRP Steel Structures Reinforcement in Civil Engineering," *Procedia Engineering,* vol. 153, pp. 124-130, 2016.
[4] J. Teng, T. Yu, and D. Fernando, "Strengthening of steel structures with fiber-reinforced polymer composites," *Journal of Constructional Steel Research,* vol. 78, pp. 131-143, 2012.
[5] H. Draganić, G. Gazić, and D. Varevac, "Experimental investigation of design and retrofit methods for blast load mitigation – A state-of-the-art review," *Engineering Structures,* vol. 190, pp. 189-209, 2019.
[6] P. A. Buchan and J. F. Chen, "Blast resistance of FRP composites and polymer strengthened concrete and masonry structures – A state-of-the-art review," *Composites Part B: Engineering,* vol. 38, no. 5, pp. 509-522, 2007.
[7] A. P. Mouritz, "Advances in understanding the response of fibre-based polymer composites to shock waves and explosive blasts," *Composites Part A: Applied Science and Manufacturing,* vol. 125, p. 105502, 2019.
[8] X. Zhao and R. Al-Mahaidi, "A brief review of CRP-metal systems subjected to impact and blast loading," presented at the 23rd Australasian Conference on the Mechanics of Structures and Materials (ACMSM23), Byron Bay, Australia, 9-12 December 2014, 2014.
[9] M. R. Bambach, X. L. Zhao, and H. Jama, "Energy absorbing characteristics of aluminium beams strengthened with CFRP subjected to transverse blast load," *International Journal of Impact Engineering,* vol. 37, pp. 37-49, 2010.
[10] D. J. Stevens, J. M. H. Puryear, N. Smith, and M. H. Gomez, "Effects of close-in charges on pipeline components," in *Structures Congress 2012*, pp. 102-113.
[11] M. Mokhtari and A. Alavi Nia, "The application of CFRP to strengthen buried steel pipelines against subsurface explosion," *Soil Dynamics and Earthquake Engineering* vol. 87, pp. 52-62, 2016.

[12] F. Dolce, M. Meo, A. Wright, M. French, and M. Bernabei, "Structural response of laminated composite plates to blast load," *Plastics, Rubber and Composites,* vol. 39, no. 3-5, pp. 180-188, 2010.
[13] D. Karagiozova, G. S. Langdon, G. N. Nurick, and S. Chung Kim Yuen, "Simulation of the response of fibre–metal laminates to localised blast loading," *International Journal of Impact Engineering,* vol. 37, no. 6, pp. 766-782, 2010.
[14] E. Sitnikova, Z. W. Guan, and W. J. Cantwell, "The analysis of the ultimate blast failure modes in fibre metal laminates," *Composites Science and Technology,* vol. 135, pp. 1-12, 2016.
[15] J. Li, C. Huang, T. Ma, X. Huang, W. Li, and M. Liu, "Numerical investigation of composite laminate subjected to combined loadings with blast and fragments," *Composite Structures,* vol. 214, pp. 335-347, 2019.
[16] N. Aquelet and M. Souli, "2D to 3D ALE mapping," in *10th International LS-DYNA Users Conference*, 2008, pp. 23-34.
[17] "LS-DYNA Keyword User's Manual," Livermore Software Technology Corporation (LSTC), Livermore, California, United States of America2017.
[18] T. P. Slavik, "A Coupling of Empirical Explosive Blast Loads to ALE Air Domains in LS-DYNA®," *IOP Conference Series: Materials Science and Engineering,* vol. 10, no. 1, p. 012146, 2010.
[19] T. Erhart, "A New Feature to Model Shell-Like Structures with Stacked Elements," in *10th European LS-DYNA Conference 2015*, Würzburg, Germany, 2015.
[20] C. Geretto, "The Effects of Different Degress of Confinement on the Deformation of Square Plates Subjected to Blast Loading," PhD Thesis, University of Cape Town, Cape Town, 2012.
[21] S. Rigby *et al.*, "Predicting the response of plates subjected to near-field explosions using an energy equivalent impulse," *International Journal of Impact Engineering,* vol. 128, pp. 24-36, 2019.
[22] S. Chung Kim Yuen, G. N. Nurick, and H. B. Brinckmann, "Response of Cylinderical Shells to Lateral Blast loads," *International Journal of Protective Structures,* vol. 4, no. 3, pp. 209-230, 2013.
[23] "ANSYS/Autodyn version 11.0, User Documentation," Century Dynamics Inc., a subsidiary of ANSYS Inc.2007.
[24] C. Reuter and T. Tröster, "Crashworthiness and numerical simulation of hybrid aluminium-CFRP tubes under axial impact," *Thin-Walled Structures,* vol. 117, no. Supplement C, pp. 1-9, 2017.
[25] F. Dolce, "Blast impact simulation on composite military armours," Master of Philosophy dissertation, University of Bath, 2009.
[26] P. Feraboli, B. Wade, F. Deleo, M. Rassaian, M. Higgins, and A. Byar, "LS-DYNA MAT54 modeling of the axial crushing of a composite tape sinusoidal specimen," *Composites Part A: applied science and manufacturing,* vol. 42, no. 11, pp. 1809-1825, 2011.
[27] S. Muflahi, G. Mohamed, and S. Hallett, "Investigation of Delamination Modeling Capabilities for Thin Composite Structures in LS-DYNA," in *Advanced Composites Centre for Innovation and Science, 13th International LS-DYNA Users Conference,* 2014, vol. 70.
[28] G. N. Nurick and G. C. Shave, "The deformation and tearing of thin square plates subjected to impulsive loads—An experimental study," *International Journal of Impact Engineering,* vol. 18, no. 1, pp. 99-116, 1996.

Numerical investigation of the hydrocyclone vortex finder depth on separation efficiency

Lesiba Mokonyama[1,*], *Thokozani Justin* Kunene[1], *and Lagouge Kwanda* Tartibu[1]

[1]University of Johannesburg, Mechanical & Industrial Engineering Technology Department, Doornfontein Campus, Johannesburg, South Africa

Abstract. Hydrocyclones are devices used in numerous chemicals, food, and mineral-related industrial sectors for the separation of fine particles. A d50 mm hydrocyclone was modelled with the use of the Computational fluid dynamics (CFD) simulation, ANSYS® Fluent 2021 R1. The vortex finder depth was varied from 20 mm, 30 mm, and 35 mm to observe the effects of pressure drop and separation efficiency from a varied vortex finder depth and characteristics of the air core. The numerical methods validated the results observed from different parameters such as volume fraction characteristics based on CFD simulations. The tangential and axial velocities increased as the vortex finder length increased. It was found that as the depth of the vortex finder is increased, particle re-entrainment time in the underflow stream increases, and separation efficiency improved.

Keywords:
Computational Fluid Dynamics, Hydrocyclone, Pressure Drop, Vortex Finder

1. Introduction

Despite its widespread use in chemical and mineral processing, determining the geometry and operating conditions of a hydrocyclone for a given application in the separation or classification of solids is still based on empirical equations or simplified theories of flow and separation mechanism within the hydrocyclone. Hydrocyclones are frequently utilized in various separation applications because of their inexpensive operating and maintenance costs, high throughput, and small footprint [1]. Various operational and structural elements determine the flow field in a hydrocyclone, which is characterized by a complicated three-dimensional whirling flow with an air core. The majority of operational factors are feed rate and slurry concentration is two factors to consider. The structural parameters are feed inlet form, vortex finder form, overflow outlet diameter, underflow outlet diameter, and cone angle. Martínez *et al* [2] used a stroboscope to measure the detailed velocity profiles in a hydrocyclone. Simulations with a hydrocyclone at a constant flow rate, overflow-to-throughput ratios, entries, and vortex finder diameters were carried out to understand the dependence of hydrocyclone performance on feed. Roping is a phenomenon that is yet primarily unknown. Because of variations in cut size and flow split, it has significant operational implications. According to Plitt and Booth [3], it is more crucial since the highest separating efficiency is attained close to the rope discharge regime. They have created response surfaces for separation efficiency as a function of classifier orifices and feed rate.

[1] Lesiba Mokonyama. Lesiba.k.mokonyama@gmail.com

Roping has been chiefly investigated in stand-alone hydro cyclones. In continuous closed grinding circuits, only a tiny amount of research has been done [3].

The form of the underflow, which is controlled by the feed circumstances, has a significant impact on hydrocyclone separation. Rope discharge with high feed solids concentrations usually results in a crude product with higher solids content and fewer particles. Spray discharge with low input solids concentration, on the other hand, results in more significant solids recovery and smaller cut sizes. As a result, ideal overall conditions may be expected at the rope/spray discharge changeover point [4][5]. Even though the impact of solids content in the feed on hydrocyclone separation efficiency has long been recognized, comprehensive study on the change in pressure drop across the hydrocyclone and the latter's separation characteristics has been absent. If all other operational parameters remain constant, an increase in feed concentration will usually result in coarser cut size, decreased separation sharpness, and increased pressure drop. Aside from the particles impeding each other's radial motion, the apex valve is limited capacity and variations in the flow field within the hydrocyclone cause more particles to be entrained by the overflow [6][7].

Raesi and Mahddahian [8] investigated air inside the air-injected deoiling hydrocyclone. Thy found that an air core forms at the centre of the hydrocyclone due to the lower density of air compared to oil and water. The volume fraction of air in the hydrocyclone's core approaches one. The majority of the air fed into the hydrocyclone escapes through the overflow. As a result, air bubble separation efficiency is 97.4 percent. There is always a need to eliminate or drastically reduce the air core formed inside the hydrocyclone to improve the separation efficiency. Sripriya et al [9] investigated a new hydrocyclone was constructed with the insertion of a solid rod at the core region of the conical section. By doing so, the air core could be successfully eliminated, and the hydrocyclone's performance could be improved. The lowering of radial and axial components of velocity and turbulence in the area near the vortex finder's entrance may cause this effect. As a result, the flow field characteristics inside the hydrocyclone without an air core become more separation-friendly. Critically to the performance in terms of air core reduction in the hydrocyclone are the feed conditions. Hararah *et al* [10] determined through their experimental results that when the feed-solids content increased, the air-core diameter shrank, which was supported by the theoretical consideration. Finally, at a critical high feed-solids level, where rope discharge of the underflow was reached, the air core vanished. The air-core diameter was associated with the underflow spray angle at low feed-solids concentration. We have also our findings can be used not only to build a thorough picture of the vortex finder's action, but also to create and control hydrocyclone's performances.

2. Model description

2.1 Hydrocyclone geometry and grid generation

The hydrocyclone's geometry was utilized in this work; numerical and experimental models are used, as seen in Figure 1. It is a standard hydrocyclone with specified dimensions for sorting mineral particles.

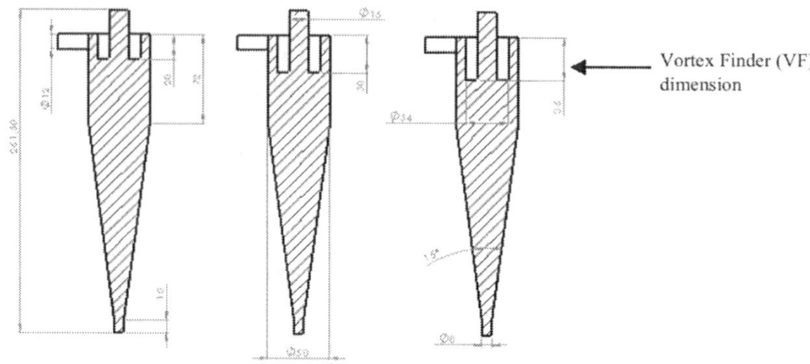

Fig. 1. Geometry of hydrocyclone of base case (unit: mm) representing three vortex finder (VF) lengths, 20 mm, 30 mm and 35 mm, respectively

Table 1 lists the parameters investigated in this study, which include air core investigation and flow structures based on the base case; variable parameters such as feed rate, cone angle, underflow outlet diameter, and overflow outlet diameter are investigated to study the influence of different parameters on the flow field inside the hydrocyclone.

Table 1: Geometry of hydrocyclone of base case (unit: mm)

Parameter	Base	Variable
Feed Rate (m/s)	3	-
Cone Angle (◦)	15	-
Underflow Diameter (mm)	8	-
Overflow Diameter (mm)	15	-
Vortex finder Depth (mm)	20	20, 30, 35

Figure 2 shows a mesh generated using the tetrahedron with seven inflation layers at the first thickness of 10^{-5} leading to the wall plus value of y+ = 5. The ANSYS® Fluent 2021 R1 (student version) has a limit of 510,000 elements. We meshed close to the limit of and archived 478 560 elements. These elements were accepted to be valid for the model based of the limitation of the commercial code.

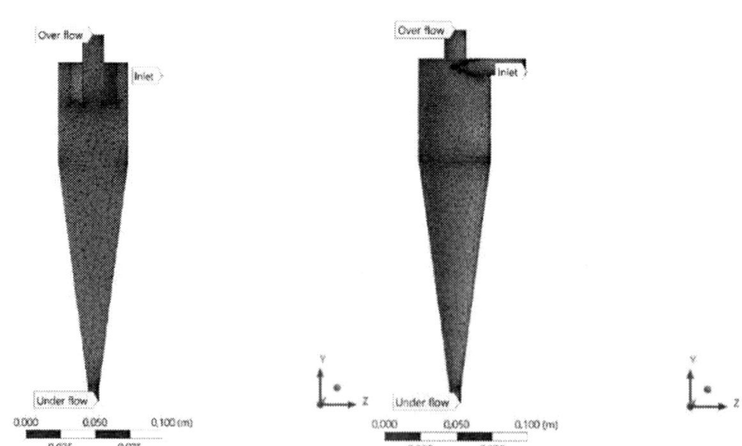

Fig. 2. Mesh density of tetrahedral elements for a computational domain a hydrocyclone.

2.2 Volume of fluid (VOF) equations

The governing equation of volume fraction in VOF model can be expressed by Li *et al.* [11] *as*:

$$\frac{\partial \alpha_q}{\partial t} + \vec{u} \cdot \nabla \alpha_q = 0 \qquad (1)$$

Where α_q denotes the volume fraction of the, i-phase fluid. When the whole space is full of a same phase, $\alpha_q = 1$; when the whole space is full of two phases, the sum of the volume fractions of two phases equal to 1.

2.3 Turbulence model

2.3.1 Shear Stress Transport (SST) model

Menter [12] developed the SST *k*–turbulence model, and the governing equations of the flow field, comprising the continuity equation, momentum equations, and transport equations of *k* and ω, can be expressed as follows.
The SST turbulence model comprises two equations: one for *k*, the specific turbulent kinetic energy (m^2 s^{-2})

$$\frac{\partial(\rho k)}{\partial t} + \frac{\partial(\rho u_j k)}{\partial x_j} = P - \beta^* \rho \omega k + \frac{\partial}{\partial x_j}\left[(\mu + \sigma_k \mu_t)\frac{\partial k}{\partial x_j}\right] \qquad (2)$$

And one for ω, the specific turbulent dissipation rate (or specific turbulent frequency) (s^{-1})

$$\frac{\partial(\rho \omega)}{\partial t} + \frac{\partial(\rho u_j \omega)}{\partial x_j} = \frac{\gamma}{v_t}P - \beta\rho\omega^2 + \frac{\partial}{\partial x_j}\left[(\mu + \sigma_\omega \mu_t)\frac{\partial \omega}{\partial x_j}\right] + 2(1 - F_1)\frac{\rho\sigma_{\omega 2}}{\omega}\frac{\partial k}{\partial x_j}\frac{\partial \omega}{\partial x_j} \qquad (3)$$

The two terms on the left side of Eqs. (2), (3) are related to the rate of change of *k* or ω and the convective transport of *k* or ω, respectively. The three terms in common on the right side of Eqs. (2), (3) represent, in this order, the transport of *k* or ω by turbulent diffusion, the rate of production of *k* or ω, and the rate of dissipation of *k* or ω. For the third term on the right side of Eq. 3,
Where:

$$P = \tau_{ij}\frac{\partial u_i}{\partial x_j};\ \tau_{ij} = \mu_t\left(2S_{ij} - \frac{2}{3}\frac{\partial u_k}{\partial x_k}\delta_{ij}\right) - \frac{2}{3}\rho k \delta_{ij};\ S_{ij} = \frac{1}{2}\left(\frac{\partial u_i}{\partial x_j} + \frac{\partial u_j}{\partial x_i}\right);\ \mu_t = \frac{\rho a_1 k}{max(a_1\omega, \Omega F_2)};\ \mu_t = \frac{\rho a_1 k}{max(a_1\omega, \Omega F_2)}$$

$$\phi = F_1\phi_1 + (1 - F_1)\phi_2;\ F_1 = \tanh(\arg_1^4);\ \arg_1 = min\left[max\left(\frac{\sqrt{k}}{\beta^*\omega d}, \frac{500v}{d^2\omega}\right), \frac{4\rho\sigma_{\omega 2}k}{\mathrm{CD}_{k\omega}d^2}\right];$$

$$\text{CD}_{k\omega} = max\left(2\rho\sigma_{\omega 2}\frac{1}{\omega}\frac{\partial k}{\partial x_j}\frac{\partial \omega}{\partial x_j}, 10^{-20}\right); \ F_2 = \tanh(\text{arg}_2^2);$$

$$\text{arg}_2 = max\left(2\frac{\sqrt{k}}{\beta^*\omega d}, \frac{500v}{d^2\omega}\right)$$

k-Ω closure: $\sigma_{k1} = 0.85, \sigma_{w1} = 0.65, \beta_1 = 0.075$

k-ε closure: $\sigma_{k2} = 1.00, \sigma_{w2} = 0.856, \beta_2 = 0.0828$

The β^* value can directly calibrate it, affecting how turbulence grows or dissipates. The final term in the ω equation (Eq. 3) represents the cross-diffusion that occurs as a result of the transformation of the ε equation into an equation for ω.

SST closure constants: $\beta^* = 0.09, a_1 = 0.31$

3. Numerical model

3.1 Solver and turbulence model

The simulation that has been carried out was carried out with a Cartesian coordinate for numerical simulation. The operating conditions were set to be transient and iterative methods to reduce the margin of error by convergence. Water properties are used along with pressure and mass fluxes for the momentum equations. The velocity kept constant PRESTO (pressure staggered option) is highly recommended for fluid swirling characteristics inside the body. For turbulence, Reynolds stress (Shear Stress Transport model) and a Volume of Fluid (VOF) model were selected to evaluate; Semi implicit pressure linked equations combine the momentum and continuity laws to obtain derivatives of the pressure. The simulations are run with a fixed time step of 5.0 x10^{-4} s. Furthermore, the effects of the controlled variables and parameters are discussed to validate experimental results. A velocity boundary condition is employed on the feed and pressure outlet boundary conditions for both the overflow and underflow, with an air backflow volume percentage of 1.0. When the air backflow volume fraction is set to 1.0, the simulation can build an air-core by drawing air via the overflow and underflow apertures when the swirl produces an area of negative pressure along the axis [13].

4. Model validation

At this reporting time, the lack of experimental data for the current study varied vortex finder lengths is only validated using similar CFD studies in the literature. Ghodrat *et al* [14] used a two-fluid model to simulate multiphase flow and hydrocyclone performance, paying specific attention to the impacts of vortex finder diameter, length, and shape over a wide range of feed solids concentrations. We selected this literature as it had the same numerical approach as we have undertaken. However, the size of the hydrocyclone is from the *d75*. We had to continue with a *d50* hydrocyclone as our objective for this study. Overall, Figure 3 shows the trends obtained in the numerical conducted using the LES are in good agreement with SST model.

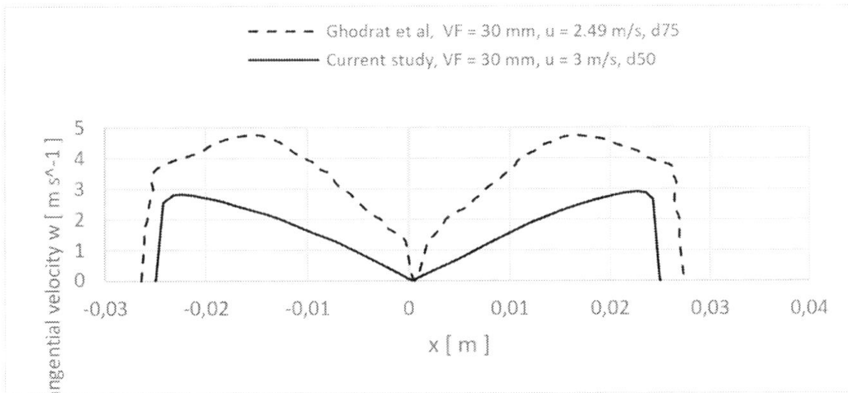

Fig. 3. Radial distribution of the tangential velocities in the horizontal plane located at 170 mm away from the underflow of the hydrocyclone Ghodrat *et al* [14].

The use of the VOF approach facilitated with the Large Eddy Simulation (LES) model should be used [14]. As it can produce results that are better comparable to experimental measurements should be used. Our observation is that LES model can better predict the tangential velocities. The trends the two approached area provide an expected velocity profile of particles experiencing centrifugal forces pushing against the wall. Our limit is also extenuated by much finer meshes that the LES model would require than the SST model. It is then computationally much more demanding at this moment. The transition to the LES model will be complimented by the acquisition of commercially available licence, using a cluster for high performance computing and conducting experiments.

5. Results and discussion

5.1 Effects of vortex finder depth on pressure drop

The vortex finder's primary function is to minimise any short circuit flow at the feed of the hydrocyclone [2]. In practice, only the difference between the inlet and the overflow pressure of the hydrocyclone is considered to calculate a pressure drop in a hydrocycle. As seen from the topology diagrams, the 35mm depth hydrocyclone produces a lot of overflow pressure. Relative to the 30mm depth hydrocyclone. This means that the pressure drop will be higher for the VF = 35mm cyclone; this will significantly impact the feed separation of the cyclone.

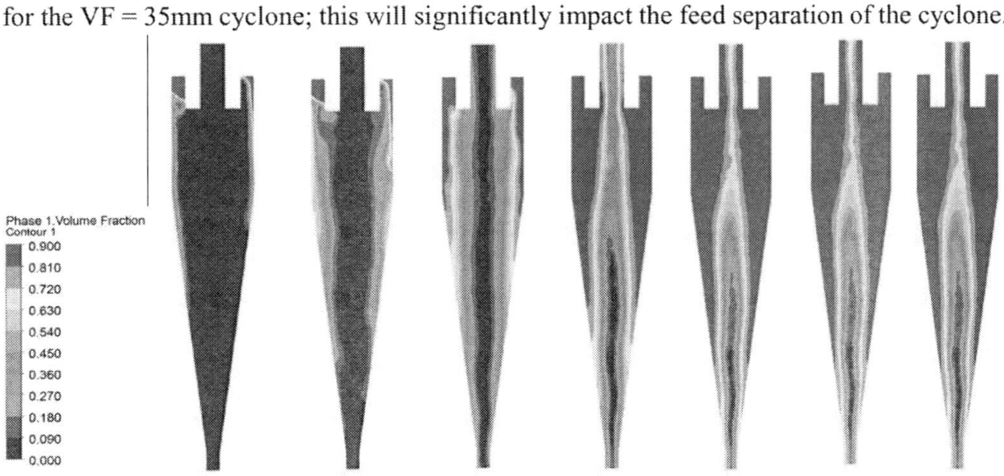

Fig. 4. Water's (Phase 1) forming process and the nature of the air core of 20 mm vortex finder over time: $t = 0.05$ seconds to $t = 4.05$ seconds viewed from left to right, at gradual time increments.

The separation of the centrifugal swirl in the cyclone's core causes the vortex to become narrower, and the friction will increase. This can be seen in Figures 4 and 5 that the walls of the 35mm hydrocyclone have a higher pressure drop. This influence on geometrical specifications on the pressure drop shows that the pressure drop is a function of the vortex finder depth. With a correct aspect ratio of the inlet/overflow diameter and vortex finder depth, a more acceptable feed separation can be produced. The separation efficiency of a hydrocyclone depends on the vortex finder depth as the particles are guided by the outside wall in the presence of the vortex finder. The particle re-entrainment duration in the underflow stream increases as the vortex finder depth increases, increasing separation efficiency [6]. Under a higher pressure, fine particles more easily moved towards the centre, which was favourable for the separation of fine particles. The particles were separated more efficiently due to high centrifugal force, and the swirling motion within the hydrocyclone increased as the feed velocity increased. The intensity of the centrifugal force is directly proportional to the tangential velocity. The main driving force in two-phase separation is centrifugal force [11].

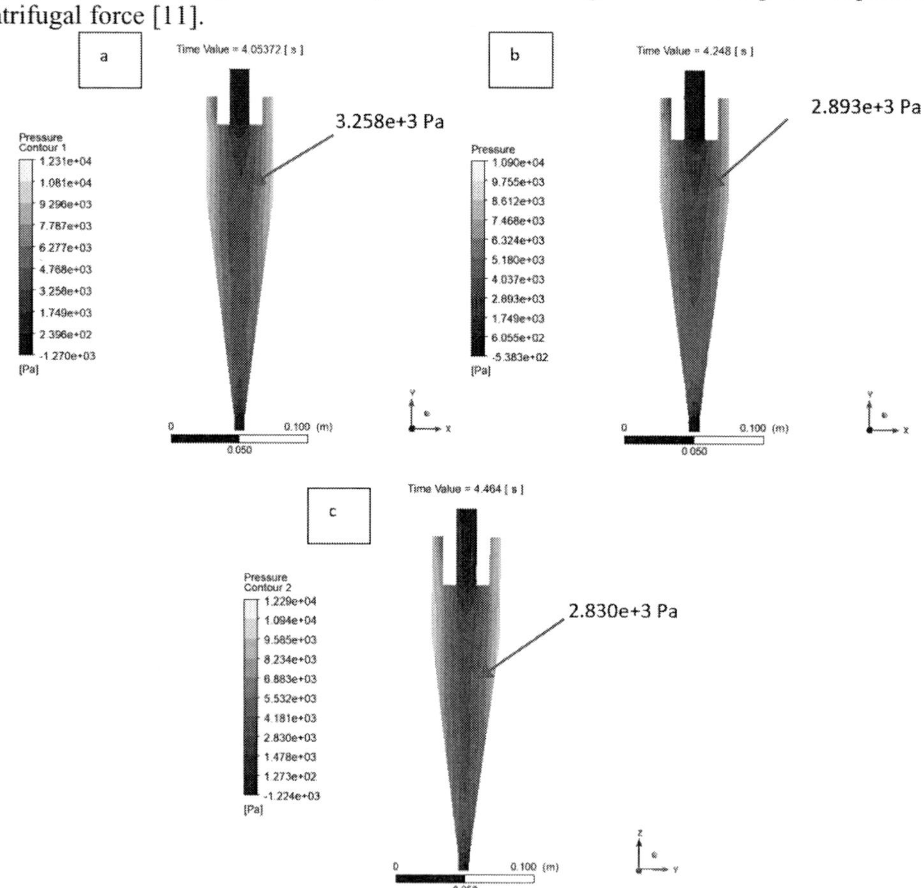

Fig. 5. Pressure gradient of the vortex finders in relation to the pressure drop across for (a) VF = 20 mm, (b) VF = 30 mm, (c) VF = 35 mm.

5.2 Effect of vortex finder depth on the air core

The prediction of the air-core profile is crucial for the flow dynamics to be described. The water is pushed against the wall by the tangential velocity, forming the air-core. As a result,

as the tangential velocity decreases, the diameter of the air-core decreases [15]. It is seen in Figure 6 that the increasing the vortex finder length has a bearing in the magnitude and behaviour of the kinetic energy of the hydrocyclone. The tangential velocity increases with the VF length thus providing a better classification of the particles.
The axial velocity determines the particle retention time and split ratio in the hydrocyclone. At a lower axial velocity, the particles lingered in the hydrocyclone for longer and separated more thoroughly [11]. Narasimha et al [16] found that increasing the feed flow rate causes an increase in the maximum tangential velocity (accounting for the increase of kinetic energy) in a given cyclone, which in turn causes a decrease of the pressure at the hydrocyclone axis near the apex. If the pressure is lowered near the apex, it is easier for the air to be sucked in through the apex. Hence the air-core diameter increases with inlet flow rate.

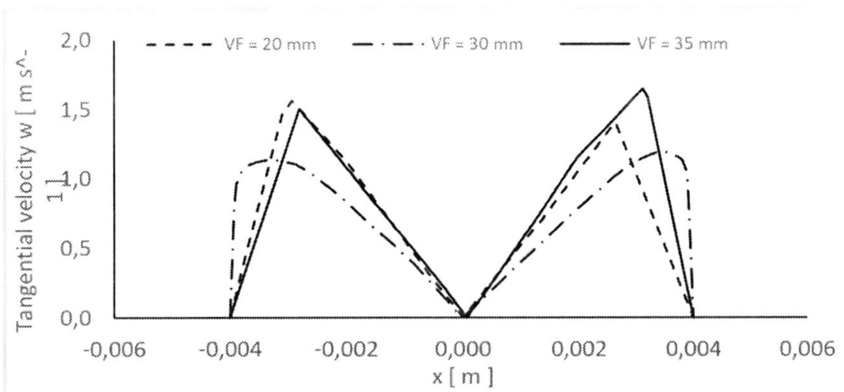

Fig. 6. The tangential velocity in the y = 0 m (underflow area) radial position.

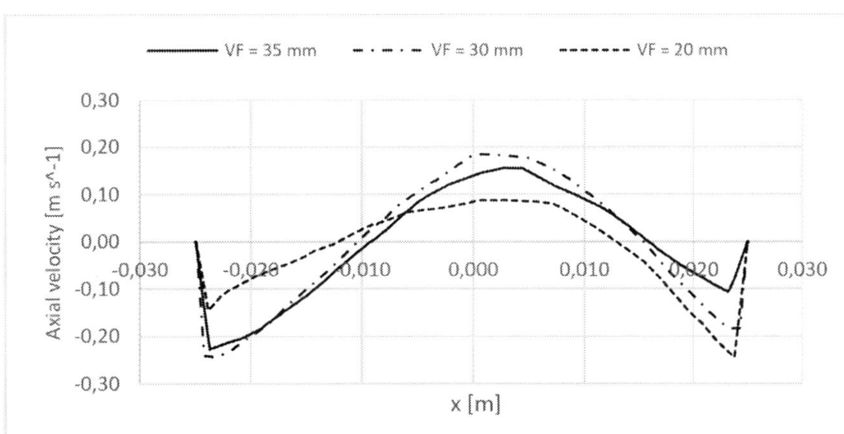

Fig. 7. The axial velocity in the y = 0.169 m (diameter line, *d*50 mm) radial position.

Neesse and Dueck [17] established an empirical formula (Eq. 4) to determine an air core diameter based on the radial pressure distribution in the hydrocyclone. The Navier–Stokes equations were used by Neesse and Dueck [17] to calculate the radius of an air core, assuming force equilibrium between pressure gradient and centrifugal force at the liquid–gas interface. This physical consideration leads to Eq. 4, which shows that the hydrocyclone shape is the primary determinant of the air core radius.

$$\frac{R_a}{R_c} = \left[1 + \frac{2n}{k_w^2}\left(\frac{R_c}{R_{in}}\right)^2\left(\frac{R_{in}}{R_o}\right)^{2m}\left(\frac{p_{in}-p_u}{p_{in}-p_o}\right)\right]^{-1/2n} \quad (4)$$

Where: dimensionless relation for the air core radius R_a as function of hydrocyclone radius R_c, inlet radius R_{in}, overflow radius R_o and the pressures in the underflow p_u and in the overflow p_o. The values of $k = 0.27$, $n = 0.8$ and $m = 1$ the empirical constants that depend on hydrocyclone geometry.

We used Eq. 4 to determine an air core diameter for a varied vortex finders. Presented in Figures 8 and 9 are the two extreme cases of our study. It is proving to be difficult to have air core formation that comes from instabilities of the multiphase flow up to the stable air core. The air diameter is still not stable along the centre hydrocyclone and the influence of the length of vortex finder as it increases the pressure drop in the overflow. It is not providing a stable air core. The air core diameter in the underflow over time is presented in figures Fig 8 and 9. It stabilizes from 1 second at a diameter of around 7.84 mm and 7.216 mm for our two extreme cases of VF of 20 mm and 35 mm, respectively.

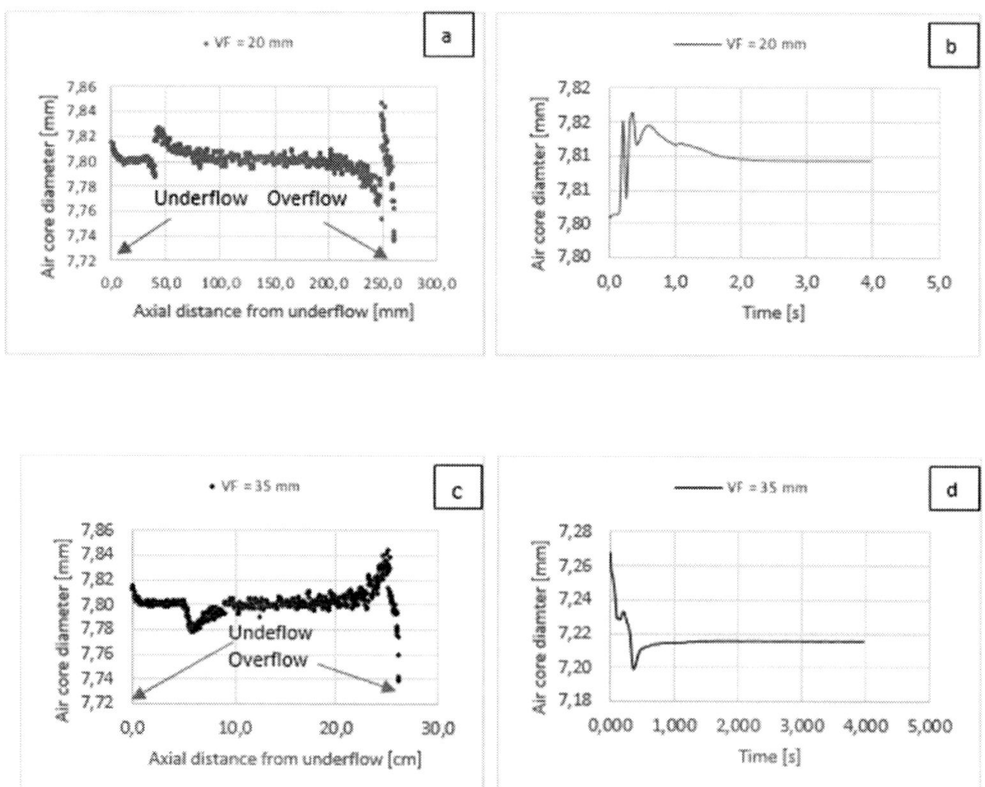

Fig. 8: (a, b) The air core along the axial distance at time 3.95 seconds (c, d) the air core stabilizing with time in the underflow region of the hydrocyclone with VF of 20 mm and 35 mm, respectively.

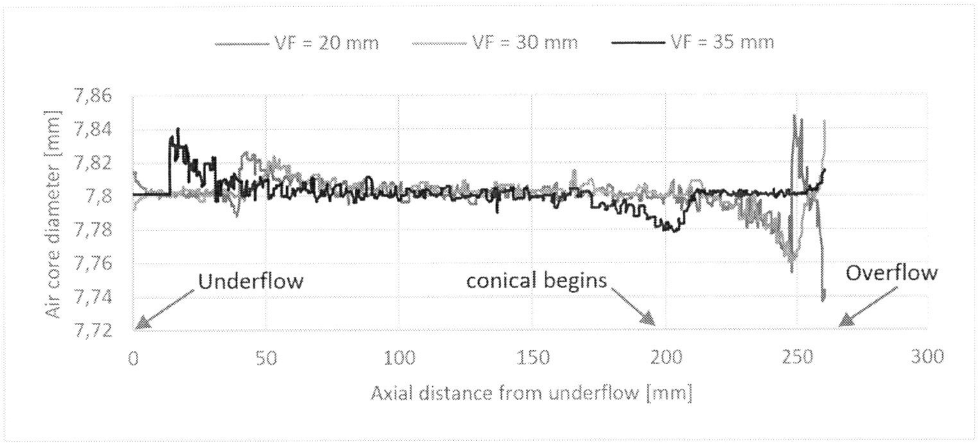

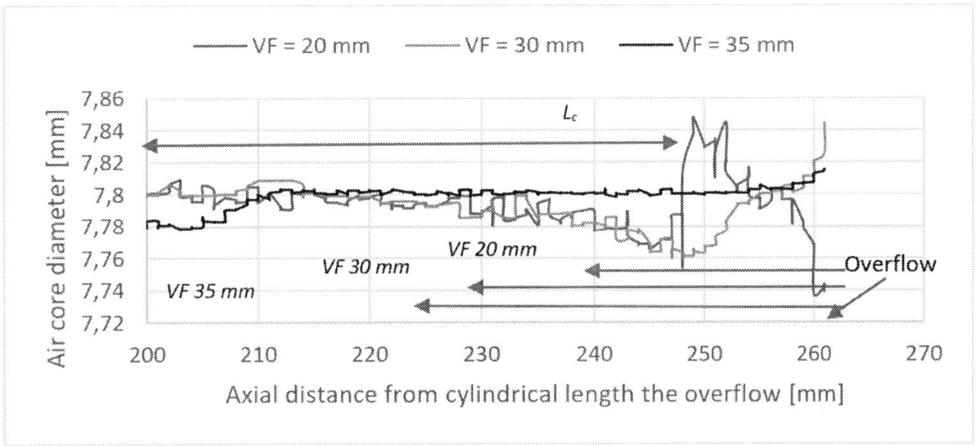

Fig. 9: Illustration of the positions of the L_c, cylindrical part length and the different vortex finders.

If the VF becomes longer and reaches the conical section, the current turbulence may force coarser particles to depart through the overflow stream rather than the intended underflow stream [2]. Our study of the VF 30mm shows an almost constant air core diameter with the cylindrical part length but starts to quickly destabilize under 210 mm or closer to the cylindrical length that ends 200 mm from the overflow. Hence, there is a greater importance to have critical length of a VF to the cylindrical length. It is prevalent in maintaining a stable air core diameter. A suitable VF like the 35 mm, as noticed in this study, proves a constant air flow through the core that leads the air to flow upwards to the overflow. The centrifugal force in the overflow pipe is minimized, preventing the formation of a stable air core. It is important to have the balance between making use of the advantages a longer VF provides like a stable air core, to the point or conical section. For the VF 35 mm, the air core diameter decreases and remains stable with increasing VF length. The change is more obvious for a longer vortex finder length, and this may be due to the strengthening tangential velocity and increased pressure drop [18]. However, extending the vortex finder shortens the natural vortex in the cyclone body, reducing the chance of small particles separating from the vortex [19].

6. Conclusion

An optimization numerical study using the SST model was conducted for the varied vortex finders of a *d*50 hydrocyclone. We find that the designs of new geometries opens a new chapter in the optimization of hydrocyclones. We have computed pressure drop is lower at vortex finders with a shorter length. The length can be increased to match the pressure drop of the standard hydrocyclone. The air core diameter decreases and remains stable with increasing under the conical section. As a result, the flow field characteristics inside the hydrocyclone without an air core become more separation-friendly. The airflow as characteristics of the air core may be used to control the hydrocyclone.

Acknowledgement

The author would like to thank supervisor Thokozani Justin Kunene and co-supervisor Lagouge Kwanda Tartibu and for the support guidance patience during the preparation of this paper. The institution University of Johannesburg for financially committing to the conference cost.

REFERENCES

[1] B. Cui, C. Zhang, D. Wei, S. Lu, and Y. Feng, "Effects of feed size distribution on separation performance of hydrocyclones with different vortex finder diameters," *Powder Technol.*, vol. 322, pp. 114–123, 2017.

[2] L. F. Martínez, A. G. Lavín, M. M. Mahamud, and J. L. Bueno, "Vortex finder optimum length in hydrocyclone separation," *Chem. Eng. Process. Process Intensif.*, vol. 47, no. 2, pp. 192–199, 2008.

[3] K. Heiskanen, "Experimental hydrocyclone roping models," *Chem. Eng. J.*, vol. 80, no. 1–3, pp. 289–293, 2000.

[4] T. Dyakowski and R. A. Williams, "Prediction of air-core size and shape in a hydrocyclone," *J. Miner. Process*, vol. 43, pp. 1–14, 1995.

[5] M. Narasimha, M. S. Brennan, and P. N. Holtham, "CFD modeling of hydrocyclones: Prediction of particle size segregation," *Miner. Eng.*, vol. 39, pp. 173–183, 2012.

[6] G. Patra, S. Chakraborty, and B. C. Meikap, "Role of vortex finder depth on pressure drop and performance efficiency in a ribbed hydrocyclone," *South African J. Chem. Eng.*, vol. 25, pp. 103–109, 2018.

[7] T. Braun and M. Bohnet, "Influence of feed solids concentration on the performance of hydrocyclones," *Chem. Eng. Technol.*, vol. 13, no. 1, pp. 15–20, 1990.

[8] R. Raesi and R. Maddahian, "Numerical investigation of air-injected deoiling hydrocyclones using population balance model," *Chem. Eng. Sci.*, vol. 248, p. 117103, 2021.

[9] R. Sripriya, M. D. Kaulaskar, S. Chakraborty, and B. C. Meikap, "Studies on the performance of a hydrocyclone and modeling for flow characterization in presence and absence of air core," *Chem. Eng. Sci.*, vol. 62, no. 22, pp. 6391–6402, 2007.

[10] M. A. Hararah, E. Endres, J. Dueck, L. Minkov, and T. Neesse, "Flow conditions in the air core of the hydrocyclone," *Miner. Eng.*, vol. 23, pp. 295–300, 2010.

[11] F. Li *et al.*, "Numerical Analysis on the Effect of Combined-Curve Tapered Segment on the Flow Field and Separation Performance of Hydrocyclones," *Arab. J. Sci. Eng.*, vol. 172, no. August, 2021.

[12] F. R. Menter, "Two-equation eddy-viscosity turbulence models for engineering applications," *AIAA J.*, vol. 32, no. 8, pp. 1598–1605, 1994.

[13] M. Narasimha, M. Brennan, and & P. N. Holtham, "A Review of CFD Modelling for Performance Predictions of Hydrocyclone," *Eng. Appl. Comput. Fluid Mech.*, vol. 1, no. 2, pp. 109–125, 2007.

[14] M. Ghodrat, S. B. Kuang, A. B. Yu, A. Vince, G. D. Barnett, and P. J. Barnett, "Numerical analysis of hydrocyclones with different vortex finder configurations," *Miner. Eng.*, vol. 63, pp. 125–138, 2014.

[15] J. A. Delgadillo and R. K. Rajamani, "Exploration of hydrocyclone designs using computational fluid dynamics," *International Journal of Mineral Processing*, vol. 84, no. 1–4. pp. 252–261, 2007.

[16] M. Narasimha, M. Brennan, and P. N. Holtham, "Large eddy simulation of hydrocyclone — prediction of air-core diameter and shape," vol. 80, pp. 1–14, 2006.

[17] T. Neesse and J. Dueck, "Air core formation in the hydrocyclone," *Miner. Eng.*, vol. 20, no. 4, pp. 349–354, 2007.

[18] B. Tang, Y. Xu, X. Song, Z. Sun, and J. Yu, "Numerical study on the relationship between high sharpness and configurations of the vortex finder of a hydrocyclone by central composite design," *Chem. Eng. J.*, vol. 278, pp. 504–516, 2015.

[19] F. Ficici, A. Vedat, and M. Kapsiz, "The effects of vortex finder on the pressure drop in cyclone separators," *Int. J. Phys. Sci.*, vol. 5, no. 6, pp. 804–813, 2010.

AUTHOR INDEX

EDP Sciences – Web of Conferences
17, Avenue du Hoggar
Parc d'Activité de Courtabouf
BP 112
F-91944 Les Ulis Cedex A
France

ISBN 978-1-7138-4077-0

CURRAN ASSOCIATES INC.
proceedings.com